BOREAL SHIELD WATERSHEDS

WATERSHEDS

Lake Trout Ecosystems
in a Changing Environment

Integrative Studies in Water Management and Land Development

Series Editor
Robert L. France

Published Titles

Handbook of Water Sensitive Planning and Design
Edited by Robert L. France

Boreal Shield Watersheds: Lake Trout Ecosystems in a Changing Environment
Edited by J.M. Gunn, R.J. Steedman, and R.A. Ryder

Forthcoming Titles

Forests at the Wildland-Urban Interface: Conservation and Management
Edited by Mary Duryea

Restoration of Boreal and Temperate Forests
Edited by John A. Stanturf

Stormwater Management for Low Impact Development
Edited by Lawrence Coffman

The Economics of Groundwater Remediation and Protection
Paul E. Hardisty, Ece Ozdemiroglu, and Jonathan Smith

BOREAL SHIELD WATERSHEDS

Lake Trout Ecosystems *in a* Changing Environment

Edited by
J.M. Gunn, R.J. Steedman, and R.A. Ryder

CRC Press
Taylor & Francis Group
Boca Raton London New York

CRC Press is an imprint of the
Taylor & Francis Group, an **informa** business

CRC Press
Taylor & Francis Group
6000 Broken Sound Parkway NW, Suite 300
Boca Raton, FL 33487-2742

First issued in paperback 2019

ISBN-13: 978-1-56670-646-9 (hbk)
ISBN-13: 978-0-367-39505-6 (pbk)
Library of Congress Card Number 2003051624

Library of Congress Cataloging-in-Publication Data

Boreal shield watersheds : lake trout ecosystems in a changing environment / edited by
   J.M. Gunn, R.J. Steedman, and R.A. Ryder.
      p. cm. — (Integrative studies in water management and land development)
   Includes bibliographical references and index.
   ISBN 1-56670-646-7 (alk. paper)
     1. Lake trout—Ecology. 2. Lake ecology—North America. I. Gunn, J.M. (John Maxwell), 1952-
II. Steedman, Robert John, 1958- III. Ryder, R.A. (Richard Alan) IV. Series.

QL638.S2B57 2003
597.5'54—dc21

2003051624

Visit the Taylor & Francis Web site at
http://www.taylorandfrancis.com

and the CRC Press Web site at
http://www.crcpress.com

Series statement: Integrative studies in water management and land development

Ecological issues and environmental problems have become exceedingly complex. Today, it is hubris to suppose that any single discipline can provide all the solutions for protecting and restoring ecological integrity. We have entered an age where professional humility is the only operational means for approaching environmental understanding and prediction. As a result, socially acceptable and sustainable solutions must be both imaginative and integrative in scope; in other words, garnered through combining insights gleaned from various specialized disciplines, expressed and examined together.

The purpose of the CRC Press series Integrative Studies in Water Management and Land Development is to produce a set of books that transcends the disciplines of science and engineering alone. Instead, these efforts will be truly integrative in their incorporation of additional elements from landscape architecture, land-use planning, economics, education, environmental management, history, and art. The emphasis of the series will be on the breadth of study approach coupled with depth of intellectual vigor required for the investigations undertaken.

Robert L. France
Series Editor
Integrative Studies in Water Management
and Land Development
Associate Professor of Landscape Ecology
Science Director of the Center for
Technology and Environment,
Harvard University
Principle, W.D.N.R.G. Limnetics
Founder, Green Frigate Books

Foreword by series editor

This volume, edited by John Gunn, Rob Steedman, and Dick Ryder, pulls together an incredibly broad mix of people and topics under a single cover. As such, it is a worthy addition to the new series from CRC Press — Integrative Studies in Water Management and Land Development — that was initiated in 2002 with publication of my own edited volume, *Handbook of Water Sensitive Planning and Design*. Books like these are rare, but they shouldn't be. Complex environmental problems can only be identified, understood, and rectified through the collective actions of a diversity of approaches from a variety of disciplines. Gunn, Steedman, and Ryder well recognize this as witness to the fact that their contributors to this volume come from many different provincial, state, and federal agencies, universities, and private consulting or research organizations. Likewise, the topics covered in these pages are truly catholic in scope: natural and cultural history, stocking and management, rehabilitation, commercial fisheries, land-use modifications, reservoir creation, nutrient inputs and transformations, lake chemistry and morphometry influences, atmospheric deposition, trace contaminant cycling, species introductions, and climatic alterations. All directed toward a single sentinel species — the lake trout of the Boreal Shield — a wonderful fish I know well as a research subject (and also as a culinary object!), and in an area of the continent of incredible sublime beauty in which I have spent much time in both recreational and scholarly pursuits.

Until some future author writes a popular account of the anthropological history of the lake trout — along the lines of, for example, John McPhee's *The Founding Fish* (about the shad), Mark Kurlonsky's *Cod: A Biography of the Fish that Changed the World*, or Richard Scheid's *Consider the Eel* — the present book, with its emphasis on the management of, and environmental influences on, this particular species of fish, should become widely read. What all of these works share is their demonstration that the true distribution for certain species of fish encompasses sociological space just as much as it does Euclidian space. Lake trout, then, are a truly integrated cultural and biological symbol of the Boreal Shield ecoregion.

Another important message that one takes away from the present book — one alluded to several times but not formally enunciated — is of a compelling challenge to our myth of "pristine nature" or "wilderness" free from human influences. When looking at a map of human inhabitation in North America (or the photo of illuminated cities shown in the first chapter), one could erroneously assume that somehow the great Boreal forest is "the true north, strong and free" from human manipulation. What we learn from this book is that the Boreal Shield ecosystem is really just as much a designed landscape as any on the planet. So, in addition to the well-known artificiality of the forests due to wildfire suppression, we now realize that since soon after glaciation, the resident relict populations of lake trout have been repeatedly poked at and prodded by us. While in the past (and even in the recent past), this has been mostly through direct tinkering such as fisheries and restocking programs, today it seems that these fish populations function as barometers of changes in both the landscape and the airscape. We would be wise to learn the lessons that these aquatic canaries might be able to tell us, and for this we should be indebted to the authors of this timely and important volume.

Robert L. France
Harvard University

Foreword: An ideal icon

The lake trout, a coldwater denizen of Boreal lakes, makes an ideal icon. The spectacular fish is a memory of its past and a vision for a desired future, an icon to stir human action on behalf of valued and relatively unspoiled Boreal lakes. These lakes are increasingly exposed to new and more intense human pressures. An icon can help foster the protection, management, and restoration of these treasured systems. Can lake trout be such an icon? Is this fish the only icon needed to stir the human passions to behave ethically for a sustainable future? In the Pacific Northwest, anadromous salmon, Douglas fir, marine mammals, and other components combine into a more general set of icons worth preserving because they are valued by different groups. Is the lake trout part of such a set of effective icons for the Boreal lake systems? My answer would be a hearty "yes."

This noble animal depends on the maintenance of a suite of aquatic, terrestrial, and aerial environments; thus it is an indicator not only of the deep, cold, oxygenated waters, but also of land at a landscape scale and of air at regional and global scales. Thus, the species integrates anthropogenic pressures on the environment giving further credibility to Barry Commoner's first law of ecology: "Everything is related to everything else." Does it seem inconsistent that the icon is also the indicator? I think not. This is often the case. This interlocking of the vision and the practice brings together excitement and technique, purpose and strength. Is the lake trout a sufficient indicator through which to judge status, function, and dynamics of Boreal lake ecosystems? I doubt it. The inshore fish community would be a great indicator, but not as good an icon. The spruce and the aspen, the moose and the wolf, and other components inform us about other facets of our influence that could influence the lakes, and mechanisms are equally or more important as indicators.

Challenges are many: overfishing and extraction, exotics and toxins, human population growth and expansion, energy use, and climate change. Some of these influences can be dealt with or fixed at the local, lake, or perhaps watershed level. Others are more provincial and linked to regional economic development that may undervalue ecosystem sustainability. Some of the pressures are continental with transboundary movement among nations of people, dollars, toxins, water, and exotics. Others are truly global, such as the generation of greenhouse gases or development of carbon storage.

As I read the chapters, it became increasingly clear that some of these Boreal lakes are more sensitive to different pressures, and that they are not all equally sensitive to the same pressure. For example, a lake sensitive to overfishing because the trout are key to the local economy or because an urban, recreational fishing population is only a short drive away may not be the same lake that is most vulnerable to climate warming or aerially borne toxics or acids. Of this the writers are well aware.

More daunting was the realization that some lakes we can protect, some we can manage to some degree, some we can restore, but others we cannot help, at least in the short term or through local action. Changes will occur, and one needs to decide how to respond to those changes. As in the medical analogy, triage should be part of any strategy. Behaviors in respect to short-term, faster-acting pressures may differ depending on the expected response of Boreal lakes to the long-term drivers. Sorting such things out among the various kinds of lakes is important to establishing short- and long-term strategies.

So from my point of view, the lake trout is certainly an icon and a tool that can help us realize the more desirable future. The species is perhaps uniquely suited to help achieve a sustainable future for Boreal lake ecosystems and the humans who love them. It cannot do it alone.

John J. Magnuson
Center for Limnology
University of Wisconsin

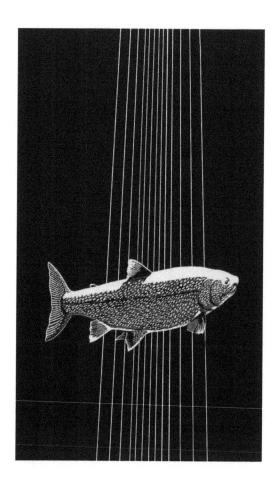

Preface: Boreal Shield ecosystems

Deep, clear Boreal Shield lakes carved from Precambrian bedrock have long defined the northern wilderness and are the ancestral home and interglacial refuge of the lake trout, *Salvelinus namaycush*. The lakes, streams, and wetlands of this ecozone are tightly linked to the austere watersheds of the north woods and are sustained by them. This land of white pine, black spruce, moose, wolf, beaver, and woodland caribou poses daunting environmental management challenges at the beginning of the 21st century. New science gleaned from these ecosystems may provide a powerful general model for those concerned about freshwater fisheries, water quality, and watershed ecosystems worldwide.

Humans have long been part of the Boreal Shield world. A few adaptive and resource-ful aboriginal peoples followed fish, game, young forests, and receding glaciers northward 5000 to 10,000 years ago. The number of people living in the Boreal forest is still small relative to those in more hospitable regions, but humans continue to move northward and exert ever-increasing demands on the Boreal landscape. Now, 200 years after the area's rich fur, fish, timber, and mineral resources first attracted the interest of Europeans, forestry and mining still form the backbone of the region's economy. The unspoiled landscape and waters have become easily accessible and support a huge tourism and recreation industry. The new wave of industry and technology in distant cities now plays a dominant role in the health of Boreal Shield ecosystems through market-driven extraction and consumption of resources, through long-range atmospheric transportation of contaminants, and through changing global climate.

This book brings together a uniquely qualified group of scientists to extend and interpret the scientific legacy of the Boreal watersheds. For the last 50 years, pristine Boreal Shield waters have served as crucibles for world-class research into impacts of water pollution, acid rain, climate change, fisheries, and watershed disturbance. This book builds on that research foundation and explores the ability to manage human interactions with these unique ecosystems at local, regional, and global scales. Our ability to sustain healthy Boreal Shield waters constitutes a crucial test of ecosystem management concepts, techniques, and commitment.

John M. Gunn
Ontario Ministry of Natural Resources
Laurentian University

Robert J. Steedman
Ontario Ministry of Natural Resources

Richard A. Ryder
RAR & Associates

xi

Acknowledgments

We would like to thank all the authors for their time and effort in producing these chapters. It was a long struggle from start to finish, and we really appreciate their patience and continued support. Special thanks to Carissa Brown and Christine Brereton, our very able editorial assistants. This project could not have been completed without them.

Many of the authors participated as peer reviewers on associated chapters. We were also fortunate to have the assistance of the following external reviewers: Chris Brousseau, Randy Eshenroder, David Evans, John Fitzsimons, Chris Goddard, John Havel, Bill Keller, Terry Marshall, Norman Mercado-Silva, Greg Mierle, George Morgan, Henk Rietveld, Helen Sarakinos, Wolfgang Schieder, Ed Snucins, Vincent St. Louis, and James Wiener.

Michael Malette, Seija Mallory, Leila Tuhkasaari, and Amanda O'Neil (Cooperative Freshwater Ecology Unit, Laurentian University, Sudbury, Ontario) compiled the lake trout data set with assistance from Rob Korver, Rod Sein, and Wayne Selinger (Ontario Ministry of Natural Resources), Michel Legault (Société de la faune et des parcs du Québec), Gary Siesennop and Mark Ebbers (Minnesota Department of Natural Resources), and Walter Kretser, Richard Costanza, Bill Gordon, and Richard Preall (Adirondack Lake Survey Corporation). Paul Morgan established the Canadian Shield Trout Scholarship Program at Laurentian University to support associated research projects. Michel Legault (Société de la faune et des parcs du Québec) and Judi Orendorff (Ontario Ministry of Natural Resources) participated in the original steering committee for this project. Ed Snucins and Vic Liimatainen provided many of the photographs.

We gratefully acknowledge the Canadian National Atmospheric Chemistry (NatChem) Database and its data-contributing agencies and organizations for the provision of the wet deposition data used to produce the 1980–1989 and 1990–1999 average annual deposition figures (Plate 6). The agencies and organizations responsible for data contributions to the NatChem Database include Environment Canada; the provinces of Ontario, Quebec, New Brunswick, Nova Scotia, and Newfoundland; the U.S. Environmental Protection Agency; and the U.S. National Atmospheric Deposition Program/National Trends Network.

Information and maps for the long-term monitoring sites were provided by John Shearer (Experimental Lakes Area), Jim Rusak (North-Temperate Lakes — Trout Lake Station), Martyn Futter (Dorset), Mark Ridgway, Trevor Midell (Harkness/Lake Opeongo), Dean Jeffries (Turkey Lakes Watershed), Bill Keller (Sudbury Lakes), Christine Brereton (Sudbury Lakes and Killarney Park), and John Gunn (Killarney Park).

Financial and logistic support for the project was provided by the Ontario Ministry of Natural Resources, Laurentian University (Cooperative Freshwater Ecology Unit), and the Sustainable Forest Management Network.

About the Editors

John M. Gunn is a senior research scientist for the Ontario Ministry of Natural Resources and heads the Cooperative Freshwater Ecology Unit at Laurentian University. During the past 25 years much of his research has focused on restoration ecology of acid-damaged ecosystems in northeastern Ontario, with particular emphasis on the recovery of stressed lake trout ecosystems. He was the recipient of several awards, including the 2000 President's Award for Conservation from the American Fisheries Society.

Robert J. Steedman is a research scientist with the Ontario Ministry of Natural Resources in Thunder Bay, where he has led long-term, interdisciplinary studies of watershed ecosystem response to forest management and provided science-based policy advice to the Province of Ontario. He is presently on assignment with the National Energy Board in Calgary, Alberta, as Professional Leader, Environment.

Richard A. Ryder is a semiretired fisheries research scientist after a 44-year career with the Ontario Ministry of Natural Resources and its predecessor, the Ontario Department of Lands and Forests. He is the recipient of numerous awards and honors, including most recently an election into the National Fisheries Hall of Excellence (1999) and the Meritorious Service Award (2001). He has served as president of the American Fisheries Society (1980–1981) and the Canadian Conference for Fisheries Research (1987–1988).

Contributors

Craig J. Allan
Department of Geography and Earth
 Sciences
University of North Carolina at Charlotte
Charlotte, North Carolina

Jean Benoît
Direction de l'aménagement de la faune de
 Lanaudière
Société de la faune et des parcs du Québec
Repentigny, Québec

Roger Bérubé
Hydraulique et Environnement
Hydro-Québec
Montréal, Québec

R.A. (Drew) Bodaly
Freshwater Institute
Department of Fisheries and Oceans
Winnipeg, Manitoba

Arthur J. Bulger
Department of Environmental Sciences
University of Virginia
Charlottesville, Virginia

Thomas J. Butler
Center for the Environment
Cornell University
Ithaca, New York

Leon M. Carl
Great Lakes Science Center
Ann Arbor, Michigan

John M. Casselman
Aquatic Research and Development
 Section
Ontario Ministry of Natural Resources
Glenora Fisheries Station
Picton, Ontario

Bev J. Clark
Dorset Environmental Science Centre
Ontario Ministry of the Environment
Dorset, Ontario

Christopher S. Cronan
Department of Biological Sciences
University of Maine
Orono, Maine

Peter J. Dillon
Environmental and Resource Studies
Trent University
Peterborough, Ontario

Charles T. Driscoll
Department of Civil and
 Environmental Engineering
Syracuse University
Syracuse, New York

Warren I. Dunlop
Southcentral Science and Information
 Section
Ontario Ministry of Natural Resources
Bracebridge, Ontario

Christopher Eagar
Northeast Forest Experiment Station
USDA Forest Service
Durham, New Hampshire

Mark P. Ebener
Great Lakes Fishery Commission
Sault Saint Marie, Michigan

Hayla E. Evans
RODA Environmental Research Limited
Lakefield, Ontario

Henri Fournier
Direction de l'aménagement de la faune de
 l'Outaouais
Société de la faune et des parcs du Québec
Hull, Québec

Robert L. France
Graduate School of Design
Harvard University
Cambridge, Massachusetts

Mike Fruetel (deceased)
Quetico Mille Lacs Fisheries Assessment
 Unit
Ministry of Natural Resources
Thunder Bay, Ontario

John M. Gunn
Cooperative Freshwater Ecology Unit
Ontario Ministry of Natural Resources
Laurentian University
Sudbury, Ontario

Karen A. Kidd
Department of Fisheries and Oceans
Freshwater Institute
Winnipeg, Manitoba

Charles C. Krueger
Great Lakes Fishery Commission
Ann Arbor, Michigan

Robert S. Kushneriuk
Ontario Ministry of Natural Resources
Centre for Northern Forest Ecosystem
 Research
Thunder Bay, Ontario

Kathleen F. Lambert
Hubbard Brook Research Foundation
Hanover, New Hampshire

Gregory B. Lawrence
Water Resources
U.S. Geological Survey
Troy, New York

Michel Legault
Direction de la recherche sur la faune
Société de la faune et des parcs
 du Québec
Québec, Québec

Nigel P. Lester
Aquatic Research and Development
 Section
Ontario Ministry of Natural Resources
Peterborough, Ontario

Gene E. Likens
Institute of Ecosystem Studies
Millbrook, New York

John J. Magnuson
Center for Limnology
University of Wisconsin, Madison
Madison, Wisconsin

Nicholas E. Mandrak
Department of Fisheries and Oceans
Burlington, Ontario

Lewis A. Molot
Faculty of Environmental Studies
York University
Toronto, Ontario

Daniel Nadeau
Direction de l'aménagement de la faune de
 l'Abitibi-Témiscamingue
Société de la faune et des parcs
 du Québec
Rouyn-Noranda, Québec

Charles H. Olver
Ontario Ministry of Natural Resources
 (retired)
Hunstville, Ontario

Roger Pitblado
Geography Department
Laurentian University
Sudbury, Ontario

Michael J. Powell
Ontario Ministry of Natural Resources
Bracebridge, Ontario

Richard A. Ryder
RAR & Associates
Thunder Bay, Ontario

David W. Schindler
Department of Biological Sciences
University of Alberta
Edmonton, Alberta

Rod Sein
Ontario Ministry of the Environment
Northern Region
Sudbury, Ontario

Brian J. Shuter
Aquatic Research and Development
  Section
Ontario Ministry of Natural Resources
Peterborough, Ontario

Robert J. Steedman
Ontario Ministry of Natural Resources
Centre for Northern Forest Ecosystem
  Research
Thunder Bay, Ontario

John L. Stoddard
U.S. Environmental Protection Agency –
  Corvallis Environmental Research
  Laboratory
Corvallis, Oregon

M. Jake Vander Zanden
Center for Limnology
University of Wisconsin, Madison
Madison, Wisconsin

Kathleen C. Weathers
Institute of Ecosystem Studies
Millbrook, New York

Chris C. Wilson
Aquatic Research and Development
  Section
Ontario Ministry of Natural Resources
Trent University
Peterborough, Ontario

Karen A. Wilson
Department of Biology
Carleton College
Northfield, Minnesota

Norman D. Yan
Biology Department
York University
Toronto, Ontario

Contents

Section V: Synthesis

Section VI

Photo Credits

Cover

- Front cover: Lake 223 of the Experimental Lakes Area in northwestern Ontario by John Shearer
- Inset: Spawning lake trout (*Salvelinus namaycush*) at Ox Narrows, Kushog Lake, by Skuli Skulason

Foreword

- Lake trout drawing by B.E. Harding, *Game Fish and Fishing in Algonquin Park*, Government of Ontario, Department of Lands and Forests, Parks Branch, Toronto, 1955; revised 1958, 1965

Section Breaks

- Section I: Vic Liimatainen
- Section II: Ed Snucins
- Section III: Vic Liimatainen
- Section IV: Ed Snucins
- Section V: Carissa Brown
- Section VI: Steve Elliott/Mark Johnston

section I

Introduction

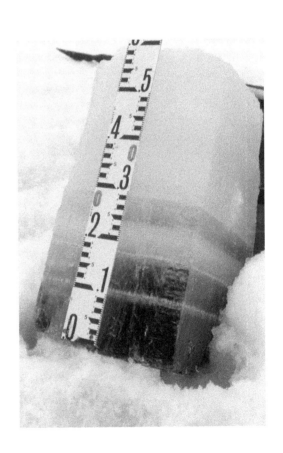

chapter one

Lake trout, the Boreal Shield, and the factors that shape lake trout ecosystems

John M. Gunn
Ontario Ministry of Natural Resources, Laurentian University
Roger Pitblado
Laurentian University

Contents

Introduction

The lake trout *Salvelinus namaycush* (also known as lake charr) is a northern species well adapted to austere conditions in the Arctic and Boreal regions of North America (Figure 1.1). For much of the last 2.5 million years, the land of the lake trout experienced various ice ages, and *S. namaycush* made its living in cold unproductive waters, lakes, and rivers near the ever-shifting margins of the continental glaciers (Winograd et al., 1997; Power, 2002; Wilson and Mandrak, Chapter 2, this volume). A winter scene on the Dog River, a Lake Superior tributary that once supported a spawning run of lake trout (Loftus, 1958), is perhaps reminiscent of some of the habitat occupied by lake trout during much of its evolutionary history (Figure 1.2).

The lake trout is well equipped to survive in such demanding and dynamic environments. It is a large and long-lived fish that produces large, well-provisioned eggs; it has a metabolism that allows movement and growth at low temperatures; it can withstand long periods of food deprivation and will eat almost any available prey item; it is a strong

(A)

(B)

Figure 1.1 Lake trout (*Salvelinus namaycush*). (A) Large adult (photo by Vic Liimatainen). (B) Age 1+ juveniles (top), age 5+ first time spawners (bottom) (photo by John Gunn).

long-distance swimmer and is able to use thermal refugia (lake hypolimion, groundwater springs; Figure 1.3) to survive weather and climate extremes (Martin and Olver, 1980; Power, 2002; Snucins and Gunn, 1995). Although extremely hardy in northern conditions, *Salvelinus namaycush* performs best in the absence of competitors and predators and is often most productive in small lakes that contain only simple fish communities (Evans and Olver, 1995; Shuter et al., 1998; Vander Zanden et al., 1999).

Ironically, this hardy northern species appears to be rather poorly adapted to deal with many "southern" phenomena associated with modern human activity. Like other

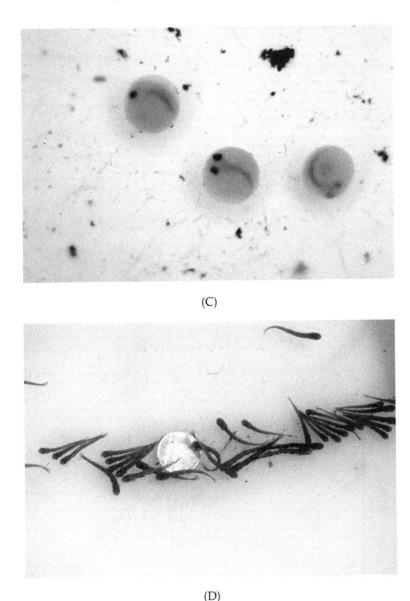

(C)

(D)

Figure 1.1 (continued) (C) Embryos ready to hatch in early March (photo by Vic Liimatainen). (D) Emergent alevins in May (photo by Rod Sein).

Arctic animals that live in unproductive ecosystems, the lake trout has evolved a life history strategy that invests in a few, long-lived, late maturing, large-bodied individuals. This makes the lake trout highly vulnerable to a suite of modern threats. These include introductions of warm water and cool water competitors; stocking of domesticated trout; increasing access, exploitation, and habitat disruption by humans; climate warming; inputs of nutrients and toxic contaminants; and hydroelectric development (Ryder and Johnson, 1972; Schindler, 1998; also see later chapters in this volume).

In North America, lake trout lakes can be found throughout the Precambrian Shield but are heavily concentrated in a great sweeping arc that extends along the Shield's southern edge (Figure 1.4). That same arc pattern coincides with or parallels other environmental phenomena (isotherms, vegetation communities, human activity, etc.) that

Figure 1.2 Denison Falls on Dog River, just above a historic spawning site for lake trout (photo by Vic Liimatainen).

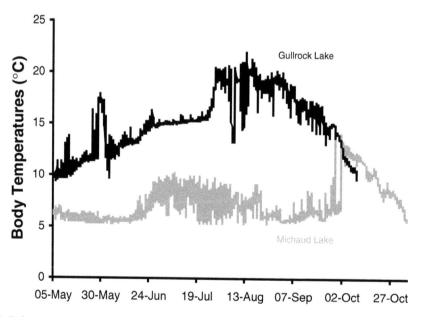

Figure 1.3 Behavioral thermoregulation in lake trout. The graph shows the core body temperature of a lake trout in a deep, cold lake (Michaud) where it maintains its preferred temperature by staying in the hypolimnion or moving up and down in the water column. In an extremely warm lake (Gullrock), the lake trout make use of a cold water seepage site during midsummer, when water temperature approaches 20°C. Note the fluctuations in the fish's body temperature in August as it moves in and out of the seepage area. From Gunn (2002).

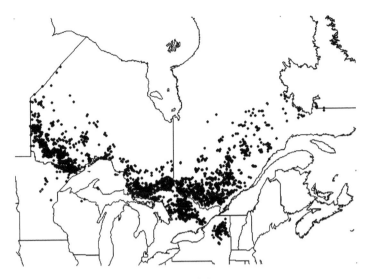

Figure 1.4 Boreal Shield lake trout lakes.

shape the distribution and dynamics of lake trout populations and their management challenges. These lakes exhibit a rather narrow range of physical and chemical characteristics (Table 1.1). They are usually rather cold, clear, deep, and dilute lakes (Martin and Olver, 1976) and have been aptly described as "swimming pools carved out of granite" (Ryder and Johnson, 1992). This book focuses on a group of about 3000 small lake trout lakes (75% of which are less than 500 ha in surface area) in forested catchments at the southern edge of the lake trout distribution range in the ecozone referred to as the Boreal Shield (Ecological Stratification Working Group, 1995). Some examples of these ecosystems are illustrated in Figure 1.5.

The Shield

In the terminology of plate tectonics, the process of plate collisions has left behind stable geological cores, which are generally located within the interiors of the continent. These shield areas are comprised of rocks that were initially formed during periods of the building of ancient mountain systems. In North America, this continental core is referred to as the Precambrian Shield, the Canadian Shield, in some places the Laurentide Plateau, or simply the Shield. Like many exposed shield areas throughout the world, it is often characterized by broadly convex surface profiles. The Shield is exposed at the surface as a vast horseshoe-shaped geologic region (about 7,000,000 square kilometers) that covers central and eastern Canada and small parts of the northern United States (Figure 1.6). Within the United States, the continental core or shield is extensive (roughly 75% of the surface area of the country), but most of it is overlain by post-Precambrian earth materials. The Shield landscape in the United States reveals itself only in the Adirondack Mountains, the Boundary Waters Area of Minnesota, and along the Great Lakes Basin.

The Precambrian Shield consists mostly of Archean rocks that are at least 2.5 billion years old. This is particularly the case in the northern two-thirds of the exposed Shield where granite, granitic gneiss, and Archean metamorphosed sedimentary and volcanic rocks are dominant. Archean rocks are also present along the southern Shield, but in these areas (for example, at the western end of Lake Superior, just north of the St. Lawrence River basin, and throughout the Adirondack Mountains) there is a greater predominance of Proterozoic or late Precambrian volcanic and sedimentary rocks in various degrees of

Table 1.1 Physical and Chemical Characteristics of Lake Trout Lakes within the Boreal Shield Study Area

	Ontario	Quebec	Adirondacks (NY)	Minnesota	Total
Surface area (ha)	553.4 ± 1066.6 median = 191.7 (1583)	459.4 ± 903.6 median = 153.0 (900)	266.6 ± 1059.8 median = 52.2 (129)	442.3 ± 1024.4 median = 103.4 (122)	504.0 ± 1015.4 median = 165.9 (2734)
Max. depth (m)	34.0 ± 17.7 median = 30.0 (1558)	36.6 ± 20.6 median = 32.0 (347)	17.9 ± 9.9 median = 18.0 (124)	32.5 ± 17.2 median = 27.6 (122)	33.4 ± 18.3 median = 29.3 (2151)
Mean depth (m)	11.0 ± 5.7 median = 9.7 (1554)	14.7 ± 9.1 median = 12.0 (162)	6.2 ± 3.0 median = 6.2 (108)	11.7 ± 4.2 median = 10.5 (13)	11.0 ± 6.2 median = 9.6 (1837)
Secchi (m)	5.3 ± 2.2 median = 4.9 (1566)	5.1 ± 1.9 median = 4.7 (318)	4.7 ± 2.2 median = 4.5 (119)	4.9 ± 1.7 median = 4.9 (119)	5.2 ± 2.2 median = 4.9 (2122)
Conductivity (µS/cm at 25°C)	49.5 ± 35.5 median = 37.0 (797)	43.7 ± 53.4 median = 30.0 (395)	36.4 ± 20.7 median = 31.6 (129)	55.2 ± 52.4 median = 39.7 (78)	47.0 ± 41.6 median = 35.0 (1399)
TDS (mg/L)	38.0 ± 29.4 median = 28.0 (1428)	38.7 ± 11.8 median = 37.3 (24)	N/A	63.2 ± 51.5 median = 48.0 (35)	38.6 ± 30.1 median = 29.0 (1487)
Ca (mg/L)	4.64 ± 4.53 median = 3.20 (556)	4.27 ± 3.80 median = 3.00 (199)	3.66 ± 2.31 median = 3.23 (113)	4.49 ± 2.37 median = 3.80 (57)	4.43 ± 4.06 median = 3.20 (925)

Mg (mg/L)	1.13 ± 0.91 median = 0.90 (533)	0.85 ± 0.50 median = 0.70 (181)	0.74 ± 0.54 median = 0.61 (107)	1.50 ± 0.55 median = 1.40 (57)	1.05 ± 0.80 median = 0.85 (878)
Total P (mg/L)	0.007 ± 0.005 median = 0.006 (183)	0.007 ± 0.004 median = 0.006 (131)	0.016 ± 0.016 median = 0.011 (101)	0.013 ± 0.006 median = 0.012 (78)	0.010 ± 0.009 median = 0.008 (493)
DOC (mg/L)	3.63 ± 1.64 median = 3.40 (362)	3.92 ± 1.47 median = 3.55 (74)	4.08 ± 1.95 median = 3.55 (107)	5.39 ± 2.00 median = 5.00 (23)	3.82 ± 1.73 median = 3.55 (566)
Alkalinity (mg/L)	12.4 ± 19.6 median = 5.1 (802)	10.9 ± 14.7 median = 6.4 (75)	129.9 ± 154.6 median = 100.6 (103)	22.4 ± 26.7 median = 14.3 (112)	24.4 ± 61.4 median = 7.0 (1092)
pH	6.69 ± 0.71 median = 6.69 (829)	6.60 ± 0.67 median = 6.53 (437)	6.66 ± 0.72 median = 6.81 (129)	7.07 ± 0.48 median = 7.07 (109)	6.69 ± 0.70 median = 6.70 (1504)

Note: Only lakes located on Precambrian Shield bedrock are included in the data set for this table. Mean ± 1 S.D., median and sample size (N) are indicated. The Laurentian Great Lakes and 33 other large lakes (>10,000 ha) were not included. We tried to limit the data set only to lakes that supported reproducing lake trout populations. Populations maintained by hatchery stocking are excluded when the information was available. For a list of the known lake trout lakes from the Boreal Shield ecozone, see Appendix 1.

(A)   (B)

(C)   (D)

Figure 1.5 Examples of lake trout ecosystems. (A) Ultraclear lakes in Killarney Park, Ontario (photo by Ed Snucins). (B) Lake 233 at the Experimental Lakes Area in northwestern Ontario (photo by John Shearer). (C) Close-up of southern part of McCulloch Lake (47°20′N, 80°42′W) near the height of land for Ontario. Lake shown again in Landsat image (Figure 1.7) (photo by Vic Liimatainen). (D) Trout Lake, Wisconsin (photo by Carl Bowser).

metamorphosis. Most Shield rocks are low in cations such as calcium and magnesium. This limits the productivity of the region's soils and provides surface waters with limited acid-neutralizing capability. Consequently, many of the lakes in these areas that are inhabited by lake trout are highly sensitive to acidic precipitation.

During the latter parts of the Precambrian, the last 500 to 600 million years, the Shield was reduced to an almost featureless surface. Then, it was gradually submerged beneath shallow seas beginning in the early Paleozoic. The sedimentary limestones, sandstones, and shales that formed beneath those seas can still be seen in areas surrounding the presently exposed Shield and in isolated pockets within the Shield itself. Later erosion stripped away these Paleozoic sedimentary rocks leaving what is today, for the most part, a tectonically stable area of rolling or undulating topography. However, as illustrated in Figure 1.7 (produced using a digital elevation model), the contemporary Shield is not flat. It has hills and mountains that in places reach elevations of between 1200 and 1700 m above sea level. Some of these areas, notably the Adirondack Mountains, were uplifted by tectonic activity that began during the Tertiary period and continues to this day. Many millions of years of precipitation and stream erosion significantly modified all of these landscapes, even before they were scraped and scoured by the last continental glaciations (Figure 1.8).

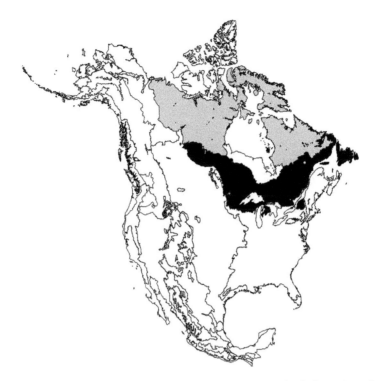

Figure 1.6 Outline map of the Precambrian Shield. The southern shaded portion of the Shield (in black) is the ecozone referred to as the Boreal Shield and is the focus of this book. *Boreal* or *northern* forest refers to the mainly coniferous forest that covers most of the northern portion of the ecozone. *Shield* refers to the exposed Precambrian Shield bedrock that extends across the entire ecozone (Ecological Stratification Working Group, 1995).

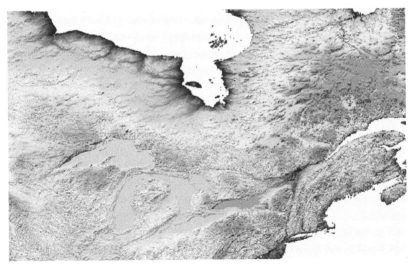

Figure 1.7 A digital elevation model highlighting the general topographic features of parts of eastern North America, following the methods of Pitblado (1992).

Figure 1.8 Bedrock polished by the glaciers (photo by David Pearson).

Continental glaciation

It has long been held that the Pleistocene epoch consisted of four continental-scale glacial stages. In North America these stages were named (from earliest to latest) Nebraskan, Kansan, Illinoian, and Wisconsinan, with the names usually representing the most southerly extent of ice movement from north to south. However, it is now recognized that there may have been as many as 10 major glacial periods during the last 1.6 million years and 40 or more minor episodes (Douglas, 1970; Winograd et al., 1997). Still, it was the advances and retreats of the Laurentide ice sheet during the most recent Wisconsinan glaciation that shaped today's lake trout waters. The three main legacies of the Wisconsin glaciers were deeply scoured lake basins, creation of complex temporary drainage systems that acted as dispersal routes and refugia for aquatic biota, and the deposition of glacial tills and lacustrine deposits, the parent materials of present-day soils.

As the 1- to 2-km-thick Laurentide ice sheet advanced southward from multiple domes in northern Canada, it stripped away the existing soils and most of the underlying weathered bedrock. Over ice-scoured plains and undulating terrain the advancing glaciers deepened existing depressions and created numerous others. These have now filled with water to form the myriad of lakes and ponds that are so characteristic of the Shield. The number of lakes that the Pleistocene glaciers left behind is not known. However, there are approximately 500,000 lakes (>1 ha) in Ontario alone (Cox, 1978), and the Precambrian Shield is thought to have close to half of the fresh water lake surface area of the world. Elsewhere, numerous glacial lakes have given rise to nicknames such as "The Land of 10,000 Lakes" for Minnesota. Figure 1.9 shows a portion of a satellite image that illustrates a Shield landscape dotted by lakes that were created by the scouring and dredging of Pleistocene glaciers.

As the continental ice sheets retreated, proglacial lakes of various sizes and longevity were formed. The earliest of these were probably Lake Maumee, which developed some 14,000 years ago in the western Erie Basin, and Lake Chicago, only a few hundred years later in the most southerly basin of what is now Lake Michigan. These early proglacial lakes in the Laurentian Great Lakes area initially drained southward through the Mississippi River system. As the glacier retreated further northward, removal of the

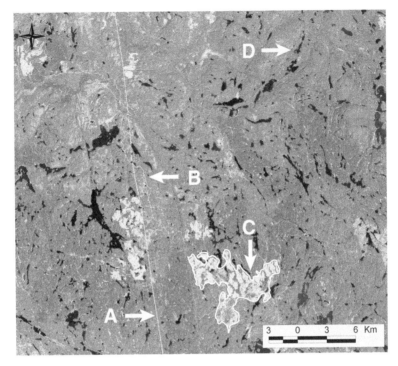

Figure 1.9 Landsat satellite image of a lake-strewn Shield landscape north of Sudbury, Canada. Electrical transmitting corridors (A), forestry roads (B), clearcut areas (C), and the location of one of the lakes in Figure 1.5 (D) are indicated.

great masses of ice allowed underlying land to rise (isostatic rebound). Drainage patterns were altered, first adding the eastward outlets of the Mohawk–Hudson River valleys; elimination of the southward flow through the Mississippi; opening of the North Bay spillway through the Mattawa and Ottawa rivers; and then finally drainage of the Great Lakes through the St. Lawrence River.

Further west and north, the largest of the glacial lakes existed, taking on many shapes and sizes from about 12,000 to 8,000 years ago. Lake Agassiz was impounded between the retreating ice margins and the Manitoba Escarpment — the present-day remnants including Lakes Manitoba, Winnipeg, Dauphin, and Winnipegosis. In addition to these Manitoban areas, at various stages of its life Lake Agassiz also extended well into Ontario and Saskatchewan in Canada and the northern parts of the states of North Dakota and Minnesota in United States. Additional proglacial lakes also appeared and disappeared north and west of Lake Agassiz as the Laurentide ice sheet retreated from its earlier coalescence with the Cordilleran ice sheet.

These lakes, and the rivers that either fed them or emanated from them, were likely ideal pathways for the movement of lake trout throughout this glacial and postglacial period (see Wilson and Mandrak, Chapter 2, this volume).

Climate, soils, and vegetation

The climate of the North American Precambrian Shield can generally be described as "continental," with long cold winters and short warm summers. However, the principal ecoregions (Bailey et al., 1985; Ecological Stratification Working Group, 1995; Bailey, 1996) of the southern portion of the Shield (Figure 1.10) are buffeted by alternating cold and dry polar air masses from the north and warmer moist maritime air masses, generally

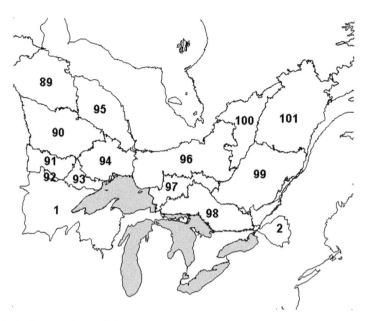

Figure 1.10 Principal ecoregions of the southern portions of Shield (see Table 1.2 for the mapping keys to these ecoregions).

Table 1.2 Temperature and Precipitation within Selected Ecoregions of the Boreal Shield

ID	Ecoregion	Mean Temperatures (°C)			Total Annual Precipitation (mm)
		Annual	Summer	Winter	
89	Hayes River Upland	−4.0	11.5	−20.0	400–600
90	Lac Seul Upland	0.5	14.0	−14.5	450–700
91	Lake of the Woods	1.5	15.0	−13.0	500–700
92	Rainy River	2.0	15.5	−12.5	600–700
93	Thunder Bay-Quetico	1.5	14.0	−13.0	700–800
94	Lake Nipigon	1.5	−14.0	−13.0	700–800
95	Big Trout Lake	−2.0	12.5	−17.0	550–775
96	Abitibi Plains	1.0	14.0	−12.0	725–900
97	Lake Timiskaming Lowland	3.0	15.0	−9.0	800–1000
98	Algonquin-Lake Nipissing	3.5	15.5	−8.5	900–100
99	Southern Laurentians	1.5	14.0	−11.0	800–1600
100	Rupert River Plateau	0.0	12.5	−13.5	650–900
101	Central Laurentians	0.0	12.5	−12.5	800–1000
1	Northern Minnesota	3.6	16.7	−10.1	580–690
2	Adirondack Mountains	4.8	14.8	−7.3	1010–1220

Source: Ecological Stratification Working Group, 1995.

from the south. Annual precipitation increases from west to east in the region, and mean annual temperatures increase from north to south. However, there are large seasonal and spatial variations throughout the region, as illustrated in Table 1.2.

Advances of the Laurentide ice sheet removed most, if not all, of the soils that had developed in previous interglacial periods. It left behind vast areas of Archean and Proterozoic bedrock that were completely bare or plastered with a relatively thin (2 to 8 m) till of poorly sorted, coarse-grained, noncalcareous rock debris. In more localized areas the Laurentide ice sheet deposited unsorted morainic debris or sorted glaciofluvial

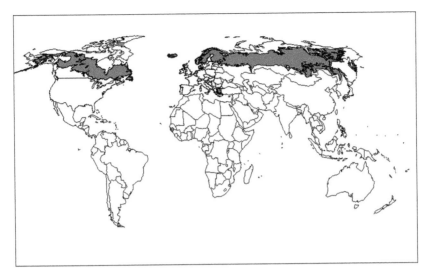

Figure 1.11 Circumpolar boreal forest. Data source: Olson et al. (2001).

sediments. The latter include a wide variety of ice-contact landforms (kames, eskers) and proglacial features. Proglacial features range from the moderately sorted sand and gravel deposits of streams emanating from the ice terminus to the highly sorted silts and clays deposited in proglacial lakes.

Broadly speaking, then, with the retreat of the Laurentide ice sheet dated at only 8,000 to 10,000 years ago, relatively little time has passed for the development of mature mineral soils, especially over the exposed bedrock areas. Thus, thin well-drained stony soils predominate, interspersed with pockets of poorly drained but deeper silt/clay and organic soils. The dominant soil-forming process is podzolization, a process whereby organic acids form in the surface horizons, leach basic elements (calcium, magnesium), iron, and aluminum from the upper layers, and then deposit these in soil horizons immediately below. Under forest cover, brunisolic soils may form with relatively poor horizonation and only slight illuviation. They are characterized by a subsoil horizon that is only slightly altered by hydrolysis, oxidation, or solution. Elsewhere, but still under forest cover, light-colored luvisols develop, distinguished from the previous two soil orders by a subsoil horizon containing clay that has been translocated from upper mineral horizons. Scattered throughout the study area are soils developed under waterlogged or very poor drainage conditions (gleysols and mesisols). Large, contiguous areas of gleysols are most often associated with extensive areas of silt and clay deposited in glacial lakes.

The vegetation of these areas characterize the southern portion of the circumpolar boreal forest (Figure 1.11), dominated by conifers, and the transition zone of mixed coniferous–deciduous woodlands of the Great Lakes–St. Lawrence forest region. Black spruce (*Picea mariana*) is the climax tree species of the more northerly ecoregions, but as one moves southward it is replaced by other conifers such as white spruce (*Picea glauca*), balsam fir (*Abies balsamea*), and eastern white pine (*Pinus strobes*). Throughout the area, warmer and drier sites are dominated by jack pine (*Pinus banksiana*) and red pine (*Pinus resinosa*); poorly drained sites are characterized by black and white spruce, balsam fir, tamarack (*Larix laracina*), eastern red cedar (*Juniperus virginiana*), and willow (*Salix* sp.). As one moves closer to the Great Lakes, the dominant vegetation is mixedwood forest of sugar maple (*Acer saccharum*), yellow birch (*Betula alleghaniensis*), eastern hemlock (*Tsuga canadensis*), and eastern white pine, with beech (*Fagus* sp.) appearing on warmer sites. Dry sites are dominated by red and eastern white pine and red oak (*Quercus rubra*). Wetter sites support red maple (*Acer rubrum*), black ash (*Fraxinus nigra*), white spruce, tamarack, and eastern white cedar (*Thuja occidentalis*).

Where winters are milder and summers warm, broadleaf deciduous species become dominant (maples, oaks, beech). This is especially true as one moves into southern Ontario and northern New York. However, as one continues into the Adirondacks, the northern hardwoods give way to spruce–fir vegetation types with increasing elevations (including localized alpine meadows).

Human activity

Humans have long been part of the Shield ecosystems (Wright, 1981). Paleo Indians, descendants of the people that first entered North America from Asia, arrived on the Shield at least 7000 years ago. These first colonists moved north following the receding glaciers and meltwater lakes, occupying what was then a vast barren tundra along the edge of the ice sheet. In this harsh land they survived as small bands of nomadic hunters, relying primarily on large game animals such as the barren ground caribou (*Rangifer tarandus groenlandicus*) (Gillespie, 1981). In the southern forested areas the primary game was woodland caribou (*Rangifer tarandus caribou*) and moose (*Alces alces*). Small mammals, plants, waterfowl, and fish were also harvested, but the high caloric needs for subsistence could be met only when the large game were present. Among the fish species, whitefish (*Coregonus clupeaformis*) and sturgeon (*Acipenser fulvescens*) were particularly important. Lake trout were also taken when they were available and were primarily captured with bone hooks or spears (Rogers and Smith, 1981).

The demographics, life style, and resource use of people on the Shield appears to have remained remarkably uniform throughout most of the postglacial period. This vast area contained only a few tens of thousands of people (Rogers and Smith, 1981), and they were almost entirely of the same language groups — Algonquin (Wright, 1981).

The constancy of the past is in sharp contrast to the rapid recent changes in human population patterns and resources use of the Shield. Although still low in population density (Figure 1.12), the Shield now supports over two million people (Urquizo et al., 1998), and their individual and collective demands on the aquatic and terrestrial resources have changed greatly. The subsistence lifestyle of the native people is now largely gone, but even before it disappeared it underwent some important changes. For example, fish increased in importance as a subsistence food after European contact (Gillespie, 1981). The improved fishing techniques that the Europeans brought, such as metal tools for cutting ice, or various twines for making lines and nets, made capturing fish far easier. The gillnet was one of these improvements. Gillnets appear to have been used in Shield lakes only after European contact, even though they were in use for thousands of years in other parts of North America. In the postcontact period the total demand for fish for food also increased. Big game began to decline, and the need for fish as food for sled dogs increased with the development of the fur trade.

The technological changes in native fishing techniques and demand were substantial during the past few hundred years, but they pale in comparison to the explosion in harvest of fish from small inland lakes in recent decades, a period when subsistence use of aquatic resources was largely replaced by recreational use by a wide variety of local and nonresident sport anglers. Snowmobiles, light aircraft, and all-terrain vehicles now make all lakes accessible throughout the year. Power augers, downriggers, and fish finders add further to the efficiency of the modern angler. With these changes, not only has the harvest of the scarce resources of the Shield increased dramatically (Post et al., 2002), but the degradation of various atmospheric, terrestrial, and aquatic components of these ecosystems by other

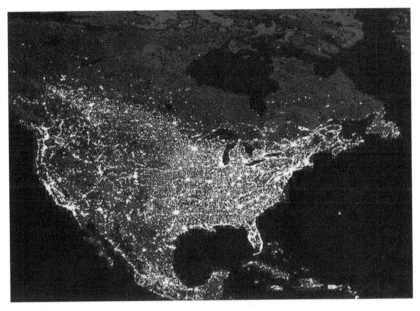

Figure 1.12 Today's population distribution and density can be inferred from this image of North America's city lights. It was created with data from the United States Defense Meteorological Satellite Program Operational Linescan System (source: visibleearth.nasa.gov/Human_ Dimensions/Population/index_2.html).

Figure 1.13 Lake trout have helped sustain humans on the Shield for thousands of years.

human activities has generally reduced the productive capacity of the systems. Humans have gone from being a small component, or even what we might consider a "product" of these Shield ecosystems (Figure 1.13), to being one of the largest influences on the landscape. Illustrating these trends and discussing ways of better managing our effects on these ecosystems will be the topic of later chapters.

Acknowledgments

We would like to thank Geoff Power for inspiration and information regarding the glacial history of lake charr, and Rob Steedman for his excellent editing efforts. Data on lake trout lakes were kindly provided by agency personnel in Ontario, Quebec, Minnesota, New York, Michigan, and Wisconsin. Mike Malette compiled the data, and the Ontario Ministry of Natural Resources provided financial support.

References

Bailey, R.G., 1996, *Ecosystem Geography*, Springer-Verlag, New York

Bailey, R.G., Zoltai, S.C., and Wiken, E.B., 1985, Ecological regionalization in Canada and the United States, *Geoforum.* 116(3): 265–275.

Cox, E.T., 1978., *Counts and Measurements of Ontario Lakes*. Ontario Ministry of Natural Resources, Toronto.

Douglas, R.J.W. (Ed.), 1970, *Geology and economic minerals of Canada. Economic Geology Report No.1. Map 1250A, A Geological Map of Canada at scale of 1:5 000 000*, Geological Survey of Canada, Department of Energy, Mines and Resources Canada, Ottawa.

Ecological Stratification Working Group, 1995, *A National Framework for Canada.* Cat. No. A42–65/1996E, Agriculture and Agri-Food Canada and Environment Canada, Ottawa.

Fulton, R.J. (compiler), 1995, *Surficial materials of Canada. Map 1880A, Scale 1:5 000 000*, Geological Survey of Canada, Natural Resources Canada, Ottawa.

Evans, D.O. and Olver, C.H., 1995, Introduction of lake trout (*Salvelinus namaycush*) to inland lakes of Ontario, Canada: factors contributing to successful colonization, *Journal of Great Lakes Research.* 21(Suppl 1): 30–53.

Gillespie, B.C., 1981, Major fauna in the traditional economy. In *Handbook of the North American Indian*, Vol. 6, edited by W.E. Sturtevant, Smithsonian Institution, Washington, D.C., pp. 15–18.

Gunn, J.M., 2002, Impact of the 1998 El Nino event on a lake charr, *Salvelinus namaycush* population recovering from acidification. *Environmental Biology of Fishes.* 64: 343–351.

Loftus, K.H., 1958, Studies on river spawning populations of lake trout in eastern Lake Superior. *Transactions of the American Fisheries Society.* 87: 259–277.

Martin, N.V. and Olver, C.H., 1976, *The distribution and characteristics of Ontario lake trout lakes. Fish and Wildlife Research Report. 97*, Ontario Ministry of Natural Resources, Toronto.

Martin, N.V. and Olver, C.H., 1980, The lake charr, *Salvelinus namaycush*. In *Charrs: Salmonid Fishes of the Genus* Salvelinus, edited by E.K. Balon, Dr. W. Junk, The Hague, pp. 205–277.

Olson, D.M., Dinerstein, E., Wikramanayake, E.D., Burgess, N.D., Powell, G.V.N., Underwood, E.C., D'Amico, J.A., Strand, H.E., Morrison, J.C., Loucks, C.J., Allnutt, T.F., Lamoreux, J.F., Ricketts, T.H., Itoua, I., Wettengel, W.W., Kura, Y., Hedao, P., and Kassem, K., 2001, Terrestrial ecoregions of the world: a new map of life on Earth, *BioScience.* 51(11): 933–938.

Pitblado, J.R., 1992, Landsat views of Sudbury (Canada) area acidic and nonacidic lakes, *Canadian Journal of Fisheries and Aquatic Sciences.* 49 (Suppl. 1): 33–39.

Post, J.R., Sullivan, M., Cox, S., Lester, N.P., Walters, C.J., Parkinson, E.A., Paul, A.J., Jackson, L., and Shuter, B.J., 2002, Canada's recreational fisheries: the invisible collapse, *Fisheries.* 27(1): 6–17.

Power, G., 2002, Charrs, glaciation and seasonal ice, *Environmental Biology of Fishes.* 64: 17–35

Rogers, E.S. and Smith, J.G.E., 1981, Environment and culture in the Shield and Mackenzie Borderlands. In *Handbook of North American Indian*, Vol. 6, edited by W.E. Sturtevant, Smithsonian Institution, Washington, D.C., pp. 130–145.

Ryder, R.A. and Johnson, L., 1972, The future of salmonid communities in North American oligotrophic lakes, *Journal Fisheries Research Board of Canada.* 29: 941–949.

Schindler, D.W., 1998, A dim future for boreal waters and landscapes. *BioScience.* 48: 157–164.

Shuter, B.J., Jones, M.L., Korver, R.M., and Lester, N.P., 1998. A general life history based model for regional management of fish stocks: The inland lake trout (*Salvelinus namaycush*) fisheries in Ontario, *Canadian Journal of Fisheries and Aquatic Sciences.* 55: 2161–2177.

Snucins, E.J. and Gunn, J.M., 1995, Coping with a warm environment: behavioural thermoregulation by lake trout, *Transactions of the American Fisheries Society.* **124**: 118–123.

Urquizo, N., Brydges, T., and Shear, H., 1998, *Ecological Assessment of the Boreal Shield Ecozone*, Environment Canada, Ottawa.

Vander Zanden, M.J., Casselman, J.M., and Rasmussen, J.R., 1999, Stable isotope evidence for the food web consequences of species invasions in lake, *Nature.* **401**: 464–467.

Wilson, C. and Mandrak, N., 2003, History and evolution of lake trout in shield lakes: past and future challenges. In *Boreal Shield Watersheds: Lake Trout Ecosystems in a Changing Environment*, edited by J.M. Gunn, R.J. Steedman, and R.A. Ryder, Lewis/CRC Press, Boca Raton, in this volume.

Winograd, I.J., Landwehr, J.M., Ludwig, K.R., Coplen, T.B., and Riggs, A.C., 1997. Duration and structure of the past four interglaciations. Quaternary Research. **48**: 141–154.

Wright, J.V., 1981, Prehistory of the Canadian Shield. In *Handbook of North American Indian*, Vol. 6, edited by W.E. Sturtevant, Smithsonian Institution, Washington, D.C., pp. 86–96.

chapter two

History and evolution of lake trout in Shield lakes: past and future challenges

Chris C. Wilson
Ontario Ministry of Natural Resources
Nicholas E. Mandrak
Department of Fisheries and Oceans

Contents

Introduction

Evidence suggests that lake trout likely evolved in response to environmental changes caused by glaciation events during the Pleistocene era. Following the retreat of the last ice sheets, lake trout recolonized their present range from multiple refugia. Populations in Boreal Shield lakes are descended from at least six refugial groups, with the greatest contributions from Mississippian and Atlantic sources. Secondary contact among these separate lineages was a significant source of genetic diversity, enabling adaptation to local environments.

Surprisingly little is known about the diversity and adaptive structure of inland lake trout populations. There is some evidence of local adaptation among inland populations and stocking strains as well as differential fitness between native and stocked lake trout.

However, many issues regarding lake trout adaptation and fitness remain unresolved. In addition, the biological characteristics of lake trout make them extremely vulnerable to natural and anthropogenic disturbances, which present significant management challenges to ensure their persistence.

Lake trout have often been described as a "glacial relict," a holdover from the Ice Age that lives primarily in deep, cold-water habitats. The species has certainly been intimately associated with glacial events and is perhaps better adapted to the past than the present. The fascination of anglers and biologists with lake trout may in part stem from the prehistoric mystique of this giant, primitive-appearing freshwater salmonid. In fact, however, lake trout are anything but ancient and primitive, having evolved no more than 3 million years ago and being specialized for habitats that did not exist before the Pleistocene glaciations.

Based on morphologic systematic and genetic data, lake trout are thought to have diverged from other species of *Salvelinus* approximately 1 to 3 million years ago (Behnke, 1972; Grewe et al., 1990; Phillips et al., 1992), roughly coincident with the onset of the Pleistocene glaciations (Dawson, 1992). Lake trout have the most extensive freshwater distribution of any salmonine: the species' natural distribution closely matches the North American limits of the recent (Wisconsinan) glaciation (Lindsey, 1964), where it is largely restricted to lakes carved out by glacial scouring. Lake trout also differ from all other members of the genus by lacking the ability to live in salt water, suggesting their evolution in a purely freshwater environment. Taken together, these facts suggest that the species evolved in response to altered habitat and ecological conditions caused by Pleistocene glacial events.

The purpose of this chapter is to describe the historical influences that underpin the evolutionary biology of lake trout populations in small lakes on the southern Canadian Shield. Beginning with the historical origins of lake trout (when they first appeared and where they came from), and postglacial history (how they arrived where they are today), we examine the influence of these events on modern populations. As well as reconstructing past events that influenced the current structure and diversity of lake trout, this chapter attempts to summarize both what is known and what needs to be learned about the present biology and potential future of these populations. As well as developing some suggested research themes, some recommendations for sustainable management are presented.

Evolutionary history

The early details of lake trout evolution will likely remain unknown. The close relationship between lake trout distribution and formerly glaciated areas (Lindsey, 1964; Scott and Crossman, 1973) suggests that fossil remains preceding the Wisconsinan glaciation were destroyed long ago by advancing glaciers. However, the roughly coincident divergences of the three subgenera of *Salvelinus* (*Salvelinus, Cristivomer*, and *Baione*) approximately 2 to 3 million years ago (Behnke, 1972, 1980; Grewe et al., 1990; Phillips et al., 1992) indicate a short burst of evolutionary diversification and selection, potentially in response to major environmental or biotic pressures. The placement of lake trout in its own subgenus (*Cristivomer*) indicates the extent of its divergence from other charr species in response to Pleistocene events (Behnke, 1972).

The nature of glacial disturbances makes it impossible to reconstruct the complete Pleistocene history of lake trout. Each glaciation effectively bulldozed continental landscapes under a wall of ice, wiping clean the geologic and biologic traces of preceding events. Because of this, the exact number of Pleistocene glaciations in North America is uncertain, with estimates ranging between 4 and 20 cycles (Pielou, 1991; Dawson, 1992; Gunn and Pitblado, Chapter 1, this volume). Each of these undoubtedly had major impacts

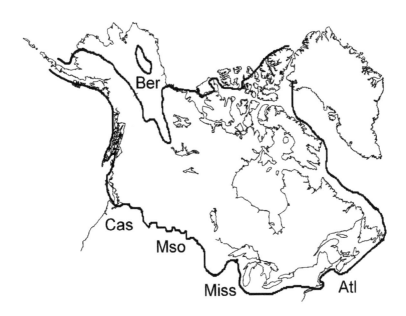

Figure 2.1 Present-day distribution of lake trout, *Salvelinus namaycush,* in relation to geographic coverage of glacial ice at the maximum of the Wisconsinan glaciation, approximately 23,000 years ago. Recognized glacial refugia for northern fishes are shown along the glacial margins (Atl = Atlantic; Ber = Beringia; Cas = Cascadia; Miss = Mississippi; Mso = Missouri). Glacial coverage taken from Dyke and Prest (1987b); lake trout distribution from Lee et al. (1980); glacial refugia from Crossman and McAllister (1986).

on the distribution and genetic structure of lake trout, based on the species' present distribution in relation to glacial coverage (Figure 2.1). During each glacial interval, fish populations were extirpated or displaced, in some cases by thousands of kilometers. Those fish that survived were limited to a handful of refugial habitats along the periphery of the ice sheets.

Four main southern and one northern Wisconsinan refugia have been hypothesized (Bailey and Smith, 1981; Crossman and McAllister, 1986; Figure 2.1). The Atlantic Coastal refugium existed along the coastal plain east of the Appalachians, south of present-day Long Island. Smaller Atlantic Coastal Uplands and Grand Banks refugia were also present along the east coast of North America (Schmidt, 1986). The Mississippi basin, south of the ice sheets, was the largest Wisconsinan refugium and was separated from the Missourian refugium by a lobe of ice. Several small refugia (e.g., Nahanni, Banff-Jasper, Waterton) also existed between the Cordilleran and Laurentide ice sheets (Crossman and McAllister, 1986; Lindsey and McPhail, 1986). The Pacific Coastal (or Cascadia) refugium existed along the coastal plain west of the Rockies, south of present-day Vancouver Island. Part of present-day Alaska and land now covered by the Bering Sea formed the Beringian refugium in northwestern North America (Figure 1.1).

Very little is known about the habitat or ecological conditions in these peripheral glacial refuges, including their exact locations. On a local scale, ice movements and shifting glacial margins would have changed the locations of meltwater impoundments. The weight of the ice sheets depressed the landscape, so that water draining downhill would have been dammed by the glaciers themselves (Pielou, 1991). These periglacial aquatic habitats, which were both dammed and fed by the ice sheets, have no parallel today. Pielou (1991) gives an excellent description of ice-dammed lakes to the south of the glaciers, with icebergs and deep, cold water next to the ice walls, gradually becoming shallower to the south. These proglacial lakes probably stratified in summer, and received a variety of substrates from rock flour to boulders from the ice and meltwater. As a result, these lakes may have contained a variety of habitats that supported diverse aquatic communities. Although lake trout probably thrived under these conditions, they were nonetheless confined to a small number of fringe habitats or regional metapopulations for tens of thousands of years during each glaciation (Bailey and Smith, 1981).

As the glaciers retreated at the end of each glacial cycle, huge volumes of meltwater were released, creating giant lakes that dwarfed any lake that exists today. These lakes were only temporary, however, and constantly changed their size, shape, and location in response to the retreating ice (Dyke and Prest, 1987b). As the glaciers receded and the newly exposed landscape rose up through isostatic rebound, the lakes followed the retreating glacial margins. Further retreat uncovered progressively lower drainage outlets, often causing dramatic drainage events and emptying or connection of water bodies (Mandrak and Crossman, 1992). The size, mobility, and interconnectedness of these proglacial lakes provided unprecedented opportunities for dispersal and colonization of the newly exposed landscape. These postglacial drainage events permitted movement of lake trout over hundreds and perhaps thousands of kilometers, potentially within a single generation.

Ecological conditions within the proglacial lakes changed considerably over time and may have varied significantly within each lake as well. For example, the Champlain Sea — a marine incursion of the St. Lawrence Lowlands between 11,800 and 8800 years ago — underwent dramatic ecological change, going from a fresh, meltwater lake (glacial Lake Vermont) to an inland sea to fresh again (Lampsilis Lake) in just a few thousand years (Dyke and Prest, 1987a; Pielou, 1991). There is also some evidence for spatial variation in habitat conditions within the major lakes, which is not surprising considering their enormous sizes. Thermal gradients, glacial till, and varying depth as a result of crustal deformation may have effectively segregated species within these lakes based on their ecological requirements.

The huge volumes of cold-water habitat and diverse prey community would have been ideal for lake trout, and it may be assumed that lake trout populations expanded rapidly in these proglacial habitats. As the glaciers continued to retreat northward, new drainage outlets were exposed, which enabled the meltwater lakes to drain over newly deglaciated terrain. This provided new connections and colonization opportunities among the proglacial lakes but also decreased the remaining habitat volume. As the volume of meltwater declined with continued glacial retreat, the proglacial lakes gradually disappeared, leaving most of the modern drainage systems in place by about 5000 to 6000 years ago (Dyke and Prest, 1987a).

Most of the present lake trout lakes on the Shield began as basins excavated by the glaciers. Many of these depressions were covered by glacial lakes, which facilitated extensive colonization. Other basins that were isolated from the large glacial lakes provided reduced opportunities for colonization. As a result, the timing and extent of colonization were highly variable and strongly influenced by lake elevation, distance from source pools (proglacial lakes), and type of access (lake source, outlet, postglacial connections) (Mandrak, 1995). As a result of their different origins and histories, lakes on the Shield were isolated for varying lengths of time, with considerably variable colonization opportunities. Lakes covered by proglacial meltwaters were likely colonized by more species and by greater numbers within species than uncovered lakes, leading to variation in the composition and size of the founding gene pools and assemblages. Conversely, areas that were not covered by the meltwater lakes had reduced colonization opportunities, which led to lower species diversity, smaller founding numbers, and reduced biodiversity at community, population, and genetic levels. As water levels dropped and colonization opportunities ended, the immigrant populations and assemblages were forced to adapt to local abiotic and biotic conditions in the diverse mosaic of smaller inland lakes that exists today.

Glacial refugia and historical genetics: peering into the past

The number and locations of refugia used by lake trout during the last (Wisconsinan) glacial event have been debated for half a century. The many, often-conflicting views have been summarized elsewhere (Martin and Olver, 1980; Crossman and McAllister, 1986; Wilson and Hebert, 1996, 1998) and need not be repeated here. The debate over the species' Quaternary history has largely been resolved by recent genetic evidence, which indicates that lake trout colonized their modern range from multiple refugia. Ihssen et al. (1988) detected two genetic groups of lake trout in Ontario and Manitoba based on allozyme data, which they interpreted as being Mississippian and Atlantic in origin. In addition, they inferred a third, distinct source for several lake trout populations in the Haliburton area of eastern Ontario (Ihssen et al., 1988). Analysis of mitochondrial DNA (mtDNA) among eastern hatchery strains similarly revealed three distinct lineages, which were interpreted as having Atlantic, Mississippian, and Beringian origins (Grewe and Hebert, 1988). Geographic surveys of mtDNA variation among native lake trout populations across the species' range confirmed the existence of multiple refugial groups (Wilson and Hebert, 1996, 1998). In addition to the three major lineages identified by Grewe and Hebert (1988), genetic and geographic substructure within these lineages suggested their dispersal from multiple glacial refugia (Figure 2.2; Wilson and Hebert, 1996, 1998). Based on the geographic distribution of the different mtDNA lineages, it appears that lake trout survived the Wisconsinan glaciation in at least six different refugia: two Beringian, two Mississippian, one Atlantic, and one in southern Alberta or northern Montana (Wilson and Hebert, 1996, 1998). The postglacial scenario described here largely relies on these latter studies while also drawing on data from earlier studies and complementary sources (Dyke and Prest, 1987a,b; Mandrak and Crossman, 1992).

Although much of the species' genetic history and diversity has undoubtedly been lost because of prehistoric (Pleistocene) events, the mode and tempo of genetic differentiation among the different lineages was largely influenced by the timing and extent of glacial cycles. The mtDNA C lineage mostly likely diverged from other lake trout during the mid-Pleistocene, at least 600,000 years ago (Figure 2.2; Wilson and Hebert, 1998). Separation of fish with A mtDNA types from groups B and D probably occurred roughly 300,000 years ago, and the latter two groups diverged during subsequent glaciations (Wilson and Hebert, 1998). Divergence within the C mtDNA group was probably caused by the displacement of ancestral populations into separate Beringian, Montana, and Mississippian refugia during the Wisconsinan glaciation, where they subsequently diverged (Wilson and Hebert, 1998). Similar processes within each glacial cycle (population displacement and subdivision by advancing glaciers, allopatric divergence in separate refugia, and subsequent expansion/recolonization) were probably responsible for much of the geographic genetic structure that exists among lake trout populations today.

Postglacial dispersal

The major mtDNA lineages show clear differences in their geographic distribution despite considerable overlap in some regions (Figure 2.3). Populations in Quebec, eastern Ontario, the maritime provinces, and New England were primarily colonized by lake trout from an Atlantic refuge, with some Atlantic fish reaching as far west as Lake Superior and Lake Nipigon. Fish from two Mississippian sources (lineages A and C1) colonized areas south

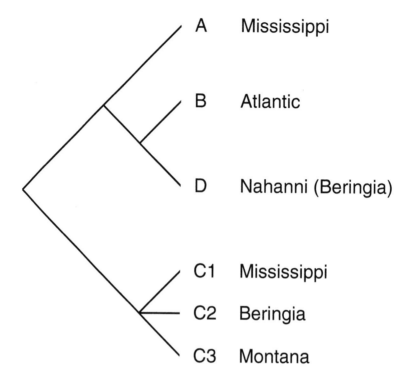

Figure 2.2 Simplified cladogram showing *S. namaycush* mitochondrial DNA lineages (A–D) detected by Wilson and Hebert (1996, 1998) and their association with glacial refugia during the most recent glaciation. Branch lengths are not proportional to genetic divergence among lineages.

of the Great Lakes, virtually all of Ontario, parts of Quebec, and northwestward through central Canada to Great Slave Lake. Lake trout from two separate Beringian refugia dispersed south and east but primarily contributed to areas west of the Canadian Shield. For southern Shield lakes, Beringian lake trout contributed to populations in western Ontario and Quebec as well as the upper Great Lakes. Lake trout from the Montana refuge also made minor contributions to western Ontario populations, but the Mississippian and Atlantic refugia were the major contributing sources for establishing populations in most of the study area (Figure 2.3).

The detailed geologic record left by the glaciers and proglacial lakes has enabled the reconstruction of postglacial drainage and colonization events (Figure 2.4; see also Mandrak and Crossman, 1992). Fish communities in the study area were greatly influenced by the formation of giant proglacial lakes in and near what is now the Great Lakes basin between 14,500 and 7900 years ago (Dyke and Prest, 1987a; Mandrak and Crossman, 1992). Lake Agassiz, which lasted for most of this period, covered much of Manitoba and large portions of Saskatchewan, North Dakota, Minnesota and Ontario during its existence, providing dispersal and colonization opportunities for many species that were present in the Mississippian and Missourian refugia (Teller and Clayton, 1983; Mandrak and Crossman, 1992).

Lake Agassiz drained southward into the Mississippi basin, briefly to the northwest, then southeastward into the Superior basin, and finally eastward into Lake Barlow-Ojibway (Mandrak and Crossman, 1992). Lake Barlow-Ojibway extended across northeastern Ontario and northwestern Quebec, and drained into the Ottawa River before abruptly emptying into James Bay 7900 years ago (Dyke and Prest, 1987b). A series of proglacial and glacial lakes also drained into Lake Agassiz from the west, facilitating immigration of lake trout from western refugia (Wilson and Hebert, 1998).

The fossil evidence suggests that lake trout were able to closely follow the retreating glacial margins: the species was present in northern Wisconsin (Hussakof, 1916) 12,500 to 16,000 years ago (Lindsey 1964) and in eastern Ontario approximately 10,000 years ago (Gruchy, 1968, in Crossman and McAllister, 1986). This close connection among lake trout movements, glacial retreat, and proglacial lake positions and connections is also supported by the genetic data (Wilson and Hebert, 1996, 1998). Colonization of the study area by lake trout with Mississippian and Atlantic origins occurred initially through glacial lakes in the lower Great Lakes basin, roughly 14,000 to 13,500 years ago (Wilson and Hebert, 1996). Although dispersal from the Atlantic refuge may have been hindered by the Champlain Sea, eastern lake trout were probably able to effectively colonize southern Shield lakes by 10,000 years ago (Wilson and Hebert, 1996). A series of proglacial lakes (Lakes Peace, McConnell, Agassiz, and Barlow-Ojibway) extending from the Yukon southeast through to Quebec allowed colonization by lake trout with Atlantic, Beringian, Mississippian, and, to a lesser extent, Montanan affinities (Wilson and Hebert, 1996, 1998; Figure 2.4). Lake trout from these western refugia did not reach the Canadian Shield until 9500 years ago (Wilson and Hebert, 1996). It is possible that these groups did not make greater contributions because of early colonization by Mississippian and Atlantic fish (Wilson and Hebert, 1996, 1998). Early populations would have expanded rapidly, creating a strong priority effect. Alternatively, isolation in separate refugia may have resulted in differences in fitness among lineages under varying conditions, although this has not been tested.

Secondary contact among refugial groups

Secondary contact among the different refugial groups occurred throughout the central portion of the species range, particularly within areas covered by the proglacial lakes (Figures 2.3 and 2.4). In the principal study area, contact was most extensive between the Mississippian and Atlantic refugial groups, with both lineages occurring through southern Quebec and Ontario as well as northern New York State (Wilson and Hebert, 1996).

This extensive contact and interbreeding among refugial groups accounts for some disagreements or conflicts among earlier studies of lake trout postglacial history. This is clearly illustrated by previous morphologic analyses: although differences among geographic populations were recognized by Lindsey (1964) and quantified by Khan and Qadri (1971), the extensive contact among refugial groups prevented discrimination of postglacial history among most populations examined (Crossman and McAllister, 1986). It is also possible that adaptation to local conditions resulted in morphologic differentiation, further confounding historical analyses.

Other data sources also show evidence of secondary contact among refugial groups. The parasitic nematode *Cystidicola stigmatura* uses lake trout and the opossum shrimp *Mysis relicta* as its primary and alternate hosts and occurs in many lake trout populations in Ontario and southwestern Quebec (Black, 1983). Although originally used to infer dispersal from a Mississippian refuge (Black, 1983), the distribution of *C. stigmatura* likely also reflects colonization events by other lake trout that came into contact with infected Mississippian-origin hosts. Mixed-origin ancestry of Shield lake trout populations was also shown by Ihssen et al. (1988) using allozyme data. Although their data were initially interpreted as showing the allopatric distribution of lake trout from Atlantic versus Mississippian refugia (Ihssen et al., 1988), it actually provides detailed information on the extent of two contact zones in Ontario. The majority of populations surveyed east of 82°W show shared ancestry from Mississippian and Atlantic fish, whereas Atlantic-refuge fish

Figure 2.3 Geographic distributions of *S. namaycush* refugial groups, as evidenced by distribution of mitochondrial DNA lineages (modified from Wilson and Hebert 1996, 1998).

are largely absent from more western populations (Ihssen et al., 1988). Rather than representing purely Mississippian fish, however, these western populations resulted from contact of Mississippian lake trout with those from the three western refugia (Figures. 2.3 and 2.4). Ironically, the "glacial relict" populations of lake trout identified by Ihssen et al. near Haliburton, Ontario appear to be isolated descendants of Mississippian fish that did not come into contact with lake trout from other refugia (Wilson and Hebert, 1996), probably because of the brief time window for colonization as the glaciers retreated (Mandrak and Crossman, 1992).

Genetic structure of modern populations

Although historical events played a major role in structuring the genetic and phenotypic diversity of lake trout, populations in southern Shield habitats have been adapting to local conditions for over a thousand generations and now reflect the joint influence of historical and ecological factors. The low levels of genetic diversity that have been observed in lake trout (Ihssen et al., 1988; Wilson and Hebert, 1996, 1998) are similar to those observed in other species from northern climes (Bernatchez and Dodson, 1991; Billington and Hebert, 1991; Wilson et al., 1996; Bernatchez and Wilson, 1998). This is most likely a result of constrained or bottlenecked refugial populations during glaciations (Avise et al., 1984). In addition, it is possible that genetic diversity of lake trout has been further reduced by demographic constraints on populations such as large-bodied, long-lived top predators that generally live in low productivity environments.

Levels of diversity within populations are likely constrained by a combination of historical and ecological factors. For example, lake trout populations that occur within areas formerly covered by proglacial lakes have relatively high levels of mtDNA diversity as a result of secondary contact of refugial groups. Conversely, populations outside the

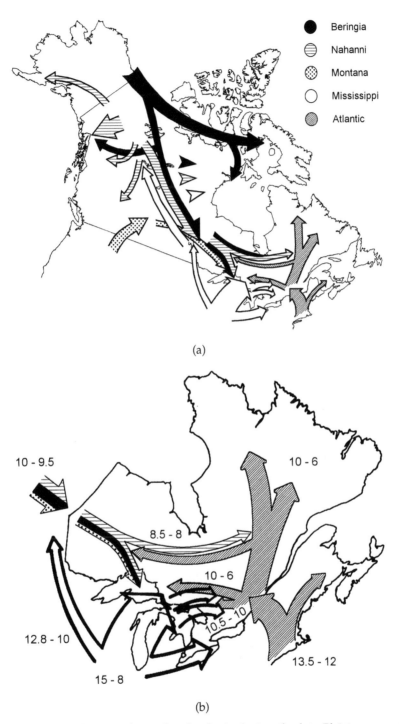

(a)

(b)

Figure 2.4 Dispersal of lake trout from glacial refugia during the late Pleistocene and Holocene (18,000 years ago to the present). (a) Recolonization of the species range; (b) timing of colonization events in the principal study area. Modified from Wilson and Hebert (1996, 1998).

former lake borders have very low levels of genetic (mtDNA) diversity, which likely reflect the more restricted opportunities for colonization and smaller sizes of founding populations (Wilson and Hebert, 1998).

In addition, genetic diversity of populations may be limited by volume of suitable habitat (lake area vs. diversity) and habitat age (time since deglaciation vs. diversity). This is largely speculative, although Ihssen et al. (1988) observed a significant correlation between lake area and the proportion of allozyme loci that were polymorphic in the resident lake trout population ($p < 0.01$). Larger lakes in their survey also had more alleles per locus than did small inland lakes, although this may have been partly confounded by historical (colonization) factors. Ihssen et al. attributed their findings to population size effects, as large populations may be expected to contain greater genetic diversity than smaller populations.

Although many existing populations of lake trout have been in place for less than 10,000 years, there have been ample opportunities for local adaptation to occur. The areas of secondary contact among the different refugial groups contained huge potential for selection and local adaptation, as the coming together of lake trout that had been isolated and exposed to different ecological conditions for tens of thousands of years would have provided plenty of raw material for natural selection to act on. This would have been compounded as the different lineages began to interbreed, as their offspring would have possessed entirely new combinations of genes and adaptive abilities. Recent evidence that significant evolutionary change can occur in short ecological time spans in some species (Thompson, 1998) may indicate that selection and adaptation to local environments occurred within the first few centuries following colonization. This is supported by the residual genetic structure of lake trout in Lac des Chasseurs, Quebec, where the entire population carries an mtDNA haplotype that may have provided a selective advantage in the early stages of population formation (Wilson and Bernatchez, 1998).

There is anecdotal evidence for multiple locally adapted stocks within the Great Lakes, although many of these disappeared as lake trout stocks declined (Brown, 1981; Goodier et al., 1981). For populations in inland lakes, there is less evidence for local adaptation. It must be emphasized, however, that this lack of evidence reflects the different focus of past research efforts and the difficulties involved in measuring local adaptation, rather than the lack of local adaptation itself. Ecological data such as life history, growth rates, and spawning preferences certainly suggest that inland populations are well adapted to their particular surroundings (Martin and Olver, 1980). Most evidence of local adaptation is indirect, such as examples of native populations having higher fitness than stocked lake trout (described below). One clear example of local specialization was the subspecies *Salvelinus namaycush huronicus*, which was described by Hubbs (1929) from Rush Lake in northern Michigan. This dwarf subspecies or morph coexisted with normal lake trout by inhabiting deeper waters and living as a benthic feeder (Hubbs, 1929). In addition to being at most one-quarter the size of the more pelagic regular form, these fish had broader, blunt heads and large fins (Hubbs and Lagler, 1959). Hubbs (1929) also reported that *huronicus* spawned in mid- to late summer, at a depth of about 120 feet. Hubbs was of the opinion that the two forms did not interbreed and were very distinct ecologically. Sadly, this unique subspecies is now presumed to be extinct (Hubbs and Lagler, 1959).

There is surprisingly little evidence dealing with adaptive variation of lake trout populations from small inland lakes despite abundant anecdotal data about individual populations. Given the diverse history, sources, and length of isolation of many Shield lakes, however, adaptation to local conditions has most likely led to considerable genotypic and phenotypic variation among resident populations. Numerous color variants have been identified in Shield lake trout populations, including black fins (e.g., Redstone Lake) and pale or dark background coloration (e.g., Lost Dog Lake and Canisbay Lake in Algonquin Park). One unique population that is both visually and genetically distinct from neighboring populations is Kingscote Lake in southern Algonquin Park: lake trout native to this lake lack any spotting pattern and have a uniform silvery-gold body coloration (C. Wilson, unpublished data). In general, however, it would be inadvisable to assign too much importance to color variants because coloration of lake trout may be influenced by diet (Scott and Crossman, 1973; Martin and Olver, 1980). Greater variability is exhibited in other morphologic characteristics such as pyloric caecae and head morphology (Martin and Olver, 1980; P. Ihssen, Ontario Ministry of Natural Resources, unpublished data). It is likely that lake trout populations exhibit substantial genetic variation at both neutral marker systems and functional gene loci as the result of their varying histories and subsequent adaptation. Identifying and assessing the significance of potentially adaptive genetic diversity both within and among populations is an urgent need for both research and management.

In contrast to other salmonids, lake trout show very little tendency to diversify into specialized sympatric morphs or subspecies, although the Rush Lake *S. namaycush huronicus* and the siscowet, *S. namaycush siscowet*, are notable exceptions. Interestingly, mtDNA from at least four refugial groups has been observed in siscowet, indicating that this morph has evolved from mixed origins within the last 14,000 years, since the last glacial retreat (Burnham-Curtis, 1993; Wilson and Hebert, 1996, 1998). As an aside, it is worth noting that both examples of subspecific differentiation in lake trout come from extremely deep lakes that may allow for vertical partitioning of predator niches. It may be that what are generally considered normal conditions for lake trout (cold, low productivity, well-oxygenated lakes) provide stable ecological environments that limit or constrain ecological diversification and differentiation within the species.

For many years it was thought that lake trout had preprogrammed developmental tendencies for piscivory or planktivory, and many populations were classified as large-growing (piscivorous) or small (planktivory). The contrast between lake trout in Lake Opeongo and those in Lake Louisa in Algonquin Park provided a widely touted example of this (Martin and Olver, 1980). Recent data, however, indicate that either feeding/growth type may be expressed by the same population under different circumstances. Using stable isotope analysis, Vander Zanden et al. (1999) demonstrated that formerly piscivorous lake trout populations switched to planktivory after their lakes were invaded by rock bass.

Although lake trout have been intensively studied for decades, recent studies may indicate that we do not yet understand its ecological niche. The strong historical focus on Great Lakes populations has created a perception of lake trout as having narrow ecological tolerances for ecological factors such as temperature and oxygen levels (e.g., Behnke, 1972). Work done on lakes in Ontario's Experimental Lakes Area, however, has demonstrated that thermal characteristics of utilized habitats can vary considerably among lakes (Sellers et al., 1998). Similarly, Snucins and Gunn (1995) have shown that lake trout may have adaptive behavioral mechanisms and thermal tolerances that do not fall within stereotypic confines. An intriguing possibility is that Great Lakes and inland lake trout have become qualitatively different organisms over the past 10,000 years as a result of divergent selective and evolutionary trajectories despite their recent (in evolutionary time) shared history. If so, a great deal of what we think we understand about lake trout biology may need to be revisited. It may also be that the historical niche of lake trout in inland lakes was broader than we now recognize but that most "unusual" populations such as those described above disappeared soon after human colonization because of exploitation and eutrophication. Models that attempt to describe the niche and ecological characteristics of lake trout, such as Ryan and Marshall (1994) and Shuter et al. (1998), should broaden our understanding and provide greater appreciation for the species' adaptive abilities.

Species interactions and anthropogenic disturbances

The close relationship between the distribution and biology of lake trout and Pleistocene glacial events highlights the interplay between historical and ecological processes in structuring modern populations. It should be noted that many communities or species assemblages in formerly glaciated areas are not in equilibrium in terms of species composition (Mandrak, 1995). Instead, many of these communities represent subsets of the species assemblages that were present just before the drainage or disappearance of the giant proglacial lakes. As communities became constrained to smaller and more restrictive environments, ecological suitability to the available habitats and interactions between species have resulted in the species assemblages now present.

Perhaps the best known set of species associations for lake trout is the deepwater "glaciomarine relict" community (deepwater sculpin, *Mysis relicta*, and several other invertebrate species) (Dadswell, 1974). More recently, Evans and Olver (1995), in an analysis of lake trout introductions among 183 inland populations, found that resident fish communities had a significant effect on the success or failure of introductions. Examination of the Ontario Ministry of Natural Resources Lake Inventory database has also suggested the possibility of both positive and negative associations between lake trout and a surprising number of species (N. Mandrak, unpublished data). Many of these potential relationships, however, have yet to be experimentally or critically examined. Recent alarming data from rock bass range expansions indicate that lake trout can be quite sensitive to community perturbations (Vander Zanden et al., 1999; J. Casselman, Ontario Ministry of Natural Resources, personal communication). In addition, data from recovering acidified lakes suggest that differential species response to recovery may prevent or deter lake trout from reassuming their role as dominant top predator (Mills et al., 2000). Resolving the biologic relationships between inland lake trout and co-occurring species continues to be a significant research and management need.

The geographic and genetic structures of existing lake trout populations have been strongly influenced by historical and ongoing human activities. Past activities such as overexploitation, habitat degradation, stocking, and transplants have undoubtedly had major impacts on the structure and fitness of native populations (OMNR, 1991). Although such activities sometimes resulted in short-term socioeconomic gains, they also regularly incurred significant long-term ecological and economic costs such as habitat rehabilitation, population reestablishment and restoration, and long-term stocking programs to compensate for the absence of naturally reproducing native fish. Given the lake trout's ecological characteristics (top predator, long generation time, restrictive habitat requirements, etc.), we strongly recommend that the species be managed as conservatively as possible.

Of the major impacts that humans have had on lake trout populations (overexploitation, habitat loss/degradation, species introductions, and stocking), stocking has perhaps had the most insidious effect. Although each of the other factors has long been known to cause populations to disappear, stocking has only recently been recognized as having the potential to do the same (Evans and Willox, 1991). The luxury of having native populations with no stocking history is becoming scarce, and realistic management decisions must be made in a complex multiuser environment (Waples, 1991). Stocking has and will continue to be a potent management tool for a variety of purposes (Powell and Carl, this volume). For the dwindling number of native lake trout populations on the Shield, however, it carries considerable genetic risks, particularly in small lakes where the native population may have been progressively fine-tuned to local conditions over literally hundreds or thousands of generations. It is also worth noting that these same populations may be less adaptable than when they were first founded, as they no longer have a massive gene pool (mixed-origin founding population) from which to begin.

Research needs

Many important questions remain to be addressed, such as potential fitness differences between populations and/or ecotypes under different conditions. Some particular knowledge gaps that need to be filled are to:

- Characterize the genetic structure and diversity of Boreal Shield populations to assess effective population sizes and historical demographics from lakes of different sizes and ecological attributes
- Build a "genetic map" of lake trout populations in high-use areas to identify native, mixed, and hatchery populations in order to establish mixed-use management plans that balance conservation with resource use
- Assess the adaptive potential of populations from different histories and habitat sizes in terms of plasticity in life history traits, growth and maturation rates, thermal tolerance, etc.
- Assess the effects of exploitation and/or other anthropogenic stressors on long-term diversity and viability of inland lake trout populations

Conclusion

The long-term sustainability of wild lake trout populations requires that we conserve not only enough fish to catch but enough that populations can continue to evolve and adapt to future climatic and ecological conditions. The preservation of resident biodiversity is essential for ensuring the long-term viability of populations as well as maintaining their heritage value. This is particularly important for deepwater species such as lake trout, where isolated Shield populations are unlikely to be rescued or rejuvenated by immigration from other lake populations (Meffe, 1995). Given the species' ecological vulnerability, lake trout management must be governed from the perspective of resource health rather than being harvest-oriented. With changing times, this has largely been achieved in most of the jurisdictions within the Boreal Shield, but continuing angler demands, species invasions, and decreasing habitat quality and quantity will continue to make this a significant challenge.

References

Avise, J.C., Neigel, J.E., and Arnold, J., 1984, Demographic influences on mitochondrial DNA lineage survivorship in animal populations, *Journal of Molecular Evolution* 20: 99–105.

Bailey, R.M. and G.R. Smith, 1981. Origin and geography of the fish fauna of the Laurentian Great Lakes. *Canadian Journal of Fisheries and Aquatic Sciences* 38: 1539–1561.

Behnke, R.J., 1972, The systematics of salmonid fishes of recently glaciated lakes, *Journal of the Fisheries Research Board of Canada* 29: 639–671.

Behnke, R.J., 1980, A systematic review of the genus *Salvelinus*. In *Charrs: Salmonid Fishes of the Genus Salvelinus*, edited by E.K. Balon, Dr.W. Junk, The Hague, pp. 441–480.

Bernatchez, L. and Dodson, J.J., 1991, Phylogeographic structure in mitochondrial DNA of the lake whitefish (*Coregonus clupeaformis*) and its relation to Pleistocene glaciations, *Evolution* 45: 1016–1035.

Bernatchez, L. and Wilson, C.C., 1998, Comparative phylogeography of Nearctic and Palearctic fishes, *Molecular Ecology* 7: 431–452.

Billington, N. and Hebert, P.D.N., 1991, Mitochondrial DNA diversity in fishes and its implications for introductions, *Canadian Journal of Fisheries and Aquatic Sciences* Suppl. 1: 80–94.

Black, G.A., 1983, Origin, distribution, and postglacial dispersal of a swimbladder nematode, *Cystidicola stigmatura*, *Canadian Journal of Fisheries and Aquatic Sciences* 40: 1244–1253.

Burnham-Curtis, M.K., 1993, Intralacustrine speciation of *Salvelinus namaycush* in Lake Superior: an investigation of genetic and morphological variation and evolution of lake trout in the Laurentian Great Lakes, Ph.D. dissertation, University of Michigan.

Crossman, E.J. and McAllister, D.E., 1986, Zoogeography of freshwater fishes of the Hudson Bay drainage, Ungava Bay and the Arctic Archipelago, In *The Zoogeography of North American Freshwater Fishes*, edited by C.H. Hocutt and E.O. Wiley, John Wiley & Sons, New York, pp. 53–104.

Dadswell, R.J., 1974, *Distribution, ecology, and postglacial dispersal of certain crustaceans and fishes in eastern North America*. National Museum of Natural Science Publication 11.

Dawson, A.G., 1992, *Ice Age Earth*, Routledge Press, London.

Dyke, A.S. and Prest, V.K., 1987a, Late Wisconsinan and Holocene history of the Laurentide ice sheet, *Géographie Physique et Quaternaire* **41**: 237–263.

Dyke, A.S. and Prest, V.K., 1987b, Paleogeography of northern North America, 18,000–5,000 years ago *Geological Survey of Canada, Map 1703A*.

Evans, D.O. and Olver, C.H., 1995, Introduction of lake trout (*Salvelinus namaycush*) to inland lakes of Ontario, Canada — factors contributing to successful colonization, *Journal of Great Lakes Research* **21** (Suppl. 1): 30–53, 1995.

Evans, D.O. and Willox, C.C., 1991, Loss of exploited, indigenous populations of lake trout, *Salvelinus namaycush*, by stocking of non-native stocks. *Canadian Journal of Fisheries and Aquatic Sciences*, **48** (Suppl. 1): 134–147.

Goodier, J.L., 1981, Native lake trout (*Salvelinus namaycush*) stocks in the Canadian waters of Lake Superior prior to 1955. *Canadian Journal of Fisheries and Aquatic Sciences* **38**: 1724–1737.

Grewe, P.M. and Hebert, P.D.N., 1988, Mitochondrial DNA diversity among brood stocks of the lake trout, *Salvelinus namaycush, Canadian Journal of Fisheries and Aquatic Sciences* **45**: 2114–2122.

Grewe, P.M., Billington, N., and Hebert, P.D.N., 1990, Phylogenetic relationships among members of *Salvelinus* inferred from mitochondrial DNA divergence, *Canadian Journal of Fisheries and Aquatic Sciences* **47**: 984–991.

Hubbs, C.L., 1929, The fishes. In *The Huron Mountains*, Huron Mountain Club, Michigan, pp. 153–164.

Hubbs, C.L. and Lagler, K.F., 1959, *Fishes of the Great Lakes Region*, University of Michigan Press, Ann Arbor.

Hussakof, L., 1916, Discovery of the great lake trout, *Cristivomer namaycush*, in the Pleistocene of Wisconsin, *Journal of Geology* **24**: 685–689.

Ihssen, P.E., Casselman, J.M., Martin, G.W., and Phillips, R.B., 1988, Biochemical genetic differentiation of lake trout (*Salvelinus namaycush*) stocks of the Great Lakes region, *Canadian Journal of Fisheries and Aquatic Sciences* **45**: 1018–1029.

Khan, N.Y. and Qadri, S.U., 1971, Intraspecific variations and postglacial distribution of lake char (*Salvelinus namaycush*), *Journal of the Fisheries Research Board of Canada* **28**: 465–476.

Lee, D.S., Gilbert, C.R., Hocutt, C.H., Jenkins, R.E., McAllister, D.E., and Stauffer, J.R., Jr., 1980, *Atlas of North American Freshwater Fishes*. North Carolina Biological Survey Publication 1980–12.

Lindsey, C.C., 1964, Problems in zoogeography of the lake trout *Salvelinus namaycush, Journal of the Fisheries Research Board of Canada* **21**: 977–994.

Lindsey, C.C. and McPhail, J.D., 1986, Zoogeography of fishes of the Yukon and Mackenzie basins. In *The Zoogeography of North American Freshwater Fishes*, edited by C.H. Hocutt and E.O. Wiley, John Wiley & Sons, Toronto, pp. 639–674.

Mandrak, N.E., 1995, Biogeographic patterns of fish species richness in Ontario lakes in relation to historical and environmental factors, *Canadian Journal of Fisheries and Aquatic Sciences* **52**: 1462–1474.

Mandrak, N.E. and Crossman, E.J., 1992, Postglacial dispersal of freshwater fishes into Ontario, *Canadian Journal of Zoology* **70**: 2247–2259.

Martin, N.V. and Olver, C.H., 1980, The lake charr, *Salvelinus namaycush*. In *Charrs: Salmonid Fishes of the Genus* Salvelinus, edited by E.K. Balon, Dr. W. Junk, The Hague, pp. 205–277.

Meffe, G.K., 1995, Genetic and ecological guidelines for species reintroduction programs — application to Great Lakes fishes, *Journal of Great Lakes Research* **21** (Suppl. 1): 3–9, 1995.

Mills, K.H., Chalanchuk, S.M., and Allan, D.J., 2000, Recovery of fish populations in Lake 223 from experimental acidification, *Canadian Journal of Fisheries and Aquatic Sciences* **57**: 192–204.

Ontario Ministry of Natural Resources, 1991, Anthropogenic stressors and diagnosis of their effects on lake trout populations in Ontario lakes, Ontario Ministry of Natural Resources Lake Trout Synthesis.

Phillips, R.B., Pleyte, K.A., and Brown, M.R., 1992, Salmonid phylogeny inferred from ribosomal DNA restriction maps, *Canadian Journal of Fisheries and Aquatic Sciences* 49: 2345–2353.

Pielou, E.C., 1991, *After the Ice Age: The Return of Life to Glaciated North America,* University of Chicago Press, Chicago.

Ryan, P.A. and Marshall, T.R., 1994, A niche definition for lake trout (*Salvelinus namaycush*) and its use to identify populations at risk, *Canadian Journal of Fisheries and Aquatic Sciences* **51**: 2513–2519.

Schmidt, R.E. 1986. Zoogeography of the northern Appalachians. In *The Zoogeography of North American Freshwater Fishes,* edited by C.H. Hocutt and E.O. Wiley, John Wiley & Sons, New York, pp. 137–159.

Scott, W.B. and Crossman, E.J., 1973, *Freshwater Fishes of Canada,* Bulletin of the Fisheries Research Board of Canada 184.

Sellers, T.J., Parker, B.R., Schindler, D.W., and Tonn, W.M., 1998, Pelagic distribution of lake trout (*Salvelinus namaycush*) in small Canadian Shield lakes with respect to temperature, dissolved oxygen, and light, *Canadian Journal of Fisheries and Aquatic Sciences* 55: 170–179.

Shuter, B.J., Jones, M.L., Korver, R.M., and Lester, N.P., 1998, A general, life history based model for regional management of fish stocks — the inland lake trout (*Salvelinus namaycush*) fisheries of Ontario, *Canadian Journal of Fisheries and Aquatic Sciences* 55: 2161–2177.

Snucins, E.J. and Gunn, J.M., 1995, Coping with a warm environment — behavioral thermoregulation by lake trout, *Transactions of the Amererican Fisheries Society* 124: 118–123.

Teller, J.T. and Clayton, L. (Eds.), 1983, *Glacial Lake Agassiz,* Geological Association of Canada, Special Paper 26.

Thompson, J.N., 1998, Rapid evolution as an ecological process, *Trends in Ecology and Evolution* **13**: 329–332.

Vander Zanden, M.J., Casselman, J.M., and Rasmussen, J.B., 1999, Stable isotope evidence for the food web consequences of species invasions in lakes, *Nature* **401**: 464–467.

Waples, R.S., Genetic interactions between hatchery and wild salmonids: lessons from the Pacific Northwest. *Canadian Journal of Fisheries and Aquatic Sciences,* **48** (Suppl. 1): 124–133.

Wilson, C.C. and Bernatchez, L., 1998, The ghost of hybrids past: fixation of arctic charr (*Salvelinus alpinus*) mitochondrial DNA in an introgressed population of lake trout (*S. namaycush*), *Molecular Ecology* 7: 127–132.

Wilson, C.C. and Hebert, P.D.N., 1996, Phylogeographic origins of lake trout (*Salvelinus namaycush*) in eastern North America, *Canadian Journal of Fisheries and Aquatic Sciences* 53: 2764–2775.

Wilson, C.C. and Hebert, P.D.N., 1998, Phylogeography and postglacial dispersal of lake trout (*Salvelinus namaycush*) in North America, *Canadian Journal of Fisheries and Aquatic Sciences* 55: 1010–1024.

Wilson, C.C., Hebert, P.D.N., Reist, J.D., and Dempson, J.B., 1996, Phylogeography and postglacial dispersal of arctic charr (*Salvelinus alpinus* L.) in North America. *Molecular Ecology* 5: 187–198.

Winter, B.D. and Hughes, R.M., 1995, AFS draft position statement on biodiversity, *Fisheries* **20**(4): 20–26.

chapter three

Rehabilitation of lake trout in the Great Lakes: past lessons and future challenges

Charles C. Krueger
Great Lakes Fishery Commission
Mark Ebener
Great Lakes Fishery Commission

Contents

Lake trout *Salvelinus namaycush*, before colonization by European peoples, were native to each of the five Great Lakes, occurring throughout Lake Superior, Lake Michigan, Lake Huron, and Lake Ontario, and in the eastern basin of Lake Erie. Life history characteristics of Great Lakes populations span the range for the species. For example, spawning typically occurs in the fall, but some populations may spawn as early as August (Hansen et al.,

1995) or even in the spring (Bronte, 1993). Age at first maturity is also highly variable across the lakes. For some shallow-water populations in Lake Superior, age at first maturity can be as late as age 9 or beyond, and some individuals may not spawn every year (e.g., Rahrer, 1967; Swanson and Swedburg, 1980). This maturity schedule is comparable to those of some populations from Precambrian Shield lakes (Martin and Olver, 1980) or northern Arctic lakes. On the other extreme, lake trout in lakes Erie and Ontario at the southern edge of their native range mature at age 3, and all fish are mature at age 5 or 6 (Cornelius et al., 1995; Elrod et al., 1995). These life history differences can dramatically affect the speed with which populations respond to management actions such as fishery regulations and stocking.

The native lake trout of the Great Lakes used a diversity of semiisolated spawning habitats from shallow near-shore reefs to deep offshore shoals to rivers, and this habitat diversity yielded a variety of lake trout forms. Aboriginal peoples, Jesuit missionaries, French *voyageurs*, commercial fishermen, and naturalists identified several types of lake trout in the Great Lakes (e.g., Agassiz, 1850; Roosevelt, 1865; Goodier, 1981; Jordan and Evermann, 1911). Three general forms or morphotypes — leans, siscowet, and humpers (also known as bankers) — were recognized based on fat content, morphology, location caught, and spawning condition and timing (Figure 3.1). Within these three forms, several other types were often described (Krueger and Ihssen, 1995). For example, some of the types of lean trout recognized by commercial fishermen were called Mackinaw, yellowfin, redfin, moss, sand, and racer trout (Goodier, 1981; Brown et al., 1981). Whether these lean types all represented significant genetic differentiation is unknown. The differences between the deepwater (sicowet) and shallow-water (lean) forms are known to be heritable phenotypic differences, and these differences probably represent important ecological adaptations for the habitats they use. Deepwater forms appear to be adapted for rapid vertical migration (Eshenroder et al., 1995a) because adults have high fat content and therefore are nearly neutrally buoyant without gas in their swim bladders (Crawford, 1966; Henderson and Anderson, 2002). This level of differentiation and adaptation within the Great Lakes stands in considerable contrast to the similarity of lake trout within and among small Precambrian Shield lakes (Wilson and Mandrak, Chapter 2, this volume).

Unfortunately, the settlement of the Great Lakes basin by Europeans in the 1800s and 1900s threatened the rich diversity of life histories, forms, and adaptations expressed by Great Lakes lake trout. This loss also jeopardized use of lake trout as a food and sport fish. Lake trout populations declined catastrophically and, in some lakes, were lost because of the stresses caused by commercial fisheries; the construction of canals for shipping; and the timber industry. By the late 1950s, native lake trout were gone from Lake Ontario, Lake Erie, and Lake Michigan; nearly gone in Lake Huron; and seriously depleted at most near-shore locations in Lake Superior. Finally, when the remaining populations were threatened by predation from the proliferating non-native sea lamprey *Petromyzon marinus*, management programs began in earnest to protect and rehabilitate lake trout (Hansen, 1999).

This chapter reviews the history of the species in the Great Lakes during the past century, describes the progress made toward lake trout rehabilitation, and identifies management lessons and challenges.

The lake trout fishery and its management: 1800–1950s

The history of Lake Superior and its lake trout provides a useful case history for the species in the basin. Fishing for lake trout by native aboriginal people around Lake Superior was ongoing when Louis Agassiz and his expedition made their journey to record the natural history along the north shore (Agassiz, 1850). Lake trout were used as a subsistence food fish and were bartered with the members of the expedition. The earliest records of

Figure 3.1 Siscowet (top), lean (middle), and humper (bottom) morphotypes are examples of the phenotypic diversity of lake trout from the Great Lakes. (Photo courtesy of Gary Cholwek and Seth Moore of the USGS Ashland Biological Station.)

commercial fisheries in Lake Superior are from the 1830s, when the Hudson's Bay Company and the American Fur Company shipped salted fish in barrels from Lake Superior (Bogue, 2000). Seines were used in the earliest days of the commercial fisheries. Later, settlers to the region were quick to recognize the value of lake trout as a food fish. By 1875, 0.75 million kilograms of lake trout were caught from the lake. Efficiency of catching fish increased with technological improvements. Pound nets were first used in the 1860s, and steam tugs introduced in the early 1870s (Goodier, 1989). The efficiency of fishing gear further expanded as gas- and diesel-powered fishing tugs were introduced after World War I. About 1930, cotton gill nets replaced those made of linen, and in the late 1940s multifilament nylon gill nets replaced cotton (Goodier, 1989; Hansen et al., 1995). When gill net fisheries converted from cotton to multifilament nylon, their nets became much better at catching fish (Pycha, 1962). Between 1913 and 1950, harvest of lake trout averaged 2.0 million kilograms per year (Baldwin et al., 1979). During this same period, sport fishing, though a minor contributor to the total catch, was culturally important at locations such as Duluth, Minnesota, and Munising, Michigan. Anglers, while trolling for lake trout, often used copper line to get lures deep. Overfishing was clearly evident when fishing effort increased sharply after World War II and yield did not increase, even though

Figure 3.2 Sea lamprey colonized the Great Lakes by using shipping canals to pass by barriers such as Niagara Falls.

gear efficiency had increased further (Hile et al., 1951; Pycha and King, 1975). Populations at this point had declined by 50%.

While overfishing of Lake Superior and other Great Lakes populations was occurring, the sea lamprey invaded upstream from Lake Ontario. The sea lamprey gained access to the upper lakes sometime in the late 1910s by passing around Niagara Falls via the Welland Canal and associated feeder canal, built for the shipping industry (Eshenroder and Burnham-Curtis, 1999). Sea lampreys were first recorded in Lake Erie in 1921, in Lake Michigan in 1936, in Lake Huron in 1937, and in Lake Superior in 1938 (Smith and Tibbles, 1980). By the 1950s, sea lampreys were very abundant in all the Great Lakes. Sea lamprey attached to lake trout with their circular, suctorial mouth and fed on their blood and other body fluids (Figure 3.2). Lake trout suffered serious mortality from these attacks. In Lake Superior, stocks of lake trout were already declining due to overfishing before the sea lamprey invaded; however, the collapse of the stocks was undoubtedly accelerated by the added mortality caused by sea lamprey (e.g., Hile et al., 1951; Coble et al., 1990). Although the exact roles of overfishing and sea lamprey predation in the collapse of lake trout in the other Great Lakes have been a source of debate (Coble et al., 1992; Eshenroder, 1992; Eshenroder et al., 1995b), general agreement exists that both sources of mortality were important in the ultimate demise of the species (Hansen, 1999).

Besides overfishing and sea lamprey predation, other environmental factors may have also played a role in the decline of lake trout. Exotic species other than sea lamprey, such as alewife *Alosa pseudoharengus* and smelt *Osmerus mordax*, invaded the Great Lakes, altered food webs, and replaced native coregonines as a primary forage species for lake trout. These species may also have directly controlled natural recruitment of lake trout through competition with, and predation on, juvenile lake trout. In addition, a variety of toxic substances, specifically organochlorine compounds, can become concentrated in lake trout eggs, fry, and the environment. These compounds reduce the hatching success of lake trout in the laboratory, but under natural conditions the relationship between toxic substances and lake trout mortality remains unclear (Zint et al., 1995). Quarrying operations may have destroyed spawning habitat in some areas. During the late 1800s and early 1900s, a fleet of 45 schooners quarried cobble for more than 40 years from Lake Ontario along 100 km of shoreline between Burlington and Whitby (Whillans, 1979). The timber industry also may have affected some near-shore populations. Sawmills deposited large amounts of woody debris and sawdust in the lakes (Lawrie and Rahrer, 1973; Bogue, 2000). This organic debris may have covered some near-shore spawning reefs. Logging drives downriver may have affected river-spawning lake trout. Dams associated with hydroelectric development such as on the Montreal River, a tributary to eastern Lake Superior, must also have contributed to the decline in river-spawning lake trout.

What happened by 1960? Natural reproduction failed to sustain lake trout populations in Lake Ontario, Lake Erie, and Lake Michigan. Wild lake trout were eliminated from these systems. In Lake Huron only two small populations remained, one in Parry Sound and the other in Iroquois Bay, both located on the eastern shore of the lake (Berst and Spangler, 1973; Reid et al., 2001). Near-shore populations in Lake Superior were decimated, although remnants of a few populations, such as the one adjacent to Gull Island Shoal, persisted (Schram et al., 1995). Offshore lake trout in Lake Superior were comparatively unaffected, especially the humper and deepwater siscowet forms, which continued to support limited fisheries through the 1960s and 1970s (Peck et al., 1974). Distance offshore and deep water may have provided some protection to these fish from commercial fisheries and sea lamprey predation. The sequence of the collapse of lake trout stocks in the upper lakes was Lake Huron first, Lake Michigan next, and Lake Superior last (see review in Hansen, 1999). Another important commercial species in the upper Great Lakes, lake whitefish *Coregonus clupeaformis* also declined to their lowest point in the late 1950 and early 1960s because of the effects of sea lampreys, other non-native species, and overfishing. By 1960, the lake trout and whitefish fisheries of the Great Lakes were devastated.

Rehabilitation management: 1950s to present

As lake trout populations declined precipitously through the 1950s in the Great Lakes, the governments in Canada and the United States embarked on a program of fishery rehabilitation focused on lake trout that has continued to the present. The two federal governments undertook many of the early management actions used to restore the lake trout and whitefish fisheries of the Great Lakes. For example, though the states have authority for fishery management, essentially the U.S. Bureau of Commercial Fisheries managed U.S. waters during the 1950s and early 1960s as a result of an absence of state interest. Over time, however, the lake trout rehabilitation program shifted toward, and has been sustained by, the eight Great Lakes states, the Province of Ontario, and the U.S. aboriginal tribal authorities. In 1955, the two federal governments formed the international Great Lakes Fishery Commission for the purpose of developing a program of sea lamprey control, conducting fishery research, and promoting coordinated management (Figure 3.3). The commission explicitly adopted lake trout population rehabilitation as one of its goals and has always encouraged fishery management agencies to restore lake trout populations (Great Lakes Fishery Commission, 2001).

A variety of management actions have been implemented since the late 1950s to rehabilitate the fisheries and overcome the obstacles faced by lake trout in the Great Lakes. Approximately 330 million yearling and fingerling lake trout were released into the Great Lakes from 1950 to 2001, primarily into near-shore areas (Table 3.1). In Lake Superior,

Figure 3.3 The convention between Canada and the United States that formed the Great Lakes Fishery Commission was fully ratified in 1955.

117 million lake trout were stocked, and of these 86% were yearlings. Lake trout fishing was closed in 1962 in Lake Superior and restricted in the other lakes during the early years of the rehabilitation program.

To reduce sea lamprey predation, selective lamprey toxicants (lampricides) were used to kill sea lamprey larvae in streams. The first stream treatments began experimentally in Lake Superior in 1958, and routine applications were extended to the upper Great Lakes shortly thereafter (Smith and Tibbles, 1980). Sea lamprey control measures were implemented later in Lake Ontario (1971) and in Lake Erie (1986). A suite of techniques is now used to control the lamprey, including lampricides, adult trapping, the release of sterile males, and the placement of barriers to block access to spawning streams. Beginning in 1966, management activities were coordinated through lake committees organized by the Great Lakes Fishery Commission. These committees included a representative from each fishery management authority on a lake. The lake committees developed lake trout management plans and helped to coordinate management actions such as stocking and fishery regulations within each Great Lake.

During the 1960s, lake trout survived, and the abundance of subadults and adults increased in the upper three Great Lakes (Superior, Huron, and Michigan) in response to sea lamprey control, regulation of fishery harvests, and stocking. Most of these lake trout were hatchery-origin fish, the survivors of past stockings. Similar increases in lake trout abundance occurred in Lake Ontario in the late 1970s and in Lake Erie in the late 1980s and early 1990s.

Unfortunately, rehabilitation management became increasingly complicated in the mid-1960s because new sport fisheries began, particularly in Lake Huron, Lake Michigan, and Lake Ontario, and were focused on catching non-native Pacific salmon (*Oncorhynchus* sp.) (Bence and Smith, 1999; Kocik and Jones, 1999). Alewives had become enormously abundant and were experiencing massive die-offs that fouled beaches and clogged the water-intake pipes of cities. Salmon were stocked initially to control these fish and to create new sport fisheries. Salmon stocking was successful, and the fisheries that resulted provided an economic stimulus for coastal communities after the loss of the commercial fishing

Table 3.1 Number of Fingerling and Yearling Lake Trout Stocked into the Great Lakes, 1950–2001

Year	Superior	Huron	Michigan	Erie	Ontario	Total
1950	50,000	0	0	0	0	50,000
1952	312,000	0	0	0	0	312,000
1953	472,000	0	0	0	0	472,000
1954	500,000	0	0	0	0	500,000
1955	164,000	0	0	0	0	164,000
1956	201,000	0	0	0	0	201,000
1958	1,020,000	0	0	0	0	1,020,000
1959	635,000	0	0	0	0	635,000
1960	1,050,000	0	113,000	0	0	1,163,000
1961	1,201,000	0	95,000	0	0	1,296,000
1962	1,852,000	0	73,000	0	0	1,925,000
1963	2,311,000	0	0	0	181,000	2,492,000
1964	2,651,000	0	0	0	111,000	2,763,000
1965	1,825,000	0	1,274,000	0	0	3,099,000
1966	3,279,000	0	1,766,000	0	0	5,046,000
1967	3,289,000	0	2,424,000	0	0	5,713,000
1968	3,375,000	0	1,876,000	0	0	5,251,000
1969	2,890,000	0	2,000,000	17,000	0	4,907,000
1970	2,785,000	0	1,960,000	0	0	4,745,000
1971	2,016,000	0	2,344,000	0	0	4,359,000
1972	2,103,000	0	2,926,000	0	0	5,029,000
1973	1,904,000	1,110,000	2,509,000	0	66,000	5,589,000
1974	2,527,000	793,000	2,397,000	26,000	1,163,000	6,907,000
1975	2,149,000	1,053,000	2,613,000	184,000	385,000	6,383,000
1976	2,453,000	1,024,000	2,548,000	202,000	531,000	6,757,000
1977	2,509,000	1,658,000	2,418,000	125,000	586,000	7,295,000
1978	3,076,000	1,262,000	2,539,000	236,000	1,243,000	8,357,000
1979	2,740,000	2,171,000	2,497,000	709,000	887,000	9,004,000
1980	3,156,000	2,164,000	2,791,000	507,000	1,577,000	10,194,000
1981	3,643,000	2,117,000	2,642,000	41,000	1,531,000	9,973,000
1982	4,017,000	2,295,000	2,746,000	235,000	1,650,000	10,944,000
1983	4,102,000	2,808,000	2,241,000	222,000	1,469,000	10,842,000
1984	4,772,000	2,998,000	1,565,000	176,000	1,538,000	11,049,000
1985	5,073,000	4,075,000	3,782,000	154,000	1,911,000	14,995,000
1986	5,171,000	3,770,000	3,297,000	199,000	2,234,000	14,671,000
1987	4,818,000	3,236,000	1,998,000	205,000	2,313,000	12,570,000
1988	4,776,000	4,132,000	2,546,000	203,000	2,285,000	13,942,000
1989	2,516,000	3,147,000	5,377,000	273,000	982,000	12,295,000
1990	2,805,000	1,428,000	1,317,000	349,000	2,054,000	7,954,000
1991	3,445,000	2,496,000	2,779,000	326,000	2,083,000	11,129,000
1992	3,653,000	4,053,000	3,435,000	277,000	1,736,000	13,154,000
1993	1,936,000	3,163,000	2,697,000	258,000	1,066,000	9,120,000
1994	2,034,000	3,945,000	3,854,000	200,000	507,000	10,540,000
1995	1,971,000	3,280,000	2,265,000	160,000	500,000	8,175,000
1996	1,496,000	4,144,000	2,115,000	83,000	350,000	8,187,000
1997	1,291,000	3,288,000	2,235,000	120,000	500,000	7,435,000
1998	1,560,000	4,385,000	2,302,000	98,000	426,000	8,771,000
1999	1,371,000	3,401,000	2,348,000	199,000	476,000	7,794,000
2000	1,357,000	4,655,000	2,260,000	135,000	489,000	8,896,000
2001	234,000	1,217,000	2,382,000	120,000	500,000	4,452,000

industry. Little contribution to the salmon fisheries came from natural recruitment, and the fisheries were dependent on stocking.

In the decades that followed, mortality of adult hatchery-origin lake trout increased to levels at which too few fish survived to either their first or second spawning. Angling effort, in general, increased in the Great Lakes because of the salmon fisheries, and anglers often caught lake trout. State and provincial agencies, in most cases, allowed anglers to keep hatchery-origin lake trout even though these fish had been stocked for rehabilitation purposes. Lake trout were harvested as by-catch in the salmon fisheries, but sometimes lake trout were targeted by anglers. Gill-net fisheries grew as aboriginal peoples exercised treaty rights in some U.S. waters. These fisheries harvested substantial numbers of lake trout incidental to targeting lake whitefish (Brown et al., 1999; Hansen, 1999). Also, sea lamprey in the 1960s and 1970s expanded into new spawning habitats because water quality improved as a result of new environmental laws and policies such as the Great Lakes Water Quality Agreement (e.g., Peshtigo River, Moore and Lychwick, 1980). In spite of these new challenges, large, abundant stocks of hatchery-origin lake trout became established in several localized areas of the Great Lakes. Surprisingly, little reproduction and natural recruitment was observed in any of the lakes other than Lake Superior. The lack of natural recruitment in the Great Lakes stands in sharp contrast to the frequent establishment of naturally reproducing populations after stocking lake trout into the small inland lakes of the Province of Ontario (Evans and Olver, 1995) and suitable western lakes such as Yellowstone Lake (Kaeding et al., 1996). Why then was there so little successful reproduction in the Great Lakes?

Causes of the slow recovery

Several hypotheses have been offered to explain why lake trout rehabilitation has been slow in the Great Lakes (Eshenroder et al., 1984, 1999; Selgeby et al., 1995), and some of these are described below. Among the explanations proposed, empirical data support them all, but none accounts fully for the slow recovery. For example, one cause proposed for the slow recovery is that too few fish survive to spawning age after stocking, and thus natural reproduction is so low as not to be detectable or capable of sustaining a population. Although lake trout at some locations have had difficulty attaining spawning age because of excessive fishing and sea lamprey mortality, at many other locations catch rates of spawning-age lake trout in gill nets have been comparable to, or exceeded, those observed for wild populations in Lake Superior (Krueger et al., 1986; Elrod et al., 1995; Hansen, 1999).

Another hypothesis is that lake trout do not reproduce successfully because they cannot find spawning grounds or mates because of the absence of proper cues (olfactory homing or pheromones). Hatchery-origin lake trout have spawned over inappropriate substrates at some locations. Moreover, stocking often has occurred at locations where little spawning habitat exists in the immediate vicinity. However, hatchery-origin lake trout are also known to locate and readily spawn over clean natural substrate that has been deposited along the shoreline, such as in Tawas Bay, Lake Huron (Foster and Kennedy, 1995), as well as over natural reef substrate such as at Stony Island reef in Lake Ontario (Perkins and Krueger, 1995).

A third proposed cause is that lake trout gametes are infertile because of toxic chemical contamination and a thiamine nutritional deficiency. Toxic substances such as organochlorines and their effects on gamete viability have been suspected of being one of the causes of lake trout reproductive failure (Zint et al., 1995). A serious disease known as early mortality syndrome (EMS) also occurs in adult lake trout from the Great Lakes, apparently because of consumption of alewives, which are non-native (Fitzsimons et al., 1998). If lake

Figure 3.4 Alewives invaded and colonized Lake Ontario around 1870.

trout feed heavily on alewives, adult females become thiamine deficient and their eggs are not viable. However, Brown et al. (1998) reported that not all females captured from Lake Ontario produced families that showed this syndrome. Also, lake trout gametes collected from Lake Ontario in the 1980s from hatchery-origin adults were propagated successfully in hatcheries. More than a million fish from this source were stocked back into the lake (Elrod et al., 1995).

A fourth cause is that the wrong genetic types of lake trout were stocked, and these fish were maladapted for survival and reproduction. The shallow-water, lean morphotype has been the only form of lake trout stocked into the Great Lakes during the past 50 years, and this form may be poorly adapted for colonizing the extensive offshore or deepwater habitats (Krueger and Ihssen, 1995). Nevertheless, some sources such as the Superior strain originating from the Apostle Islands region of Lake Superior and the Seneca strain from the Finger Lakes region are known to successfully reproduce at some locations (e.g., Grewe et al., 1994).

A fifth cause is that predation on lake trout eggs and fry by non-native and/or native species inhibits natural recruitment. For example, predation by the non-native alewife on lake trout fry has been implicated in preventing natural recruitment at some near-shore locations (Figure 3.4; Johnson and VanAmberg, 1995; Krueger et al., 1995a). This source of mortality, however, would be comparatively unimportant at offshore locations such as the midlake reef in Lake Michigan or Six Fathom Bank in Lake Huron or in Lake Superior and Lake Erie, where alewife abundance is apparently minimal.

Although evidence exists to support each of these hypotheses, no single one explains the general lack of natural reproduction by lake trout. Probably, varying combinations of each plus other causes account for the slow recovery of lake trout in the Great Lakes.

Status of rehabilitation

Lake Superior

Lake trout were declared rehabilitated along most of the Lake Superior shoreline in 1996 (Schreiner and Schram, 1997). Wild lake trout have continued to increase in abundance since that time and may be more abundant now than at any time during the last century in many areas of the lake (Wilberg et al., 2003). Substantial natural reproduction has been

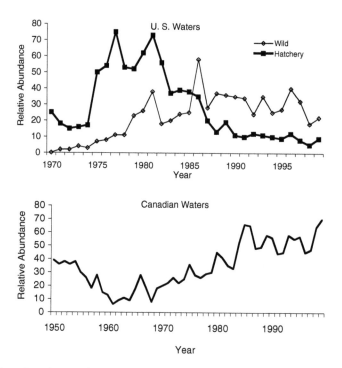

Figure 3.5 Relative abundance of wild and hatchery lake trout caught in gill net surveys in U.S. waters of Lake Superior and of all lake trout caught in waters <70 m deep in Canadian waters, 1950–1999. (Data from C.R. Bronte, U.S. Fish and Wildlife Service.)

noted at most locations in the lake. Population recovery was first noted at areas where remnant stocks were present, such as at Gull Island Shoal (Swanson and Swedburg, 1980), Isle Royale, and Standard Rock, and recruitment, presumably from hatchery-origin lake trout, was noted later in areas where wild lake trout were absent (Figure 3.5; Hansen et al., 1995). Recovery at one location, Devils Island Shoal, was aided by the stocking of fertilized eggs in artificial-turf incubators (Bronte et al., 2002). Rehabilitation has been slowest along the Minnesota shoreline. Abundant and widely distributed spawning habitat, remnant stocks, effective harvest regulation, and few non-native species such as the alewife probably all contributed to the success in Lake Superior. Supplemental stocking is no longer required at most locations. Sea lampreys continue to cause lake trout mortality, and control efforts must be maintained to protect the wild populations. Abundance of the siscowet or deepwater form appears high and may be increasing (Bronte, C.R., U.S. Fish and Wildlife Service, personal communication, 2003).

Lake Huron

Natural reproduction of lake trout has been detected at several sites in Lake Huron; but, at only one site, Parry Sound in Ontario waters, has a self-sustaining population been established (Reid et al., 2001). The success at Parry Sound appears to be caused by reduced fishing mortality because of restrictive angling regulations, successful sea lamprey control, and the establishment of a refuge. A remnant population of wild lake trout was also present in Parry Sound and may have speeded the recovery. Naturally reproduced lake trout have also been caught from an artificial reef in Tawas Bay (Foster and Kennedy, 1995) and from natural sites in South Bay (Anderson and Collins, 1995), Thunder Bay

(Johnson and VanAmberg, 1995), Gravelly Bay, Six Fathom Bank, and Iroquois Bay (Woldt et al., in press). Assessment data from these sites indicate that natural reproduction is not sufficient to sustain populations. Although these successes are encouraging, rehabilitation has not occurred lake-wide. Excessive fishery harvests, sea lamprey-caused mortality, and low lake-wide stocking rates are the major obstacles to lake trout rehabilitation in Lake Huron (Eshenroder et al., 1995b). Recent efforts to apply lampricides in the St. Mary's River may help to reduce sea lamprey abundance in Lake Huron, especially in northern waters. Increased stocking of lake trout could also help reestablish northern populations.

Lake Michigan

Mature hatchery-origin lake trout have been abundant for many years at several spawning locations but have failed to produce detectable natural recruitment. Naturally produced fry were collected from rock rubble deposited at two locations in Grand Traverse Bay (Wagner, 1981), from rock covering a power-plant water-intake pipe in southeastern Lake Michigan (Jude et al., 1981), and from rock deposited at Burns Waterway Harbor in Indiana (Marsden, 1994). Viable fertilized eggs have been recovered from several locations on the east and west shorelines as well as in Traverse Bay (Holey et al., 1995). Wild yearling and older lake trout of the 1976, 1981, and 1983 year classes were caught from Grand Traverse Bay and nearby Platte Bay (Rybicki, 1991). Excessive fishing mortality at these two locations during the mid- to late 1980s may have eliminated the wild year classes and the hatchery-origin adults responsible for the natural reproduction (Hansen, 1999). No wild fry or yearling lake trout have been recovered recently, but few studies are under way to catch these life stages. Predation by round gobies *Neogobius melanostomous* on lake trout eggs and by alewives on fry is believed to be an important block to successful recruitment. Zebra mussel *Dreissena polymorpha* and quagga mussel *D. bugensi* colonization of some reefs may increase damage to and mortality of lake trout eggs after deposition (Marsden and Chotkowski, 2001).

Lake Erie

No detectable natural recruitment of lake trout has occurred in Lake Erie. This lake has the least lake trout habitat of any of the Great Lakes, and suitable waters are restricted to the eastern basin of the lake. Lake Erie was the last Great Lake to be managed for lake trout rehabilitation, so the lack of success may not be surprising because of the short time since the program began. Lake trout stocking has exceeded 150,000 yearlings each year since 1982, and sea lamprey control began in 1986 (Cornelius et al., 1995). Since control began, hatchery trout have survived better, and maximum age now exceeds 14. Fitzsimons and Williston (2000) sampled three natural spawning reefs in late autumn of 1994 and 1995 for lake trout eggs. Eggs were found at two reefs along the New York shoreline but not at one reef on the north shore near Port Dover, Ontario. Egg densities at the two sites ranged from 4.8 to 62.5 eggs/m². Concentrations of mature lake trout in the fall have not been detected at Brocton Shoal, located on the south shore. This reef has the requisite characteristics for lake trout spawning (Edsall et al., 1992) and may be one of the best reefs in the area. The New York State Department of Environmental Conservation plans, in cooperation with U.S. Fish and Wildlife Service, to stock 30,000 yearlings of the humper morphotype at this location in 2004. These deepwater fish will be propagated from a broodstock that originated from Klondike Reef in eastern Lake Superior. This planting will be the first time that a morphotype other than the lean shallow-water form has been stocked in the Great Lakes.

Lake Ontario

Limited natural recruitment has been detected on a lake-wide basis since 1993. Fertilized eggs have been caught at both the eastern and western ends of the lake (Perkins et al., 1995). Evidence of successful reproduction by stocked lake trout was documented in 1982 with the capture of a single lake trout fry by the New York State Department of Environmental Conservation and by a collection of 75 fry off the north end of Stony Island in 1986 (Marsden et al., 1988). Lake trout fry were captured from Stony Island reef every year that surveys were conducted, 1986–1993 (Marsden and Krueger, 1991; Krueger et al., 1995a). Nevertheless, no detectable naturally produced year classes of age-1 and older lake trout occurred until the early 1990s. Since then, a few age-1 or older wild lake trout have been caught (approximately 175 fish). The lake trout population in Lake Ontario continues to depend on hatchery stocking as its primary source of recruitment.

New management approaches

The slow recovery of lake trout populations prompts the consideration of different management approaches than have been used in the past. For decades, the typical management practice was to stock near-shore areas with yearling lake trout, conduct sea lamprey control, and provide a modest level of protection from fisheries. The widespread lack of success (except in Lake Superior) supports the notion of experimentally trying some different approaches. One alternative approach that appeared to show some success was the stocking of fertilized eggs instead of lake trout yearlings at Devils Island Shoal in Lake Superior. Fertilized eggs were placed in bundles of artificial turf and anchored over Devils Island Shoal in Lake Superior from 1981 to 1995 (Bronte et al., 2002). Approximately 17 million eggs were stocked in this way. Stocking this life stage was based on the concept that the hatched fry would leave the turf incubators, reside in the rock rubble of the shoal, imprint to the site, and increase the number of adults homing to the site. Beginning in 1992, the presence of mature adult lake trout without evidence of being of hatchery origin (no fin clips) rose sharply in the samples obtained by assessment gill netting (Bronte et al., 2002). The recruitment of unclipped spawners from 1985 to 1997 showed a closer relationship to the number of eggs planted in previous years than to the number of adult lake trout that had been observed on the reef during the spawning season.

Three other "new" ideas are described below that are also somewhat different from the typical approach used in the past.

Pulse stocking

First, we propose stocking all available hatchery-reared lake trout allocated to a lake in a given year into one selected locality that has an abundance of spawning habitat, leaving other areas of the lake unstocked. In the next year all lake trout would be stocked in a different area of abundant spawning habitat. This process would continue each year until all comparable areas have received a pulse of hatchery lake trout, and then repeat the cycle. This approach is proposed for two reasons. First, the best survival of stocked fish always occurs early in a rehabilitation program (e.g., Elrod et al., 1995; Hansen et al., 1994). Survival of stocked lake trout consistently declines 8 to 10 years after stocking is initiated, when adults become abundant. Poor survival of hatchery yearlings may result from cannibalism of newly stocked fish by adult hatchery-origin lake trout. Our approach should minimize this problem. Second, if reproduction should occur, wild recruits should have a better chance for survival without competition from stocked hatchery yearlings. Pulse stocking should minimize competition.

Creation of spawning areas

Second, we propose that large areas of rock rubble be deposited in near-shore locations to create spawning areas where mature hatchery lake trout congregate in the fall. This approach was developed from the universal observation in the Great Lakes that rock structures, if placed in suitable near-shore lake trout habitat, are quickly discovered by adult hatchery-origin fish and used for spawning (e.g., Wagner, 1981; Marsden, 1994; Marsden et al., 1995). In some cases, lake trout spawning over such areas have successfully produced measurable recruitment of age-1+ lake trout (Peck, 1986; Rybicki, 1991). Spawning shoals would be most appropriately placed in areas where severe loss of near-shore habitat has occurred. These reefs are to create areas for spawning, not fishing. Shoal placement should be combined with refuge or no-fishing regulations that protect fish from all forms of fishing mortality within a sizable area around the reef. The primary rationale for this approach is simply that past experience shows it will work. Artificial spawning reefs could be the key catalyst for establishment of new near-shore populations. New populations established through the use of this approach could serve as surrogates for remnant populations. Rehabilitation success most often has been observed where remnant populations survived (Lake Superior and Parry Sound). New wild populations could, through straying of individual fish, help to colonize other areas of the lake. Careful design and placement of man-made spawning shoals, when combined with assessment, will reveal much about spawning-site selection and demonstrate its potential as a rehabilitation strategy. Artificial spawning shoals could be designed and created as sets of experiments to improve understanding of lake trout spawning-site selection, reproductive success, and adult homing.

Reefs would need to be placed in waters deep enough to avoid processes that could cause mortality of eggs, such as ice scour and high wave energy. Sites should also be selected to avoid areas subject to severe long-shore transport and deposition of sediment. Employment of this approach presumes that survival from egg deposition to the yearling life stage is possible and not precluded by egg and fry predators and that sea lamprey control, stocking, and harvest regulation continue. The creation of new spawning reefs must be on a spatial scale that produces enough juveniles to sustain populations. Based on past observations of egg deposition and survival by hatchery trout on a natural reef in Lake Ontario (Perkins and Krueger, 1995), a 2-ha spawning area could generate from 64,000 to 131,000 yearlings and, assuming 60% survival thereafter, could in the first generation establish a population of 3,000 to 6,000 age-7 spawners.

Transplantation of adults

Third, we suggest experimentally transplanting wild adult lake trout from Lake Superior into isolated offshore areas possessing suitable spawning habitats such as the midlake reef complex in Lake Michigan or the Six Fathom Bank in Lake Huron. This approach addresses three problems observed in lake trout rehabilitation. First, hatchery-origin lake trout have difficulty finding and using off-shore areas of suitable spawning habitat (e.g., Krueger et al., 1986). This approach attempts to circumvent this problem by placing spawning-condition adults directly on a suitable reef. Because of the impending urgency of spawning, mature lake trout placed directly on offshore spawning areas in late September or October may spawn on these reefs. After the first spawning, they may remain in the area or return in subsequent spawning seasons. Second, lake trout rehabilitation takes a long time because lake trout require 5 to 7 years or more to attain maturity. Stocking eggs, fry, or yearlings requires managers to wait until fish mature before they can assess the success or failure of their actions. The proposed approach will accelerate this process by as much as 5 to 7 years because the lake trout are already mature. Natural seeding of the reef with eggs would immediately occur, and these eggs would have the rearing advantages associated with natural substrates (e.g., fry imprinting). If these events took place, populations could become established rapidly.

Third, nearly all lake trout currently stocked may suffer from domestication effects (Reisenbichler et al., 2003) because they have been raised in a hatchery for more than 1 year and are the products of hatchery brood stocks. Because of the small numbers of adults transplanted, areas stocked should have complete protection from fishing. Transplantation of wild adults avoids the problem of reduced fitness in the wild caused by domestication because this approach does not use hatcheries. Adult transfers have proven amazingly successful in the management and restoration of fish (Kerr et al., 1996; Lasenby and Kerr, 2000) and wildlife populations. For example, the stocking of domestic and domestic-wild turkeys failed to reestablish populations within their native range during the 1920s to 1940s. With the development and use of live-capture and transfer techniques in the 1950s, successful reestablishment of turkey populations was widespread across North America (Hewitt, 1967). Similarly, reestablishment of self-sustaining populations of other birds and mammals was based on trapping and transfer of adults (Wolf et al., 1996).

The slow recovery of lake trout in the Great Lakes indicates that new management approaches should be tried and tested. In the situation where fertilized eggs were stocked in artificial turf in Lake Superior (described above), stocking and its assessment was undertaken for 20 years, and this persistence was a critical element that fostered learning. Implementation of new strategies must have a long-term commitment and must incorporate carefully designed assessment studies.

Management lessons and future challenges

Much has been learned during five decades of lake trout rehabilitation efforts in the Great Lakes. Certainly, one can conclude that this effort has been much more difficult and complicated than anyone dreamed back in the late 1950s when it began. The successful rehabilitation of Lake Superior and of Parry Sound in Lake Huron and the capture of naturally produced juveniles from Lake Michigan and Lake Ontario clearly indicate the feasibility of the rehabilitation goal. Described below are lessons that future managers need to consider if the mistakes of the past are not to be repeated (see Krueger et al., 1995b).

1. *Fishing mortality must be controlled to rehabilitate populations.* Populations cannot be reestablished if excessive mortality of subadults and adults is permitted. A refuge (no harvest) and severe angling restrictions were believed to be key management practices that led to the successful restoration of the Parry Sound population (Reid et al., 2001). If population rehabilitation is the management goal, it does not make sense to have directed fisheries on the species before establishment of self-sustaining populations. Lenient harvest regulations are inconsistent with the goal of rehabilitation. Control of all forms of mortality, such as from sea lampreys, is important, but fishery harvest is typically the portion of mortality that managers have the most ability to control.

2. *Non-native species can seriously impede rehabilitation.* New invasions from non-native species pose an unpredictable, serious threat to the success of lake trout rehabilitation in the Great Lakes. Further invasion of non-native aquatic organisms must be prevented. Non-native organisms can be serious predators on lake trout eggs and fry (Krueger et al., 1995a; Marsden, 1997; Chotkowski and Marsden, 1999), can alter habitat (Marsden and Chotkowski, 2001), and have community level

effects (O'Gorman et al., 2000). Fishery and environmental agencies must immediately take action to prevent any further colonization by exotic aquatic organisms.

3. *Survival and reproduction of stocked fish is related to their genetic origins.* Some strains
    of lake trout survive and reproduce better than others. Yet, little consideration has
    been given to matching lake trout types or strains to particular habitats (e.g.,
    Krueger et al., 1981). The predominant and formerly most productive lake trout
    habitats in the Great Lakes are deepwater areas and offshore shoals (Eshenroder
    et al., 1995c). These areas, with few exceptions, have been virtually ignored when
    considering the sources of lake trout to stock. The lake trout reintroduced to these
    areas should be those that have the genetic adaptations required for survival and
    reproduction in these habitats. Siscowet lake trout should be ideal for deep-water
    areas, and humpers are suited for offshore shoals. More consideration needs to be
    given to restoring the full complement of phenotypic diversity of lake trout to the
    Great Lakes (Krueger et al., 1995b).

4. *Long-lived species take a long time to restore.* Though this lesson may seem obvious,
    the problem of time has both ecological and social implications. Wild, self-sustain-
    ing, naturally reproducing populations of lake trout in the Great Lakes typically
    had 13 or more adult age classes (age 7 to 20 or more). Thus, the buildup of age
    classes typical of wild populations takes decades, but they are fragile and vulner-
    able and can be disrupted by a few short years of over harvest (see lesson 1!). One
    brief lapse of protection can set back for decades the possibility of success for
    rehabilitation. Fish management agencies and the public must be vigilant to protect
    fish and be patient for success. Unfortunately, the time frame required for lake trout
    rehabilitation spans well beyond the careers of managers, biologists, and politicians.
    Even more serious is sustaining public support for such long-term programs. An
    extraordinary level of commitment is required from members of the public who
    may not experience any benefits in their lifetimes from a rehabilitation effort.

5. *Coordinated management is essential to management success in multijurisdictional waters.*
    Jurisdictions share authority for management on every Great Lake. Coordinated
    management actions consistent with agreed-on goals are critical for success in lake
    trout rehabilitation. The annual meetings of the Great Lakes Fishery Commission's
    lake committees have provided the forum where fish management agencies discuss
    and decide on common strategies and actions to accomplish their goals. Without
    this approach, the successes experienced so far in lake trout rehabilitation would
    not have occurred.

Conclusion

What does the future hold for lake trout rehabilitation? We don't know! If it was the mid-
1980s, we could make a confident prediction that rehabilitation successes are likely if
mortality (fishing and sea lamprey) is controlled and intensive stocking included a variety
of lake trout strains. Unfortunately, in the 2000s major ecological disruptions to the food
web are occurring rapidly as a result of new invasions of aquatic species from Eurasia.
High levels of ecological instability currently characterize the Great Lakes food web. The
possibility exists that critical native components of the food web such as the amphipod
Diporeia or the opossum shrimp *Mysis relicta* may be lost through interactions with non-
native species. Yet-to-come invading species could be even more effective predators of
lake trout eggs and fry than the presently established alewife and round goby. Stemming
the biological pollution caused by the flow of invading species to the Great Lakes is critical
if progress in lake trout rehabilitation is not to be lost and new successes are to be gained.

Acknowledgments

Shawn Riley, Heather Kirshman, and William Alguire assisted with the review of literature used in this chapter. C. I. Goddard and R. L. Eshenroder provided many helpful suggestions on earlier drafts of this chapter.

References

Agassiz, L., 1850, *Lake Superior: Physical Character, Vegetation, and Animals, Compared with Those of Other and Similar Regions*. Gould, Kendall and Lincoln, Boston.

Anderson, D.M. and Collins, J.J., 1995, Natural reproduction by stocked lake trout (*Salvelinus namaycush*) and hybrid (backcross) lake trout in South Bay, Lake Huron, *Journal of Great Lakes Research* **21**(Suppl. 1):260–266.

Baldwin, N.S., Saalfeld, R.W., Ross, M.A., and Buettner, H.J., 1979, *Commercial Fish Production in the Great Lakes 1867–1977*, Great Lakes Fishery Commission Technical Report 3, Ann Arbor, Michigan.

Bence, J.R. and Smith, K.D., 1999, An overview of recreational fisheries of the Great Lakes, In *Great Lakes Fishery Policy and Management. A Binational Perspective*, edited by W.W. Taylor and C.P. Ferreri, Michigan State University Press, East Lansing, pp. 259–306.

Berst, A.H. and Spangler, G.R., 1973, Lake Huron: effects of exploitation, introductions, and eutrophication on the salmonid community, *Journal of the Fisheries Research Board of Canada* **29**:877–887.

Bogue, M.B., 2000, *Fishing the Great Lakes*, The University of Wisconsin Press, Madison.

Bronte, C.R., 1993, Evidence of spring spawning lake trout in Lake Superior, *Journal of Great Lakes Research* **19**:625–629.

Bronte, C.R., Schram, S.T., Selgeby, J.H., and Swanson, B.L., 2002, Reestablishing a spawning population of lake trout in Lake Superior with fertilized eggs in artificial turf incubators, *North American Journal of Fisheries Management* **22**:796–805.

Brown, E.H., Jr., Eck, G.W., Foster, N.R., Horrall, R.M., and Coberly, C.E., 1981, Historical evidence for discrete stocks of lake trout (*Salvelinus namaycush*) in Lake Michigan, *Canadian Journal of Fisheries and Aquatic Sciences* **38**:1747–1758.

Brown, S.B., Fitzsimons, J.D., and Palace, V.P., 1998, Thiamine and early mortality syndrome in lake trout, *American Fisheries Society Symposium* **21**:18–25.

Brown, R., Ebener, M., and Gorenflo, T., 1999, Great Lakes commercial fisheries: historical overview and prognosis for the future. In *Great Lakes Fishery Policy and Management. A Binational Perspective*, edited by W.W. Taylor and C.P. Ferreri, Michigan State University Press, East Lansing, pp. 417–453.

Chotkowski, M.A. and Marsden, J.E., 1999, Round goby and mottled sculpin predation on lake trout eggs and fry: field predictions from laboratory experiments, *Journal of Great Lakes Research* **25**:26–35.

Coble, D.W., Bruesewitz, R.E., Fratt, T.W., and Scheirer, J.W., 1990, Lake trout, sea lampreys, and overfishing in the upper Great Lakes: a review and reanalysis, *Transactions of the American Fisheries Society* **119**:985–995.

Coble, D.W., Bruesewitz, R.E., Fratt, T.W., and Scheirer, J.W., 1992, Decline of lake trout in Lake Huron, *Transactions of the American Fisheries Society* **121**:548–554.

Cornelius, F.C., Muth, K.M., and Kenyon, R., 1995, Lake trout rehabilitation in Lake Erie: a case history, *Journal of Great Lakes Research* **21**(Suppl. 1):65–82.

Crawford, R.H., 1966, *Buoyancy regulation in lake trout*, doctoral dissertation, University of Toronto.

Edsall, T.A., Brown, C.L., Kennedy, G.W., and French, J.R.P., III, 1992, *Surficial Substrates and Bathymetry of Five Historical Lake Trout Spawning Reefs in Nearshore Waters of the Great Lakes*. Great Lakes Fishery Commission Technical Report 58, Ann Arbor, Michigan.

Elrod, J.H., O'Gorman, R., Schneider, C.P., Eckert, T.H., Shaner, T., Bowlby, J.N., and Schleen, L.P., 1995, Lake trout rehabilitation in Lake Ontario, *Journal of Great Lakes Research* **21**(Suppl. 1):83–107.

Eshenroder, R.L., 1992, Decline of lake trout in Lake Huron, *Transactions of the American Fisheries Society* **121**:548–554.

Eshenroder, R.L., Bronte, C.R., and Peck, J.W., 1995c, Comparison of lake trout-egg survival at inshore and offshore and shallow-water and deepwater sites in Lake Superior, *Journal of Great Lakes Research* **21**(Suppl. 1):313–322.

Eshenroder, R.L. and Burnham-Curtis, M.K., 1999, Species succession and sustainability of the Great Lakes fish community. In *Great Lakes Fishery Policy and Management. A Binational Perspective*, edited by W.W. Taylor and C.P. Ferreri, Michigan State University Press, East Lansing, pp. 145–184.

Eshenroder, R.L., Crossman, E.J., Meffe, G.K., Olver, C.H., and Pister, E.P., 1995a, Lake trout rehabilitation in the Great Lakes: an evolutionary, ecological, and ethical perspective, *Journal of Great Lakes Research* **21**(Suppl. 1):518–529.

Eshenroder, R.L., Payne, N.R., Johnson, J.E., Bowen, C., and Ebener, M.P., 1995b, Lake trout in rehabilitation in Lake Huron. *Journal of Great Lakes Research* **21**(Suppl. 1):108–127.

Eshenroder, R.L., Peck, J.W., and Olver, C.H., 1999, *Research priorities for lake trout rehabilitation in the Great Lakes: a 15-year perspective*. Great Lakes Fishery Commission Technical Report 64, Ann Arbor, Michigan.

Eshenroder, R.L., Poe, T.P., and Olver, C.H., 1984, *Strategies for rehabilitation of lake trout in the Great Lakes: proceedings of a conference on lake trout research, August 1983*. Great Lakes Fishery Commission Technical Report 40, Ann Arbor, Michigan.

Evans, D.O. and Olver, C.H., 1995, Introduction of lake trout, *Salvelinus namaycush*, to inland lakes of Ontario, Canada: factors contributing to successful colonization, *Journal of Great Lakes Research* **21**(Suppl. 1):30–53.

Fitzsimons, J.D., Brown, S.B., and Vandenbyllaardt, L., 1998, Thiamine levels in food chains of the Great Lakes, *American Fisheries Society Symposium* **21**:90–98.

Fitzsimons, J.D. and Williston, T.B., 2000, Evidence of lake trout spawning in Lake Erie, *Journal of Great Lakes Research* **26**:489–494.

Foster, N.R. and Kennedy, G.W., 1995, Patterns of egg deposition by lake trout and lake whitefish at Tawas artificial reef, Lake Huron, In *The Lake Huron Ecosystem: Ecology, Fisheries and Management*, edited by M. Munawar, T. Edsall, and J. Leach, SPB Academic Publishing, Amsterdam, pp. 191–206.

Goodier, J.L., 1981, Native lake trout (*Salvelinus namaycush*) stocks in the Canadian water of Lake Superior prior to 1955, *Canadian Journal of Fisheries and Aquatic Sciences* **38**:1724–1737.

Goodier, J.L., 1989, Fishermen and their trade on Canadian Lake Superior: one hundred years, *Inland Seas* **45**:284–306.

Great Lakes Fishery Commission, 2001, *Strategic Vision of the Great Lakes Fishery Commission for the First Decade of the New Millennium*. Great Lakes Fishery Commission, Ann Arbor, Michigan.

Grewe, P.M., Krueger, C.C., Marsden, J.E., Aquadro, C.F., and May, B., 1994, Hatchery origins of naturally produced lake trout fry captured in Lake Ontario: temporal and spatial variability based on allozyme and mitochondrial DNA data, *Transactions of the American Fisheries Society* **123**:309–320.

Hansen, M.J., 1999, Lake trout in the Great Lakes: basinwide stock collapse and binational restoration. In *Great Lakes Fishery Policy and Management. A Binational Perspective*, edited by W.W. Taylor and C.P. Ferreri, Michigan State University Press, East Lansing, pp. 417–453.

Hansen, M.J., Ebener, M.P., Schorfhaar, R.G., Schram, S.T., Schreiner, D.R., and Selgeby, J.H., 1994, Declining survival of lake trout stocked during 1963–1986 in U.S. waters of Lake Superior, *North American Journal of Fisheries Management* **14**:395–402.

Hansen, M.J., Peck, J.W., Schorfhaar, R.G., Selgeby, J.H., Schreiner, D.R., Schram, S.T., Swanson, B.L., MacCallum, W.R., Burnham-Curtis, M.K., Curtis, G.L., Heinrich, J.W., and Young, R.R., 1995, Lake trout (*Salvelinus namaycush*) populations in Lake Superior and their restoration in 1959–1993, *Journal of Great Lakes Research* **21**(Suppl. 1):152–175.

Henderson, B.A. and Anderson, D.M., 2002, Phenotypic differences in buoyancy and energetics of lean and siscowet lake charr in Lake Superior, *Environmental Biology of Fishes* **64**:203–209.

Hewitt, O.H., 1967, *The Wild Turkey and Its Management*, The Wilderness Society, Washington, D.C.

Hile, R., Eschmeyer, P.H., and Lunger, G.F., 1951, Status of the lake trout fishery in Lake Superior, *Transactions of the American Fisheries Society* **80**:278–312.

Holey, M.E., Rybicki, R.W., Eck, G.W., Brown, E.H., Jr., Marsden, J.E., Lavis, D.S., Toneys, M.L., Trudeau, T.N., and Horrall, R.M., 1995, Progress toward lake trout restoration in Lake Michigan, *Journal of Great Lakes Research* **21**(Suppl. 1):128–151.

Johnson, J.E. and VanAmberg, J.P., 1995, Evidence of natural reproduction of lake trout in western Lake Huron, *Journal of Great Lakes Research* **21**(Suppl. 1):253–259.

Jordan, D.S. and Evermann, B.W., 1911, A review of the salmonoid fishes of the Great Lakes, with notes on the whitefishes of other regions, *Bulletin of the United States Bureau of Fisheries* 29:41.

Jude, D.J., Klinger, S.A., and Enk, M.D., 1981, Evidence of natural reproduction by planted lake trout in Lake Michigan, *Journal of Great Lakes Research* 7:57–61.

Kaeding, L.R., Boltz, G.D., and Carty, D.G., 1996, Lake trout discovered in Yellowstone Lake threaten native cutthroat trout, *Fisheries* 21(3):16–20.

Kerr, S.J., Corbett, B.W., Flowers, D.D., Fluri, D., Ihssen, P.E., Potter, B.A., and Seip, D.E., 1996, *Walleye Stocking as a Management Tool. Percid Community Synthesis*, Ontario Ministry of Natural Resources, Peterborough, Ontario.

Kocik, J.F. and Jones, M.L., 1999, Pacific salmonines in the Great Lakes basin. In *Great Lakes Fishery Policy and Management. A Binational Perspective*, edited by W.W. Taylor and C.P. Ferreri, Michigan State University Press, East Lansing, pp. 455–488.

Krueger, C.C., Gharrett, A.J., Dehring, T.R., and Allendorf, F., 1981, Genetic aspects of fishery rehabilitation programs, *Canadian Journal of Fisheries and Aquatic Sciences* 38:1877–1881.

Krueger, C.C. and Ihssen, P.E., 1995, Review of genetics of lake trout in the Great Lakes: history, molecular genetics, physiology, strain comparisons, and restoration management, *Journal of Great Lakes Research* **21**(Suppl. 1):348–363.

Krueger, C.C., Jones, M.L., and Taylor, W.W., 1995b, Restoration of lake trout in the Great Lakes: challenges and strategies for future management, *Journal of Great Lakes Research* **21**(Suppl. 1):547–558.

Krueger, C.C., Perkins, D.L., Mills, E.L., and Marsden, J.E., 1995a, Predation by alewives on lake trout fry in Lake Ontario: role of an exotic species in preventing restoration of a native species, *Journal of Great Lakes Research* **21**(Suppl. 1):458–469.

Krueger, C.C., Swanson, B.L., and Selgeby, J.H., 1986, Evaluation of hatchery-reared lake trout for reestablishment of populations in the Apostle Islands region of Lake Superior, 1960–84. In *Fish Culture in Fisheries Management*, edited by R.H. Stroud, American Fisheries Society, Bethesda, pp. 93–107.

Lasenby, T.A. and Kerr, S.J., 2000, *Bass Stocking and Transfers: An Annotated Bibliography and Literature Review.* Ontario Ministry of Natural Resources, Peterborough, Ontario.

Lawrie, A. and Rahrer, J.F., 1973, *Lake Superior: A Case History of the Lake and Its Fisheries.* Great Lakes Fishery Commission Technical Report 19, Ann Arbor, Michigan.

Marsden, J.E., 1994, Spawning by stocked lake trout on shallow, near-shore reefs in southwestern Lake Michigan, *Journal of Great Lakes Research* 20:377–384.

Marsden, J.E., 1997, Carp diet includes zebra mussels and lake trout eggs, *Journal of Freshwater Ecology* 12:491–492.

Marsden, J.E. and Chotkowski, M.A., 2001, Lake trout spawning on artificial reefs and the effect of zebra mussels: fatal attraction? *Journal of Great Lakes Research* 27:33–43.

Marsden, J.E. and Krueger, C.C., 1991, Spawning by hatchery-origin lake trout (*Salvelinus namaycush*) in Lake Ontario: data from egg collections, substrate analysis, and diver observations, *Canadian Journal of Fisheries and Aquatic Sciences* 48:2377–2384.

Marsden, J.E., Perkins, D.L., and Krueger, C.C., 1995, Recognition of spawning areas by lake trout: deposition and survival of eggs on small, man-made rock piles, *Journal of Great Lakes Research* **21**(Suppl. 1):330–336.

Marsden, J.E., Krueger, C.C., and Schneider, C.P., 1988, Evidence of natural reproduction by stocked lake trout in Lake Ontario, *Journal of Great Lakes Research* 14:3–8.

Martin, N.V. and Olver, C.H., 1980, The lake charr, *Salvelinus namaycush*. In *Charrs: Salmonid Fishes of the Genus* Salvelinus, edited by E.K. Balon, Dr. W. Junk Publishers, The Hague, pp. 205–277

Moore, J.D. and Lychwick, T.J., 1980, Changes in mortality of lake trout (*Salvelinus namaycush*) in relation to increased sea lamprey (*Petromyzon marinus*) abundance in Green Bay, 1974–1978, *Canadian Journal of Fisheries and Aquatic Sciences* 37:2052–2056.

O'Gorman, R., Elrod, J.H., Owens, R.W., Schneider, C.P., Eckert, T.H., and Lantry, B.F., 2000, Shifts in depth distribution of alewives, rainbow smelt, and age-2 lake trout in southern Lake Ontario following establishment of dreissenids, *Transactions of the American Fisheries Society* **129**:1096–1106.

Peck, J.W., 1986, Dynamics of reproduction by hatchery lake trout on a man-made spawning reef, *Journal of Great Lakes Research* **12**:293–303.

Peck, J., Schorfhaar, R., and Wright, A.T., 1974, Status of selected fish stocks in Lake Superior and recommendations for commercial harvest. In *Status of Selected Fish Stocks in Michigan's Great Lakes Waters and Recommendations for Commercial Harvest,* edited by J.D. Bails and M.H. Patriarche, Michigan Department of Natural Resources, Technical Report 70-33, Fisheries Division, Lansing, pp. 1–70.

Perkins, D.L., Fitzsimons, J.D., Marsden, J.E., Krueger, C.C., and May, B., 1995, Differences in reproduction among hatchery strains of lake trout at eight spawning areas in Lake Ontario: genetic evidence from mixed-stock analysis, *Journal of Great Lakes Research* **21**(Suppl. 1):364–374.

Perkins, D.L. and Krueger, C.C., 1995, Dynamics of reproduction by hatchery-origin lake trout (*Salvelinus namaycush*) at Stony Island reef, Lake Ontario, *Journal of Great Lakes Research* **21**(Suppl. 1):400–417.

Pycha, R.L., 1962, The relative efficiency of nylon and cotton gill nets for taking lake trout in Lake Superior, *Journal of the Fisheries Research Board of Canada* **19**:1085–1094.

Pycha, R.L. and King, G.R., 1975, *Changes in the Lake Trout Population of Southern Lake Superior in Relation to the Fishery, the Sea Lamprey, and Stocking, 1950–70.* Great Lakes Fishery Commission Technical Report 28, Ann Arbor, Michigan.

Rahrer, J.F., 1967, Growth of lake trout in Lake Superior before the maximum abundance of sea lampreys, *Transactions of the American Fisheries Society* **96**:268–277.

Reid, D.M., Anderson, D.M., and Henderson, B.A., 2001, Restoration of lake trout in Parry Sound, Lake Huron, *North American Journal of Fisheries Management* **21**:156–169.

Reisenbichler, R.R., Utter, F.M., and Krueger, C.C., 2003, Genetic concepts and uncertainties in restoring fish populations and species. In *Strategies for Restoring River Ecosystems: Sources of Variability and Uncertainty in Natural and Managed Systems,* edited by R.C. Wissmar and P.A. Bisson, American Fisheries Society, Bethesda, Maryland, pp. 149–183.

Roosevelt, R.B., 1865, *Superior Fishing: or, the Striped Bass, Trout, and Black Bass of the Northern States,* Carleton Publishers, New York.

Rybicki, R.W., 1991, *Growth, Mortality, Recruitment and Management of Lake Trout in Eastern Lake Michigan,* Michigan Department of Natural Resources, Fisheries Research Report 1979, Ann Arbor, Michigan.

Schram, S.T., Selgeby, J.H., Bronte, C.R., and Swanson, B.L., 1995, Population recovery and natural recruitment of lake trout at Gull Island Shoal, Lake Superior, 1964–1992, *Journal of Great Lakes Research* **21**(Suppl. 1):225–232.

Schreiner, D.R. and Schram, S.T., 1997, Lake trout rehabilitation in Lake Superior, *Fisheries (Bethesda)* **22**:12–14.

Selgeby, J.H., Eshenroder, R.L., Krueger, C.C., Marsden, J.E., and Pycha, R.L. (Eds.), 1995, International conference on restoration of lake trout in the Laurentian Great Lakes, *Journal of Great Lakes Research* **21**(Suppl. 1):1–564.

Smith, B.R. and Tibbles, J.J., 1980, Sea lamprey (*Petromyzon marinus*) in lakes Huron, Michigan and Superior: history of invasion and control, 1936–78, *Canadian Journal of Fisheries and Aquatic Sciences* **37**:1780–1801.

Swanson, B.L. and Swedburg, D.V., 1980, Decline and recovery of the Lake Superior Gull Island Reef lake trout (*Salvelinus namaycush*) population and the role of sea lamprey (*Petromyzon marinus*) predation, *Canadian Journal of Fisheries and Aquatic Sciences* **37**:2074–2080.

Wagner, W.C., 1981, Reproduction of planted lake trout in Lake Michigan, *North American Journal of Fisheries Management* **1**:159–164.

Whillans, T.H., 1979, Historic transformation of fish communities in three Great Lakes bays, *Journal of Great Lakes Research* **5**:195–215.

Wilberg, M.J., Hansen, M.J., and Bronte, C.R., 2003, Historic and modern abundance of wild lean lake trout in Michigan waters of Lake Superior: implications for restoration goods. *North American Journal of Fisheries Management* **23**:100–108.

Wilson, C.C. and Mandrak, N., 2003, History and evolution of lake trout in Shield lakes: past and future challenges, In *Boreal Shield Watersheds: Lake Trout Ecosystems in a Changing Environment*, edited by J.M. Gunn, R.J. Steedman, and R.A. Ryder, Lewis/CRC Press, Boca Raton, Florida, chap. 2.

Woldt, A.P., Reid, D.M., and Johnson, J.E., 2003, Status of the open-water predator community. In *The State of Lake Huron in 1999*, edited by M.P. Ebener, Great Lakes Fishery Commission Special Publication, Ann Arbor, Michigan, in press.

Wolf, C.M., Griffith, B., Reed, C., and Temple, S.A., 1996, Avian and mammalian translocations: update and reanalysis of 1987 survey data, *Conservation Biology* **10**:1142–1154.

Zint, M.T., Taylor, W.W., Carl, L., Edsall, C.C., Heinrich, J., Sippel, A., Lavis, D., and Schaner, T., 1995, Do toxic substances pose a threat to rehabilitation of lake trout in the Great Lakes? A review of the literature, *Journal of Great Lakes Research* **21**(Suppl. 1):530–546.

Environmental factors that affect Boreal Shield ecosystems

chapter four

Land, water, and human activity on Boreal watersheds

Robert J. Steedman
Ontario Ministry of Natural Resources
Craig J. Allan
University of North Carolina at Charlotte
Robert L. France
Harvard University Graduate School of Design
Robert S. Kushneriuk
Ontario Ministry of Natural Resources

Contents

Introduction: land–water linkages in managed Boreal landscapes

This chapter reviews ecological linkages between Precambrian Shield waters and the Boreal forest that sustains them. Two key uncertainties emerge and are partially addressed by recent science: (1) Has human activity threatened the identity or sustainability of aquatic ecosystems on the Boreal Shield? (2) Will new disturbances and environmental changes

Table 4.1 Boreal Forest Land Capabilities
in Ontario, 1991 (%) (OEAB, 1994)

Land Type	% Area
Production forest[1]	58
Private, federal lands	17
Water	13
Nonforested	9
Protection forest[2]	3

[1] Production forest is Crown land with no obvious limitations on the ability to practice timber management.

[2] Protection forest cannot normally support timber management because of steep slopes or thin soils over bedrock

exacerbate or counteract existing ecological impacts? Research to answer these questions must be designed and interpreted in the context of historical watershed disturbances such as wildfire, insects, disease, and climate change. Through this review we found confirmation of some "best guesses" used by land and water managers over the last 20 years. However, we also found evidence of unexpected ecological resilience and underestimated ecological threats.

Forests cover most of the Boreal Shield (Table 4.1) and are notable for their tendency to undergo and recover from frequent stand-destroying disturbances. Before 20th-century fire suppression programs, forests on the Boreal Shield were disturbed by wildfire, windthrow, and insect infestation at intervals of 50 to 100 years (Cwynar, 1978; Bonan and Shugart, 1989; Kuusela, 1990). Under contemporary fire suppression regimens, the predicted fire recurrence interval in protected areas of Boreal Shield forests has increased to as long as 500 to 600 years in some areas (Li, 2000). Data from the last 20 to 30 years suggest that most wildfires are less than about 1 km² in size. However, a relatively small number of large (10- to 50-km²) fires are responsible for most of the area burned (Li, 2000).

The Boreal Shield has thin soils, rugged terrain, thousands of lakes, and valuable ore bodies. Forestry is the dominant land use, and the density of human settlement is low. Agriculture, urbanization, and mining also occur and are associated with long-term changes to forests and soils, although at smaller spatial scales (Table 4.2). Outdoor recreation and tourism are important components of the regional culture and economy and depend to a large extent on attractive and healthy forest landscapes (Haider and Hetherington, 2000).

Logging now accounts for most forest disturbance on the managed portion of the Boreal Shield. In the 500,000-km² Boreal forest management zone of Ontario, the percentage of the area logged each year increased from about 0.1% in the decade starting in 1951 to about 0.4% in the first half of the decade starting in 1991. This value is typical of other managed forests in Canada. Forest disturbance by wildfire remained at about 0.1% per year over this 45-year period (Perera and Baldwin, 2000). Insects kill about the same volume of wood as logging, and disease (mainly fungi) kills about twice that (Ontario Environmental Assessment Board, 1994, p. 26; Fleming et al., 2000).

Assessment of aquatic ecosystem response to contemporary Boreal forest disturbance is complicated. It is not uncommon for a lake and its watershed to have been subjected in recent decades to a variety of potentially interacting disturbances. These may include upland and shoreline wildfire, logging or permanent deforestation, shoreline restructuring, nutrient enrichment, acidification, persistent contaminants, increased angling pressure, increased water residence time and ultraviolet light penetration, introduction of

Table 4.2 Speculative Comparison of Contemporary Land Uses on the Boreal Shield

Effect	Land Use			
	Forestry	Agriculture	Urbanization	Mining
% of Boreal Shield surface area directly affected	60%	<1%	<<1%	<<1%
Severity of soil disturbance	Low on cutovers, high on road networks	High and recurring on cultivated fields; high on road networks	High to extreme during development; topsoil may be removed entirely	Locally extreme; soil is removed from mined surface
Duration of deforestation	Years to decades; 10 years is a common estimate of time required for "free-to-grow" status in Ontario; repeated at rotation intervals of 50 to 100 years	Decades to centuries; as long as the land is cultivated regularly	Decades to centuries; until paving and drainage is removed, topsoil is restored, and forest cover is re-established	Decades to centuries; depends on site characteristics such as tailings toxicity and landscape rehabilitation
Extent of artificial drainage networks	Low density of trails, roads, ditches	Medium density of roads, ditches and drains	High density of roads, ditches, sewers, roof tops	Locally high impact from roads, ditches, diversions, mine drainage
Water yield	Moderate increases (40%) at small scales; negligible at large scales	Moderate	Extreme	Negligible to moderate; baseflow may be increased when groundwater is intercepted or pumped from pits and shafts
Peak flows	Moderate, primarily on small streams during snowmelt	Moderate	Extreme	Negligible to low
Destruction of riparian habitat	Low to moderate, primarily associated with water crossings and machine trespass around smaller streams and lakes	Often extensive, involving cultivation or grazing to the water's edge, particularly on smaller streams	Often extensive due to development of shorelands on both small and large streams and lakes	Low-extreme (see below)
Destruction of aquatic habitat	Low; generally associated with water crossings	Moderate; often involving livestock in and around stream channels, nutrient enrichment and sedimentation	Almost always extensive, due to storm runoff, nutrient enrichment, contaminants, sedimentation, and channelization	Low to extreme; may involve infilling of lakes, or permanent flooding of pits (creation of habitat)
Alteration of water quality	Low to moderate for up to 10 yrs. After logging, with some chronic sedimentation associated with road networks	Continuously moderate-high	Very high during urban development, then stabilizing at somewhat lower level	Low to extreme, depending on nature of ore, tailings disposal and treatment methods

Note: With regard to potential for harmful aquatic impacts, these very general overviews take into account road networks and common mitigative measures.

Sources: Gregory (1977); Swank and Crossley (1988); Naiman (1992); Statistics Canada (1994); OEAB (1994); Waters (1995); OMA (1998); Carignan and Steedman (2000).

exotic zooplankton and fish species, and changes in water level caused by dam construction and beaver trapping. Several forms of disturbance may be active simultaneously, and the different disturbances may cause similar symptoms of stress in the aquatic ecosystem, making diagnosis of cause and effect difficult (Loftus and Regier, 1972; Rapport et al., 1985; Steedman and Regier, 1987).

Watersheds: the terrestrial link to Boreal Shield waters

A watershed (also known as catchment, basin, or drainage area) can be defined by the topography of a landscape or drainage network. It is the maximum area of land that may contribute surface runoff to an arbitrary point on the drainage network. The surface drainage network in a watershed includes hillslopes, wetlands, temporary and permanent stream channels, and lakes. Subsurface or groundwater drainage may or may not mirror surface runoff, depending on local geology and soil characteristics. Hydrologists recognize that the entire watershed does not always contribute runoff to streams. The "active" portion of the watershed expands and contracts depending on rainfall and soil moisture, and is almost always smaller than the topographic watershed (Hibbert and Troendle, 1988). Most of the time, when upland soils are not saturated with water, runoff is generated relatively low in the drainage network, and the smallest headwater streams channels are dry. Realistic models of aquatic ecosystem responses to land use must therefore reflect the fact that the intensity of the land–water linkage is often most intense near a water body and decreases with distance from the water (Marsh and Luey, 1982; Steedman, 1988; Håkanson and Peters, 1995). Exceptions to this generalization include landslides or overland flows that bypass normal drainage networks during severe storms and wash material directly into lakes and streams (Swanson et al., 1987). Concerns about long-range atmospheric transport and deposition of acid-generating substances (e.g., sulfate, nitrate; Figure 4.1) and of contaminants (e.g., mercury, PCBs) on land and water led to the concept of an "airshed," the atmospheric analog of the watershed (Gorham and Gordon, 1960; Schindler et al., 1972; Eisenreich, 1982).

The watershed is not the only part of the world that influences a lake, stream, or wetland, but it is a very important one. Small watersheds have long served as convenient, quantitative model systems for terrestrial and aquatic phenomena (Black, 1970; Schindler et al., 1976; Lotspeich, 1980; Likens, 1984). The study of watersheds was led by hydrologists interested in the factors that influence river systems, including channel pattern, channel form, water yield, and flow extremes (Leopold and Maddock, 1953; Strahler, 1957; Chorley, 1962). This science strongly influenced ecologists interested in effects of land use and human activity on biogeochemistry and aquatic biota (Likens et al., 1970; Hammer, 1972; Swank and Crossley, 1988). The role of "riparian" (shoreline or floodplain) forest was examined from several perspectives, including "allochthonous litter inputs" (Kaushik and Hynes, 1971), the river continuum concept (Vannote et al., 1980; Minshall et al., 1985), nutrient spiraling (Elwood et al., 1983), and the ecology of floodplain rivers (e.g., Welcomme, 1985; Regier et al., 1989). The marriage of hydrologic and ecological watershed science traditions can be recognized in contemporary watershed-based analysis and planning for urban, agricultural, and forested landscapes (Hornbeck and Swank, 1992; Frissell and Bayles, 1996; France, 2002a).

Hydrologic pathways through Boreal Shield watersheds

The Precambrian Shield is primarily metamorphosed crystalline bedrock. Glacial scour patterns and till deposits create a disjointed drainage system characterized by numerous lakes and wetlands. The landscape is moderately hilly, with relief typically less than 50 m.

Figure 4.1 Typical potzolic soil profile. High acid deposition increases the solubilization and release of base cations (Ca^{++}, Mg^{++}, K^+) and metals such as Al from such soils. Hypotheses for the recent observation that base cations are declining in many lakes and streams include: depletion of easily leached base cation pools, lower weathering rates because of recent declines in acid deposition, decreases in atmospheric dust, and removal of base cations by forest biomass harvest. For additional details, see Chapter 10, this volume. (Photo by G. Spiers.)

Surface waters are low in dissolved constituents (<100 mg L^{-1} total dissolved solids), soft, neutral to slightly acidic, often colored but generally not turbid. Groundwater hydrology of the Shield has not been studied extensively, primarily because there is generally ample high-quality surface water for human use (Farvolden et al., 1988).

On its way downhill through a Boreal Shield watershed, snowmelt or rainfall may pass through a series of lichen patches on exposed bedrock, through the organic litter and soils of coniferous or deciduous forests and wetlands, and into intermittent and perennial stream channels and beaver ponds (Figure 4.2). These various environments modify the dissolved and particulate load of the water by removing or altering some substances and adding others. The relative proportions of slow hydrologic pathways (e.g., wetlands, groundwater) and fast hydrologic pathways (e.g., channel flow) in a watershed affect runoff chemistry and lake productivity. This can be indexed by an integrative measure such as average watershed gradient, which appears to influence the amount of carbon, nutrients, and other substances exported in runoff. Export of dissolved organic carbon, total phosphorus, chlorophyll, calcium, and magnesium are all negatively related to watershed slope, whereas nitrate and ammonium export are positively related to slope (Engstrom, 1987; Rasmussen et al., 1989; Dillon et al., 1991; D'Arcy and Carignan, 1997). Phytoplankton biomass in Boreal Shield lakes may be predicted with reasonable certainty from a combination of watershed slope, drainage area, and wetland area (D'Arcy and Carignan, 1997).

Small wetlands are common on the Shield and are important modifiers of stream chemistry even when they cover a relatively small portion of a watershed. Wetlands are

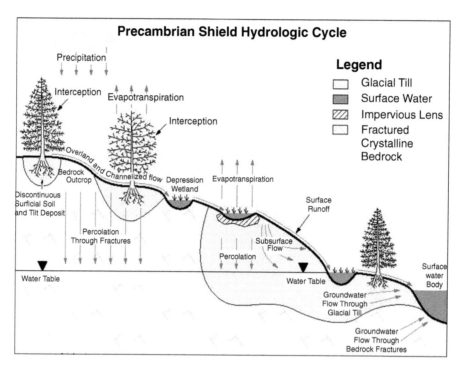

Figure 4.2 Hydrologic pathways on the Precambrian Shield (modified from Grey, 1970, *Handbook of Principles of Hydrology*, National Research Council of Canada, Ottawa).

characterized by low gradients and seasonally saturated organic soils and can have flat to highly irregular ground surfaces formed by boulders and bedrock depressions. Wetland vegetation may consist of relatively vigorous black spruce *Picea mariana* and white cedar *Thuja occidentalis* in conifer swamps or bog-like associations with abundant ericaceous shrubs over peat moss *Sphagnum* (Newmaster et al., 1997). The biogeochemical behavior of a given wetland is difficult to predict, but has been shown to be related to soil and overburden depth over bedrock. Wetlands on deep overburden are more likely to experience continuous groundwater inputs from upland areas and are less likely to undergo episodic drying, oxidation, and export of P, N, and sulfate (Devito and Dillon, 1993; Devito and Hill, 1997). Extensive Shield wetland systems are found on ancient glacial lake sediments such as the "claybelt" of northeastern Ontario and the Red Lake Peatlands in northern Minnesota.

Well-defined stream channels generally don't develop on upland portions of Shield watersheds. Instead, upland runoff commonly moves as shallow subsurface flow through surficial organic soil layers and shallow mineral soils (Renzetti et al., 1992; Allan and Roulet, 1994). Overland flow is often generated from exposed bedrock outcrops and may also contribute a significant portion of the unchannelized upland runoff. Even though travel times and soil–water contact times may be brief in these areas, upland forest soils efficiently sequester atmospheric inputs of N, P, Ca, and K. Organic acids from decomposing litter are exported, and contribute to the acidic, aluminum-rich runoff typical of conifer-dominated headwater watersheds on the Shield (Allan et al., 1993). Downstream, as watershed size and soil depth increase, groundwater contributions to runoff may become substantial. Perennial streams with recognizable channels develop where sufficient headwater runoff accumulates via the various pathways described above. On the Boreal Shield, this typically occurs where drainage area exceeds about 25 ha.

Figure 4.3 Wetlands and beaver ponds have a very significant effect on the hydrology and bio-geochemistry of many Shield lakes. (Photo by E. Snucins.)

In stream channels, dissolved and particulate material may be altered through inter-actions of sediment and water and biotic uptake by algae, macrophytes, and invertebrates. Streams on the Shield are commonly impounded by beavers (Figure 4.3), creating depo-sitional environments with deep organic substrates. Beaver ponds have highly seasonal behavior with regard to nutrient retention and export (Devito and Dillon, 1993). The ponds tend to retain P and N during the growing season when the biota are active and pond waters are ice-free and oxygenated. Severe anoxia develops in water and sediments under winter ice, mobilizing nutrients that are exported downstream with spring runoff. In a study on the Precambrian Shield of eastern Quebec, Naiman (1982) found that headwater streams and associated aquatic biota retain and process most of the forest litter that falls, blows, or is washed into the channel. Unlike particulate sediment, which was exported primarily during brief high-flow periods (10–80 kg ha^{-1} yr^{-1}), dissolved organic carbon export (3–50 kg ha^{-1} yr^{-1}, depending on stream order) was only weakly associated with stream discharge. Northwestern Ontario Boreal streams transport material primarily in dissolved form (87%) rather than as suspended load (10%) or bedload (3%) (Beaty, 1994).

Lakes with basins scoured out of bedrock are common on the northern part of the Boreal Shield. Such lake basins generally have weak groundwater connections with their watersheds. Investigations by the Canadian Nuclear Fuel Waste Management Program (CNFWMP) have shown that localized, shallow groundwater flow systems controlled by surface topography and fractured bedrock are widespread on the Shield, where the major-ity of lake trout lakes are found (Dugal et al., 1981; Lee et al., 1983). There are distinct areas of groundwater recharge and discharge as well as connections between groundwater in surficial deposits and fractured bedrock to a depth of at least 150 m. The elevation of the groundwater table follows surface topography in a general manner. Porosity of the bedrock is low even when it is fractured, and groundwater inputs to Shield lakes are likely to occur only in areas with significant surficial deposits of sand and gravel. Shallow Shield groundwaters are similar in composition to surface waters, with dissolved constituents increasing with depth. Where glacial lake sediments and carbonate materials are found, Na^+, Ca^{2+}, and HCO_3^- and hardness increase (Farvolden et al., 1988). Lakes formed in deep deposits of glacial sand and gravel are common on the southern fringe of the Shield and may be strongly influenced by groundwater (Frape and Patterson, 1981). Groundwater seepage often accounts for most of the water movement through these lakes and is generally concentrated in a narrow band near the shoreline (Lee, 1977).

Figure 4.4 Aerial photo of Lake 42, facing northwest, during 1996 experimental watershed logging at the Coldwater Lakes Experimental Watersheds, 70 km NW of Atikokan, Ontario. (Photograph by M. Friday.)

Watershed and shoreline disturbance on the Boreal Shield

Recent studies suggest that forest disturbance by wildfire or logging produces relatively subtle responses in aquatic biota of Shield lakes (Figure 4.4). Minor or equivocal changes in relative abundance have been reported for phytoplankton (Planas et al., 2000; Knapp et al., 2002), zooplankton (Patoine et al., 2000, 2003), and fish (Rask et al., 1998; St-Onge and Magnan, 2000; Marcogliese et al., 2001; Steedman, 2003; Tonn et al., in review). These findings are consistent with studies from other regions of North America (Moring and Lantz, 1975; Hartman and Scrivener, 1990; Rutherford et al., 1992; Stone and Wallace, 1998; Young et al., 1999; Williams et al., 2002). In contrast, historical mining and smelting practices often caused severe watershed disturbance or deforestation, including massive soil erosion, metal poisoning, and acidification of receiving waters. Watersheds disturbed in this way take many decades, and perhaps centuries, to recover healthy forests and waters (Gunn, 1995).

Amphibian populations in streams appear to be less resilient to forest disturbance (Corn and Bury, 1989). Shaded forest streams often show temporarily increased production of algae, invertebrates, and fish after forest disturbance that removes riparian cover (Murphy and Hall, 1981; Behmer and Hawkins 1986). However, this general response may be modified or suppressed by other factors such as nutrient availability, disturbance of large wood debris, sediment inputs and temperature preferences of resident biota (Moring and Lantz, 1975; Murphy et al., 1986; Culp, 1988; Hartman and Scrivener, 1990; Young et al., 1999). Drainage, practiced primarily in Fennoscandia to lower the water table in wet forests, appears to be particularly disruptive to stream biota (Holopainen et al., 1991; Vulori et al., 1998).

Forest management may influence aquatic biota and habitat through both watershed and riparian mechanisms. The relative importance of these depends on the amount of

logging in the watershed and on the location and design of clearcuts, roads, water crossings, and machinery operations in and around riparian areas. Other forest management impacts on aquatic ecosystems include road networks and wastewater discharge from processing mills.

Hydrologic response

Extensive watershed forest disturbance by wildfire, logging, blowdown, or disease may increase water yield and peak flow by temporarily reducing forest evapotranspiration (Bethlahmy, 1974; Schindler et al., 1980; Verry, 1986; Nicolson, 1988; Dubé et al., 1995; Stednick, 1996). Additional impacts may also occur as a result of localized soil disturbance, rutting, and compaction, which alter infiltration capacity, microtopography, and soil biogeochemical processes (Grigal, 2000). Increases in water yield or peak flows on small forested watersheds are generally not detected unless at least 25 to 50% of the watershed has been disturbed (Plamondon and Oeullet, 1980; Bosch and Hewlett, 1982; Cheng, 1989).

The increase in water yield from disturbed or deforested watersheds depends on several factors, including climate, watershed physiography, pretreatment vegetative cover, the amount and type of remaining vegetation, and the type and intensity of the harvesting practice. Postlogging increases in water yield are mainly expressed in the growing season, when the intact forest would have exerted its greatest evapotranspiration demands, and are associated with higher peak flows from individual rain events and higher wetland and groundwater levels (Nicolson et al., 1982; Verry, 1986). Complete disturbance of small Shield watersheds appears to increase annual water yield by about 40% (Plamondon and Oeullet, 1980; C. Allan, unpublished data). Spring runoff and peak flows may be particularly sensitive to watershed disturbance because of increased snow accumulation and earlier snowmelt after removal of coniferous canopies (Buttle et al., 2000) and enhanced drainage efficiency in roads and ditches. Recent analyses suggest that in large watersheds, flow routing along roads can be more important than logging in controlling the size and timing of peak flows (Jones and Grant, 1996).

In summary, present levels of temporary watershed disturbance associated with Boreal forest management generally appear to be low and unlikely to cause unnatural or harmful impacts on water yield or peak flows, particularly at the landscape scale (Buttle and Metcalfe, 2000; Verry, 2000). Extensive synchronous forest disturbance by clearcutting (i.e., 25 to 50% or more of a watershed within a 5-year period) is most likely to occur on relatively small watersheds (i.e., generally much smaller than 1 km^2) under modern sustainable forestry scenarios. In contrast, wildfire occasionally causes much larger disturbances.

Water quality

At any watershed scale, extensive forest disturbance causes significant but temporary changes in groundwater, stream, and lake water quality, including increased concentrations of dissolved material such as organic carbon, cations (particularly potassium), and plant nutrients (N and P). These temporary water quality changes are driven by watershed-scale changes in forest hydrology and soil chemistry, involve normal groundwater, wetland, and streamflow pathways, and typically peak in the first or second year after disturbance. Soil disturbance and warming, ash deposition, and decomposition of damaged vegetation after fire or logging are the primary sources of excess nutrient export after forest disturbance. Nutrient export increases as a result of increased runoff volume, increased concentrations in runoff, or both.

The risk to water quality after forest disturbance has generally been assessed from landscape properties related to the hydraulic energy and transportation of sediment eroded from the catchment. These risk factors include precipitation intensity, slope and slope position, soil exposure, compaction, and texture (coarse soils with low organic content are more erodible) (Trimble and Sartz, 1957; Dyrness, 1967; Beaty, 1994; France et al., 1998). According to these landscape risk factors, Shield lakes should generally be at lower risk from increased sedimentation and nutrient enrichment than lakes in areas such as the Pacific Northwest, which have more runoff, steeper slopes, and erodible soils (Krause, 1982).

However, in spite of this generalization there are areas on the Shield where topography, vegetation, and soils may interact to produce locales at relatively high risk of erosion. Following windthrow and wildfire, stream bedload in a northwestern Ontario stream increased by 20-fold (Beaty, 1994). Forest disturbance may temporarily increase the erosion of surface duff by raindrop impact, with the greatest potential for increased soil erosion near lake shorelines, where litterfall is lower and soils are often thinner than in upland areas (France, 1997b; France et al., 1998). Aeolian (wind-driven) soil erosion has been observed on new cutovers in the Boreal forest where mineral soil has been exposed on skid trails, roads, and landings and can reach nearby lakes (Steedman and France, 2000). Most water-borne sediment erosion during Boreal forest management originates from roads and associated structures and disturbances (e.g., ditches, culverts). On the Shield, overland transport of road sediment in storm runoff appears to be attenuated within about 40 m on both clearcuts and undisturbed forest floor, even on steep slopes (France, 2002b).

Most watershed studies in temperate areas that have shown altered water budgets and nutrient dynamics after watershed disturbance have also shown significant recovery trends within 5 to 10 years. The rate of recovery after watershed disturbance is closely associated with the rate and extent of herbaceous and forest regeneration (Hibbert, 1966; Bayley et al., 1992). Hall and Smol (1993) investigated historical lake responses to the widespread dieoff 4,800 years ago of hemlock (*Tsuga canadensis*) forests on the southern Boreal Shield. All five lakes that they studied showed shifts in their algal communities and minor changes in trophic status as inferred from historical diatom communities. One lake with a large, steep watershed showed doubling in total P (from predieoff levels of about 14 µg L^{-1}) following the hemlock dieoff, followed by a decrease in total P of 11 µg L^{-1} as the forests recovered.

Water quality impacts are strongly modulated by watershed characteristics, drainage position, and morphometry of the water body. Lakes and streams both respond to forest disturbance in similar ways, but the interplay of watershed and lake basin morphology may cause water quality impacts to be delayed or reduced because of dilution of runoff in lake volume. Lakes with long water renewal times (i.e., relatively large lake volume and relatively small watersheds) may be relatively insensitive to temporary watershed disturbance by wildfire or forestry (Carignan et al., 2000; Steedman, 2000). For the same watershed disturbance, streams will show the greatest change in water quality, shallow lakes with short water renewal times an intermediate response, and deep lakes with long water renewal times the smallest response (Nicolson, 1975; Schindler et al., 1980; Bayley et al., 1992; Rask et al., 1993, 1998; Jewett et al., 1995).

Clearcut watersheds appear to export more dissolved organic carbon (DOC) and less nitrate and bioavailable phosphorus than burnt watersheds (Carignan et al., 2000). Forest disturbance, like forest flooding, is also associated with increased movement of terrestrially deposited mercury into aquatic biota (Garcia and Carignan, 1999, 2000). Water color is strongly influenced by DOC and may have a significant influence on surface heating and stratification of small lakes (Fee et al., 1996). In northwestern Ontario, water clarity in three lakes declined by 25% after partial clearcutting of their watersheds. Late-summer thermoclines were about 1 m shallower after logging in two of the lakes, but it was not possible to exclude weather as a factor (Steedman and Kushneriuk, 2002).

Phosphorus is generally the limiting nutrient for primary production in most Shield lakes (Schindler, 1974) because of its scarcity in bedrock and overburden on Shield watersheds and efficient retention by upland forests and wetlands (Allan, 1993, Devito et al., 1989). Where lake nutrient budgets have been estimated for undisturbed Shield lakes, precipitation and dry aerial deposition directly on the lake surface are generally found to be dominant sources of P (Schindler et al., 1976; Dillon et al., 1993). In contrast, watershed disturbance by agriculture and urbanization is associated with significant increases in P export (Dillon and Kirchner, 1975). Severe shoreline trampling and soil erosion associated with heavily used hiking trails were implicated in P enrichment of a Quebec lake (Dickman and Dorais, 1977). Although N is commonly exported from disturbed forests, it is not likely to have significant effects on algal growth in Shield waters. It is possible that increased short-term nutrient inputs to lakes following forest disturbance (e.g., from airborne dust, rill erosion, or groundwater) may be offset to some extent by decreased inputs of riparian litterfall (France et al., 1996).

Nutrient enrichment after watershed disturbance could be expected to modify the volume of lake trout habitat (i.e., cool, highly oxygenated water) in deep, stratified lakes. Intermediate mechanisms would be involved, including decreased water clarity (caused by increased loadings of dissolved organic carbon [DOC], nutrients, and sediment) and increased wind-induced mixing. However, these expected changes were not apparent in an experiment designed to test such linkages (Steedman and Kushneriuk, 2000).

In shallow lakes, forest disturbance may increase the frequency or severity of winterkill brought about by increased primary production and winter oxygen depletion. However, inter-annual variation in timing of ice cover of shallow lakes is also an important factor in overwinter fish survival (Danylchuk and Tonn, 2003).

Although water quality is influenced by both wildfire and forest management, long-term precipitation patterns may have a greater effect on watershed erosion and lake productivity. Studies of recent sediments in northwestern Ontario revealed pronounced declines in sedimentation rates after 1970. The decline in sedimentation was positively correlated with the size of the watersheds but was unrelated to disturbance history and coincided with a 60% decrease in regional runoff and precipitation (Blais et al., 1998; Schindler, 1998). Analysis of chrysophyte (planktonic algae) remains from the same sediment cores (Paterson et al., 1998) indicated that there were significant differences in the composition of chrysophyte communities among the various lakes and that, in general, these chrysophyte communities had not changed much in the last 50 years or so, even for lakes with burned or logged watersheds. However, there were significant differences in chrysophyte communities corresponding to the recent climate-driven reduction in sedimentation rates.

Riparian disturbance

In the following discussion, "riparian" refers to the area of wet soils and distinctive vegetation immediately adjacent to streams, rivers, and lakes. In low-relief Boreal landscapes, riparian areas commonly take the form of wide, fringing wetlands dominated by herbaceous or shrubby wetland vegetation. In areas of higher relief, or where stream channels are incised, the riparian zone may be narrow and behave more like an upland forest, providing shade, forest litterfall, and wood debris inputs to aquatic habitats. We use the word shoreline in a more inclusive way to refer to both upland and riparian areas in the vicinity of a water body, typically within 100 m or so. Shoreline areas are typically subject to special forest management policy such as "riparian buffer strips," which are often comprised of both upland forest and riparian vegetation.

Unlike watershed-scale impacts, riparian disturbance may directly influence aquatic habitats through shade loss, physical disturbance, or changes to forest litter inputs, particularly along streams. Shorelines with upland vegetation are more likely to be disturbed by wildfire than shorelines with broad wetlands. In managed forests, riparian disturbance is often associated with water crossings or machine activity too close to streams or lakes.

Shoreline forest disturbance is not likely to influence midlake wind speeds, stratification, and mixing on most lake trout lakes. Clearcut logging around three 30- to 40-ha dimictic northwestern Ontario lakes was associated with increases of 5% or less in midlake wind speed and no measurable changes in mixing or duration of stratification (Steedman and Kushneriuk, 2000). However, average wind velocities close to lee shores probably increase after wildfire or logging and may slightly increase average littoral turbulence and wave energy. Only small lakes in relatively flat terrain, with fetch less than a few hundred meters, should experience significantly increased effective wind exposure after shoreline forest disturbance. This has been shown for a 0.4-ha lake in Finland (Rask et al., 1993) and for a 4-ha lake basin in Michigan (Scully et al., 2000).

In streams, shoreline forest disturbance is associated with increased summer water temperature and temperature variability. These increases are usually attributed to decreased shading, although other factors including wind, stream discharge, channel form, and groundwater inputs can all be important (Brown and Krygier, 1967; Holtby, 1989). Boreal stream temperature response to riparian shade loss is probably similar to that reported elsewhere (Krause, 1982) but will be influenced by slope, vegetation, and groundwater inputs. Increased exposure to ultraviolet light (UVB) after riparian canopy removal has been shown to influence benthic algae and invertebrates in small streams, but this effect may be counteracted to some extent by increased DOC in runoff from logged watershed (Bothwell et al., 1993; Kelly et al., 2001).

Lake water temperature is influenced by a complex and interacting suite of factors, including air temperature, solar energy, wind, water clarity, lake depth, and morphology. Different factors may dominate at different locations and at different spatial or temporal scales. Shade associated with shoreline forest influences only a small portion of the water surface on all but the smallest lakes. Shoreline logging did not significantly increase average littoral water temperatures in two small Boreal forest lakes in northwestern Ontario but was associated with increases of 1–2°C in maximum littoral water temperature, and increases of 0.3–0.6°C in average diurnal temperature range, compared with undisturbed shorelines or shorelines with 30-m riparian reserves (Steedman et al., 1998, 2001). However, these small, transient impacts did not influence whole-lake thermal regimes (Steedman and Kushneriuk, 2000) and are unlikely to affect the distribution or growth of aquatic biota. In very small lakes and ponds, shoreline logging may increase wind exposure, mixing, and dissolved oxygen (Rask et al., 1993; Scully et al., 2000; Steedman and Kushneriuk, 2000). Removal of riparian vegetation has been shown to stimulate aquatic primary production in light-limited streams (Gregory et al., 1987), but similar studies have not been conducted on lakes.

Riparian (shoreline) forests contribute energy and nutrients to lake biota (Likens, 1985). As with forests along headwater streams, these inputs are comprised of large wood debris (boles, branches, twigs) and particulate forest litter such as leaves, flowers, pollen, terrestrial insects, and insect frass (fecal pellets). Riparian energy and nutrient inputs are probably more important on small lakes or lakes with complex shorelines, where the ratio of shoreline length to lake surface area is relatively large.

Lake trout (*Salvelinus namaycush*) in small Shield lakes depend on a combination of forest litter and littoral foodwebs to satisfy their energy requirements (France and Steedman, 1996). Nitrogen isotope ratios in the flesh of juvenile lake trout indicate that they are relatively omnivorous, exhibiting $\delta^{15}N$ values (deviations of the $^{15}N/^{14}N$ ratio from an isotopic standard) consistent with predation on the opossum shrimp *Mysis relicta* (41% of samples), zooplankton (35% of samples), and littoral macroinvertebrates (25% of samples). Carbon isotope ratios indicate that these juvenile lake trout obtain on average about half of their carbon from terrestrial sources. Littoral macroinvertebrates colonize sunken forest litter, probably to graze on biofilm (a mixture of periphyton, fungal hyphae, bacteria, and organic debris) or to avoid predators (France, 1997a). However, because forest litter decomposes slowly in lakes and is regularly resupplied during forest regeneration, this important habitat resource is probably rarely in limited supply (France, 1998).

Airborne (primarily coniferous) forest litter input to four small lakes in northwestern Ontario was estimated to be about 30 g dry weight per meter of forested shoreline per year (France and Peters, 1995). Rough estimates of litter input to lakes in other, primarily deciduous regions range up to 350 g dry weight per meter of shoreline (Jordan and Likens, 1975; Gasith and Hasler, 1976; Hanlon, 1981). Cole et al. (1990) measured midsummer P inputs (mostly ants, spiders, and leaf fragments) to a small oligotrophic lake in New Hampshire and found that these terrestrial inputs were 50 to 70 times greater than rainfall and stream P inputs during the same period.

The amount of forest litter that reaches the lake surface declines with distance beyond about 3 m from shore and is related to the height of the riparian forest and its proximity to the shoreline (France and Peters, 1995). Lateral transport of forest-floor litter by wind or water (as opposed to direct litter inputs from the forest canopy) represents less than 6% of the total forest litter supply to small Boreal lakes (France, 1995). A model developed from these studies indicated that in oligotrophic headwater lakes, forest litter can contribute up to 15% of the total carbon supply and up to 10% of the total P supply (France and Peters, 1995).

Clearcut logging of shoreline forests in northwestern Ontario removes conifers and allows small deciduous trees to dominate the riparian forest for several decades during the time of riparian forest regrowth. Inputs of leaves and small woody debris to the littoral zone may be locally reduced by over 90% for several years after shoreline logging (France et al., 1997b). Based on estimates of airborne litter input from forested and clearcut shorelines and laboratory measurement of leaf leachate, France et al. (1996) determined that riparian forest disturbance could reduce DOC inputs from about 18 to less than 1 g per meter of shoreline per year and reduce total P inputs from about 3 to less than 1 g per meter of shoreline per year. In the absence of increased soil erosion and transport, such reductions could decrease primary production of lake plankton communities by as much as 9% and reduce respiration by as much as 17%. These effects could last more than 10 years in northwestern Ontario, depending on the rate of riparian forest regeneration, and would tend to counteract the runoff nutrient pulse occurring after watershed forest disturbance.

Little is known about abundance or recruitment of wood debris into Boreal lakes and streams. Wood may enter Boreal streams and lakes through natural forest disturbances, primarily windthrow of dead or damaged trees, particularly after shoreline wildfire. Such wood inputs are probably episodic and highly variable. Beavers also fell or drag significant amounts of wood into Boreal lakes and streams. Natural wood inputs were augmented through much of the 20th century in some Boreal Shield lakes and rivers by losses from log booms and log drives (e.g., Kelso and Demers, 1993). Because of the relatively low topographical relief and a smaller, younger forest, abundance of wood in Boreal waters may be lower than that reported for other regions.

Submerged tree trunks and branches provide habitat complexity and nutrients for littoral biota, including fish, invertebrates, algae, and decomposer organisms. Fallen tree trunks and large branches may persist for centuries when submerged in fresh water (Guyette and Cole, 1999). In streams, large wood debris creates and maintains complex stream channels, including deep pools important to fish production (Swanson et al., 1982; Murphy et al., 1986; Bisson et al., 1987). Large wood debris may have a similar role in Boreal lakes, particularly in littoral waters (Mallory et al., 2000). In three northwestern Ontario lake trout lakes, littoral waters contained 100 to 200 pieces of large wood (>10 cm diameter, >1 m length) per kilometer of shoreline (Ontario Ministry of Natural Resources/Centre for Northern Forest Ecosystem Research, unpublished data). Shoreline cottage development has been associated with a reduction in the amount of large wood debris in the littoral zone of northern lakes (Christensen et al., 1996).

Depending on configuration and timing, shoreline forest harvesting has the potential to reduce the future supply of large wood debris unless appropriate measures are taken. To contribute to aquatic habitat, retained trees must be close enough to the shoreline to have a reasonably high probability of falling into the water, i.e., within $1/2$ tree height of the shoreline, or about 7 m in the Boreal forest. Partial shoreline logging may decrease wood inputs to lakes (through reduced tree abundance) or increase inputs (through increased windthrow).

Mitigation of impacts

Boreal Shield management agencies administer a variety of policies to protect lake trout lakes from watershed and shoreline disturbance caused by forestry, road construction, agriculture, or residential development. These policies generally involve three approaches: (1) "best management" practices to reduce erosion associated with stream crossings and road construction near water bodies; (2) protection of riparian vegetation and soils during forestry activities; and (3) restrictions on shoreline development for cottages and homes. Although rare, some management agencies on the Shield restrict development at the watershed scale when this is believed to be degrading water quality or fish habitat in lake trout lakes. For example, on Clearwater Bay, Lake of the Woods, Ontario, the Ontario Ministry of Natural Resources issued in 1990 a "Restricted Area Order" that limits development of new housing in the watershed of Clearwater Bay (Ontario Ministry of Natural Resources, 1994).

Roads and water crossings

Road construction almost always accompanies human activity on the Shield and in recent decades has frequently been implicated in occurrences of destructive erosion and sedimentation in forested landscapes (Megahan and Kidd, 1972; Duck, 1985; Ontario Environmental Assessment Board, 1994, p. 126; Swift, 1988). The potential for erosion is high during and after construction of roads, culverts, and bridges because these structures require extensive disturbance to vegetation and soils, including excavation, filling, and compaction for ditches, roadbeds, and embankments, and creation of artificial drainage channels. Poor road layout and maintenance and inappropriate selection of crossing structures can potentially degrade fish habitat. Habitat degradation may be caused by physical disruption or burial of stream channels and lakeshores, by sedimentation, or by introduction of hanging culverts or other structures that limit fish movement or migration (Warren and Pardew, 1998). Abandoned or poorly maintained road networks may also pose a long-term threat to aquatic habitat (Jones et al., 2000). When culverts fail or roads wash out, large volumes of sediment are introduced into streams and lakes. However, proper construction and maintenance techniques developed in the last 20 years have the potential to greatly reduce the frequency and intensity of erosion and sedimentation. Most

Figure 4.5 Forest buffer strips surrounding lakes. The effectiveness of such forest management practices are under review in some jurisdictions. (Top photo by OMNR, bottom photo by Jim Buttle.)

jurisdictions now employ systems of mandatory design standards, good practice guidelines, and mitigation techniques designed to minimize the risk of damage to water quality and aquatic habitats during road construction and maintenance (e.g., Ontario Ministry of Natural Resources, 1990; Québec Ministère des Forêts, 1992). However, the benefits of these systems are difficult to quantify (Park et al., 1994), and road failures still occur as a result of improper construction techniques and extreme precipitation or runoff events.

Easy forest access facilitated by new roads almost invariably leads to increased exploitation of fish populations, intentional and unintentional introduction of exotic species, and pressure for cottage development in previously inaccessible areas. Angling pressure may be more of a threat to Boreal fish populations than habitat degradation associated with the road construction itself (Gunn and Sein, 2000). Non-native fishes may be introduced to newly accessed waters accidentally through use of live bait or intentionally to provide angling opportunities. Such introductions have reduced native fish biodiversity in Boreal Shield lakes (MacRae and Jackson, 2001).

Riparian protection

Riparian buffer strips are reserves of undisturbed shoreline forest designed to protect lakes and streams from forestry or other land uses, and are widely prescribed by forest managers

(Figure 4.5). Fish habitat and water quality protection are common design objectives, but shoreline visual esthetics and terrestrial wildlife habitat are also frequently addressed through this approach. Preservation of shoreline forest is thus intended to provide a range of benefits to aquatic habitats. These include shade, bank stability, inputs of leaves, branches, and boles important as aquatic food and habitat, retention of sediments and nutrients from upland runoff, and protection of shoreline esthetics. Forested riparian buffer strips of at least 15 m width are generally sufficient to prevent stream temperature increases after logging, but the other benefits have proven more difficult to quantify (Clinnick, 1985; Ribe, 1989; Norris, 1993; Castelle et al., 1994; Haider and Hetherington, 2000). Riparian buffer strips do not appear to be effective in preventing water quality changes associated with watershed forest disturbance (Norris, 1993; Steedman, 2000; Carignan et al., 2000). In northwestern Ontario, a forested shoreline buffer strip prevented increases in midlake wind speed but did not prevent declines in water clarity and thermocline depth (Steedman and Kushneriuk, 2000). Windthrow is known to be a chronic problem with riparian buffer strips because nonwindfirm trees from interior forest become exposed to higher wind velocities on edges of clearcuts (Gratkowski, 1956; Steinblums et al., 1984; Quine et al., 1995). Recent research from the Boreal Shield suggests that local topographic control of wind velocity is more important than buffer width in predicting windthrow in buffer strips (Ruel, 2000).

Administrative jurisdictions on the Boreal Shield have adopted a variety of approaches to protect riparian areas during forest management. For example, in Ontario, foresters must identify "areas of concern" around lakes and streams, where special protective measures may be considered. The area of concern ranges from 30 m width (measured from high water) on slopes up to 15% to 90 m width on slopes up to 60%. On lake trout lakes forestry operations are severely restricted within the area of concern, so that there may be no road or landing construction, no tree cutting (except for selection cutting under certain circumstances), and no site preparation (Ontario Ministry of Natural Resources, 1988). Additional operational measures to minimize riparian soil disturbance are also specified in a "Riparian Code of Practice" (Ontario Ministry of Natural Resources, 1991). Proposed revisions to these guidelines, based in part on recent science findings reviewed above, may in the future allow some logging to the shoreline where this is consistent with esthetic values, wildlife habitat, and emulation of regional wildfire patterns, and provided that no riparian disturbance or sedimentation occurs. In the Boreal forest of Ontario, most "no-cut" reserves are for shoreline protection and account for about 10% of potential timber harvest in managed forests (Puttock, 1985; Ontario Environmental Assessment Board, 1994, p. 182; France et al., 2002). In Quebec, a 20-m-wide "wooded strip" (measured from the tree line beside the water's edge) must be preserved around lakes and water courses during forestry operations. If the slope is less than 40%, and tree basal area (stocking) exceeds 60%, one-third of trees 10 cm or more in diameter must be harvested from the wooded strip (Québec Ministère des Forêts, 1992). In Minnesota, most lake trout lakes are in the northeastern part of the state and are wholly or partly under Federal jurisdiction in the Boundary Waters Canoe Area Wilderness, National Forests (Superior and Chippewa), or Voyageurs National Park. In these areas, legislation requires that the natural beauty of shorelines be protected, generally by prohibition of logging within 400 feet (122 m) of the waterline. Special riparian management zones of various widths (50–300 feet or 15–91 m) are also recognized in the National Forests and allow forest management that is consistent with protection of riparian values.

Harvest and regeneration of some shoreline forests may become a component of landscape-scale emulation of natural disturbance in managed Boreal Shield forests, as it has in the Pacific northwestern and eastern deciduous forests of North America (Gregory, 1997; Palik et al., 2000). Although wildfire is known to naturally disturb shorelines of Boreal forest lakes, there are uncertainties regarding environmental consequences associated with shoreline forest management. Boreal forest managers in Ontario are now required by law to "emulate natural disturbances and landscape patterns while minimizing adverse effects on plant life, animal life, water, soil, air and social and economic values..." (Ontario Crown Forest Sustainability Act 2[3]2). Shoreline forests are not excluded from the Crown Forest Sustainability Act (CFSA) requirement to emulate natural disturbance and landscape patterns during forest management. In the Boreal forest of Ontario, efficient fire suppression is combined with management policies that routinely prescribe 30 to 90 m of forested shoreline reserves around most lakes and streams. However, many foresters believe that extensive, protected shoreline reserves are producing shoreline forests dominated by shade-tolerant tree species not representative of natural Boreal forest conditions, in unnatural linear patterns highly susceptible to blowdown and wildfire, contrary to the intent of the CFSA. Research reviewed in this chapter suggests that shoreline forestry operations, if carefully conducted, need not be associated with deleterious effects to water quality and aquatic habitat in Boreal lakes. If this is a general and reliable result, it implies that some shoreline reserves could be harvested and regenerated without significant risk to water quality or fish habitat. Remaining uncertainties regarding the generality and reliability of these research findings can be resolved only through operational adaptive-management trials, coupled with appropriate quantitative monitoring.

Shoreline development

Because of their clear waters and rocky shorelines, lake trout lakes on the Shield are popular for seasonal and year-round residential shoreline development. Whether as cottages or permanent homes, this development brings sewage effluent (usually via private septic tanks and associated tile fields), riparian and upland deforestation, and littoral habitat modifications such as bathing beaches, landings, docks, and boathouses. Elevated P inputs to Shield lakes, usually from municipal sewage or agricultural runoff, have been associated with algal blooms, hypolimnetic oxygen depletion, fouling of spawning habitat, and impaired visual esthetics (e.g., Dillon and Rigler, 1975; Dillon et al., 1986; Molot et al., 1992).

Ontario's Trophic Status Model (Dillon et al., 1986; Hutchinson et al., 1991) provides a methodology for predicting phosphorus concentration in lakes subjected to various forms of shoreline or catchment development. The Trophic Status Model is based on a series of empirical and semiempirical relationships among lake morphometry, catchment characteristics, and natural and anthropogenic sources of phosphorus, and is a refinement of the Dillon and Rigler (1975) model. The main application of this methodology has been the regulation of shoreline "cottage" development on Shield lakes in Ontario. In practice, small headwater lakes with intensive shoreline development or lakes with significant upstream or catchment phosphorus sources were the most likely candidates for development restrictions under this approach. From 1970 to 1998 Ontario's policy was to regulate shoreline development so that lakes with total phosphorus concentrations less than 10 µg L^{-1} would not be permitted to rise above 10 µg L^{-1}, while lakes with total phosphorus concentrations above 10 µg L^{-1} would not be permitted to rise above 20 µg L^{-1} (Ministry of the Environment, 1984).

Shoreline cottage development may not yet have significantly affected the productivity of Shield lakes. Analyses of diatoms in sediment cores from 54 south-central Ontario lakes with "moderate" cottage densities suggested that present-day total P has not increased relative to pre-1850 levels (Hall and Smol, 1996). Paleoecological evidence suggests that mesotrophic lakes in that study appear to have experienced decreases in total P over the same period, perhaps because of watershed acidification or long-term forest recovery after clearcutting in the early 1900s. Because P appears to be less mobile in domestic septic system effluent than previously assumed, models may have overestimated the influence on water quality of domestic P inputs (Hutchinson, 2002). In contrast, municipal P inputs have had dramatic but reversible water quality impacts on Shield lakes. Gravenhurst Bay, which receives municipal sewage effluent and was seriously enriched before 1972 (total P of 40 to 50 µg L^{-1}), recovered steadily following improved sewage treatment and now has total P of about 13 µg L^{-1}, similar to the pre-1850 level inferred from diatom remains.

Policy implications and recommendations

Research reviewed in previous sections of this chapter provided some of the "best guesses" that land and water managers have used to protect lake trout lakes over the last 20 years. For example, evidence is accumulating regarding the need to protect the riparian zone and its complex geomorphic, geochemical, and biotic connections to lakes. However, some assumptions about the effects of watershed disturbance around lake trout lakes have not been strongly supported by recent evidence. It has long been thought that forest disturbance could place Boreal waters at risk through sedimentation, nutrient enrichment, and degradation of littoral and hypolimnetic lake trout habitat. This process appears to occur, but primarily off of the Shield, in association with extensive, long-term agricultural or residential watershed deforestation and development (e.g., Evans et al., 1996). There is little evidence that temporary forest disturbance associated with wildfire or forestry has produced detrimental changes of this type on the Shield.

The research reviewed in this chapter has a number of implications for management of Boreal Shield waters:

1. Strategic, long-term ecosystem monitoring and experimental policy assessment are required for sustainable management of irreplaceable resources such as healthy Boreal waters and aquatic biota.
2. Management agencies must be conservative users of science (Steedman, 1994). Although recent findings regarding the aquatic effects of watershed disturbance are reasonably compelling, researchers are less certain that all relevant phenomena have been reliably assessed. It is difficult and expensive to measure subtle land-use effects, and we don't know how far we can extrapolate detailed findings from case studies.
3. Present approaches used by Boreal forest managers for the protection of water quality and aquatic habitat during forest management will likely prove conservative and reliable for most aquatic values associated with Shield lakes.
4. Conceptual integration of riparian land-use policy is required. Fish habitat (broadly defined, as in Canada's Fisheries Act) may not always be a limiting factor for sustainable management of shoreline forests. However, other values such as shoreline esthetics and historical landscape pattern may be more sensitive in some locations. Shield lakes appear to be relatively insensitive to temporary catchment forest disturbance. This may alleviate some concerns associated with forest management designs intended to more closely reflect historical forest disturbance by wildfire.

Acknowledgments

Thanks to Gary Siesennop of the Minnesota Department of Natural Resources for researching and summarizing shoreline protection policy for Minnesota lake trout lakes. Thanks to David Schindler and Peter Dillon for constructive criticism of an early draft.

References

Allan, C.J., Roulet N.T., and Hill, A.R., 1993, The biogeochemistry of pristine, headwater Precambrian shield watersheds: an analysis of material transport within a heterogeneous landscape, *Biogeochemistry* **22**:37–79.

Allan, C.J. and Roulet, N.T., 1994, Runoff generation in zero order Precambrian Shield catchments: the stormflow response of a heterogeneous landscape, *Hydrological Processes* **8**:369–388.

Bayley, S.E., Schindler, D.W, Beaty, K.G., Parker, B.R, and Stainton, M.P., 1992, Effect of multiple fires on nutrient yields from streams draining boreal forest and fen watersheds: nitrogen and phosphorus, *Canadian Journal of Fisheries and Aquatic Sciences* **49**:584–596.

Beaty, K.G., 1994, Sediment transport in a small stream following two successive forest fires, *Canadian Journal of Fisheries and Aquatic Sciences* **51**:2723–2733.

Behmer, D.J. and Hawkins, C.P., 1986, Effects of overhead canopy on macroinvertebrate production in a Utah stream, *Freshwater Biology* **16**:287–300.

Bethlahmy, N., 1974, More streamflow after a bark beetle epidemic, *Journal of Hydrology* **23**:185–189.

Bisson, P.A., Bilby, R.E., Bryant, M.D., Dolloff, C.A., Grette, G.B., House, et al., 1987, Large woody debris in forested streams: Past, present and future, In *Streamside Management: Forestry and Fishery Interactions*, Contribution Number 57, Institute of Forest Resources, edited by E.O. Salo and T.W. Cundy, University of Washington, Institute of Forest Resources, Seattle, pp. 143–190.

Black, P.E., 1970, The watershed in principle, *Water Resources Bulletin* **6**:153–162.

Blais, J.M., France, R.L., Kimpe, L.E., and Cornett, R.J., 1998, Climatic changes have had a greater effect on erosion and sediment accumulation than logging and fire: evidence from ^{210}Pb chronology in lake sediments, *Biogeochemistry* **43**:235–252.

Bonan, G.B. and Shugart, H.H., 1989, Environmental factors and ecological processes in boreal forests, *Annual Review of Ecology and Systematics* **20**:1–28.

Bosch, J.M. and Hewlett, J.D., 1982, A review of catchment experiments to determine the effect of vegetation changes on water yield and evapotranspiration, *Journal of Hydrology* **55**:3–33.

Bothwell, M.L., Sherbot, D., Roberge, A., and Daley, R.J., 1993, Influence of natural UV-radiation on lotic periphytic diatom community growth, biomass accrual and species composition: Short-term versus long-term effects, *Journal of Phycology* **29**:24–35.

Brown, G.W. and Krygier, J.T., 1967, Changing water temperatures in small mountain streams, *Journal of Soil and Water Conservation* **22**:242–244.

Buttle, J.M. and Metcalfe, R.A., 2000, Boreal forest disturbance and streamflow response, northeastern Ontario, *Canadian Journal of Fisheries and Aquatic Sciences* **57**(Suppl. 2):5–18.

Buttle, J.M., Creed, I.F., and Pomeroy, J.W., 2000, Advances in Canadian forest hydrology, *Hydrological Processes* **14**:1551–1578.

Carignan, R., D'Arcy, P., and Lamontagne, S., 2000, Comparative impacts of fire and forest harvesting on water quality in boreal shield lakes, *Canadian Journal of Fisheries and Aquatic Sciences* **57**(Suppl. 2):105–117.

Carignan, R. and Steedman, R.J., 2000, Impacts of major watershed perturbations on aquatic ecosystems. Introduction to *Canadian Journal of Fisheries and Aquatic Sciences*, Volume 57, Supplement S2, *Canadian Journal of Fisheries and Aquatic Sciences* **57**(Suppl. 2):1–4.

Castelle, A.J., Johnson, A.W., and Conolly, C., 1994, Wetland and stream buffer size requirements — a review, *Journal of Environmental Quality* **23**:878–882.

Cheng, J.D., 1989, Streamflow changes after clear-cut logging of a pine beetle-infested watershed in southern British Columbia, Canada, *Water Resources Research* **25**(3):449–456.

Chorley, R.J., 1962, *Geomorphology and general systems theory. United States Geological Survey Professional Paper* 500-B: 10 pp.

Christensen, D.L., Herwig, B.R., Schindler, D.E., and Carpenter, S.R., 1996, Impacts of lakeshore residential development on coarse woody debris in north temperate lakes, *Ecological Applications* **6**(4):1143–1149.

Clinnick, P.F., 1985, Buffer strip management in forest operations: a review, *Australian Forestry* **48**(1):34–45.

Cole, J.J., Caraco, N.F., and Likens, G.E., 1990, Short-range atmospheric transport: a significant source of phosphorus to an oligotrophic lake, *Limnology and Oceanography* **35**:1230–1237.

Culp, J.M., 1988, The effect of streambank clearcutting on the benthic invertebrates of Carnation Creek, British Columbia, In *Applying 15 Years of Carnation Creek Results,* Proceedings of a workshop held January 13–15, 1987, Nanaimo, B.C., edited by T.W. Chamberlin, Carnation Creek Steering Committee, c/o Pacific Biological Station, Nanaimo, B.C., pp. 87–92.

Cwynar, L.C., 1978, Recent history of fire and vegetation from laminated sediment of Greenleaf Lake, Algonquin Park, Ontario, *Canadian Journal of Botany* **56**:10–21.

Danylchuk, A.J. and Tonn, W.M., 2003, Natural disturbances and fish: local and regional influences on winterkill of fathead minnows, in boreal lakes, *Transactions of the American Fisheries Society,* **132**:289–298.

D'Arcy, P. and Carignan, R., 1997, Influence of catchment topography on water chemistry in south-eastern Québec shield lakes, *Canadian Journal of Fisheries and Aquatic Sciences* **54**:2215–2227.

Devito, K.J., Dillon, P.J., and Lazerte, B.D., 1989, Phosphorus and nitrogen retention in five Precambrian shield wetlands, *Biogeochemistry* **8**:185–204.

Devito, K.J. and Dillon, P.J., 1993, Importance of runoff and winter anoxia to the P and N dynamics of a beaver pond, *Canadian Journal of Fisheries and Aquatic Sciences* **50**:2222–2234.

Devito, K.J. and Hill, A.R., 1997, Sulphate dynamics in relation to groundwater–surface water interactions in headwater wetlands of the southern Canadian Shield, *Hydrological Processes* **11**:485–500.

Dickman, M. and Dorais, M., 1977, The impact of human trampling on phosphorus loading to a small lake in Gatineau Park, Québec, Canada, *Journal of Environmental Management* **5**:335–344.

Dillon P.J. and Kirchner, W.B., 1975, The effects of geology and land use on the export of phosphorus from watersheds, *Water Research* **9**:135–148.

Dillon, P.J. and Rigler, F.H., 1975, A simple method for predicting the capacity of a lake for development based on lake trophic status, *Journal of the Fisheries Research Board of Canada* **32**:1519–1531.

Dillon, P.J., Nicolls, K.H., Scheider, W.A., Yan, N.D., and Jeffries, D.S., 1986, *Lakeshore Capacity Study — Trophic Status,* Ontario Ministry of Municipal Affairs Report, Toronto, Ontario.

Dillon, P.J., Molot, L.A., and Scheider, W.A., 1991, Phosphorus and nitrogen export from forested stream catchments in central Ontario, *Journal of Environmental Quality* **20**(4):857–864.

Dillon, P.J., Reid, R.A., and Evans, H., 1993, The relative magnitude of phosphorus sources for small, oligotrophic lakes in Ontario, Canada, *Internationale Vereinigung fur Theoretische und Angewandte Limnologie* **25**:355–358.

Dubé, S., Plamondon, A.P., and Rothwell, R.L., 1995, Watering up after clear-cutting on forested wetlands of the St. Lawrence lowland, *Water Resources Research* **31**(7):1741–1750.

Duck, R.W., 1985, The effect of road construction on sediment deposition in Loch Earn, Scotland, *Earth Surface Processes and Landforms* **10**:401–406.

Dugal, J.J.B., Pearson, R., and Stone, D., 1981, *Hydrogeologic Testing and Fracture Analysis of the Eye-Dashwa Lakes Granitic Pluton at Atikokan, Ontario,* Atomic Energy of Canada Limited Paper TR-7363, Ottawa, pp. 8.4.1–8.4.16.

Dyrness, C.T., 1967, Erodibility and erosion potential of forest watersheds, In *Forest Hydrology,* edited by W.E. Sopper and H.W. Lull, Pergamon Press, Oxford. pp. 599–611.

Eisenreich, S.J., 1982, Overview of atmospheric inputs and losses from films, *Journal of Great Lakes Research* **8**(2):241–242.

Elwood, J.W., Newbold, J.D., O'Neill, R.V., and Van Winkle, W., 1983, Resource spiralling: an operational paradigm for analyzing lotic ecosystems. In *Dynamics of Lotic Ecosystems,* edited by T.D. Fontaine, III and S.M. Bartell, Ann Arbor Science, Ann Arbor, MI. pp. 3–27.

Engstrom, D.R., 1987, Influence of vegetation and hydrology on the humus budgets of Labrador lakes, *Canadian Journal of Fisheries and Aquatic Sciences* **44**:1306–1314.

Evans, D.O., Nicholls, K.H., Allen, Y.C., and McMurty, M.J., 1996, Historical land use, phosphorus loading, and loss of fish habitat in Lake Simcoe, Canada, *Canadian Journal of Fisheries and Aquatic Sciences* **53**(Suppl. 1):194–218.

Farvolden, R.N., Pfannkuch, O., Pearson, R., and Fritz, P., 1988, Region 12, Precambrian Shield, In *Hydrogeology: Geology of North America,* vol. O-2, edited by W. Back, J.S. Rosenshein, and P.R. Seaber, Geological Society of America, Boulder, Colorado, pp. 101–114.

Fee, E.J., Hecky, R.E., Kasian, S.E.M., and Cruikshank, D.R., 1996, Effects of lake size, water clarity, and climatic variability on mixing depths in Canadian Shield lakes, *Limnology and Oceanography* **41**:912–920.

Fleming, R.A., Hopkin, A.A., and Candau, J.-N., 2000, Insect and disease disturbance regimes in Ontario's forests, In *Ecology of a Managed Terrestrial Landscape: Patterns and Processes of Forest Landscapes in Ontario,* edited by A.H. Perera, D.J. Euler, and I.D. Thompson, UBC Press, Alberta, pp. 141–162

France, R.L., 1995, Empirically estimating the lateral transport of riparian leaf litter to lakes, *Freshwater Biology* **34**:495–499.

France, R.L., 1997a, Macroinvertebrate colonization of woody debris in Canadian Shield lakes following riparian clearcutting, *Conservation Biology* **11**(2):513–521.

France, R.L., 1997b, Potential for soil erosion from decreased litterfall due to riparian clearcutting: implications for boreal forestry and warm- and cool-water fisheries, *Journal of Soil and Water Conservation* **52**(6):452–455.

France, R.L., 1998, Colonization of leaf litter by littoral macroinvertebrates with reference to successional changes in boreal tree composition expected after riparian clearcutting, *American Midland Naturalist* **140**:314–324.

France, R.L. (Ed.), 2002a, *Handbook of Water Sensitive Planning and Design,* Lewis Publishers, Boca Raton.

France, R.L., 2002b, Factors influencing sediment transport from logging roads near boreal trout lakes (Ontario, Canada), In *Handbook of Water Sensitive Planning and Design,* edited by R.L. France, Lewis Publishers, Boca Raton. pp. 635–644.

France, R., Culbert, H., and Peters, R., 1996, Decreased carbon and nutrient input to boreal lakes from particulate organic matter following riparian clear-cutting, *Environmental Management* **20**(4):579–583.

France, R.L. and Peters, R.H., 1995, Predictive model of the effects on lake metabolism of decreased airborne litterfall through riparian deforestation, *Conservation Biology* **9**:1578–1586.

France, R., Peters, R., and McCabe, L., 1998, Spatial relationships among boreal riparian trees, litterfall and soil erosion potential with reference to buffer strip management and coldwater fisheries, *Annals Botonici Fennici* **35**:1–9.

France, R. and Steedman, R.J., 1996, Energy provenance for juvenile lake trout in small Canadian Shield lakes as shown by stable isotopes, *Transactions of the American Fisheries Society* **125**:512–518.

France, R., Felkner, J.S., Flaxman, M., and Rempel, R., 2002, Spatial investigation of applying Ontario's timber management guidelines: GIS analysis for riparian areas of concern, In *Handbook of Water Sensitive Planning and Design,* edited by R.L. France, Lewis Publishers, Boca Raton. pp. 601–612

Frape, S.K. and Patterson, R.J., 1981, Chemistry of interstitial water and bottom sediments as indicators of seepage patterns in Perch Lake, Chalk River, Ontario, *Limnology and Oceanography* **26**:500–517.

Frissell, C.A. and Bayles, D., 1996, Ecosystem management and the conservation of aquatic biodiversity and ecological integrity, *Water Resources Bulletin* **32**(2):229–240.

Gasith, A. and Hasler, A.D., 1976, Airborne litterfall as a source of organic matter in lakes, *Limnology and Oceanography* **21**:253–258.

Gorham, E. and Gordon, A.G., 1960, The influence of smelter fumes upon the chemical composition of lake waters near Sudbury, Ontario, and upon the surrounding vegetation, *Canadian Journal of Botany* **38**:477–487.

Gratkowski, H.J., 1956, Windthrow around staggered settings in old-growth Douglas-fir, *Forest Science* **2**(1):60–74.

Gregory, K.J. (Ed.), 1977, *River Channel Changes,* John Wiley & Sons, Chichester.

Gregory, S.V., 1997, Riparian management in the 21st century, In *Creating a Forestry for the 21st Century: The Science of Ecosystem Management,* edited by K.A. Kohm and J.F. Franklin, Island Press, Washington, D.C., pp. 69–86.

Gregory, S.V., Lamberti, G.A., Erman, D.C., Koski, K.V., Murphy, M.L., and Sedell, J.R., 1987, Influence of forest practices on aquatic production, In *Streamside Management: Forestry and Fishery Interactions*, edited by E.O. Salo and T.W. Cundy, University of Washington Institute of Forest Resources Contribution No. 57, Seattle, pp. 233–255.

Grey, D.M., 1970, *Handbook of Principles of Hydrology*. National Research Council of Canada, Ottawa.

Grigal, D.F., 2000, Effects of extensive forest management on soil productivity, *Forest Ecology and Management* **138**:167–185.

Gunn, J.M., 1995, *Restoration and Recovery of an Industrial Region*, Springer-Verlag, New York.

Guyette, R.P. and Cole, W.G., 1999, Age characteristics of coarse woody debris (*Pinus strobus*) in a lake littoral zone, *Canadian Journal of Fisheries and Aquatic Sciences* **56**:496–505.

Haider, W. and Hetherington, J., 2000, Effects of forest regeneration practices on resource-based tourism and recreation. In *Regenerating the Canadian Forest: Principles and Practice for Ontario*, edited by R.G. Wagner and S.J. Colombo, Fitzhenry & Whiteside, Markam, Ontario, pp. 557–570

Håkanson, L. and Peters, R.H., 1995, *Predictive Limnology: Methods for Predictive Modelling*, Academic Publishing bv, Amsterdam.

Hall, R.I. and Smol, J.P., 1993, The influence of catchment size on lake trophic status during the hemlock decline and recovery (4800 to 3500 BP) in southern Ontario lakes, *Hydrobiologia* **269/270**:371–390.

Hall, R.I. and Smol, J.P., 1996, Paleolimnological assessment of long-term water-quality changes in south-central Ontario lakes affected by cottage development and acidification, *Canadian Journal of Fisheries and Aquatic Sciences* **53**:1–17.

Hammer, T.R., 1972, Stream channel enlargement due to urbanization, *Water Resources Research* **8**:1530–1540.

Hanlon, R.D.G., 1981, Allochthonous plant litter as a source of organic material in an oligotrophic lake (Llyn Frongoch), *Hydrobiologia* **80**:257–261.

Hartman, G.F. and Scrivener, J.C., 1990, Impacts of forestry practices on a coastal stream ecosystem, Carnation Creek, British Columbia, *Canadian Bulletin of Fisheries and Aquatic Sciences* 223.

Hibbert, A.R., 1966, Forest treatment effects on water yield, In *Proceedings of a National Science Foundation Advanced Science Seminar: International Symposium of Forest Hydrology;* August 1965, University Park, Pennsylvania, Pergamon Press, New York, pp. 725–736.

Hibbert, A.R. and Troendle, C.A., 1988, Streamflow generation by variable source area, In *Forest Hydrology and Ecology at Coweeta*, edited by W.T. Swank, and D.A. Crossley, Jr., Springer-Verlag, New York, pp. 111–127

Holopainen, A.-L., Huttunen, P., and Ahtianen, M., 1991, Effects of forestry practices on water quality and primary productivity in small forest brooks, *Internationale Vereinigung fur Theoretische und Angewandte Limnologie* **24**:1760–1766.

Holtby, L.B., 1989, Changes in the temperature regime of a valley-bottom tributary of Carnation Creek British Columbia over-sprayed with the herbicide roundup (glyphosate), In *Proceedings of the Carnation Creek Herbicide Workshop*, edited by P.E. Reynolds, Forestry Canada and B.C. Ministry of Forests, Alberta, pp. 212–223.

Hornbeck, J.W. and Swank, W.T., 1992, Watershed ecosystem analysis as a basis for multiple-use management of eastern forests, *Ecological Applications* **2**(3):238–247.

Hutchinson, N.J., Neary, B.P., and Dillon, P.J., 1991, Validation and use of Ontario's trophic status model for establishing lake development guidelines, *Lake Reservoir Management* **7**:13–23.

Hutchinson, N.J., 2002, Limnology, plumbing and planning: evaluation of nutrient-based limits to shoreline development in Precambrian Shield watersheds, In *Handbook of Water Sensitive Planning and Design*, edited by R.L. France, Lewis Publishers, Boca Raton, pp. 647–680.

Jewett, K.J., Daugharty, D., Krause, H.H., and Arp, P.A., 1995, Watershed responses to clear-cutting: effects on soil solutions and stream water discharge in central New Brunswick, *Canadian Journal of Soil Science* **75**:475–490.

Jones, J.A. and Grant, G.E., 1996, Peak flow responses to clear-cutting and roads in small and large basins, western Cascades, Oregon, *Water Resources Research* **32**(4):959–974.

Jordan, M. and Likens, G.E., 1975, An organic carbon budget for an oligotrophic lake in New Hampshire, U.S.A., *Internationale Vereinigung fur Theoretische und Angewandte Limnologie* **19**:994–1003.

Kaushik, N.K. and Hynes, H.B.N., 1971, The fate of dead leaves that fall into streams, *Archiv für Hydrobiologie* **68**:465–515.

Kelly, D.J., Clare, J.J., and Bothwell, M.L., 2001, Attenuation of solar ultraviolet radiation by dissolved organic matter alters benthic colonization patterns in streams, *Journal of the North American Benthological Society* **29**(1):96–108.

Kelso, J.R.M. and Demers, J.W., 1993, *Our Living Heritage: the Glory of the Nipigon,* Mill Creek, Echo Bay, Ontario.

Knapp, C.W., Graham, D.W., Steedman, R.J., and deNoyelles, F., Jr., 2003, Short-term impact of experimental deforestation on deep phytoplankton communities in four small boreal forest lakes, *Boreal Environment Research,* **8**:9–18.

Krause, H.H., 1982, Effect of forest management practices on water quality — a review of Canadian studies, In *Proceedings of the Canadian Hydrology Symposium '82 — Hydrological Processes of Forested Areas,* National Research Council, Fredericton, N.B.

Kuusela, K., 1990, *The Dynamics of Boreal Coniferous Forests,* Finnish National Fund for Research and Development (SITTRA) 112, Helsinki, Finland.

Lamontagne, S., Carignan, R., D'Arcy, P., Prairie, Y.T., and Paré, D., 2000, Element export in runoff from eastern Canadian Boreal Shield drainage basins following forest harvesting and wild-fires, *Canadian Journal of Fisheries and Aquatic Sciences* **57**(Suppl. 2):118–128.

Lee, D.R., 1977, A device for measuring the seepage flux in lakes and estuaries, *Limnology and Oceanography* **22**: 140–147.

Lee, P.K., Pearson, R., Leech, R.E.J., and Dickin, R., 1983, Hydraulic testing of deep fractures in the Canadian Shield, *International Association of Engineering Geologists Bulletin* **26–27**:461–465.

Leopold, L.B. and Maddock, T., Jr., 1953, *The Hydraulic Geometry of Stream Channels and Some Physiographic Implications,* United States Geological Survey Professional Paper 252.

Li, C., 2000, Fire regimes and their simulation with reference to Ontario, In *Ecology of a Managed Terrestrial Landscape: Patterns and Processes of Forest Landscapes in Ontario,* edited by A.H. Perera, D.J. Euler, and I.D. Thompson, UBC Press, Alberta, pp. 115–140.

Likens, G.E., 1984, Beyond the shoreline: a watershed-ecosystem approach, *Internationale Vereinigung fur Theoretische und Angewandte Limnologie* **22**:1–22.

Likens, G.E., 1985, *An Ecosystem Approach to Aquatic Ecology: Mirror Lake and Its Environment,* Springer-Verlag, Berlin.

Likens, G.E., Bormann, F.H., Johnson, N.M., Fisher, D.W., and Pierce, R.S., 1970, Effects of forest cutting and herbicide treatment on nutrient budgets in the Hubbard Brook watershed-ecosystem, *Ecological Monographs* **40**:23–47.

Loftus, K.H. and Regier, H.A. (Eds.), 1972, Proceedings of the 1971 Symposium on Salmonid Communities in Oligotrophic Lakes (SCOL), *Journal of the Fisheries Research Board of Canada* **29**:611–986.

Lotspeich, F.B., 1980, Watersheds as the basic ecosystem: this conceptual framework provides a basis for a natural classification system, *Water Research Bulletin* **16**:581–586.

MacDonald, G.M., Larsen, C.P.S., Szeicz, J.M., and Moser, K.A., 1991, The reconstruction of boreal forest fire history from lake sediments: a comparison of charcoal, pollen, sedimentological, and geochemical indices, *Quaternary Science Reviews* **10**:53–71.

Mallory, E.C., Ridgway, M.S., Gordon, A.M., and Kaushik, N.K., 2000, Distribution of woody debris in a small headwater lake, central Ontario, Canada, *Archiv für Hydrobiologie* **148**:589–606.

Marsh, P.C. and Luey, J.E., 1982, Oases for aquatic life within agricultural watersheds, *Fisheries* **7**(6):16–19,24.

Megahan, W.F. and Kidd, W.J., 1972, Effects of logging and logging roads on erosion and sediment deposition from steep terrain, *Journal of Forestry* **80**:136–141.

Minshall, G.W., Cummins, K.W., Peterson, R.C., Cushing, C.E., Bruns, D.A., Sedell, J.R., and Vannote, R.J., 1985, Developments in stream ecosystem theory, *Canadian Journal of Fisheries and Aquatic Sciences* **42**:1045–1055.

Ministry of the Environment, 1984, Water management goals, policies, objectives and implementation procedures for the Ministry of the Environment. November 1978 (revised May 1984), Ontario Ministry of the Environment, Toronto.

Molot, L.A., Dillon, P.J., Clark, B.J., and Neary, B.P., 1992, Predicting end-of-summer oxygen profiles in stratified lakes, *Canadian Journal of Fisheries and Aquatic Sciences* **49**:2363–2372.

Moring, J.R. and Lantz, R.L., 1975, *The Alsea Watershed Study: Effects of Logging on the Aquatic Resources of Three Headwater Streams of the Alsea River, Oregon, Part I. Biological studies. Fishery Research Report Number 9*, Oregon Department of Fish and Wildlife, Corvallis, Oregon.

Murphy, M.L. and Hall, J.D., 1981, Varied effects of clear-cut logging on predators and their habitat in small streams of the Cascade Mountains, Oregon, *Canadian Journal of Fisheries and Aquatic Sciences* **38**:137–145.

Murphy, M.L., Heifetz, J., Johnson, S.W., Koski, K.V., and Thedinga, J.F., 1986, Effects of clear-cut logging with and without buffer strips on juvenile salmonids in Alaskan streams, *Canadian Journal of Fisheries and Aquatic Sciences* **43**:1521–1533.

Naiman, R.J., 1982, Characteristics of sediment and organic carbon export from pristine boreal forest watersheds, *Canadian Journal of Fisheries and Aquatic Sciences* **39**:1699–1718.

Naiman, R.J. (Ed.), 1992, *Watershed Management*, Springer-Verlag, New York.

Newmaster, S.G., Harris, A.G., and Kershaw, L.J., 1997, *Wetland Plants of Ontario*, Lone Pine Publishing and Queen's Printer for Ontario, Edmonton, Alberta.

Nicolson, J.A., 1975, Water quality and clearcutting in a boreal forest ecosystem, In Canadian Hydrology Symposium — 75 Proceedings, Winnipeg, Manitoba, August 11–14, 1975, National Research Council of Canada, pp. 734–738.

Nicolson, J.A., 1988, Alternate strip clearcutting in upland black spruce V. The impact of harvesting on the quality of water flowing from small basins in shallow-soil boreal ecosystems, *The Forestry Chronicle* **64**:52–58.

Nicolson J.A., Foster, N.W., and Morrison, I.K., 1982, Forest harvesting effects on water quality and nutrient status in the boreal forest, In *Proceedings of the Canadian Hydrology Symposium '82, Hydrological Processes of Forested Areas*, National Research Council of Canada, Fredericton, New Brunswick, pp. 71–89.

Norris, V., 1993, The use of buffer zones to protect water quality: a review, *Water Resources Management* **7**:257–272.

Ontario Environmental Assessment Board, 1994, *Reason for Decision and Decision, Class Environmental Assessment by the Ministry of Natural Resources for Timber Management on Crown Lands of Northern Ontario*, Ontario Environmental Assessment Board, Toronto.

Ontario Mining Association, 1998, *Providing for the Future*, Ontario Mining Association, Toronto.

Ontario Ministry of Natural Resources, 1988, *Timber Management Guidelines for the Protection of Fish Habitat*, Ontario Ministry of Natural Resources, Toronto.

Ontario Ministry of Natural Resources, 1990, *Environmental Guidelines for Access Roads and Water Crossings*, Ontario Ministry of Natural Resources, Toronto.

Ontario Ministry of Natural Resources, 1991, *Code of Practice for Timber Management Operations in Riparian Areas*, Ontario Ministry of Natural Resources, Toronto.

Ontario Ministry of Natural Resources, 1994, *Development Guidelines, Clearwater Bay Restricted Area*, Ontario Ministry of Natural Resources, Kenora District.

Palik, B.J., Zasada, J.C., and Hedman, C.W., 2000, Ecological principles for riparian silviculture, In *Riparian Management in Forests*, edited by E.S. Verry, J.W. Hornbeck, and C.A. Dolloff, Lewis Publishers, Boca Raton, pp. 233–254

Park, S.W., Mostaghimi, S., Cooke, R.A., and McClennan, P.W., 1994, BMP impacts on watershed runoff, sediment, and nutrient yields, *Water Resources Bulletin* **30**(6):1011–1023.

Paterson, A.M., Cumming, B.F., Smol, J.P., Blais, J.M., and France, R.L., 1998, Assessment of the effects of logging and forest fires on lakes in northwestern Ontario: a 30-year paleolimnological perspective, *Canadian Journal of Forest Research* **28**:1546–1556.

Perera, A.H. and Baldwin, D.J., 2000, Spatial patterns in the managed forest landscape of Ontario, In *Ecology of a Managed Terrestrial Landscape: Patterns and Processes of Forest Landscapes in Ontario*, edited by A.H. Perera, D.J. Euler, and I.D. Thompson, UBC Press, Alberta, pp. 74–99.

Plamondon, A.P. and Ouellet, D.C., 1980, Partial clearcutting and streamflow regime of russeau des Eaux-Volées experimental basin, In *The Influence of Man on the Hydrological Regime with Special Reference to Representative and Experimental Basins.* Proceedings of the Helsinki Symposium, June 1980, IAHS Publication No. 130. pp. 129–136.

Puttock, G.D., 1985, Evaluation of the impact of forest reserves on harvesting costs and provincial revenues in Ontario, *Environmental Management* 9(1):83–88.

Québec Ministère des Forêts, 1992, *Modalités d'Intervention en Milieu Forestier*, Les Publications du Québec, Montreal.

Quine, C.P., Coutts, M.P., Gardiner, B.A., and Pyatt, D.G., 1995, *Forests and Wind: Management to Minimise Damage*, Bulletin 114, HMSO, London.

Rapport, D.J., Regier, H.A., and Hutchinson, T.C., 1985, Ecosystem behavior under stress, *American Naturalist* 125:617–640.

Rask M., Arvola, L., and Salonen, K., 1993, Effects of catchment deforestation and burning on the limnology of a small forest lake in southern Finland, *Internationale Vereinigung fur Theoretische und Angewandte Limnologie* 25:525–528.

Rask, M., Nyberg, K., Markkanen, S.-L., and Ojala, A., 1998, Forestry in catchments: effects on water quality, plankton, zoobenthos and fish in small lakes, *Boreal Environment Research* 3:75–86.

Rasmussen, J.B., Godbout, L., and Schallenberg, M., 1989, The humic content of lake water and its relationship to watershed and lake morphometry, *Limnology and Oceanography* 34(7):1336–1343.

Regier, H.A., Welcomme, R.L., Steedman, R.J., and Henderson, H.F., 1989, Rehabilitation of degraded river ecosystems, *Canadian Special Publication of Fisheries and Aquatic Sciences* 106:86–97.

Renzetti, V.E., Taylor, C.H., and Buttle, J.M., 1992, Subsurface flow in a shallow soil Canadian Shield watershed, *Nordic Hydrology* 23:209–226.

Ribe, R.G., 1989, The aesthetics of forestry: what has empirical preference research taught us? *Environmental Management* 13(1):55–74.

Ruel, J.-C.D., 2000, Factors influencing windthrow in balsam fir forests: from landscape studies to individual tree studies, *Forest Ecology and Management* 135:169–178.

Rutherford, D.A., Echelle, A.A., and Maughan, O.E., 1992, Drainage-wide effects of timber harvesting on the structure of stream fish assemblages in southeastern Oklahoma, *Transactions of the American Fisheries Society* 121:716–728.

Schindler, D.W., 1974, Eutrophication and recovery in experimental lakes: implications for lake management, *Science* 184:897–898.

Schindler, D.W., 1998, A dim future for boreal waters and landscapes, *BioScience* 48(2):157–164.

Schindler, D.W., Brunskill, G.J., Emerson, S., Broecker, W.S., and Peng, T.-H., 1972, Atmospheric carbon dioxide: its role in maintaining phytoplankton standing crops, *Science* 177:1192–1194.

Schindler, D.W., Newbury, R.W., Beaty, K.G., and Campbell, P., 1976, Natural water and chemical budgets for a small Precambrian lake basin in central Canada, *Journal of the Fisheries Research Board of Canada* 33:2526–2543.

Schindler, D.W., Newbury, R.W., Beaty, K.G., Prokopowich, J., Ruszczynski, T., and Dalton, J.A., 1980, Effects of a windstorm and forest fire on chemical losses from forested watersheds and on the quality of receiving streams, *Canadian Journal of Fisheries and Aquatic Sciences* 37:328–334.

Scully, N.M., Leavitt, P.R., and Carpenter, S.R., 2000, Century-long effects of forest harvest on the physical structure and autotrophic community of a small temperate lake, *Canadian Journal of Fisheries and Aquatic Sciences* 57(Suppl. 2):50–59.

Smith, I.R., 1979, Hydraulic conditions in isothermal lakes, *Freshwater Biology* 9:119–145.

Smol, J.P., 1992, Paleolimnology: an important tool for effective ecosystem management, *Journal of Aquatic Ecosystem Health* 1:49–58.

Statistics Canada, 1994, *Human Activity and the Environment 1994*, Statistics Canada National Accounts and Environment Division, Minister of Industry, Science and Technology, Ottawa.

Stednick, J.D., 1996, Monitoring the effects of timber harvest on annual water yield, *Journal of Hydrology* 176:79–95.

Steedman, R.J., 1988, Modification and assessment of an index of biotic integrity to quantify stream quality in southern Ontario, *Canadian Journal of Fisheries and Aquatic Sciences* 45:492–501.

Steedman, R.J., 1994, Ecosystem health as a management goal, *Journal of the North American Bentho-logical Society* **13**(4):605–610.

Steedman, R.J., 2000, Effects of experimental clearcut logging on water quality in three small boreal forest lake trout (*Salvelinus namaycush*) lakes, *Canadian Journal of Fisheries and Aquatic Sciences* **57**(Suppl. 2):92–96.

Steedman, R.J., 2003, Littoral fish response to experimental logging around small Boreal Shield lakes, *North American Journal of Fisheries Management,* 23:393–403.

Steedman, R.J. and Kushneriuk, R.S., 2000, Effects of experimental clearcut logging on thermal stratification, dissolved oxygen, and lake trout (*Salvelinus namaycush*) habitat volume in three small boreal forest lakes, *Canadian Journal of Fisheries and Aquatic Sciences* **57**(Suppl. 2):82–91.

Steedman, R.J. and France, R.L., 2000, Origin and transport of aeolian sediment from new clearcuts into boreal lakes, northwestern Ontario, Canada. *Water, Air, and Soil Pollution* **122**:139–152.

Steedman, R.J. and Regier, H.A., 1987, Ecosystem science for the Great Lakes: perspectives on degradative and rehabilitative transformations, *Canadian Journal of Fisheries and Aquatic Sciences* **44**(Suppl. 2):95–103.

Steedman, R.J., France, R.L., Kushneriuk, R.S., and Peters, R.H., 1998, Effects of riparian deforestation on littoral water temperatures in small boreal forest lakes, *Boreal Environment Research* 3:161–169.

Steedman, R.J., Kushneriuk, R.S., and France, R.L., 2001, Littoral water temperature response to experimental shoreline logging around small boreal forest lakes, *Canadian Journal of Fisheries and Aquatic Sciences* **58**:1638–1647.

Steinblums, I.J., Froelich, H.A., and Lyons, J.K., 1984, Designing stable buffer strips for stream protection, *Journal of Forestry* **92**:49–52.

Stone, M.K. and Wallace, J.B., 1998, Long-term recovery of a mountain stream from clear-cut logging: the effects of forest succession on benthic invertebrate community structure, *Freshwater Biology* **39**:151–169.

Strahler, A.N., 1957, Quantitative analysis of watershed geomorphology, *Transactions of the American Geophysical Union* 38:913–920.

Swank, W.T. and Crossley, D.A., Jr. (Eds.), 1988, *Forest Hydrology and Ecology at Coweeta*, Springer-Verlag, New York.

Swanson, F.J., Gregory, S.V, Sedell, J.R., and Campbell, A.G., 1982, Land–water interactions: the riparian zone, In *Analysis of Coniferous Forest Ecosystems in the Western United States, US/IBP Synthesis Series 14*, edited by R.L. Edmunds, Hutchinson Ross Publishing Co., Stroudsburg, Pennsylvania, pp. 267–291.

Swanson, F.J., Benda, L.E., Duncan, S.H., Grant, G.E., Megahan, W.F., Reid, L.M., and Ziemer, R.R., 1987, Mass failures and other processes of sediment production on Pacific Northwest forest landscapes, In *Streamside Management: Forestry and Fisheries Interactions*, edited by E.O. Salo and T.W. Cundy, University of Washington, Institute of Forest Resources, Seattle, pp. 9–38.

Swift, L.W., Jr., 1988, Forest access roads: design, maintenance, and soil loss, In *Forest Hydrology and Ecology at Coweeta*, edited by W.T. Swank and D.A. Crossley, Jr., Springer-Verlag, New York, pp. 313–324.

Tonn, W.M., Paszkowski, C.A., Scrimgeour, G.J., et al., 2003, Effects of harvesting and forest fire on fish assemblages in Boreal Plains lakes: a reference condition approach, *Transactions of the American Fisheries Society,* 132:514–523.

Trimble, G.R. and Sartz, R.S., 1957, How far from a stream should a logging road be located? *Journal of Forestry* 55(5)339–341.

Vannote, R.L., Minshall, G.W., Cummins, K.W., Sedell, J.R., and Cushing, C.E., 1980, The river continuum concept, *Canadian Journal of Fisheries and Aquatic Sciences* 37:130–137.

Verry, E.S., 1986, Forest harvesting and water: the lake states experience, *Water Resources Bulletin* **22**(6):1039–1047.

Vulori, K.-M., Joensuu, I., Latvala, J., Jutila, E., and Ahvonen, A., 1998, Forest drainage: a threat to benthic biodiversity of Boreal headwater streams? *Aquatic Conservation* **8**(6):745–760.

Warren, M.L., Jr. and Pardew, M.G., 1998, Road crossings as barriers to small-stream fish movement, *Transactions of the American Fisheries Society* **127**:637–644.

Waters, T.F., 1995, *Sediment in Streams: Sources, Biological Effects, and Control, American Fisheries Society Monograph 7*, American Fisheries Society, Bethesda, Maryland.

Welcomme, R.L., 1985, *FAO Fisheries Technical Report 262, River Fisheries.* Food and Agriculture Organization of the United Nations, Rome.

Williams, L.R., Taylor, C.M., Warren, M.L., Jr., and Clingenpeel, J.A., 2002, Large-scale effects of timber harvesting on stream systems in the Ouachita mountains, Arkansas, USA. *Environmental Management* **29**(1):76–87.

Young, K.A., Hinch, S.G., and Northcote, T.G., 1999, Status of resident coastal cutthroat trout and their habitat twenty-five years after riparian logging, *North American Journal of Fisheries Management* **19**:901–911.

chapter five

Impact of new reservoirs

Michel Legault
Société de la faune et des parcs du Québec
Jean Benoît
Société de la faune et des parcs du Québec
Roger Bérubé
Hydraulique et Environnement, Hydro-Québec

Contents

Introduction

Even though there are many reasons for creating reservoirs (such as drinking water supply, irrigation, or stream flow control), production of hydroelectricity (Figure 5.1) is the main reason for the existence of reservoirs on the Precambrian Shield. Exploitation of Canadian

Figure 5.1 Reservoir for production of hydroelectricity.

hydroelectric potential began in the last century and has advanced northward like a wave that is now cresting in the midnorthern latitudes (Rosenberg et al., 1987). Already, the total area of boreal reservoirs in North America is similar to that of Lake Ontario (Rudd et al., 1993). The environmental impacts that result from the creation of reservoirs are numerous and include accumulation of sediments, shore erosion, and accumulation of mercury in fish and other organisms.

There are upwards of 100 reservoirs in Québec, 68 of which are known to support lake trout populations. These represent 7.4% of the province's lake trout lakes. In Ontario, the areas with the most concerns about reservoirs and water level variations are the Algonquin and Eastern regions (Lewis et al., 1990). Together they comprise nearly 22% of Ontario's lake trout lakes.

The lake trout *Salvelinus namaycush* is among the most studied species in North America with regard to reservoir management. Lake trout have great difficulty adapting to lakes with regulated water levels or to reservoirs used for hydroelectric generation purposes (Martin, 1955; Wilton, 1985; Gendron and Bélanger, 1993; Benoît et al., 1997). The change in water level, the resulting surface area and type of substrate of the flooded lands, the seasonal drawdown regime employed, and the lotic or lentic origin of the reservoir are some of the many factors affecting lake trout's adaptation in reservoirs (Machniak, 1975; Evans et al., 1991). Among the foregoing factors, drawdown is often pinpointed as the cause of the lake trout's adaptation problems, as it has repercussions on two aspects of the species' vital cycle: reproduction success and larvae survival (Martin, 1955; Wilton, 1985; Gendron and Bélanger, 1993).

This chapter reviews major modifications made to lake trout ecosystems by the creation of reservoirs on the Precambrian Shield. It also discusses impacts of drawdown on the reproduction of lake trout and suggests management alternatives.

Transformations to ecosystems

As a result of reservoir creation and management practices, temporary and permanent transformations to ecosystems will occur. These transformations can be of greater or lesser significance depending on the reservoir's intended use (source of drinking water, electrical energy generation, flow regulation) and its particularities: shape and mean depth, retention time, surface area of flooded lands, density and nature of vegetation flooded, as well as length of impoundment period (Baxter and Glaude, 1980). Without a doubt the most spectacular and complex changes, from both physical and biological standpoints, are brought about by the creation of a reservoir on a river. Such a reservoir entails permanent transformations to ecosystems, as vast tracts of lands are flooded and river stretches are turned into lakes.

A lake can also be transformed into a reservoir by regulating its flow. In such cases, the transformations to ecosystems are less extensive and, in some instances, only temporary. The following section examines major modifications to lake trout ecosystems brought about by the creation and management of reservoirs located on the Precambrian Shield.

Water quality

The first phenomenon to occur as a result of reservoir impoundment is the leaching of flooded soil and vegetation (Baxter and Glaude, 1980). Mineral salts and nutrients present in the soil are released in the water, aided by the wave action of the rising water shredding the forest floor. While decomposition has yet to begin in the first stages of impoundment, these initial phenomena appear responsible for the sharp rise in total phosphorus and drop in pH (Figure 5.2) (Chartrand et al., 1994).

The decomposition of flooded organic matter then follows. This phenomenon leads to consumption of dissolved oxygen, mainly in the deeper strata of the reservoir, lower pH, and release of CO_2, CH_4, and nutrients such as phosphorus (Figure 5.2). The low oxygen levels and change in the chemical conditions of deep-zone waters are factors likely to limit the habitat of young lake trout, increase the threat of predation by adults, and ultimately reduce recruitment to the population.

However, overall water quality within the reservoir is barely influenced by these benthic processes. Because the volume of water rich in decomposition by-products near the bottom is very small relative to the total volume of the reservoir, the concentrations of these products throughout the reservoir remain low following spring overturn.

All of these changes are temporary and subside as time passes (Baxter and Glaude, 1980). For example, in the Robert-Bourassa and Opinaca reservoirs (Québec) as a whole (Chartrand et al., 1994), physicochemical variations peaked quickly (1 to 4 years) following impoundment (Figure 5.2). Modifications related to the decomposition of flooded organic matter were nearly over 9 to 10 years after impoundment. With respect to the Caniapiscau reservoir, Québec (Chartrand et al., 1994), the modifications measured were of the same magnitude as those measured in other reservoirs; however, the maxima for total phosphorus and silica were reached later, between the 6th and 10th year of impoundment (Figure 5.2). In this particular instance, the return to values representative of natural environments was completed 14 years later. It appears that because impoundment occurred more gradually, over a period of 3 years rather than 6 to 12 months as with the others, the period required for the return of initial conditions was extended.

The modification period is brief largely because only a small portion of the flooded organic matter composing the forest soil and vegetation decomposes easily and rapidly. Only the leaves of trees and bushes, conifer needles, forest ground cover, and the first few centimeters of humus decompose rapidly. Most of the other flooded material (tree

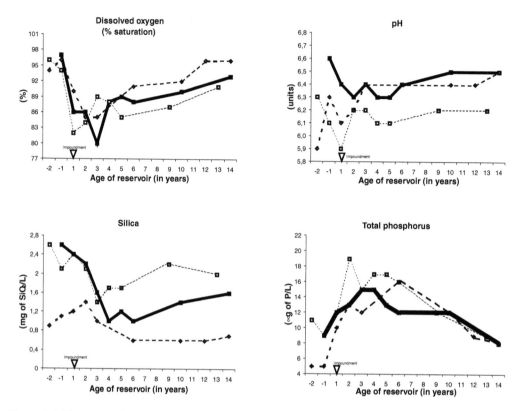

Figure 5.2 Variation of principal water quality variables before, during, and after impoundment in the major reservoirs of the La Grande complex (—— Robert-Bourassa, ... Opinaca, --- Caniapiscau). (From Chartrand et al.., 1994, Commission Internationale des Grands Barrages, Dix-huitième Congrès des Grands Barrages, Durban, Q.69-R-14, pp. 165-190.)

branches, trunks and roots, and deep soil humus) proves to be difficult to decompose and remains really intact dozens of years after impoundment.

Mercury is a widespread contaminant in freshwater fish. Lake trout frequently have high concentrations of mercury because of their position at the top of the food chain and because the Boreal lakes that they inhabit often have conditions favorable for mercury bioaccumulation (Bodaly and Kidd, Chapter 9, this volume).

Several studies have demonstrated that impoundment brings about a rapid increase in fish mercury levels (Schetagne et al., 1997). The extent of the increase in bioavailability of mercury for aquatic wildlife in reservoirs depends on many factors: the land area flooded, filling time, water residence time, volume of water, proportion of flooded land in shallow environment (where biotransfer is at its maximum), water quality, food web of the flooded environment, fish population dynamics, etc. (Jones et al., 1986; Brouard et al., 1990; Doyon et al., 1996).

Research conducted at the La Grande complex (Québec) (Schetagne et al., 1997) showed that depending on the fish species and reservoir considered, maximum mercury concentrations were 3 to 7 times higher than those measured in natural environments. In nonpiscivorous species, mercury levels stop increasing significantly 4 to 5 years after impoundment, and the return to concentrations representative of natural environments is well under way 10 to 15 years after flooding. In piscivorous species, maximum values were attained later than for nonpiscivorous species. Maximum concentrations for walleye

Stizostedion vitreum, northern pike *Esox lucius*, and lake trout were reached between 9 and 13 years after impoundment, depending on the reservoir and species.

A gradual decrease in fish mercury levels is observed once mercury release activities, including decomposition of flooded organic matter, as well as the erosion and resuspension of flooded organic matter along banks exposed to wave action have declined. Subsequently, this erosion of organic matter helps accelerate the decrease in fish mercury levels by reducing the area of shallow zones in reservoirs that still have organic matter available. It is in these shallow zones rich in organic matter where most biotransfer occurs (Shetagne et al., 1997). In the majority of reservoirs, a significant decrease begins to be apparent, for all these species, 14 to 15 years after reservoir creation.

The data collected at the La Grande complex, as well as in other reservoirs located in the Precambrian Shield, show that fish mercury levels in reservoirs return to values similar to those measured in natural environments after a period that may range from 15 to 25 years for nonpiscivorous species and from 20 to 30 years for piscivorous species (Shetagne et al., 1997)

Water impoundment also affects other aspects of water quality and because the changes result from major alterations in the reservoir's shape and management, they are permanent. For example, the creation of a reservoir on a river enables the particles that were suspended in the river's running waters to settle to the bottom more easily. This sedimentation process reduces turbidity and increases light penetration into the water (Chartrand et al., 1994).

Plankton

A rise in nutrients, particularly phosphorus, which generally limits phytoplankton production on the Precambrian Shield, can cause an increase in phytoplankton quantity. Phytoplankton abundance had increased by five times in Lake Minnewanka, Alberta 3 years after its level had been raised (Cuerrier, 1954) and had returned to initial levels 8 years later. The rise in nutrients noted in all reservoirs at the La Grande complex (Chartrand et al., 1994), particularly phosphorus, resulted in a threefold increase in chlorophyll *a* concentration (Figure 5.3). At the Robert-Bourassa and Opinaca reservoirs, maximum concentration was reached 3 to 5 years after impoundment. Once easily decomposed organic matter was depleted, nutrient levels dropped to initial levels and chlorophyll *a* concentration returned to values similar to those prevailing before impoundment. The return to initial values occurred 9 and 10 years, respectively, after impoundment of those reservoirs. At the Caniapiscau reservoir, maximum values were attained some 10 years after impoundment, and the return to initial values was nearly completed after 14 years.

Zooplankton abundance and biomass in reservoirs are influenced by water enrichment and availability of organic matter, produced by flooded vegetation and forest soils, as well as increased retention time. At the Robert-Bourassa reservoir (Chartrand, 1994), zooplankton density and biomass reached maximum values in the fourth summer after impoundment and dropped off slowly afterwards (Figure 5.3). The maximum values were attained a year after chlorophyll *a* (phytoplankton biomass) (Figure 5.3) and phosphorus concentrations had peaked (Figure 5.2).

Benthos

Benthic organisms are divided into several groups according to the way they feed: some filter the water for suspended particles, some grind organic waste such as leaves falling to the bottom of the water, and some prey on other small organisms dwelling on the bottom. Benthic organisms must adapt to the great physical transformations caused by

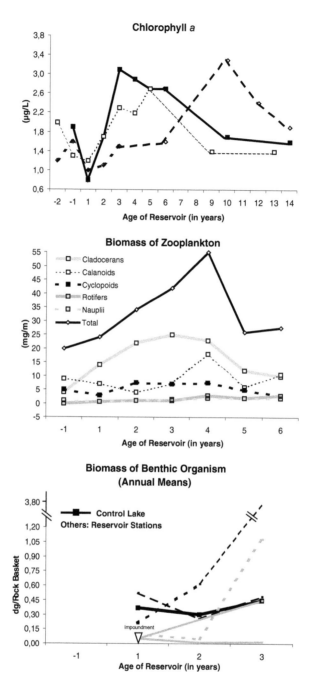

Figure 5.3 Chlorophyll *a* variation in the major reservoirs of the la Grande complex (— Robert-Bourassa, ... Opinaca, --- Caniapiscau) and evolution of zooplankton and benthic organisms biomass at Robert-Bourassa reservoir following impoundment. (From Chartrand et al.., 1994, Commission Internationale des Grands Barrages, Dix-huitième Congrès des Grands Barrages, Durban, Q.69-R-14, pp. 165-190.)

the reservoir creations, most notably the flooding of forest and the formation of huge lakes from rivers.

After the initial decrease in benthos abundance and change in their composition following the water level rise in lake Minnewanka (Cuerrier, 1954), benthos increased in abundance over the next 10 years, showing adaptation to new conditions. However, the littoral zone, which is subjected to drawdown 5 months per year, is not productive. Significant variations in water level hinder the growth of vegetation in the riparian zone and consequently limit the abundance of benthic fauna.

In the reservoirs of the La Grande complex (Chartrand et al., 1994) benthos diversity decreased, mainly in the first years. While species with poor mobility or adapted to running water became scarce, those with greater mobility or requiring little dissolved oxygen rapidly took over the new aquatic habitats. The great number of anchoring points offered by the flooded vegetation increased the surface area of feeding grounds, leading as a result to measurements of greater benthos densities and biomass than in natural lakes (Figure 5.3).

Analysis of the stomach contents of reservoir fish and monitoring of fish populations revealed that benthos diversity and quantity were sufficient to sustain major increases in growth rate and condition factors of fish feeding on benthos. These fish included lake whitefish (*Coregonus clupeaformis*), and their predators, such as the northern pike (*Esox lucius*).

Fish abundance

It is known that fish populations are often numerous in the first years of a reservoir's existence (Ellis, 1941, in Baxter and Glaude, 1980). In some cases, reservoir creation can also help increase the fisheries resources of a region (Baxter and Glaude, 1980). The rapid increase in abundance of certain fish species often observed in new reservoirs may have occurred for a number of reasons. Among these could be increased reproduction rate brought about by secure spawning grounds and protection of fry afforded by flooded vegetation. Increased food availability is another.

At the time the La Grande complex reservoirs were impounded (DesLandes et al., 1995), overall abundance for all species dropped significantly but then rose over the next 3 years before decreasing slightly up to the 10th year. The drop noted in the first year could have been related to rising water levels and the dilution effect this entails. The subsequent increase in abundance varied in range and duration depending on the species. While abundance of species such as the northern pike and lake whitefish greatly increased, the abundance of longnose sucker *Catostomus catostomus*, white sucker *Catostomus commersoni*, walleye, and cisco *Coregonus artedi*, declined sharply following the initial increase.

The northern pike is a species that generally does well in reservoirs. In the first years of a reservoir's creation, its abundance and growth usually increase (Machniak, 1975). When the La Grande 2 (DesLandes et al., 1995) and Caniapiscau (Belzile et al., 2000) reservoirs were created, the relative abundance of northern pike rose sharply and remained steady up to 10 and 17 years, respectively, following impoundment.

Although few data are available on lake trout response to the trophic pulse following reservoir creation, the ecosystem changes do not seem favorable to lake trout (DesLandes et al., 1995).

Lake trout populations are generally low in Québec's reservoirs despite the fact that reservoirs present abiotic and biotic factors considered very good for the species (Lacasse and Gilbert, 1992; Gendron and Bélanger, 1993). Québec's reservoirs containing lake trout populations tend to have much larger surface area than most natural lake trout lakes in the region (Table 5.1). This pattern is also reflected to a lesser extent in the mean and maximum depth. The physicochemical properties of many of the reservoirs on the Precambrian Shield provide good life-sustaining conditions for lake trout populations: near-neutral pH, high concentration of dissolved oxygen, and cold temperatures (Table 5.1) (Gendron and Bélanger, 1993; Benoît et al., 1997). However, adverse effects on fish popu-

Table 5.1 Characteristics of Québec's Natural Lakes and Reservoirs having a Lake Trout Population

Parameters	Natural lakes	Reservoirs
Surface area (ha)	766 ± 7332	28421 ± 73187
	N = 906	N = 67
Mean depth (m)	14.1 ± 8.6	22.2 ± 14.5
	N = 171	N = 35
Maximum depth (m)	36.7 ± 26.1	64.4 ± 54.6
	N = 359	N = 45
Secchi disk transparency (m)	5.2 ± 1.9	5.7 ± 2.9
	N = 283	N = 14
Conductivity (µS/cm at 25°C)	46.8 ± 56.7	33.0 ± 31.8
	N = 337	N = 11
pH	6.64 ± 0.71	6.63 ± 0.60
	N = 379	N = 38

Note: Mean ± SD; N: number of lakes or reservoirs

lations spawning in shallow waters can occur if reservoir drawdown exposes spawning grounds during egg incubation or larva development.

In Québec, generally speaking, the most abundant lake trout populations are found in deep (maximum depth >30 m), small area (<1,500 ha) reservoirs with an annual drawdown below 1.6 m (except for the Mitis reservoir: 3.0 m). Conversely, populations are less abundant in large upstream reservoirs (>25,000 ha) with strong annual and interannual drawdowns (7.8 m), where impoundment resulted in a sharp rise in water level (Gendron and Bélanger, 1993). However, even though the habitat appears suited to the species and drawdowns are relatively low, in reservoirs created from rivers, lake trout populations are scarce.

More recent and specific studies of five Québec reservoirs harboring lake trout populations (Benoît et al., 1997; Doyon, 1997) revealed that the populations in four of the five reservoirs had been decimated and exhibited significant recruitment problems, most likely in response to drawdown effects.

Climatic warming and reservoirs

Another concern about the creation of reservoirs is their release of carbon dioxide (CO_2) and methane (CH_4), both greenhouse gases, into the atmosphere. As we saw previously in this chapter, these gases are the major end products of the microbial decomposition of flooded organic material. Studies of CO_2 and CH_4 fluxes from existing reservoirs (Duchemin et al., 1995; Kelly et al., 1997) have demonstrated that reservoirs are sources of these gases to the atmosphere. However, the net effect of reservoir creation relative to other electric generation options (more specifically: gas, oil, coal), in terms of greenhouse gas production, is controversial.

Moreover, predictions of a warmer and drier climate resulting from greenhouse gas accumulation might shift the balance between evaporation and precipitation, which in turn will lead to overall declines in both river flows and lake levels (Magnuson et al., 1997). Under this scenario, reductions in runoff will negatively impact hydroelectric power generation, thus creating a demand for new dams. Building reservoirs in Precambrian Shield would flood more wetlands and terrestrial soils, thus further contributing to climatic warming by increasing greenhouse gas fluxes to the atmosphere.

Finally, potential impacts of climate warming on reservoirs are reductions in nutrient loading and recycling for many lakes on the Precambrian Shield (Schindler and Gunn,

Chapter 8, this volume). The change of thermal regime would also cause shrinkage of summer habitats for cold-water fish species such as lake trout.

Impacts of reservoirs use on lake trout reproduction

The following section first examines the impacts on lake trout reproduction by raising the water level during reservoir impoundment. Second, it reviews possible effects of various water level management practices on the quality of reproduction sites and on egg and fry mortality.

Characteristics of reproduction sites

The lake trout is a fish that reproduces almost exclusively in lakes, although reproductive activities have been documented in the rivers of Ontario and Québec (Loftus, 1958; Vincent and De Serres, 1963; Séguin and Roussell, 1970). The spawning period varies with the latitude. For the Precambrian Shield, this usually means that it takes place in October. Incubation extends over a period of 4 to 5 months, depending on water temperature (Martin and Olver, 1980). The great majority of eggs hatch around March 1, but hatching may occur as early as the end of January or as late as the beginning of April (Chabot and Archambault, 1981; Pariseau, 1981). Lake trout embryos can move extensively within and above the substrate immediately after hatching (Baird and Krueger, 2000). After their yolk sac is fully resorbed, approximately 2 months after hatching, it appears that the young fish immediately migrate to deep waters (Martin and Olver, 1980).

To date, it has been impossible to determine whether lake trout return to their birthplace to reproduce. However, it is clear that specific sites are used by spawning stock, sometimes year after year (Gunn, 1995). Lake trout spawning grounds in Québec's Mauricie region are generally located near the shore at a depth of less than 2 m. The substrate is composed mainly (>90%) of cobbles and boulders (40 to 500 mm) without sand or silt, and with numerous and deep interstices. The sites are subject to strong wave action, have a relatively steep slope (>20%), and are located near a deep zone (>30 m) (Benoît et al., 1999). These characteristics are similar to those of many lake trout spawning grounds observed in Ontario (MacLean et al., 1990).

For the lake trout, spawning in very deep zones appears more of an exception than the rule because in Ontario lakes (excluding the Great Lakes), 98% of the spawning grounds are less than 4.5 m deep. The average overall depth is 1.4 m (MacLean et al., 1990). The reason for this is that the occurrence of coarse substrate decrease with depth (Chabot and Archambault, 1981). As a result, most of the substrate at depths of greater than 3 to 5 m generally consists of fine particles (sand and silt). However, some spawning grounds at depths below 5 m have been reported (Machniak, 1975).

Spawning depth seems to result from a compromise between the forces needed to keep the substrate clear of fine particles and the forces that can cause egg disturbance or mortality. A significant relationship between spawning depth and lake size was obtained using data from 24 lakes (Fitzsimons, 1994) (Figure 5.4):

$$\text{Depth (m)} = 0.07 + 0.93 \log \text{ surface area (km}^2) \quad (R^2 = 0.79)$$

Effects of raised water level

Few studies have focused on the lake trout's reproductive behavior following reservoir creation, but raising the water level appears to have less of an impact than lowering the level in winter.

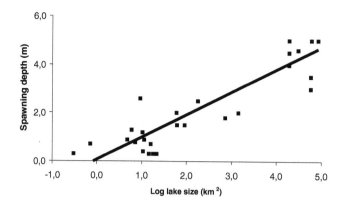

Figure 5.4 Relation between lake size (km², log 10 scale) and lake trout spawning depth (m). (From Fitzsimons, J.D., 1994, *Canadian Technical Report of Fisheries Aquatic Sciences*, No. 1962.)

At the Minnewanka reservoir, Cuerrier (1954) observed spawning as usual at traditional sites that had maintained a suitable spawning substrate despite the raising of the water level by 23 m. However, spawning took place over a wider vertical range than before construction of the dam because the lake trout were able to use new rocky areas down to a depth of about 9.5 m. On the contrary, when the water level in Bark Lake, Ontario was raised 11 m in the late 1930s, the lake trout stopped using traditional sites (Wilton, 1985). Inventories carried out from 1966 to 1972 revealed that the lake trout spawned at depths of less than 3 m during that period.

Whether it is in terms of substrate or depth, the characteristics of reproduction sites found in reservoirs (Lacasse and Gilbert, 1992; Bélanger and Gendron, 1993; Martin, 1955) are similar to those of sites in natural environments (Dumont et al., 1982). When the water level is raised, fine particles eventually cover traditional reproduction sites (P.G. Sly, personal communication in Evans et al., 1991), rendering them less attractive to breeders because of reduced substrate permeability and cleanliness. However, wave action can clear substrate, presenting characteristics favorable to reproduction and hence creating new sites in shallow areas. McAughey and Gunn (1995) demonstrated that the species was clearly capable of seeking alternative spawning sites when traditional ones were destroyed. Hence, the lake trout will readily abandon traditional reproduction sites and select new, more favorable ones in the vicinity.

Effects of water level fluctuations

Management of artificial bodies of water designed for hydroelectric production purposes usually requires that the reservoir be filled when water inputs are high and emptied gradually during periods of heavy energy demand. On the Precambrian Shield, three major steps are involved in dam management. The first step, characterized by a rather stable but high water level, covers the summer season (May to August included). The second step, known as drainage, sometimes begins after a slight fall high water stage and lasts until spring thaw. This step makes it possible to supply power plants throughout the winter when energy demand is heavy and water inputs are low. It rests nearly entirely on water loads accumulated during the previous spring thaw or drawn from groundwater tables. Winter drawdowns (October to April) are typical of this step. In the third and last step the reservoir, fed by the spring thaw, fills again at a rapid and constant pace. Figure 5.5 shows typical water level fluctuations in a reservoir used for hydroelectric production purposes. A similar management pattern is observed in reservoirs used to regulate spring flooding.

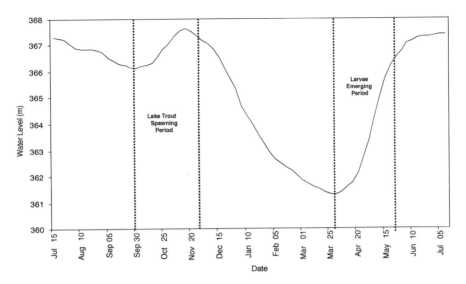

Figure 5.5 Typical water level fluctuations of reservoir in the Precambrian Shield used for hydro-electric production purposes; critical period for lake trout reproduction is indicated.

A survey of various Ministry of Natural Resources districts of Ontario (Lewis et al., 1990) determined the pattern and importance of drawdowns in 85 lake trout lakes. A large number (50) of these lakes had drawdowns occurring after lake trout spawning. Drawdowns occurred before lake spawning on 39 lakes and during spawning on 14 lakes (some lakes had drawdowns occurring at more than one time). The modal drawdown depth was 0.5 to 1.0 m for all three periods (before, during, and after spawning), with drawdown depths varying from 0.5 m to more than 9.5 m. In Québec, the modal drawdown depth was 0.1 to 0.5 m, with depths varying from 0.1 to 11.2 m (Gendron and Bélanger, 1993).

The impacts on lake trout of fluctuations in water level are many and have been extensively documented (Machniak, 1975; Evans et al., 1991; Gendron and Bélanger, 1993). Because lake trout spawn in fall, generally in shallow littoral zones, and the fry leave the spawning grounds towards late spring, the species is vulnerable to winter drawdowns over a period of 3.5 to 8 months (Figures 5.5 and 5.6).

Water management effects on lake trout populations seem to vary according to reservoir management parameters, especially the extent of winter drawdowns. Some reservoirs where winter drawdowns are rather extensive nonetheless manage to support healthy lake trout populations. For example, the Kempt (Benoît et al., 1997) and Mitis (Gendron and Bélanger, 1992) reservoirs in Québec are characterized by significant winter drawdowns (2.0 and 3.0 m) but harbor relatively abundant and stable lake trout populations. A large proportion of spawning grounds used by lake trout in those lakes are deep enough to be unimpacted by the effect of drawdown.

Fluctuations in water level create shoreline erosion and stir up significant quantities of sediment. Spatial distribution of sediment particles is strongly influenced by drawdowns (Baxter and Glaude, 1980), and, depending on shoreline instability and seasonal variations in water use, a decline in the quality of spawning grounds can occur as a result. Sand and silt deposits on these grounds can seriously disrupt lake trout reproduction, as they often result in highly reduced interstitial oxygen concentrations.

Among the factors likely to modify the lake trout's reproductive behavior is the disappearance of reproduction sites as a result of changes in their chemistry and physical attributes caused by the adverse effects of water level fluctuations. The species may have

Figure 5.6 Lake trout spawning area affected by drawdown.

the capacity to seek alternative sites for reproduction (Benoît and Legault, 2002; McAughey and Gunn, 1995), but the need to look for new spawning grounds or reproduce at less favorable sites could have long-term repercussions on the reproductive capacity and recruitment success of a population. Assessing the true impact is difficult, as few studies have been conducted on the subject.

Several studies demonstrating the effects of winter drawdowns on the survival of the spawn of lake trout populations are found in the literature. As shown by a recent summary of studies on Québec reservoirs as a whole and of scientific studies available on the lake trout (Gendron and Bélanger, 1993), the effects of drawdowns are usually highly linked to the depth of the spawning grounds. Highlights of these studies are given in Table 5.2. In the majority of cases, the spawning grounds are located in the shallow littoral zones, and the impacts of drawdowns on egg survival vary greatly. For example, in Thompson Lake and Cold Stream Pond, Maine (DeRoche 1969), the grounds are between 0.2 and 1.2 m deep and the majority of eggs are laid at a depth of 0.3 m. In these reservoirs, a drawdown depth of as little as 0.3 m could cause high egg mortality. Martin (1955) mentions that a drawdown depth of 1.5 m in Hay Lake causes the majority of lake trout eggs to die. In Mary Lake, where two spawning areas have been located — the first between 0.15 and 0.6 m, the second between 1.0 and 1.3 m — a mean drawdown depth of 0.8 m causes the destruction of the majority of eggs laid in the first area (Wilton, 1985).

In some reservoirs, the depth range of potential spawning areas is large relative to water level fluctuations. In Opeongo Lake, Ontario, for instance, spawning grounds are found at depths varying from 0.9 to 3.7 m; hence, the mean drawdown depth of 0.6 m does not put the survival of the eggs in jeopardy. At Wanapitei reservoir, Ontario, Whitfield (1950) estimated that a drawdown depth of 2.4 m resulted in the loss of 26.3% of the eggs. Rawson (1945) mentioned that in Alberta's Minnewanka reservoir, lake trout reproduced at depths varying from 1.5 to 9.4 m. He concluded that a drawdown depth of 3.0 m in this reservoir could maintain a natural lake trout population but that the actual mean drawdown depth of 9.1 m caused the total destruction of the eggs. At Bark reservoir the original spawning ground, located at a depth of 11 m after impoundment, was abandoned by the trout, which then spawned at sites located at less than 3 m in depth. The minimum drawdown depth of 3.8 m used between 1969 and 1972 impeded recruitment during that

Table 5.2 Depth of Spawning Grounds and Observed Mortality Rate as a Function of Drawdown Depth in Each Reservoir

Name and location of reservoirs	Depth of spawning grounds (m)	Winter drawdown (m)	Egg mortality rate	Reference
Bark, Ontario	1.5–3.0	3.8 (minimum) 5.1 (mean)	100%	Wilton 1985
Bays, Ontario	1–3 (99%) 6 (1%)	1.0	28% 0%	Tarandus As. 1988
Cold Stream Pond, Maine	0.2–3.0 0.3 (major)	0.3	90% (forecast)	DeRoche and Bond 1957
David, Québec	< 1.0	0.25+	60%	Pariseau 1981
Hay, Ontario	< 1.8	0.8 (mean) 1.9 (maximum)	Low High	Martin 1955
Kempt, Québec	3.4–4.5	1.8 (mean) 2.7 (maximum)	0%	GDG Envir. 1998
Mary, Ontario	0.2–0.6 (major) 1.0–1.3	0.8 (maximum)	Majority Low	Wilton 1985
Minnewanka, Alberta	1.5–6.2	9.1 11 (maximum)	100%	Rawson 1945 in Wilton 1985
Mitis, Québec	2.0–4.0 (12%) 4.5–6.0 (88%)	3.0	50% 12%	Bélanger and Gendron 1993
Mondonac, Québec	0.5–1.5	1.75	99%	Lacasse and Gilbert 1992
Opeongo, Ontario	0.9–3.7	0.6 (mean) 0.8 (maximum)	Low	Martin 1955
Saint-Patrice, Québec	0.25–3.0 6	0.6	52% 0%	Chabot and Archambault 1981
Smoke, Ontario	< 6.0	1.0 – 1.5	Low	Franck Hicks in Lacasse and Benoît 1995
Thompson, Maine	0.2–1.2 (mostly 0.3)	0.3	90% (forecast)	DeRoche 1969
Wanapitei, Ontario		2.4	26.3%	Whitfield 1950

period. During the 2-year study on the impacts of drawdown on lake trout reproduction in the Mitis reservoir (Bélanger and Gendron, 1993), the majority of eggs (88%) were seen at depths between 4.5 and 6.0 m at the main spawning ground. At six other sites, eggs were found mostly between 2.0 and 4.0 m. It was determined that a drawdown depth of 4.5 m in the winter of 1992–1993 left only 12% of the eggs potentially alive at the principal sites; about 50% of the eggs at the other sites would have been stranded.

As demonstrated by these studies, a given drawdown depth can have very different effects on lake trout populations depending on the yearly fluctuations in water level and location of spawning sites. This situation is attributable to the fact that hydrodynamic conditions required for the maintenance of conditions suited to the development of eggs and fry vary from one reservoir to another, the shallower sites generally being more favorable.

A number of factors may enable a greater number of eggs to survive in drawdown conditions. Backflow caused by wet snow weighing down on the ice cover in the pelagic zone is one of them. The rise in water level that occurs as a result had reached 35 cm in the Mondonac reservoir during a winter inspection of spawning sites (Lacasse and Gilbert, 1992). Had it not been for this backflow, the eggs inventoried earlier would probably not have survived. It may be that in some years this rise in water level is more important because of greater snowfalls, which enable a larger proportion of eggs to survive drawdown.

A second factor is interannual fluctuation of drawdown depth. In some years the reservoirs do not reach the minimum management level, leaving less shoreline out of water and enabling a greater proportion of lake trout eggs to survive winter drawdown. Reasons for the interannual fluctuation include major water input in winter and low demand for electricity.

Lake trout eggs may hatch towards the end of winter or in early spring, usually between February and April depending on developmental conditions. The fry move about very little and remain hidden among the rocks until they reach the "swim-up" stage, usually when water temperature rises to 8–10°C (DeRoche, 1969). As long as their yolk sac has not undergone full resorption, which occurs at least 2 months after hatching, they are vulnerable to drops in water level. This means that lake trout eggs and fry in their first stages of development are vulnerable to water level fluctuations as late as May, when they finally become mobile enough to swim away from the shoreline. At this stage, to ensure good recruitment success, bountiful food and adequate protection against predators are likely as important as quality spawning grounds.

Changes in water quality near spawning grounds can affect the survival of eggs and fry. For example, eggs exposed to low oxygen levels produce either crippled or incompletely developed fry (Garside, 1959). The increase in shore erosion following regulation might also affect the success of lake trout fry through reduced feeding and growth, thus making them more susceptible to predation (Daly et al., 1962 in Machniak, 1975).

A study of egg and fry survival rates in seminatural conditions at Kempt reservoir (GDG Environnement ltée, 1998) revealed that eggs and fry located in the lower drawdown zone, where wet sand is present, can survive for a short period. Under such conditions, however, the survival rate is rather low (0–12% at ice breakup) in comparison to success obtained in those parts of the spawning grounds that remain under water (40–78% survival at ice breakup).

Drawdowns can also affect fish in more subtle and less known ways. Disruption of spring primary productivity following retention of spring flood, for instance, could threaten the survival of fry, a phase critical to the future size of fish populations (Auvinen, 1988; Viljanen, 1988). Gendron and Bélanger (1991) have postulated that in certain reservoirs where drawdowns are strong, competition for planktonic and benthic organisms as a food source in spring and early summer is fierce. This competition would appear to result from the late production of these food organisms. Low spring productivity could also be caused by lowering of the water level in winter, which delays implantation of benthos in littoral zones (Tikkanen et al., 1988), and to the late warming of the waters in spring, which delays production of plankton (Patalas and Salki, 1984). The most obvious direct effect of water level changes on benthos is exposure and desiccation after drawdown. The abundance and biomass of macrobenthos in nonregulated lakes may be two to three times higher than in reservoirs. Mortality of exposed organisms undoubtedly reduces populations within the fluctuation zone and may partly explain the inverted vertical distributions of benthos observed in fluctuating reservoirs (Ploskey, 1986). Reduced benthic productivity on shoals, where we find young of many species of fish, increases competition for benthic prey. Food shortages can cause high mortality among fry of several species, thus influencing fish community structure in reservoirs. Despite the lack of documentation on the subject, it is safe to assume that very early or late hatching of lake trout larvae relative to their food supply (which may be influenced by water level regulation) will have a substantial impact on larval survival.

Changes in the seasonal temperature cycle induced by water level fluctuations can also reduce lake trout productivity by altering the thermal habitat of both juveniles and adults (Evans et al., 1991). For example, a reduction in the volume of the thermal habitat generated by a lowering of the thermocline is likely to lead to reduced productivity.

Drawdown depth may indirectly affect lake trout productivity through impacts on nutrient dynamics and production of other fish species. Although there is no documentation specific to the lake trout, Cohen and Radomski (1993) have established a relationship between maximum and minimum annual water level fluctuations and abundance of walleye, lake whitefish, and northern pike populations in two reservoirs. These fluctuations repeatedly expose and inundate shoals, and nutrients are successively oxidized and reduced. As a result, the quality of nutrients is modified. Nutrient exchanges between the littoral and pelagic zones are also modified (Kennedy and Walker, 1990). Primary productivity is consequently altered, along with the food chain, in reservoirs. The quality of the riparian habitat is also modified as a result of a reduction in the diversity and density of plant communities in the littoral zone. Drawdowns have had observed adverse effects on the summer habitat of the lake whitefish and cisco (Gaboury and Patalas, 1984), both major food items of the lake trout. The rainbow smelt *Osmerus mordax* is also likely to undergo similar variations in abundance and influence lake trout abundance (Matuszek et al., 1990).

Changes in species composition, particularly in important prey species, may have important effects on lake trout. For lake trout, change from piscivorous feeding to plankton feeding could result in a slower growth and maturation at a smaller size. Impoundment of Lake Minnewanka did indeed cause a shift in the feeding habits of lake trout (Cuerrier, 1954). Lake trout had practically ceased eating fish and were subsisting mainly on small chironomid larvae. Cuerrier suggested that after flooding immature trout were spatially separated from Rocky Mountain whitefish *Prosopium williamsoni*, their prey before flooding. The growth and survival of trout declined to the point that few large fish remained.

When recruitment of lake trout populations is already substantially impacted by drawdown, pressure from fishing can interact synergistically with drawdown to reduce fish stocks. At the Mitis reservoir, where the lake trout population is abundant despite a winter drawdown of 3 m and an observed egg mortality of 12%, Gendron and Bélanger (1992) noted that the situation could be explained, in part, by the low fishing pressure. Low fishing pressure, therefore, is a factor contributing to the maintenance of lake trout populations experiencing recruitment failure as a result of water level fluctuations. Obviously, however, imposing such a management measure is likely to cause conflicts with anglers and tourist operators.

There are few data available on the effects of water level fluctuations on lake trout reproductive behavior. In southern Québec's Kempt reservoir (GDG Environnement ltée, 1998), the lake trout was observed to spawn late in the season at a lower water temperature than that observed in neighboring lakes. In the fall of 1997, reproduction began in early November at temperatures of 6° to 7°C. Because reproduction of the species in the area usually occurs in October at temperatures of 8° to 12°C (Chabot and Archambault, 1981; Lacasse and Gilbert, 1992; Benoît and Legault, 2002), the late reproduction may be an adaptation to fall water level fluctuations. This assumption needs to be tested, however.

Management recommendations for lake trout in reservoirs

This section outlines several actions that can be taken that may improve abundance of lake trout populations in reservoirs. Some are intended to preserve the natural conditions needed to maintain lake trout populations at a high level of abundance, whereas others focus on compensating for reservoir water level management-related impacts.

Integrated management of water and wildlife

Because winter drawdown is the main factor behind the lake trout problems observed in several hydroelectric reservoirs, it is the overriding focus of any integrated management plan. Obviously, water management that calls for a stable water level during the incubation and hatching period is ideal. According to DeRoche (1969), Maine's policy recommends not lowering the water level between September 15 and the end of April. However, applying such a policy in most reservoirs within the lake trout's range is difficult; the hydroelectric reservoirs are managed to supply power plants throughout the winter when energy demand is heavy and water inputs are low.

Water level fluctuations required to help maintain a lake trout population will vary from one reservoir to the next. Findings to date indicate that the acceptable winter drawdown range varies between 0.25 and 3.5 m, depending on the reservoir. In a survey of 234 sites on 90 lakes in Ontario, mean depths of spawning shoals ranged from 0.1 to 6.0 m and averaged 1.4 m overall. Twenty-eight percent of the shoals were in water depths of less than 1.0 m, 76% were in depths of less than 2.0 m, and 92% were in depths of less than 3.0 m. Based on this survey of rather small lakes, it is recommended that drawdowns be limited to less than 1.0 m to protect >70% of the lake trout spawning shoals (J. Gunn, Laurentian University, personal communication). Moreover, in certain reservoirs it may be possible to maintain the lake trout population by irregular recruitment every 2 or 3 years, by alternating between large and small-scope drawdowns. Depending on drawdown depth, population maintenance would be less costly in terms of hydroelectric power loss. Development of a sound water management protocol rests on a thorough understanding of lake trout habitat and requirements.

When a reservoir water level is lowered before the lake trout's reproduction period, ascertaining that a quality substrate exists below the lowered level is crucial to prevent too many eggs from dying off. At Lake Shirley, Ontario, Martin (1955) noted that lake trout eggs had been preyed on because of the low quality of the lower portion of spawning areas as a result of drawdown. To counter this problem, lowering the reservoir progressively and earlier in the season to enable cleaning of the fine substrate of deep-zone spawning areas by wave action is a possibility. At the time of this writing, the efficiency of such a measure is being tested in Québec in a reservoir of the Saint-Maurice river hydroelectric complex (Benoît et al., 1999) and in the Jacques-Cartier reservoir (Fournier and Lépine, 1998).

When a lake trout lake is linked to several reservoirs, implementation of a watershed-based integrated management approach is possible without hydroelectric power loss. This type of management approach is currently being tried in the Saint-Maurice river hydroelectric complex (Benoît et al., 1999), which comprises 10 main reservoirs of which 6 support lake trout populations. At the end of the summer, the lake trout reservoirs are lowered progressively to get a minimum level before the lake trout's reproduction period. Then the water level is maintained at this level or higher through the end of the larva emergence period (Figure 5.7). As the lake trout reservoirs are drawn down, the water is accumulated in the other reservoirs. Water management constraints necessitate that early drawdowns in lake trout reservoirs are compensated by drawdowns in the other reservoirs so that the economic impact of the new water management is negligible. As a result, lake trout have access to quality spawning areas that are not dried out during winter drawdown. Lake trout populations, and ultimately other fish species, are no longer subject to major constraints by the operation of the reservoirs in the Saint-Maurice River complex because the goal is to integrate concerns about major sport species into the reservoir's operation procedure.

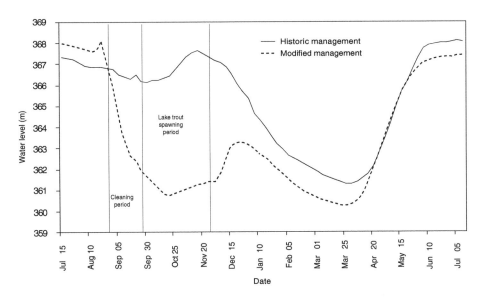

Figure 5.7 Experimental water management at reservoir Châteauvert, from the Saint-Maurice river hydroelectric complex, for improved lake trout reproduction. (Adapted from Benoît et al., 1999.)

Introduction of a strain of deep-spawning lake trout

A possible solution to the reproduction problems in reservoirs with extensive drawdowns is the use of a strain of deep-spawning lake trout. Introduction in Wisconsin's Green Lake (Hacker, 1957) of a strain from a deep-spawning subpopulation from Lake Michigan provides an interesting example. In this lake, lake trout spawn at depths of 20 to 30 m on a substrate consisting of anything from clay to fine gravel, even though rocky surfaces were available near the shore. However, this author observed that because eggs laid on this type of substrate are very vulnerable to predation, variable recruitment and a decline in the lake trout population apparently occurred. Creation of an artificial deep-zone spawning area may have provided the eggs with greater protection.

Based on studies of lake trout reproduction, especially of the species' capacity to react rapidly to the loss of reproduction habitat (McAughey and Gunn, 1995), we can assume that introduction of new stocks that would use spawning sites not subject to drawdowns is possible. In addition, egg incubation at suitable sites could encourage the return of spawning stock to these sites. For this to work, thorough knowledge of lake trout habitat and potential spawning sites is required. Site selection will need to take into account annual and seasonal drawdowns in the reservoirs concerned. As a first step and for economic reasons, the use of natural sites rather than development of artificial sites must be encouraged. A source of spawning stock that may harbor survival attributes useful in reservoirs (e.g., good health population in a reservoir with similar characteristics), from which sufficient quantities of eggs could be collected must be identified, as well as the incubation technique best suited to the sites. Also, a program to monitor the success of the techniques should be implemented.

Creating deep-zone spawning areas

The creation of artificial spawning sites for lake trout in areas located below the drawdown zone is a mitigating measure that can be applied in hydroelectric reservoirs. Past experiments have shown that such sites can be introduced with success in some reservoirs, such as those of Lake Shirley (Martin, 1955), Lake des Baies (Tarandus Associates, 1988), and Lake Mountain (Haxton, 1991) in Ontario, Green Lake in Wisconsin (Hacker, 1957), and Lake Manouane in Québec (Gendron, 2001). However, little is known of the long-term use of these artificial sites and their true efficiency in ensuring recruitment of populations (Kerr, 1998) despite the fact that their creation involves high costs.

According to Arnett (1988) and Benoît and Legault (2002), the creation of artificial spawning sites, using coarse substrate, in the deeper end of littoral zones next to natural spawning sites, resulted in the partial transfer of reproduction to the artificial sites. Substrate quality appears to have been a determining factor in the site selection process, and depth likely played a secondary role. In all probability, the factor that drove the lake trout to use artificial rather than natural sites is the greater proportion of substrate devoid of fine sediments, sand, gravel, or pebbles as well as its greater permeability to eggs, deposited in the interstices of the rock.

Two major factors must be taken into consideration in creating artificial spawning sites in deep zones, i.e., the lake trout's homing behavior to specific sites and sedimentation potential. With regard to the former it cannot be said, according to examples provided in the scientific literature, that proximity to a natural site and stocking of eggs and/or fry are prerequisites to the success of an artificial site (Lacasse and Benoît, 1995; Gunn, 1995). However, the chances that spawners will use an artificial site are likely to be improved by such management practices. In addition, vulnerability to sedimentation appears to be the factor that may limit the long-term success of artificial spawning sites.

Absence of sedimentation is a condition that is not easily found in natural deep environments, as hydrodynamic conditions are generally more favorable to the maintenance of quality egg and larva development conditions at shallow sites. Because of the limited amplitude of waves generated in most reservoirs, deep zones may not be affected by the scouring action of waves. If need be, lowering the water level of the reservoirs in the fall during the period preceding reproduction would allow the artificial sites to be cleansed by the action of the waves. Given the significance for the economy of the loss of water reserves, lowering the water level for cleaning purposes could take place every 2 or 3 years, depending on the importance of sedimentation.

High-pressure or suction hoses have been used with success to clean the substrate of lake trout spawning areas (Dumont, 1995; Kerr, 1998). Although such methods are technically feasible even on a large scale, they can be costly depending on the distance of spawning areas from the shore, their size, and depth.

Obstruction of the spawning areas in the littoral zones of Lake aux Sables using a tarp (Benoît and Legault, 2002) led lake trout to abandon these areas and reproduce in the artificial deep-water spawning areas created near the natural sites. This technique, coupled with the creation of adjacent artificial sites below the drawdown zone, may make it possible to concentrate spawning at these sites and represents a potential way of alleviating the effects of hydroelectric reservoir management. However, the long-term efficiency of this measure would be dependent on an ability to maintain the permeability and cleanliness of the artificially built substratum to continue to attract spawners. Again, sedimentation is the main factor capable of producing a countereffect to such an intervention.

Supplemental stocking

In water bodies subjected to strong pressure from sport fishing enthusiasts, regular stocking proves a practical way to artificially maintain stocks. When these water bodies are reservoirs supporting lake trout populations already stressed by water level fluctuations, users actually insist on stocking. However, maintaining a natural stock should be the goal if the lake trout's adaptation potential is to remain intact. The use of genetic strains indigenous to a water body or hydrographic basin should be the preferred option if stocking is necessary.

Depending on the context and available budgets, two stocking techniques can be used. The first, commonly used for the restoration of lake trout stocks, consists of placing incubation boxes containing layers of astroturf deep enough to avoid being dried out during drawdown time. Such a method was developed in Lake Superior (Swanson, 1982) and used with success in Québec's Jacques-Cartier reservoir (Fournier and Lépine, 1998). The mean egg-hatching rates were 78% in Lake Superior and 56% at the Jacques-Cartier reservoir. The second technique, reservoir stocking using hatchery-reared subadults (1 year +) taken from indigenous stock, represents an efficient but more costly alternative to the incubation boxes.

There are many logistic and economic constraints to the supplemental stocking of a reservoir. Stocking of large reservoirs is very costly, and egg supply problems are to be expected, particularly if genetic strains indigenous to the lake or the reservoir's hydrographic basin are to be used.

Restricting harvest

When drawdown has a substantial impact on recruitment of lake trout populations, lowering the pressure on fisheries becomes a factor propitious to the maintenance of these populations. It is obvious that such a measure is likely to conflict with fishing activities. Seeking consensus from users is necessary before such a measure can be implemented.

Conclusions

Reservoir creation brings about major transformations to ecosystems supporting lake trout. These changes are variable in importance and can be temporary or permanent.

Most of the transformations occur at reservoir creation and last but a short while, as environments tend to return to their natural and initial state of dynamic equilibrium. However, reservoir management practices, like water level fluctuations, have a permanent impact that is most adverse for lake trout.

Water level fluctuations can seriously affect lake trout populations by disrupting reproduction success. The margin of flexibility with respect to winter drawdown depth without affecting lake trout spawning areas is quite variable and ranges from 0.25 to 3.5 m. If the reproduction habitat undergoes modifications that are too extensive, reproduction of the species runs the risk of being seriously affected, causing the decline of a stock or disappearance of the species from the reservoir. Knowledge about the depth of spawning sites is therefore essential for assessing the effects of drawdown management practices in a given reservoir.

The sequence of reproduction events and scope of drawdowns are the two other major parameters that need to be taken into consideration in designing an intervention strategy aimed at ensuring the conditions required to support an abundant lake trout population, such as integrated management of water and wildlife resources.

In cases where adapting water level management of a reservoir to lake trout reproduction imperatives is impossible, alternatives have been presented. Although some of the actions proposed may be permanent (use of deep-spawning strain of lake trout, creation of spawning areas below the drawdown zone), others are recurrent (supplemental stocking). Even though such actions have been used with success in the past, their true efficacy has yet to be properly documented.

References

Arnett, G., 1988, *Comparison of spawning success on natural vs. man made shoals, Lake of Bays*, Ontario Ministry of Natural Resources, Bracebridge District, Toronto.

Auvinen, H., 1988, Factors affecting the year-class strength of vendance (*Coregonus albula* (L.)) in lake Pyhäjärvi (Karelia, SE Finland), *Finnish Fisheries Research* 9:235-243.

Baird, O.E. and Krueger, C.C., 2000, Behavior of lake trout sac fry: vertical movement at different developmental stages, *Journal of Great Lakes Research* 26:141-151.

Baxter, R.M. and Glaude, P., 1980, Les effets des barrages et des retenus d'eau sur l'environnement au Canada: expérience et perspectives d'avenir, *Canadian Bulletin of Fisheries and Aquatic Sciences* 205F:36.

Bélanger, B. and Gendron, M., 1993, *Le touladi: réservoirs Mitis et Manic-3; caractérisation et validation des sites de reproduction (1992); incubation des œufs et échantillonnage des alevins (1993)*, Le Groupe de Recherche SEEEQ ltée, pour le Service ressources et aménagement du territoire, vice-présidence Environnement, Hydro-Québec.

Belzile, L., Lalumière, R., and Doyon, J.F., 2000, *Réseau de suivi environnemental du complexe La Grande. Suivi des communautés de poissons du secteur est (1980-1999). Rapport synthèse*, Prepared by Groupe Conseil Génivar Inc. pour la Direction expertise et support technique de production, Unité hydrolique et Environnement, Hydro-Québec.

Benoît, J. and Legault, M., 2002, Assessment of the feasibility of preventing reproduction of lake trout (*Salvelinus namaycush*) in shallow areas of reservoirs affected by drawdowns, *Environmental Biology of Fishes* 64:303-311.

Benoît, J., Scrosati, J., and Dumont, D., 1997, *Situation du touladi (Salvelinus namaycush) des réservoirs Châteauvert, Kempt, Manouane et Mondonac*, Ministère de l'Environnement et de la Faune, Direction régionale Mauricie—Bois-Francs, Trois-Rivières, Rapport technique.

Benoît, J., Bérubé, R., Frigon, J.R., Milette P., and Legault, M., 1999, *La restauration du touladi des réservoirs de la Haute-Mauricie, Plan d'action 1995-1997, Rapport synthèse*, Ministère de l'Environnement et de la faune: Direction de la faune et des habitats, Direction régionale Mauricie et Hydro-Québec: Téléconduite Centre, Direction régionale Mauricie et Production des Cascades, Direction Expertise et Support technique de production.

Brouard, D., Demers, C., Lalumière, R., Schetagne, R., and Verdon, R., 1990, *Évolution des teneurs en mercure des poissons du complexe hydroélectrique La Grande, Québec (1978-1990); rapport synthèse*, Vice-présidence Environnement, Hydro-Québec et Groupe Environnement Shooner, Inc.

Chabot, J. and Archambault, J., 1981, *Quelques caractéristiques de frayères à touladi du lac St-Patrice, comté de Pontiac*, Ministère du Loisir, de la Chasse et de la Pêche du Québec, Service de l'aménagement et de l'exploitation de la faune, région de l'Outaouais.

Chartrand, N., Schetagne, R., and Verdon, R., 1994, *Enseignements tirés du suivi environnemental au complexe la Grande*, Commission Internationale des Grands Barrages, Dix-huitième Congrès des Grands Barrages, Durban, Q.69-R-14, pp. 165-190.

Cuerrier, J.-P., 1954, The history of Lake Minnewanka with reference to the reaction of lake trout to artificial changes in environment, *Canadian Fish Culturist* 15:1-9.

Cohen, Y. and Radomski, P., 1993, Water level regulations and fisheries in Rainy Lake and the Namakan reservoir, *Canadian Journal of Fisheries and Aquatic Sciences* 50:1934-1945.

DeRoche, S.E., 1969, Observations on the spawning habits and early life of lake trout, *Progressive Fish Culturist* 3:109-113.

DeRoche, S.E. and Bond, L.H., 1955, The lake trout of Cold Stream Pond, Enfield, Maine, *Transactions of the American Fisheries Society* 85:257-270.

DesLandes, J.-C., Guénette, S., Prairie, Y., Roy, D., Verdon, R., and Fortin, R., 1995, Changes in fish populations affected by the construction of the La Grande complex (phase 1), James Bay region, Québec, *Canadian Journal of Zoology* 73:1860-1877.

Doyon, J.F., 1997, *Réseau de suivi environnemental du complexe La Grande, Phase 1 (1995). Suivi des communautés de poissons et étude spéciale sur le touladi du réservoir Caniapiscau (secteur est du territoire)*, Rapport présenté par le Groupe-conseil Génivar inc. à la direction principale Communication et Environnement d'Hydro-Québec.

Doyon, J.F., Tremblay, A., and Proulx, M., 1996, *Régime alimentaire des poissons du complexe La Grande et teneurs en mercure dans leurs proies (1993-1994)*, Groupe-conseil Génivar inc. pour Hydro-Québec, vice-présidence Environnement et Collectivités.

Duchemin, E., Lucotte, M., Canuel, R., and Chamberland, A., 1995, Production of the greenhouse gases CH_4 and CO_2 by hydroelectric reservoirs of the boreal region, *Global Biogeochemical Cycles.*

Dumont, B., 1995, *Expérimentation d'une méthode de nettoyage de frayère à touladi au lac Mégantic*, Rapport de Pro Faune pour l'Association Chasse et Pêche du Lac-Mégantic inc.

Dumont P., Pariseau, R., and Archambault, J., 1982, Fraie du touladi (*Salvelinus namaycush*) en très faible profondeur, *Canadian Field-Naturalist* **96**(3):353-354.

Evans, D.O., Brisbane, J., Casselman, J.M., Coleman, K.E., Lewis, C.A., Sly, P.G., Wales, D.L., and Willox, C.C., 1991, *Anthropogenic stressors and diagnosis of their effects on lake trout populations in Ontario lakes. Lake trout synthesis*, Ontario Ministry of Natural Resources, Toronto.

Fitzsimons, J.D., 1994, An evaluation of lake trout spawning habitat characteristics and method for their detection, *Canadian Technical Report of Fisheries Aquatic Sciences*, No. 1962.

Fournier, G. and Lépine, C., 1998, *Efforts de restauration de la population de touladis (Salvelinus namaycush) du grand lac Jacques-Cartier, Réserve faunique des Laurentides*, Rapport présenté à La Fondation de la faune du Québec par Inkshuk, division, consultation faunique et environnementale, St-Raymond de Portneuf.

Gaboury, M.N. and Patalas, J.W., 1984, Influence of water level drawdown on the fish populations of Cross Lake, Manitoba, *Canadian Journal of Fisheries and Aquatic Sciences* **41**:118-125.

Garside, E.T., 1959, Some effects of oxygen in relation to temperature on the development of lake trout embryos, *Canadian Journal of Zoology* **37**:689-698.

GDG Environnement ltée, 1998, *La restauration du touladi des réservoirs de la Haute-Mauricie. Plan d'action 1995-1997. Évaluation de l'impact du marnage sur la reproduction du touladi au réservoir Kempt (Manouane «A»)*, Rapport présenté à Hydro-Québec, vice-présidence Environnement et Collectivités et région Mauricie.

Gendron, M., 2001, *La restauration du touladi des réservoirs de la Haute-Mauricie. Plan de la mise en œuvre. Étude de la reproduction du touladi dans les réservoirs Châteauvert et Manouane, automne 1999*, Rapport réalisé par Environnement Illimité inc. Présenté à Hydro-Québec, Unité Hydrolique et Environnement et région Mauricie.

Gendron, M. and Bélanger, B., 1991, *Étude de l'effet du marnage sur la faune ichtyenne des réservoirs Pipmuacan, Outardes-4 et Manic-5, travaux de recherche automne 1990*, Le Groupe de Recherche SEEEQ Ltée, pour le Service Ressources et Aménagement du Territoire, vice-présidence Environnement, Hydro-Québec.

Gendron, M. and Bélanger, B., 1992, *Évaluation de l'impact du marnage sur les sites et le potentiel de reproduction du touladi dans le réservoir Mitis*, Le Groupe de Recherche SEEEQ Ltée, pour le Service Ressources et Aménagement du Territoire, vice-présidence Environnement, Hydro-Québec.

Gendron, M. and Bélanger, B., 1993, *Étude de l'état des populations de touladi dans les réservoirs*, Le Groupe de Recherche SEEEQ Ltée, pour le Service Ressources et Aménagement du Territoire, vice-présidence Environnement, Hydro-Québec.

Gunn, J.M., 1995, Spawning behavior of lake trout: effects on colonization ability, *Journal of Great Lakes Research* **21**(Suppl. 1):323-329.

Hacker, V.A., 1957, Biology and management of lake trout in Green lake, Wisconsin, *Transactions of the American Fisheries Society* **86**:71-83.

Haxton, T., 1991, *Lake trout spawning shoal rehabilitation on Mountain lake, Minden township*, Ontario Ministry of Natural Resources, Minden District.

Jones, M.L., Cunningham, G.L., Marmorek, D.R., Stokes, P.M., Wren, C., and DeGrass, P., 1986, *Mercury release in hydroelectric reservoirs*, Canadian Electrical Association, Ottawa.

Kelly, C.A., Rudd, J.W.M., Bodaly, R.A., Roulet, N.P., St. Louis, V.L., Heyes, A., Moore, T.R., Schiff, S., Aravena, R., Scott, K.J., Dyck, B., Harris, R., Warner, B., and Edwards, G., 1997, Increases in fluxes of greenhouse gases and methyl mercury following flooding of an experimental reservoir, *Environmental Science and Technology* **31**:1334-1344.

Kennedy, R.H. and Walker, W.W., 1990, Reservoir nutrient dynamics, In *Reservoir Limnology: Ecological Perspective*, edited by K.W. Thornton, B.L. Kimmel, and F.E. Payne, pp. 109-132, John Wiley & Sons, New York.

Kerr, S.J., 1998, *Enhancement of lake trout spawning habitat: a review of selected projects*, Technical Report TR-108, Southcentral Sciences Section, Ontario Ministry of Natural Resources, Kemptville, Ontario.

Lacasse, S. and Benoît, J., 1995, *Évaluation de la faisabilité d'induire la reproduction du touladi (Salvelinus namaycush) sur les frayères situées en profondeur — Application à la problématique des réservoirs de la Haute-Mauricie*, Ministère de l'Environnement et de la Faune, Direction de la faune et des habitats et Direction régionale Mauricie — Bois-Francs. Rapport technique.

Lacasse, S. and Gilbert, L., 1992, *Évaluation de l'impact du marnage sur la reproduction du touladi au lac Mondonac*, GDG Environnement ltée, pour le service Activités d'exploitation, vice-présidence Environnement, Hydro-Québec.

Lewis, C.L., Cunningham, G.L., and Chen, T., 1990, *Analysis of Questionnaire on Stresses Acting on Lake Trout Lakes*, Lake Trout Synthesis, Ontario Ministry of Natural Resources, Toronto.

Loftus, K.H., 1958, Studies on river-spawning populations of lake trout in eastern Lake Superior, *Transactions of the American Fisheries Society* **87**:259-277.

Machniak, K., 1975, *The effects of hydroelectric development on the biology of northern fishes (reproduction and population dynamics). IV. Lake trout* Salvelinus namaycush *(Walbaum). A literature review and bibliography*, Fisheries and Marine Service Research and Development, Technical Report 530.

MacLean, N.G., Gunn, J.M., Hicks, F.J., Ihssen, P.E., Malhiot, M., Mosindy, T.E., and Wilson, W., 1990, *Environmental and genetic factors affecting the physiology and ecology of lake trout*, Ontario Ministry of Natural Resources, Lake Trout Synthesis, Physiology and Ecology Working Group.

Magnuson, J.K.J., Assel, R.A., Bowser, C.J., Dillon, P.J., Eaton, J.F., Evans, H.E., Fee, E.J., Hall, R.I., Mortsch, L.R., Schindler, D.W., Quinn, F.H., and Webster, K.E., 1997, Potential effects of climate changes on aquatic systems: Laurentian Great Lakes and Precambrian shield region, *Hydrobiological Processes* **11**:221-249.

Martin, N.V., 1955, The effect of drawdowns on lake trout reproduction and the use of artificial spawning beds, *Transactions of the North American Wildlife Conference* **20**:263-271.

Martin, N.V. and Olver, C.H., 1980, The lake charr, *Salvelinus namaycush*, In *Charrs, Salmonid Fishes of the Genus* Salvelinus, edited by E.K. Balon, pp. 205-277, Dr. W. Junk Publishers, The Hague.

Matuszek, J.E., Shuter, B.J., and Casselman, J.M., 1990, Changes in Lake trout growth and abundance after introduction of Cisco into Lake Opeongo, Ontario, *Transactions of the American Fisheries Society* **119**:718-729.

McAughey, S.C. and Gunn, J., 1995, The behavioral response of lake trout to a loss of traditional spawning sites, *Journal of Great Lakes Research* **21**(Suppl. 1):375-383.

Pariseau, R., 1981, *Inspection d'hiver de la frayère à touladi du lac David (Pontiac)*, Ministère du Loisir, de la Chasse et de la Pêche, Service de l'Aménagement et de l'Exploitation de la Faune, Région de l'Outaouais.

Patalas, K. and Salki, A., 1984, Effects of impoundment and diversion on the crustacean plankton of Southern Indian Lake, *Canadian Journal of Fisheries and Aquatic Sciences* **41**:613-637.

Ploskey, G.R., 1986, Effects of water level changes on reservoir ecosystems, with implications for fisheries management, In *Reservoir Fisheries Management: Strategies for the 80s*, edited by G.E. Hall and M.J. Van Den Avyle, Reservoir Committee, Southern Division American Fisheries Society, Bethesda.

Rawson, D.W., 1945, *Further investigation of the effect of power development on the fisheries of lake Minnewanka, Banff*, Canadian Department of Mines Resources, Report to the Natural Parks Bureau.

Rosenberg, D.M., Bodaly, R.A., Hecky, R.E., and Newbury, R.W., 1987, The environmental assessment of hydroelectric impoundments and diversions in Canada, *Canadian Bulletin of Fisheries and Aquatic Sciences* **215**:71-104.

Rudd, J.W.M., Harris, R., Kelly, C.A., and Hecky, R.E., 1993, Are hydroelectric reservoirs significant sources of greenhouse gases? *Ambio* **22**:246-248.

Schetagne, R., Doyon, J.-F., and Verdon, R., 1997, *Summary report: Evolution of fish mercury levels at the La Grande Complex, Québec (1978-1994)*, Joint Report of the Direction principale communication et Environnement, Hydro-Québec, and Groupe-conseil Génivar Inc.

Ségin, R.L. and Roussel, Y.E., 1970, Étude de la frayère et du comportement de la truite grise (*Salvelinus namaycush*) au ruisseau des Cèdres, canton Bouchette, compté Gatineau. *Service de la faune du Québec. Rapport* **5**:127-158.

Swanson, B.L., 1982, Artificial turf as a substrate for incubating lake trout eggs on reefs in Lake Superior, *Progressive Fish-Culturist* **44**(2):109-111.

Tarandus Associates Ltd., 1988, *An Evaluation of Selected Lake Trout Spawning Shoals in Lake of Baies*, Ontario Ministry of Natural Resources, Toronto.

Tikkanen, P., Niva, T., Yrjänä, T., Kuusela, K., Hellsten, S., Kantola, L., and Alasaarela, E., 1988, Effects of regulation on the ecology of the littoral zone and the feeding of whitefish, *Coregonus* spp., in lakes in northern Finland, *Finnish Fisheries Research* **9**:457-465.

Vincent, B. and De Serres, L., 1963, Description d'une frayère de touladi (*Salvelinus namaycush*) dans un ruisseau de la région de Maniwaki. Ministère du Tourisme, de la Chasse et de la Pêche, *Service de la Faune, Rapport* **3**:225-231.

Viljanen, M., 1988, Relations between egg and larval abundance, spawning stock and recruitment in Vendace (*Coregonus albula* L.), *Finnish Fisheries Research* **9**:271-289.

Wilton, M.L., 1985, *Water drawdown and its effects on lake trout* (Salvelinus namaycush) *reproduction in three south-central Ontario lakes*, Ontario Fisheries Technical Report Series no. 20, Ontario Ministry of Natural resources, Toronto, Ontario.

Whitfield, R., 1950, *Elevations of Lake Wanapitei and their effect upon the eggs of spawning fishes*, Ontario Department of Lands and Forest, North Bay District, Manuscript Report.

chapter six

Lake trout (Salvelinus namaycush) habitat volumes and boundaries in Canadian Shield lakes

Bev J. Clark
Dorset Environmental Science Centre, Ontario Ministry of the Environment
Peter J. Dillon
Environmental and Resource Studies, Trent University
Lewis A. Molot
Faculty of Environmental Studies, York University

Contents

Introduction

Lake trout *Salvelinus namaycush* populations generally require large volumes of cold, well-oxygenated water to thrive (Martin and Olver, 1980); thus, their optimal habitat boundaries have been defined by temperatures of less than 10°C and by oxygen concentrations greater than 6 mg/L (Evans et al., 1991). Oxygen and temperature criteria have been used to define other classes of lake trout habitat; for example, Evans et al. (1991) defined a "usable" habitat as one with $O_2 > 4$ mg/L and temperature <15°C. There is increasing evidence that some populations can be successful at higher temperatures under some circumstances (Snucins and Gunn, 1995), but in most cases, examining the potential for success or failure of lake trout populations requires, as a minimum, a measurement of suitable habitat volumes for the lakes in question (Christie and Regier, 1988; MacLean et al., 1990; Evans et al., 1991). Because these boundaries are measured easily using portable field equipment (temperature/oxygen meters), they present a relatively simple means of assessing habitat suitability.

Habitat volume

Lake trout habitat volumes are at their minimum each year immediately before the time when surface waters have cooled enough (to <15°C) to be usable. This occurs before thermal destratification in the fall. At this time, oxygen concentrations in the hypolimnion will be at their minimum, and the depth of the metalimnion as well as the depth of the 10°C isotherm, Z_{10}, will be at their greatest. These dates are determined each year by the rate of cooling of the mixed layer and are therefore subject to external forces associated with weather. The most appropriate time for these measurements is therefore the late summer period when minimum hypolimnetic oxygen concentrations and maximum temperatures, i.e., worst-case conditions with respect to lake trout habitat, usually are found in dimictic lakes (Evans et al., 1991).

Evans et al. (1991) used habitat volumes "standardized" to August 31, and Molot et al. (1992) standardized end-of-summer oxygen profiles to September 1. Although optimal habitat often will continue to decrease after September 1, it is difficult to obtain data sets that are closely spaced (temporally) to model the exact minimum habitat volumes or to predict the dates that these occur because of between-year variation in the turnover dates. Measured Z_{10} (i.e., the depth of the 10°C isotherm) depression rates (Dillon et al., in press) during the late summer and fall ranged from 0.013 to 0.051 m/d and were correlated with transparency, measured as DOC. This means that Z_{10} would not likely depress more than *ca.* 1.5 additional meters after September 1 for most lakes if conditions continued to deteriorate for the entire month. Oxygen concentrations in the bottom waters may continue to diminish; potentially this would present more severe problems for lake trout, especially in those cases where habitat volumes are minimal earlier in the year, which can be the situation in lakes that are relatively shallow. However, the greatest stress on populations occurs after all of the optimal habitat has disappeared; hence, we believe that the optimal habitat estimated for September 1 describes close to minimal volumes. By the end of September, temperatures are almost always such that the optimal habitat for lake trout (Evans et al., 1991) has started to increase through cooling of the mixed layer.

Models that have been developed to estimate habitat volumes usually express the volume as a percentage of the total lake volume. In addition, most existing optimal habitat models acknowledge the link between habitat volume and some measure of both lake morphometry and nutrient status. For example, the optimal habitat model used for lake trout management in inland lakes in southeastern Ontario (Ontario Ministry of the Environment and Ontario Ministry of Natural Resources, 1993) uses mixing ratios and mean summer chlorophyll *a* to estimate the proportion of the habitat that is optimal. Ryan and Marshall (1994) presented a rapid diagnostic method based on mean depth and habitat quality using phosphorus, Secchi depth, and chlorophyll. Models that estimate habitat based on specific oxygen and temperature boundaries are discussed later in the chapter

Volume-weighted habitat

Management guidelines that are based on habitat volumes may be difficult to interpret because these volumes vary considerably between lakes. Similar volumes, for example, may have very different meanings when expressed as a percent of the total lake volume, and vice versa. It also may be difficult to assess the importance of habitat loss, especially in bottom waters where the loss may represent a relatively small proportion of the total volume. The use of a volume-weighted hypolimnetic oxygen concentration (VWHO) would eliminate many of these problems. A standardized (end-of-summer) VWHO allows the use of a single number to compare conditions among lakes. These lakes may otherwise show seasonal and spatial variability with respect to O_2 concentrations and often different orders of magnitude with respect to habitat volumes. It is suggested for lake trout that VWHO should not be allowed to drop below 7 mg/L (Evans, 1999). Calculating VWHO requires morphometry data and at least one end-of-summer oxygen and temperature profile. Ideally the means of several oxygen and temperature profiles would be used to reflect long-term conditions. Temperature profiles are used only to establish the upper limit of the hypolimnion, which is defined as the lower depth of the first 1-m interval where the temperature change is less than 1°C per meter. Once the thermal hypolimnetic boundaries are established, the volumes for each stratum are calculated by:

$$V = m[A_t + A_b + v(A_t \, {}^*A_b)]/3 \qquad (6.1)$$

where

>   V is volume (m^3)
> A_t is the area of the top of the stratum (m^2)
> A_b is the area of the bottom of the stratum (m^2)
>   m is the depth (thickness) of the stratum (m)

VWHO is calculated as the summed products of the measured dissolved oxygen concentration in each stratum and the proportion of the hypolimnetic volume represented by that stratum. It can also be estimated by volume-weighting the oxygen concentrations (at a given depth) that are predicted from the hypolimnetic oxygen profile model developed by Molot et al. (1992). More details with respect to modeling oxygen concentrations are given in the following section.

Estimating habitat boundaries

Models that predict the percentage optimal habitat in lakes or calculations that give VWHOs offer no specific information about the location of the habitat boundaries defined by temperature and oxygen. Furthermore, proportional volumes alone cannot provide information regarding how environmental change might affect the habitat boundaries. For example, it has long been realized that increases in the nutrient status of a lake will deplete hypolimnetic oxygen (Molot et al., 1992) and thereby reduce lake trout habitat by raising the lower (oxygen) boundary in the lake. On the other hand, determining the effect of transparency on habitat volumes is of recent interest because of work linking climate change as well as other regional-scale stresses such as acid rain to changes in transparency (Dillon et al., 1987; Schindler, 1997; Schindler et al., 1990, 1996b; Fee et al., 1996; Molot and Dillon, 1997). For example, Fee et al. (1996) found that mixing depths (which have a strong effect on the determination of upper habitat boundaries) were best identified using extinction coefficients (converted to percentage transmission), which principally are functions of dissolved organic carbon (DOC) except in those cases where algal levels are high (i.e., eutrophic lakes).

Two simple models were previously developed and combined to estimate long-term average optimal habitat boundaries for lake trout defined by temperature and oxygen criteria. A hypolimnetic oxygen profile model (Molot et al., 1992; Clark et al., 2002) was used to define the mean long-term lower boundary ($Z_{6\,mg/L}$) of the optimal habitat. Because this model uses lake morphometry and total phosphorus (TP) to predict hypolimnetic oxygen profiles for the late summer period (Sept. 1), it can be used to estimate the depth of the specific O_2 concentration of interest.

Oxygen concentrations at each stratum z are determined by:

$$\log_{10}O_2(f)_z = 1.83 - 1.91/VSA_z - 7.06/O_2(i)_z - 0.0013TP^2_{so} \tag{6.2}$$

where

$O_2(f)_z$ = the end-of summer oxygen concentration at depth z (mg/L)
$O_2(i)_z$ = the oxygen concentration at depth z at spring turnover (mg/L)
TP_{so} = the total phosphorous concentration at spring turnover (µg/L)
VSA_z = the ratio of the stratum volume (V)/sediment surface area (SA) at depth z (m)

Spring turnover oxygen concentrations $[O_2(i)]$ are measured directly or determined for each stratum by:

$$\log_{10}O_2(i)_z = 0.99 - 5.74/A_o + 0.64/z \tag{6.3}$$

when the maximum distance from shore to shore at z_{max} (= MD) < 1.4 km or by:

$$\log_{10}O_2(i)_z = 1.07 - 6.95/A_o - 0.0043z/MD \tag{6.4}$$

when MD > 1.4 km.

A second recently developed model (Dillon et al., in press) predicts the depth at which 10°C occurs (Z_{10}) for the same late summer period. For the set of 37 lakes used for the model development, transparency was the primary determinant of Z_{10}, with the morphometric or lake size parameters playing a secondary role. A water clarity parameter, either Secchi depth or 1/DOC (the reciprocal of the dissolved organic carbon concentration) together with either lake area (A_0) or MD, was used to generate a linear relationship with Z_{10}. The combination of 1/DOC and A_0 gave the best fit as:

$$Z_{10} = 3.52 + 11.3/DOC + 0.139* \sqrt{A_0} \tag{6.5}$$

(r^2 = 0.88, p < 0.01, std. error of estimate = 0.87).

This was somewhat better than the fit obtained using Secchi depth as the transparency parameter and MD as the lake size parameter:

$$Z_{10} = 3.35 + 0.956*Secchi + 0.33*MD \tag{6.6}$$

(r^2 = 0.78, p < 0.01, std. error of estimate = 1.17)

These two modeled boundary depths (i.e., Molot et al., 1992, and Dillon et al., in press) were combined with measured morphometric data (strata volumes) to calculate long-term optimal lake trout habitat volumes standardized to the end of summer i.e., Sept. 1 (Dillon et al., in press). These authors suggest that the combined models could be used to predict the effects of changes in trophic status or transparency on the upper and lower optimal habitat boundaries for a given lake. In a similar fashion, a modeling approach analogous to those described here could be used to delineate other oxygen or temperature boundaries and thus define usable lake trout habitat, or any other class of habitat.

Effects of external stresses on habitat volume

It is clear that because these oxygen and temperature models use TP, Secchi depth (or DOC), and physical factors (morphometry, MD, or A_0), changes in either the nutrient status or transparency of a lake will impact directly on the optimal habitat volumes. These relationships, in fact, allow us to predict the loss in optimal habitat volume that would result from projected changes in water quality. It should be noted that, in some instances, the changes in conditions controlling Z_{10} and $Z_{6mg/L}$ may in part, have counteracting effects on optimal habitat volume. Increasing nutrient levels (TP), for example, will increase oxygen deficits near the bottom (Molot et al., 1992), resulting in a shallower lower boundary for optimal habitat. However, the same change in TP may result in reduced Secchi depth because of increased chlorophyll *a* concentrations (Dillon and Rigler, 1974), a shallower Z_{10}, and therefore increased habitat volume. In general, however, water clarity in lake trout lakes is controlled by DOC rather than TP, so the effects of an increase in nutrient levels will be largely transmitted through the decrease in the lower oxygen boundary rather than the increase in the upper temperature boundary.

Many lakes will likely become more transparent as a consequence of changes in climate (Magnuson et al., 1997; Schindler et al., 1990, 1996a, 1997). It has been proposed that this will be the result of declining DOC production in watersheds during prolonged periods of above-normal temperatures and below-normal precipitation (Dillon et al., 1996; Dillon and Molot, 1997; Schindler et al., 1997). This change in transparency in combination with the direct effects of increased air temperatures will result in deeper Z_{10}s (Snucins and Gunn, 2000) and decreased optimal lake trout habitat volumes. Fee et al. (1996) estimated that there is the potential for the epilimnia in smaller lakes to increase in thickness by 1 to 2 m as a result of a twofold increase in atmospheric CO_2 levels. Acidification of lakes by the input of mineral acids also results in increased transparency (Effler et al., 1985), again via removal of DOC (Schindler et al., 1996b). For example, the Secchi depth measured in Plastic Lake increased by *ca.* 2 m (Dillon et al., 1987) during the lake's last 7 or 8 years of acidification and probably by much more over the total period in which acidification occurred.

For the 37 lakes in the Dorset area used to test/develop their model, Dillon et al. (in press) calculated that a 1-m increase in transparency would result in an increase in Z_{10} of *ca.* 1 m. Following a 2-m increase in transparency, optimal habitat volumes in the same lakes would be reduced by between 8 and 100%. These calculations are based on the assumption that clarity changes result from loss of DOC and that there is no change in TP concentrations that would affect the lower optimal habitat boundaries ($Z_{6mg/L}$). Any such changes resulting from changes in TP would be much less significant than those resulting from DOC-induced changes in clarity. The proportional amount of optimal habitat that would be lost for any lake is greatest when the optimal habitat volume is relatively small. Lakes having high percentages (>50%) of their total volume as optimal habitat would lose only 10 to 20% of their optimal habitat, whereas lakes with only 10 to 25% of their volume as optimal habitat could lose 50% or more of the remaining habitat, putting the continued success of the lake trout populations in jeopardy.

There are other factors that were not included in either of the two models defining the habitat boundaries, which may be important. For example, France (1997) established linkages between lake thermocline depth and riparian deforestation, which would indicate that the Z_{10} for lakes of similar size, clarity, and nutrient status might vary with differences in riparian cover or surrounding topography. We cannot quantify the influence of this parameter on the variation observed between predicted and measured boundary depths because such data are not available.

Finally, recent studies show that the effects of logging on the mixing depths of adjacent lakes may be minimal (Steedman and Kushneriuk, 2000) or that one consequence of increasing temperature may be a reduction in mixed layer depths that results from rapid stratification caused by high air temperatures in the first part of the ice-free season (Snucins and Gunn, 2000). It has recently been reported that in high-DOC lakes, the depth of the summer mixed layer is decreased by the indirect effects of climate warming events. The latter will have the opposite effect on habitat volumes (i.e., volumes will increase) to that proposed by Fee et al. (1996).

References

Christie, G.C. and Regier, H.A., 1988, Measures of optimal thermal habitat and their relationship to yields for four commercial fish species, *Canadian Journal of Fisheries and Aquatic Sciences* **45**:301–314.

Clark, B.J., Dillon, P.J., Molot, L.A., and Evans, H.E., 2002, Application of a hypolimnetic oxygen profile model to lakes in Ontario, *Lake and Reservoir Management* **18**:32–43.

Dillon, P.J., Clark, B.J., Molot, L.A., and Evans, H.E., in press, Predicting optimal habitat boundaries for lake trout (*Salvelinus namaycush* Walbaum) in Canadian Shield lakes, *Canadian Journal of Fisheries and Aquatic Sciences*.

Dillon, P.J. and Rigler, F.H., 1974, The phosphorus–chlorophyll relationship in lakes, *Limnology and Oceanography* **19**:767–773.

Dillon, P.J., Reid, R.A., and deGrosbois, E., 1987, The rate of acidification of aquatic ecosystems in Ontario, Canada, *Nature* **329**:45–48.

Dillon, P.J., Molot, L.A., and Futter, M., 1996, The effect of El Nino-related drought on the recovery of acidified lakes, *Environmental Monitoring and Assessment* **46**:105–111.

Dillon, P.J. and Molot, L.A., 1997, Effect of landscape form on export of dissolved organic carbon, iron and phosphorus from forested stream catchments, *Water Resources Research* **33**:2591–2600.

Effler, S.W., Schafran, G.C., and Driscoll, C.T., 1985, Partitioning light attenuation in an acidic lake, *Canadian Journal of Fisheries and Aquatic Sciences* **42**:1707–1711.

Evans, D.O., 1999, *Metabolic scope-for-activity of juvenile lake trout and the limiting effect of reduced dissolved oxygen: defining a new dissolved oxygen criterion for the protection of lake trout habitat.* Community Dynamics and Habitat Unit manuscript report 1999–1.

Evans, D.O., Casselman, J.M., and Willox, C.C., 1991, Effects of exploitation, loss of nursery habitat, and stocking on the dynamics and productivity of lake trout populations in Ontario lakes, In *Lake Trout Synthesis, Response to Stress Working Group*, Ontario Ministry of Natural Resources, Toronto, Ontario.

Fee, E.J., Hecky, R.E., Kasian, S.E.M., and Cruikshank, D.R., 1996, Effects of lake size, water clarity, and climate variability on mixing depths in Canadian Shield Lakes, *Limnology and Oceanography* **41**:912–920.

France, R., 1997, Land–water linkages: influences of riparian deforestation on lake thermocline depth and possible consequences for cold stenotherms, *Canadian Journal of Fisheries and Aquatic Sciences* **54**:1299–1305.

MacLean, N.G., Gunn, J.M., Hicks, F.J., Ihssen, P.E., Malhiot, M., Mosindy, T.E., and Wilson, W., 1990, Genetic and environmental factors affecting the physiology and ecology of lake trout, In *Lake Trout Synthesis*, Ontario Ministry of Natural Resources, Toronto.

Magnuson, J.J., Webster, K.E., Assel, R.A., Bowser, C.J., Dillon, P.J., Eaton, J.G., Evans, H.E., Fee, E.J., Hall, R.I., Mortsch, L.R., Schindler, D.W., and Quinn, F.H., 1997, Potential effects of climate changes on aquatic systems: Laurentian Great Lakes and Precambrian Shield region, *Hydrological Processes.* **11**:828–871.

Martin, N.V. and Olver, C.H., 1980, The lake charr, *Salvelinus namaycush*, In *Charrs: Salmonid Fishes of the Genus Salvelinus, Perspectives in Vertebrate Science*, Vol. 1, edited by E.K. Balon, Dr. W. Junk, The Hague, pp. 209–277.

Molot, L.A., Dillon, P.J., Clark, B.J., and Neary, B.P., 1992, Predicting end-of-summer oxygen profiles in stratified lakes, *Canadian Journal of Fisheries and Aquatic Sciences* **49**:2363–2372.

Molot, L.A. and Dillon, P.J., 1997, Photolytic regulation of dissolved organic carbon in northern lakes, *Global Biogeochemical Cycles* **11**:357–365.

Ontario Ministry of the Environment (OMOE) and Ontario Ministry of Natural Resources (OMNR), 1993, *Inland Lake Trout Management in Southeastern Ontario*, Ontario Ministry of the Environment and Ontario Ministry of Natural Resources. MS Report.

Ryan, P.A. and Marshall, T.R., 1994, A niche definition for lake trout (*Salvelinus namaycush*) and its use to identify populations at risk, *Canadian Journal of Fisheries and Aquatic Sciences* **51**:2513–2519.

Schindler, D.W., 1997, Widespread effects of climatic warming on freshwater ecosystems in North America, *Hydrological Processes* **11**:1043–1067.

Schindler, D.W., Beaty, K.G., Fee, E.J., Cruikshank, D.R., DeBruyn, E.D., Findlay, D.L., Linsey, G.A., Shearer, J.A., Stainton, M.P., and Turner, M.A., 1990, Effects of climate warming on lakes of the central boreal forest, *Science* **250**:967–970.

Schindler, D.W., Bayley, S.E., Parker, S.E., Beaty, K.G., Cruikshank, D.R., Fee, E.J., Schindler, E.U., and Stainton, M.P., 1996a, The effects of climate warming on the properties of boreal lakes and streams at the Experimental Lakes Area, northwestern Ontario, *Limnology and Oceanography* **41**:1004–1017.

Schindler, D.W., Curtis, P.J., Parker, B.R., and Stainton, M.P., 1996b, Consequences of climate warming and lake acidification for UV-B penetration in North American boreal lakes, *Nature* **379**:705–708.

Schindler, D.W., Curtis, P.J., Bayley, S.E., Parker, B.R., Beaty, K.G., and Stainton, M.P., 1997, Climate-induced changes in the dissolved organic carbon budgets of boreal lakes, *Biogeochemistry* **36**:9–28.

Snucins, E.J. and Gunn, J., 1995, Coping with a warm environment: behavioral thermoregulation by lake trout, *Transactions of the American Fisheries Society* **124**:118–123.

Snucins, E.J. and Gunn, J., 2000, Interannual variation in the thermal structure of clear and colored lakes, *Limnology and Oceanography* **45**:1639–1646.

Steedman, R.J. and Kushneriuk, R.S., 2000, Effects of experimental clearcut logging on thermal stratification, dissolved oxygen, and lake trout (*Salvelinus namaycush*) habitat volume in three small boreal forest lakes, *Canadian Journal of Fisheries and Aquatic Sciences* **57**(Suppl. 2):82–91.

chapter seven

The effects of phosphorus and nitrogen on lake trout (Salvelinus namaycush) production and habitat

Peter J. Dillon
Environmental and Resource Studies, Trent University
Bev J. Clark
Dorset Environmental Science Centre, Ontario Ministry of the Environment
Hayla E. Evans
RODA Environmental Research Limited

Contents

Introduction

The relationships between the levels of nutrients, particularly phosphorus (TP), in lakes and lake trout *Salvelinus namaycush* are complex. This is a consequence of the fact that phosphorus may affect lake trout in two opposite but interrelated ways. On one hand, phosphorus is the nutrient that controls algal biomass in almost all lakes in the Boreal ecozone of Canada and most lakes directly to the south in the Laurentian Great Lakes region; i.e., in almost all lakes in North America that are inhabited by lake trout (Dillon and Rigler, 1974; Schindler, 1977). Phytoplankton is a very important component of the food web in all lakes, and algal abundance may strongly influence the biomass and productivity of higher trophic levels, including fish. On the other hand, high algal biomass is associated with oxygen depletion in the hypolimnia of lakes; an adverse effect on the amount of suitable habitat available to species including lake trout that prefer or require cold, well-oxygenated waters is therefore expected when algal biomass is high.

In this chapter, we discuss the two conflicting roles that the limiting nutrient, phosphorus, plays in lake trout biology. Considerably less information is available with respect to the role of nitrogen, but we have included as much as is available. We first review the literature with respect to relationships between nutrients and fish production in general and then review those relationships pertaining to phosphorus and salmonids in particular. We then characterize known lake trout lakes by their phosphorus and nitrogen regimes. We discuss the role of morphometric factors in determining lake trout habitat and then very briefly consider how a relatively new model that can be used to predict lake trout habitat is affected by the combination of phosphorus concentrations and morphometric factors. The question of nutrient–morphometry–habitat relationships is discussed in detail by Clark et al. (Chapter 6, this volume).

Nutrient–phytoplankton–fish linkages

The relationship between TP in lakes and biological measurements of primary standing stock, such as chlorophyll *a* concentration, has been studied extensively (Sakamoto, 1966; Dillon and Rigler, 1974; Nicholls and Dillon, 1978; Prepas and Trew, 1983; Stockner and Shortreed, 1985; Ostrofsky and Rigler, 1987; Dillon et al., 1988; Molot and Dillon, 1991). Less well documented, however, is the association of other nutrients, such as nitrogen, with primary production (e.g., Sakamoto, 1966; Stockner and Shortreed, 1985; Prepas and Trew, 1983; Molot and Dillon, 1991) and also the relationship between TP and higher levels (e.g., zooplankton and fish) of the aquatic food web (Moyle, 1956; Hanson and Leggett, 1982; Jones and Hoyer, 1982; Downing et al., 1990).

As mentioned above there are two reasons TP, and perhaps other nutrients, should be related to fish production, either directly or indirectly. First, the ecological concept of food webs suggests that there should be a direct relation between fish production and secondary production and between secondary productivity and primary productivity (McQueen et al., 1986), even if there are allochthonous inputs to the food chain that contribute significantly to fish diets (e.g., France and Steedman, 1996). If phosphorus is controlling primary production, it is reasonable to expect a relationship between TP and fish production. Secondly, we know that TP controls the trophic status of lakes (see above), and we know that phosphorus and trophic status can affect oxygen concentrations in thermally stratified lakes (Cornett and Rigler, 1979; Welch and Perkins, 1979; Molot et al., 1991). Because minimum O_2 levels can define the habitat for lake trout (Ryan and Marshall, 1994; Ranta and Lindstrom, 1998; Sellers et al., 1998) or other cold-water species, then TP should be related to cold-water fish production.

Nutrient–biomass relationships

Perhaps the best-known model used to predict fish production from a knowledge of nutrient concentrations in the lake was developed by Ryder (1965). Ryder's morphoedaphic index (MEI) was an empirically derived formula that related lake mean depth (z_{mean}) and total dissolved solids (TDS), a surrogate of phosphorus (intended as an indicator of the lake's fertility or productivity), to potential fish yield or harvest. For a number of years after its publication, Ryder's MEI spawned a plethora of papers in which the ability of the MEI to predict fish yield, fish harvest, fish production, and fish biomass was examined in North American lakes (Carlander, 1977; Adams and Olver, 1977; Oglesby, 1977; Ryan and Harvey, 1977; Matuszek, 1978; Prepas, 1983) and reservoirs (Jenkins, 1967, 1982; Henderson et al., 1973) and also in tropical systems (Regier et al., 1971; Henderson et al., 1973; Toews and Griffith, 1979). During the same period of time, methods were developed to improve the index, for example, by the introduction of scaling factors for latitude (Henderson et al., 1973; Ryder et al., 1974, Schlesinger and Regier, 1982; Kalff, 1991) and also by attempts to find better predictors for fish harvest than TDS/z_{mean}. These predictors included biotic variables such as photosynthesis (McDonnell et al., 1977), chlorophyll *a* (Jones and Hoyer, 1982), primary production (Oglesby, 1977) and benthic biomass (Matuszek, 1978; Hanson and Leggett, 1982), and also abiotic factors. Because it is generally understood that any correlate of TDS can be used as a substitute for that variable (Henderson et al., 1973), conductivity and alkalinity were substituted sometimes for TDS in the MEI (e.g., Vighi and Chiaudani, 1985).

However, evidence that nutrients, and in particular TP, may ultimately control the rate of fish production in lakes was increasing (Colby et al., 1972) and, in fact, had been provided even earlier than the publication of the MEI (Rawson, 1951, 1952; Moyle, 1956). In addition, there was a noticeable congruence of the MEI with other models incorporating morphometric and/or edaphic parameters such as TP. Ryder et al. (1974) argued that the MEI model could be related to Vollenweider's (1968) method for determining admissible and dangerous levels of phosphorus loading in lakes and also to Schindler's (1971) index for predicting TP loading to lakes from lake area, catchment area, and lake volume. Oglesby (1982) also pointed out the similarity between the MEI and Vollenweider's (1976, as cited by Oglesby, 1982) model relating lake trophic status to phosphorus loading, i.e.,

$$Chl_a = L_p(T_w + 1/T_w)/z_{mean} \qquad (7.1)$$

where

Chl_a = chlorophyll *a* concentration (mg/m^3)
L_p = areal P loading ($mg\ P/m^2/yr$)
T_w = hydraulic retention time (years)

Assuming that Chl *a* is proportional to fish yield (Jones and Hoyer, 1982) and that L_p is a surrogate for TDS, which incorporates both allochthonous and autochthonous inputs, then the resemblance of this equation to the MEI is evident.

Hanson and Leggett (1982) were perhaps the first to demonstrate conclusively that fish yields could be predicted empirically from TP concentrations in lakes. They found that the MEI consistently performed poorly when compared to other indices of fish yield/biomass. As they state, total phosphorus concentration and macrobenthos standing crop/mean depth were superior to TDS, mean depth, and the MEI as predictors of fish yield (and biomass) when comparisons were based on the same data set. Total phosphorus was the best univariate predictor of fish yield and fish biomass in two of the four data sets they examined. Furthermore, although the best multivariate predictor of fish yield in one of these data sets included z_{mean}, P, and TDS, removal of TDS in the equation did not significantly reduce the predictive efficiency of the model. They suggested that this resulted from cross correlation of TDS with phosphorus concentration (see also Vighi and Chiaudani, 1985; Chow-Fraser, 1991), which is in agreement with Henderson and co-workers' (1973) contention that any correlate of TDS should be a suitable substitute for it in the MEI.

Vighi and Chiaudani (1985) used a slightly different approach to utilize the apparent relationship between TP concentration and fish production. First they found a significant correlation between TP concentration and the MEI in 53 lakes having negligible phosphorus load as a result of anthropogenic activities. Then they plotted TP concentration versus MEI in lakes subject to cultural eutrophication and found they fell above the predicted line developed from the 53 pristine lakes. From this observation they argued that because the MEI was developed for use on unpolluted lakes (i.e., with respect to nutrient inputs), then differences between MEI-predicted fish yields and actual fish yields could be used to estimate the degree of phosphorus inputs from anthropogenic activities. Thus, indirectly, Vighi and Chiaudani used the relationship between phosphorus and fish yield to predict natural phosphorus loadings in these lakes. A similar approach was used more recently by Chow-Fraser (1991) and also by Koussouris et al. (1992).

The relationship between phosphorus concentration and fish production was explored further in a series of papers published by Downing and co-workers. Downing et al. (1990), using data collected from the literature for entire lake fish communities, demonstrated that fish production was closely correlated with mean total phosphorus concentration (r^2 = 0.67) in addition to annual phytoplankton production (r^2 = 0.79) and annual average fish standing stock (r^2 = 0.67). Similar to Hanson and Leggett (1982), they also found no correlation between fish production and the MEI. A few years later, Downing and Plante (1993) reported that the residuals from a multivariate equation (which related annual fish production to annual mean standing biomass and maximum individual biomass) indicated that fish production was positively correlated with total phosphorus concentration (in µg/L) in addition to temperature, phytoplankton production, chl a, and lakewater pH (see their Figure 4). Once again, the MEI was not a good predictor of fish production. Later the same year, Plante and Downing (1993) tested the hypothesis that there was a general positive relationship between lake trophic status (i.e., phosphorus concentration) and salmonine production. Their relationship:

$$\log \text{Production} = 0.95 \log \text{TP} - 0.47 \quad (n = 10, r^2 = 0.61) \tag{7.2}$$

where

    Production = salmonine production (kg/ha/yr)
    TP = total lake phosphorus concentration (µg/L)

This is very similar to the relation published by Hanson and Leggett (1982) over a decade earlier, i.e.:

$$\log \text{FY} = 1.021 \log \text{TP} - 1.148 \quad (n = 21, r^2 = 0.87) \tag{7.3}$$

where

    FY = fish yield (kg/ha)
    TP = total phosphorus concentration (mg/m³)

This is surprising because Hanson and Leggett's (1982) relationship was produced using many fish species, whereas Plante and Downing's (1993) relationship involves only salmonine species. Plante and Downing suggest that the similarity may indicate that the link between (abiotic) factors (such as phosphorus), which influence primary production, and salmonine populations is stronger for salmonines than for other fish species because food chains tend to be simpler in lakes in which salmonines dominate. Thus, the relationship between phosphorus and salmonine production is similar to that between phosphorus and fish community production, simply because the food chain is shorter in salmonine lakes.

The mounting evidence that TP is the nutrient limiting primary productivity and also fish (salmonine) production in lakes, together with the economic significance of salmonine fisheries, prompted some scientists/biologists to investigate the utility of adding phosphorus to oligotrophic lakes as a means of increasing salmonine production. For example, in the early 1970s (1970 to 1973 inclusive), Great Central Lake, British Columbia, was treated with ammonium nitrate and ammonium phosphate in an attempt to test the hypothesis that increasing the supply of inorganic nutrients in the lake would increase production at succeeding trophic levels (LeBrasseur et al., 1978). LeBrasseur and co-workers found that during the period when the lake was being enriched, mean summer primary production increased fivefold, zooplankton standing stock increased nine times, and the growth of age 2+ smolts, the survival of age 0+ sockeye, and the mean stock size of adult sockeye salmon *Oncorhynchus nerka* increased from <50,000 to >360,000 fish. Several years later, Hyatt and Stockner (1985) reported on the results of a more expanded fertilization experiment in which ammonium nitrate and ammonium phosphate were added to as many as 17 lakes along the British Columbia coast. Similar to the results of LeBrasseur et al. (1978), Hyatt and Stockner (1985) found that increased autotrophic and heterotrophic production resulted in larger standing stocks of zooplankton and increased in-lake growth of juvenile sockeye salmon. They suggest that the changes that occurred in the fertilized lakes may lead to increases in the harvestable surplus of sockeye adults. In Norway, Johannessen et al. (1984) also fertilized six small mountain lakes in Telemark, with ammonium phosphate, ammonium nitrate, and phosphoric acid. They found that in Lake Kanontjern, the length and weight of brown trout *Salmo trutta* increased during the three seasons of fertilization. Similarly, Johnston et al. (1990) found that whole-river fertilization of the Keogh River, British Columbia, with nitrogen and phosphorus increased the size of steelhead trout *Oncorhynchus mykiss* and coho salmon *Oncorhynchus kisutch* fry. More recently, Ashley et al. (1997) reported that phosphorus and nitrogen additions (1992–1994) to the North Arm of Kootenay Lake, British Columbia, increased the biomass of phytoplankton, zooplankton, and kokanee salmon *Oncorhynchus nerka* in the lake.

In summary, given the key role of TP in the production dynamics of freshwater ecosystems and the strong predictive relationships that have been developed previously (e.g., Hanson and Leggett, 1982; Plante and Downing, 1993), it is important that models incorporating TP and fish production continue to be produced and refined.

Characterization of Ontario lake trout lakes with respect to phosphorus and nitrogen

Because of their requirement for cold, well-oxygenated water, the classic picture of a lake suitable for lake trout is one that is oligotrophic and deep, although exceptions to this generalization are known to occur (e.g., in monomictic Pedro Lake, Ontario; see Snucins and Gunn, 1995). Notwithstanding, this notion is based on a conceptual model in which the benefits of the protection of oxygen concentration (i.e., of maintaining habitat size), derived from having low nutrient levels, exceed the advantages of having high nutrient levels and concomitant increased productivity. Survey data collected on a substantial number of lakes in Ontario support this concept. Of 1220 lakes in the province with TP data available, 293 also were classified as lake trout lakes (both native and introduced). These are distributed throughout the range of lake trout lakes in the province (Figure 7.1). The distributions of TP concentrations in lakes with and without lake trout in the data set

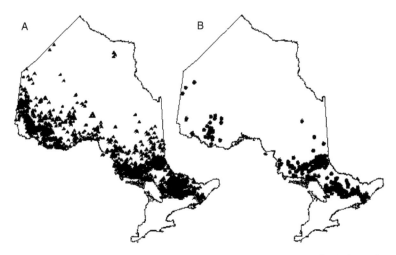

Figure 7.1 Locations of (A) 1916 known lake trout lakes in Ontario and of (B) the subset of 293 lake trout lakes from the overall set with at least one TP concentration measurement. The data were collected using a range of sampling methods and were collected at different times of the year, although almost all samples were either epilimnetic composite samples taken in the summer or spring or fall overturn samples. All analyses were carried out using identical methods at the same laboratory.

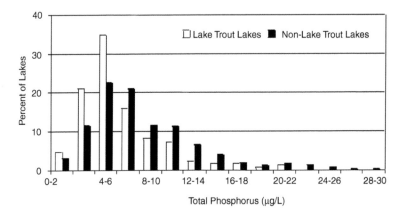

Figure 7.2 Total phosphorus concentration distributions in 293 lake trout lakes and 927 lakes with no lake trout in Ontario.

are shown in Figure 7.2. A total of 61% of the lake trout lakes surveyed had TP concentrations less than 6 µg/L, and 85% had <10 µg TP/L, the concentration that is sometimes used as a criterion for classifying lakes as oligotrophic (Dillon and Rigler, 1975). In comparison, 27% of non–lake trout lakes had TP < 6 µg/L and 62% had <10 µg TP/L. Similarly, the mean (6.9 µg/L) and median TP concentrations (6.0 µg/L) in lake trout lakes were lower than those (10.3 and 8.9 µg/L, respectively) of the non–lake trout lake set. This comparison indicates that lake trout lakes, on average, have lower TP levels than other lakes in Ontario. In fact, the data set used here may be biased to make this difference appear less than it is, as the surveys that have been compiled here into this database explicitly included almost all of the known lake trout lakes in the southernmost portion of the province where it may be expected that TP levels are higher because of anthropogenic contributions to the nutrient budgets, e.g., more shoreline development, higher atmospheric deposition of TP.

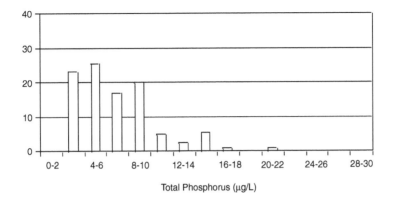

Figure 7.3 Total phosphorus concentration distributions in 125 Quebec lake trout lakes (from Prairie, 1994).

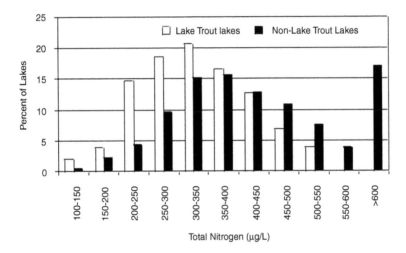

Figure 7.4 Total nitrogen concentration distributions in 102 lake trout lakes and 394 lakes without lake trout in Ontario.

The distribution of TP in lake trout lakes in Quebec (Figure 7.3) was very similar to that in Ontario (Prairie, 1994). Of 125 lakes in Quebec with TP data available, 49% had TP < 6 µg/L, and 86% had TP < 10 µg/L. The mean (7.1 µg/L) and median (6.2 µg/L) TP concentrations in Quebec lake trout lakes were virtually identical to those in Ontario lakes (6.9 and 6.0 µg/L, above). Ratios of total nitrogen to total phosphorus (TN–TP) greater than about 16 (mole ratio, equivalent to a ratio of 7 by weight) generally indicate phosphorus limitation (Wetzel, 1975); this ratio is exceeded in all but a few cases in eastern North America. As a result, nitrogen does not control trophic status and/or algal biomass in the great majority of lakes in the Boreal ecozone or the St. Lawrence Lowlands. However, nitrogen concentrations may influence species composition because lower TN–TP ratios favor nitrogen-fixing species. In Ontario lake trout lakes, total nitrogen concentrations, like TP concentrations, were lower than in non–lake trout lakes (Figure 7.4). The mean and median TN concentrations (330 and 331 µg/L, respectively) were substantially lower than those in non–lake trout lakes (495 and 394 µg/L). The median TN–TP ratio in the lake trout lakes was 56 (by weight), well above the ratio at which nitrogen may control algal biomass and higher than that in non–lake trout lakes (44 by weight).

It should be noted that the traditional view of a lake trout lake might change as the impacts of species invasions (i.e., the accidental or intentional introduction of nonnative species) into lake trout lakes are more fully elucidated. For example, it has been reported (Vander Zanden et al., 1999) that introduction of smallmouth bass *Micropterus dolomieu* and rock bass *Ambloplites rupestris* into some Canadian lakes resulted in a change in the food-web structure including both a decline in the littoral prey-fish abundance and in the trophic position of the trout. Lake trout in the invaded lakes shifted their diet toward pelagic zooplankton and reduced their dependence on littoral fish.

However, in general, the available information supports the hypothesis that lake trout are found in lakes with low TP levels and TN–TP ratios. Only a few lake trout lakes have TP concentrations that would be considered indicative of mesotrophic conditions. This suggests that the need for relatively low TP concentrations to ensure high concentrations of oxygen in the cold, deep water supersedes the benefits of increased nutrient levels that would result in increased food resources.

Characterization of Ontario lake trout lakes with respect to morphometry

In addition to nutrient levels, lake trout habitat is controlled by morphometry. The requirement for cold, well-oxygenated water means that thermally stratified lakes with large hypolimnetic volumes should be preferred. As the proportion of a lake that is composed of hypolimnetic water increases, not only is the volume of the water with suitable cold temperatures that are needed increased, but the impact of the algal nutrients and subsequent production in the trophogenic zone, which leads to oxygen-consuming degradation processes in the hypolimnion, diminishes. Thus, any consideration of the role of nutrients in lake trout biology is inextricably linked with morphometry because of the nutrient–oxygen relationship.

A comparison of the size of lake trout lakes with other lakes in Ontario is shown in Figure 7.5. There is a much lower proportion (5%) of lake trout lakes in the smallest size category (<25 ha) than is the case for all lakes in the database (34%). More than 34% of the known lake trout lakes are <100 ha in area, about half the total (63%) of all those lakes

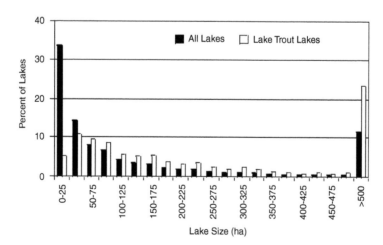

Figure 7.5 Distribution of lake surface areas for 1916 Ontario lake trout lakes and 6824 other lakes in Ontario.

in the database. However, because the database includes very few of the small lakes (<10 ha) in Ontario, the difference is certainly considerably larger than these numbers indicate. Because lake size is not normally distributed, we consider the median areas rather than the means; for lake trout lakes, the median (169 ha) is about three times greater than for all lakes (55 ha).

The requirement for cold temperatures and high oxygen concentrations suggests that, other things being equal, lake trout lakes will be relatively deep in comparison with lakes that do not have lake trout populations. This is supported by the information available in Ontario (Figure 7.6); the average mean and maximum depths of 1892 lakes with lake trout are 10.7 and 33.5 m, respectively. This is substantially greater than the comparable mean and maximum depths for 436 non–lake trout lakes with data available (4.7 and 13.0 m, respectively).

In summary, the available data support the widely held view that lake trout are more likely to be found in deep lakes; the preference for larger size (area) is probably a consequence of the fact that there is a relationship between lake area and depth. This is relevant in the context of nutrients in that an oxygen model developed for habitat (Clark et al., Chapter 6, this volume) utilizes both TP concentration and morphometric data related to hypolimnetic volume.

TP–morphometry–oxygen linkages

Recently, a model linking TP, morphometry, and water clarity (dissolved organic carbon or Secchi depth) to lake trout habitat has been proposed (Dillon et al., in press). The model addresses both the oxygen and temperature requirements of lake trout. TP and morphometry are used in the oxygen submodel, and water clarity and morphometry are utilized in the thermal submodel. The former model can be used to estimate the response of oxygen concentrations in individual hypolimnetic strata in a lake to changes in TP concentration; thus, the effects of changing nutrient levels on the volume of suitable habitat may be ascertained provided criteria for optimum and/or acceptable oxygen concentration can be established. In a similar fashion, the thermal submodel addresses the issue of habitat size from the perspective of optimal and/or acceptable temperatures. This approach to modeling optimal habitat is discussed in detail by Clark et al. (Chapter 6, this volume).

Summary

Nutrients, particularly phosphorus, play two conflicting roles in lake trout biology. Elevated nutrient levels increase productivity of at least lower trophic levels, which, in turn, may increase fish productivity. However, elevated nutrient levels also result in reduced hypolimnetic oxygen levels, i.e., reduced habitat for lake trout. There are a number of relationships in the literature between nutrients and fish production in general, but only a very few pertaining to phosphorus and salmonids in particular. The available nutrient data demonstrate that lake trout lakes are typically oligotrophic, with 85% of those in Ontario having TP concentrations below 10 µg/L. Mean and median TP values are about two-thirds of those in lakes without lake trout, and total nitrogen concentrations in lake trout lakes are also lower by about the same proportion. Nutrients, morphometric factors, and water clarity (through temperature effects) combine to determine the size of lake trout habitat; a relatively new model can be used to predict how lake trout habitat is affected by the combination of phosphorus concentrations and morphometric factors.

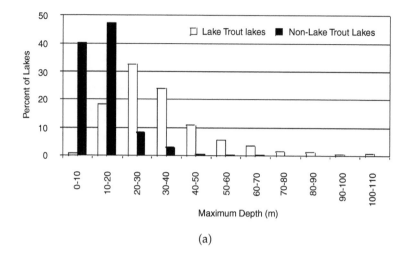

(a)

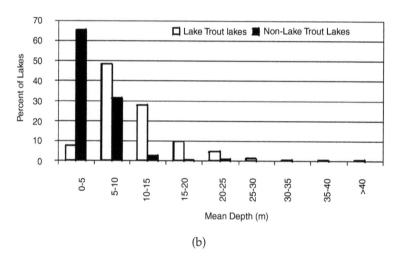

(b)

Figure 7.6 Comparison of (a) maximum and (b) mean depths between 1892 lake trout lakes and 436 other non–lake trout lakes with available data.

References

Adams, G.F. and Olver, C.H., 1977, Yield properties and structure of boreal percid communities in Ontario, *Journal of the Fisheries Research Board of Canada* **34**:1613–1625.

Ashley, K., Thompson, L.S., Lasenby, D.C., McEachern, L., Smokorowski, K.E., and Sebastian, D., 1997, Restoration of an interior lake ecosystem: the Kootenay Lake fertilization experiment, *Water Quality Research Journal of Canada* **32**:295–323.

Carlander, K.D., 1977, Biomass, production, and yields of walleye *(Stizostedium vitreum vitreum)* and yellow perch *(Perca flavescens)* in North America lakes, *Journal of the Fisheries Research Board of Canada* **34**:1602–1612.

Chow-Fraser, P., 1991, Use of the morphoedaphic index to predict nutrient status and algal biomass in some Canadian lakes, *Journal of the Fisheries Research Board of Canada* **48**:1909–1918.

Colby, P.J., Spangler, G.R., Hurley, D.A., and McCombie, A.M., 1972, Effects of eutrophication on salmonid communities in oligotrophic lakes, *Journal of the Fisheries Research Board of Canada,* **29**:975–983.

Cornett, J.J. and Rigler, F.H., 1979, Hypolimnetic oxygen deficits: their prediction and interpretation, *Science,* 205:580–581.

Dillon, P.J. and Rigler, F.H., 1974, The phosphorus–chlorophyll relationship in lakes, *Limnology and Oceanography,* 19:767–773.

Dillon, P.J. and Rigler, F.H., 1975, A simple method for predicting the capacity of a lake for development based on lake trophic status, *Journal of the Fisheries Research Board of Canada* 32:1519–1531.

Dillon, P.J., Nicholls, K.H., Locke, B.A., deGrosbois, E., and Yan, N.D., 1988, Phosphorus–phytoplankton relationships in nutrient-poor soft-water lakes in Canada, *International Association for Theoretical and Applied Limnology, Proceedings* 23:258–264.

Dillon, P.J., Clark, B.J., and Molot, L.A., 2003, Predicting optimal habitat boundaries for lake trout (*Salvelinus namaycush* Walbaum) in Canadian Shield lakes, *Canadian Journal of Fisheries and Aquatic Sciences,* in press.

Downing, J.A. and Plante, C., 1993, Production of fish populations in lakes, *Canadian Journal of Fisheries and Aquatic Sciences* 50:110–120.

Downing, J.A., Plante, C., and Lalonde, S., 1990, Fish production correlated with primary productivity, not the morphoedaphic index, *Canadian Journal of Fisheries and Aquatic Sciences* 47:1929–1936.

France, R. and Steedman, R., 1996, Energy provenance for juvenile lake trout in small Canadian Shield lakes as shown by stable isotopes, *Transactions of the American Fisheries Society* 125:512–518.

Hanson, J.M. and Leggett, W.C., 1982, Empirical prediction of fish biomass and yield, *Canadian Journal of Fisheries and Aquatic Sciences,* 39:257–263.

Henderson, H.F., Ryder, R.A., and Kudhongania, A.W., 1973, Assessing fishery potentials of lakes and reservoirs, *Journal of the Fisheries Research Board of Canada,* 30:2000–2009.

Hyatt, K.D. and Stockner, J.G., 1985, Responses of sockeye salmon (*Oncorhynchus nerka*) to fertilization of British Columbia coastal lakes, *Canadian Journal of Fisheries and Aquatic Sciences,* 42:320–331.

Jenkins, R.M., 1967, The influence of some environmental factors on standing crop and harvest of fishes in U.S. reservoirs, In: *Proceedings of the Reservoir Fisheries Resource Symposium, Southern Division,* American Fisheries Society, Bethesda, Maryland, pp. 298–321.

Jenkins, R.M., 1982, The morphoedaphic index and reservoir fish production, *Transactions of the American Fisheries Society* 111:133–140.

Johannessen, M., Lande, A., and Rognerud, S., 1984, Fertilization of 6 small mountain lakes in Telemark, southern Norway, *International Association of Theoretical and Applied Limnology, Proceedings* 22:673–678.

Johnston, N.T., Perrin, C.J., Slaney, P.A., and Ward, B.R., 1990, Increased juvenile salmonid growth by whole-river fertilization, *Canadian Journal of Fisheries and Aquatic Sciences* 47:862–872.

Jones, J.R. and Hoyer, M.V., 1982, Sportfish harvest predicted by summer chlorophyll-α concentration in midwestern lakes and reservoirs, *Transactions of the American Fisheries Society* 111:171–176.

Kalff, J., 1991, The utility of latitude and other environmental factors as predictors of nutrients, biomass and production in lakes worldwide: problems and alternatives, *International Association of Theoretical and Applied Limnology, Proceedings* 24:1235–1239.

Koussouris, T.S., Bertahas, I.T., and Diapoulis, A.C., 1992, Background trophic state of Greek lakes, *Fresenius Environmental Bulletin* 1:96–101.

LeBrasseur, R.J., McAllister, C.D., Barraclough, W.E., Kennedy, O.D., Manzer, J., Robinson, D., and Stephens, K., 1978, Enhancement of sockeye salmon (*Oncorhynchus nerka*) by lake fertilization in Great Central Lake: Summary Report, *Journal of the Fisheries Research Board of Canada* 35:1580–1596.

Matuszek, J.E., 1978, Empirical predictions of fish yields of large North American lakes, *Transactions of the American Fisheries Society* 107:385–394.

McConnell, W.J., Lewis, S., and Olson, J.E., 1977, Gross photosynthesis as an estimator of potential fish production, *Transactions of the American Fisheries Society* 105:417–423.

McQueen, D.J., Post, J.R., and Mills, E.L., 1986, Trophic relationships in freshwater pelagic ecosystems, *Canadian Journal of Fisheries and Aquatic Sciences* **43**:1571–1581.

Molot, L.A. and Dillon, P.J., 1991, Nitrogen/phosphorus ratios and the prediction of chlorophyll in phosphorus limited lakes in central Ontario, *Canadian Journal of Fisheries and Aquatic Sciences* **48**:140–145.

Molot, L.A., Dillon, P.J., Clark, B.J., and Neary, B.P., 1991, Predicting end-of-summer oxygen profiles in stratified lakes, *Canadian Journal of Fisheries and Aquatic Sciences* **49**:2363–2372.

Moyle, J.B., 1956, Relationships between the chemistry of Minnesota surface waters and wildlife management, *Journal of Wildlife Management* **20**:303–320.

Nicholls, K.H. and Dillon, P.J., 1978, An evaluation of phosphorus–chlorophyll relationships for lakes, *Internationale Revue der Gesamten Hydrobiologie* **63**:141–154.

Oglesby, R.T., 1977, Relationship of fish yield to lake phytoplankton standing crop, production, and morphoedaphic factors, *Journal of the Fisheries Research Board of Canada* **34**:2271–2279.

Oglesby, R.T., 1982, The MEI symposium — overview and observations, *Transactions of the American Fisheries Society* **111**:171–175.

Ostrofsky, M.L. and Rigler, F.H., 1987, Chlorophyll–phosphorus relationships for subarctic lakes in western Canada, *Canadian Journal of Fisheries and Aquatic Sciences* **44**:775–781.

Plante, C. and Downing, J.A., 1993, Relationship of Salmonine production to lake trophic status and temperature, *Canadian Journal of Fisheries and Aquatic Sciences* **50**:1324–1328.

Prairie, Y., 1994, *Développement de modèles prédictifs décrivant l'effet de l'eutrophisation sur l'habitat du touladi (Salvelinus namaycush)*. Report for the Quebec Ministère de l'Environnement et de la Faune, Quebec.

Prepas, E.E., 1983, Total dissolved solids as a predictor of lake biomass and productivity, *Canadian Journal of Fisheries and Aquatic Sciences* **40**:92–95.

Prepas, E.E. and Trew, D.O., 1983, Evaluation of the phosphorus–chlorophyll relationship for lakes off the Precambrian Shield in western Canada, *Canadian Journal of Fisheries and Aquatic Sciences* **40**:27–35.

Ranta, E. and Lindstrom, K., 1998, Fish yield versus variation in water quality in the lakes of Kuusamo, northern Finland, *Annales Zoologici Fennici* **35**:95–106.

Rawson, D.S., 1951, The total mineral content of lake waters, *Ecology* **32**:669–672.

Rawson, D.S., 1952, Mean depth and the fish production of large lakes, *Ecology* **33**:513–521.

Regier, H.A., Cordone, A.J., and Ryder, R.A., 1971, Total fish landings from fresh waters as a function of limnological variables, with special reference to lakes of East-Central Africa, *Food and Agriculture Organization of the United Nations FI:SF/GHA10 Fish Stock Assessment Working Paper 3*, FAO, Rome.

Ryan, P.A. and Marshall, T.R., 1994, A niche definition for lake trout (*Salvelinus namaycush*) and its use to identify populations at risk, *Canadian Journal of Fisheries and Aquatic Sciences* **51**:2513–2519.

Ryan, P.M. and Harvey, H.H., 1977, Growth of rock bass, *Ambloplites rupestris*, in relation to the morphoedaphic index as an indicator of an environmental stress, *Journal of the Fisheries Research Board of Canada* **34**:2079–2088.

Ryder, R.A., 1965, A method for estimating the potential fish production of north-temperate lakes, *Transactions of the American Fisheries Society* **94**:214–218.

Ryder, R.A., Kerr, S.R., Loftus, K.H., and Regier, H.A., 1974, The Morphoedaphic Index, a fish yield estimator — Review and evaluation, *Journal of the Fisheries Research Board of Canada* **31**:663–688.

Sakamoto, M., 1966, Primary production by phytoplankton community in some Japanese lakes and its dependence on lake depth. *Archiv fur Hydrobiologie* **62**:1–28.

Schindler, D.W., 1971, A hypothesis to explain differences and similarities among lakes in the Experimental Lakes Area, northwestern Ontario, *Journal of the Fisheries Research Board of Canada* **28**:295–301.

Schindler, D.W., 1977, Evolution of phosphorus limitation in lakes, *Science* **195**:260–262.

Schlesinger, D.A. and Regier, H.A., 1982, Climatic and morphoedaphic indices of fish yields from natural lakes, *Transactions of the American Fisheries Society* **111**:141–150.

Sellers, T.J., Parker, B.R., Schindler, D.W., and Tonn, W.M., 1998, Pelagic distribution of lake trout (*Salvelinus namaycush*) in small Canadian Shield lakes respect to temperature, dissolved oxygen, and light, *Canadian Journal of Fisheries and Aquatic Sciences* **55**:170–179.

Snucins, E.J. and Gunn, J.M., 1995, Coping with a warm environment: Behavioural thermoregulation by lake trout, *Transactions of the American Fisheries Society* **124**:118–123.

Stockner, J.G. and Shortreed, K.S., 1985, Whole lake fertilization experiments in coastal British Columbia lakes: empirical relationships between nutrient input and phytoplankton biomass and production, *Canadian Journal of Fisheries and Aquatic Sciences* **42**:649–658.

Toews, D.R. and Griffith, J.S., 1979, Empirical estimates of potential fish yield for the Lake Bangweulu system, Zambia, central Africa, *Transactions of the American Fisheries Society* **108**:241–252.

Vander Zanden, M.J., Casselman, J.M., and Rasmussen, J.B., 1999, Stable isotope evidence for the food web consequences of species invasions in lakes, *Nature* **401**:464–467.

Vighi, M. and Chiaudani, G., 1985, A simple method to estimate lake phosphorus concentrations resulting from natural background loadings, *Water Research* **19**:987–991.

Vollenweider, R.A., 1968, *The Scientific Basis of Lake and Stream Eutrophication, with Particular Reference to Phosphorus and Nitrogen as Eutrophication Factors,* Technical Report OECD, Paris, DAS/CSI/68, Vol. 27, pp. 1–182.

Welch, H.B. and Perkins, M.A., 1979, Oxygen deficit phosphorus loading relation in lakes, *Journal of the Water Pollution Control Federation,* **51**:2823–2828.

Wetzel, R.G., 1975, *Limnology,* W.B. Saunders, Philadelphia.

chapter eight

Dissolved organic carbon as a controlling variable in lake trout and other Boreal Shield lakes

David W. Schindler
Department of Biological Sciences, University of Alberta
John M. Gunn
Ontario Ministry of Natural Resources, Laurentian University

Contents

Introduction

Lake trout lakes on the Boreal Shield are usually relatively clear (Martin and Olver, 1976; Johnson et al., 1977; Marshall and Ryan, 1987). Seventy percent have Secchi depths greater than 4 m (Figure 8.1). Transparency in Shield lakes is usually largely determined by the concentration of dissolved organic matter (usually called DOC, or dissolved organic carbon, for it is usually quantified by measuring its carbon content) (Schindler, 1971; Figure 8.2), much of which originates from decomposing vegetation in wetlands and forest soils. Fulvic and tannic acids are important components of DOC, imparting a yellowish-brown color to the lakes and streams (Schindler, 1998). However, most lake trout lakes

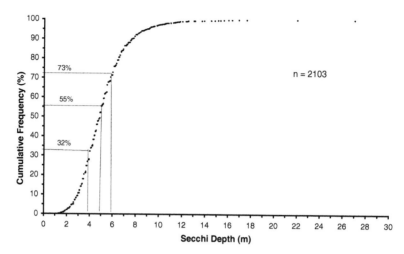

Figure 8.1 Secchi depth (m) in lake trout lakes of eastern North America. From data compiled for text edited by Gunn, Steedman, and Ryder (2003).

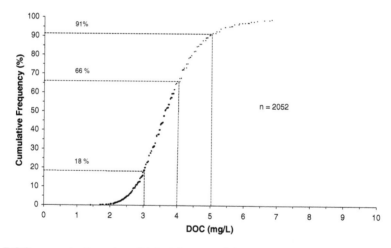

Figure 8.2 DOC concentrations (mg/L) in lake trout lakes of eastern North America. From data compiled for text edited by Gunn, Steedman, and Ryder (2003).

contain enough DOC to have important effects on optical, chemical, physical, and biological properties (Schindler et al., 1997). For example, over 80% of lake trout lakes have DOC concentrations >3 mg L^{-1} (Figure 8.2), a range where DOC attenuates enough light to reduce photosynthesis and block UV radiation. The most colored lake trout lakes tend to be small, with areas of a few hundred hectares or less. Without DOC, these Shield lakes would have Secchi depths exceeding 30 m (Figure 8.3), as are observed in the lakes on the orthoquartzite ridges of the LaCloche Mountains, along the northern shore of Lake Huron (Gunn et al., 2001).

Not all DOC is highly colored. Some is also produced by algae and other plants within the lake. Such DOC is termed *autochthonous*, meaning produced within. The highly colored DOC that originates from the watershed is termed *allochthonous*, meaning produced outside the lake. In general, the larger a lake's catchment, the higher the proportion of the basin that is covered by wetlands, and the wetter the climate, the darker the lake's color

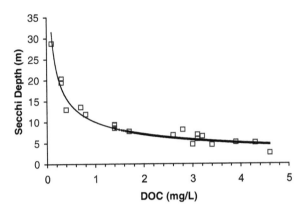

Figure 8.3 The relationship between DOC and Secchi depth in lakes of the LaCloche Mountain Lakes. From Gunn et al. (2001).

will be (Engstrom, 1987; Rasmussen et al., 1989; Urban et al., 1989; Meili, 1992; Curtis and Schindler, 1997). However, colored allochthonous DOC can be bleached by exposure to UV radiation near the surface of lakes. The degree of bleaching depends on the pH (Donahue et al., 1998) and on how long the DOC is present in a lake, which in turn is related to how rapidly a lake is flushed with water. Thus, there is a tendency for lakes with faster rates of water flushing to be darker in color.

DOC can influence the food and habitat of lake trout and other Shield lake biota in a number of very subtle ways, for it is a pivotal variable that affects many important physical, chemical, and biological properties of lakes. Below are some of its more important effects.

The effect of DOC on the physical properties of lakes

The dark color of allochthonous DOC makes it an excellent attenuator (absorber) of light. As water containing DOC absorbs solar radiation, light energy is transformed to heat. As a result, the epilimnion of a high-DOC lake tends to be warmer than that of a clear lake of the same size, subjected to the same climatic regime (Salonen et al., 1984; Perez-Fuen-tetaja et al., 1999).

Reduced penetration of light in highly colored lakes also causes the thermocline depth to be shallower than in clear lakes of equivalent size. This is particularly true of small lakes (Snucins and Gunn, 2000; Xenopoulos and Schindler, 2001), and the large majority of lake trout lakes are indeed quite small, with about 75% of them less than 500 ha (Figure 8.4). For lakes greater than about 500 ha in area, the effect of wind on the lake's surface becomes of overriding importance in determining thermocline depth (Fee et al., 1996; Figure 8.5).

Because of its effect on water temperature and thermal stratification, DOC concentration can control the summer habitat available to lake trout. Small, high-DOC lakes have higher proportions of their volumes below the thermocline, where water is cold enough to attract lake trout in summer. It has recently been recognized that DOC data can be used to predict lake trout habitat directly (Dillon et al., 2003).

It was discovered recently that DOC is also an excellent attenuator of UV radiation, serving as a sort of "sunscreen" to protect aquatic organisms (Scully and Lean, 1994; Figure 8.6). As we discuss below, climate change, increasing UV from stratospheric ozone depletion, and acid rain can potentially affect summer habitat via their effects on DOC.

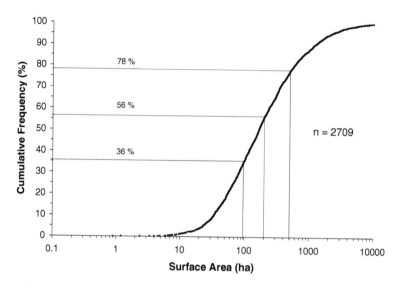

Figure 8.4 Size frequency (ha) of lake trout lakes in eastern North America. From data compiled for text edited by Gunn, Steedman, and Ryder (2003).

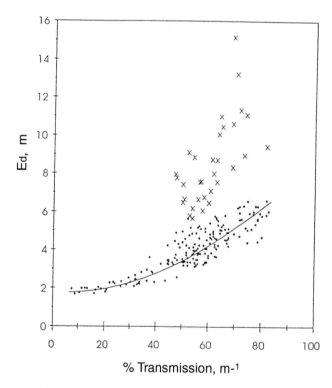

Figure 8.5 The relationship between light transmission and mixing depth (e) in large and small lakes of northwestern Ontario. Lakes <500 ha have their thermocline depth largely determined by color; larger lakes have their thermoclines determined by fetch because the fetch-determined thermocline is too deep for light penetration to have a significant effect. From Fee et al. (1996).

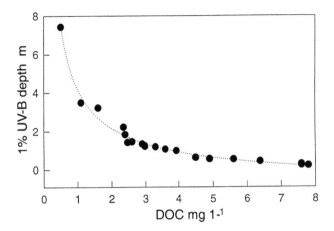

Figure 8.6 The effect of DOC on the attenuation of UV light. From Schindler et al. (1996).

The effect of DOC on the chemical properties of lakes

Many papers have been written on the chemical properties of DOC. Here, we discuss only those of relevance to lake trout and other biota of Boreal lakes.

DOC includes many substances, of varying chemical structure. Some molecules are small, in a size range ideal for uptake by organisms. However, colored allochthonous DOC is composed mostly of humic and fulvic acids and other large, charged molecules, which tend to combine with other charged chemicals, including trace metals, nutrients, and organic contaminants, to form stable colloids. DOC prevents these chemicals from participating in biological and chemical reactions as if they were in truly dissolved form but also retards further coagulation, inhibiting their removal to the sediments (Weilenmann et al., 1989).

DOC will combine with toxic trace metals, rendering most of them less toxic than if they were in dissolved ionic form. Mercury, lead, aluminum, chromium, and copper are among metals with a high affinity for DOC (Kerndorf and Schnitzer, 1980). There is a strong correlation between DOC concentration in water and mercury concentration in fish, perhaps because wetlands tend to be the most important source of both in the catchments of lakes (Urban et al., 1989; McMurty et al., 1989; St. Louis et al., 1996). For example, lake catchments containing wetlands in the Experimental Lakes Area yield 26- to 79-fold more methyl mercury than catchments without wetlands (St. Louis et al., 1994). As will be discussed later, DOC can greatly affect the toxicity of aluminum in lakes subjected to acid precipitation (Driscoll et al., 1995).

In some cases, DOC has complex effects on trace metal concentrations, biogeochemical cycles, and toxicity. For example, although wetlands supply both DOC and mercury, DOC also inhibits the methylation of mercury in lakes (Miskimmin et al., 1992).

If wetlands are flooded, as occurs when reservoirs are constructed on the Shield, DOC and methyl mercury (MeHg) concentrations increase dramatically. The latter is the result of increased activity by methanogenic bacteria (Kelly et al., 1997). As a result, concentrations of mercury in fish increase dramatically, frequently to levels that pose a hazard to humans or other species that rely on fish for food (Hecky et al., 1992; Rosenberg et al., 1995; Bodaly et al., 1998; Bodaly et al., Chapter 9, this volume).

DOC may be important in controlling the transformation of MeHg to elemental mercury in surface waters. Exposure of MeHg to UV and short-wavelength visible radiation accelerates this reaction, so that declining DOC may accelerate the process. The conversion of MeHg to elemental mercury is an important natural pathway for mercury to reach the atmosphere (Sellers et al., 1996; Amyot et al., 1997).

In well-oxygenated surface waters almost all of the iron will be bound to DOC complexes, with little remaining in dissolved ionic forms. These complexes render much of the iron unavailable to phytoplankton (Sakamoto, 1971) and also slow the removal of iron from solution (Curtis, 1993). In turn, iron–DOC complexes bind both trace metals and nutrients. Phosphorus bound to iron–DOC complexes is less available for immediate biological uptake than ionic phosphate (Jackson and Schindler, 1975).

The attenuation of solar radiation by DOC also affects the formation of transient secondary thermoclines within the epilimnion, which can last for hours or days, depending on wind conditions (Xenopoulos and Schindler, 2001). The frequency of formation of secondary thermoclines ranges from 30% to nearly 100% of summer days, with a higher incidence in lakes with higher DOC and smaller surface area (Xenopoulos and Schindler, 2001). Such thermoclines can isolate nonmotile species in near-surface layers that are less than a meter deep, causing them to be exposed to high levels of UV radiation. High exposures in near-surface layers can affect the growth and physiological state of algae and bacteria (Xenopoulos et al., 2000).

The effect of DOC on biological properties of lakes

The depth to which photosynthesis can occur is also a function of DOC, which strongly attenuates all wavelengths of solar radiation. Dark lakes are usually less productive than clear lakes with similar nutrient loadings. Carpenter et al. (1998) found that increasing DOC from 5 to 17 mg C per liter was equivalent to reducing inputs of phosphorus 10-fold, from 5 to 0.5 mg per m^2 per day. They believe that the main effect on primary production of DOC is by shading. However, complexation of phosphorus and iron by DOC may also be involved.

DOC contains energy potentially available to organisms, which may partly compensate for DOC's negative effects on photosynthesis. Bacteria, heterotrophic algae and protozoa are among the organisms that rely on DOC as an energy source. In turn, these supplement algal production, which supplies food for small zooplankton, which feed the large zooplankton and planktivorous fishes that are typically important in the diets of lake trout and other piscivorous fish. Although this has not been studied in detail in lake trout lakes, it is doubtful whether the positive effects of DOC on the microbial food chain to zooplankton would be sufficient to offset the negative effect of DOC on phytoplankton production, as discussed earlier.

The DOC that enters lakes can be quite old. In particular, DOC entering from groundwater can be quite recalcitrant, with some of it decades old (Schiff et al., 1997). However, once DOC is discharged into lakes, a variety of physical, chemical, and biological properties combine to transform molecules that have resisted decomposition in the terrestrial environment. UV radiation can cleave large, refractory DOC molecules into small molecules such as fatty acids and other substances that are directly usable by microorganisms (Wetzel et al., 1995). Allochthonous DOC tends to be rich in carbon but deficient in phosphorus, and if lake water contains phosphorus from other sources, DOC is more efficiently degraded (Schindler et al., 1992). As DOC is mineralized, CO_2 concentrations reach supersaturation, so that it is released to the atmosphere (Dillon and Molot, 1997). It is difficult to unravel the combinations and sequences of physical, chemical, and biological activity that decomposes DOC, and it is probably unrealistic to view any of the processes in isolation.

DOC can also play a role in buffering against acidification because it contains organic anions (A⁻). The amount of A⁻ is a predictable function of DOC and pH for most surface waters (Oliver et al., 1983), for carboxyl groups (COOH) are the main functional groups, varying little between regions (Jones et al., 1986). In colored lakes, DOC can contribute substantially to the ability to neutralize incoming strong acids (Lazerte and Dillon, 1984).

In the process of degradation by UV radiation, DOC can release several toxic chemicals, including hydrogen peroxide, carbon monoxide, hydroxyl radicals, and superoxides (Keiber et al., 1990; Mopper and Zhou, 1990; Cooper et al., 1994; Shao et al., 1994). Although concentrations are small, hydrogen peroxide may be present at concentrations that are toxic to microorganisms (Xenopoulos, 1997; Xenopoulos and Bird, 1997).

Attenuation of light by DOC can also potentially affect the makeup of food chains at higher levels by affecting the nutritional value of food. It is known that the zooplankton tend to be dominated by *Daphnia*, in lakes where seston (algae and detritus) has low ratios of C:P, whereas copepods tend to predominate at higher C:P ratios (Sterner et al., 1998). Higher C:P ratios in seston tend to be produced in lakes with higher penetration of light, i.e., lakes with low DOC (Hassett et al., 1997).

DOC can also directly affect trophic interactions via its effect on transparency to visible light. Sight-dependent predators are probably favored in clear lakes (O'Brien, 1987; Clark and Levy, 1988). Increased UV can cause subtle and poorly understood changes in relationships between invertebrate and fish species (Williamson, 1995). High light levels will also increase the C:P ratio of phytoplankton, lowering its nutritional value for *Daphnia* and other large grazing cladocerans (Sterner et al., 1998; Elser et al., 1998). Altered light regimes are also likely to affect the vertical migration of zooplankton (Dodson, 1990). Below, we give examples of how human activities are directly and indirectly affecting DOC, potentially changing important community interactions of several types.

Human activities that affect DOC

A number of different activities affect DOC, changing physical, chemical, and biological processes that are important to lake trout. Some of these are local, but others are regional or global in their influence, so that there is little that we can do directly to control them (Schindler, 1998).

Local influences

Although data are scarce, several types of local influences, including logging, road building, and other disruptions to the catchments of lakes can affect the inputs of DOC. In general, any activity that increases the contact between water and wetland soils, such as clearcutting, wildfire, or drainage interruptions, will increase the concentrations of DOC in runoff (Carignan et al., 2000). Similarly, raising the level of lakes even slightly can cause increased DOC if wetlands are flooded, as outlined later for reservoir construction.

Acid rain

Among regional insults, acid rain is a potent modifier of DOC. In general, lakes and streams become clearer and contain less DOC as the pH decreases below pH 5 (Schindler et al., 1996b; Yan et al., 1996). Flocculation and precipitation with aluminum (Weilenmann et al., 1989; Driscoll et al., 1995) and increased photolytic degradation (Molot and Dillon, 1996) appear to be mechanisms that remove DOC in acidifying lakes. Lakes can lose 90% or more of their natural DOC concentrations, with subsequent deepening of thermoclines and euphotic zones (Schindler et al., 1996a). Increased penetration of solar energy into the hypolimnion of a small lake could also warm it (Yan and Miller, 1984), potentially making it become too warm for lake trout (Gunn, 2001).

Loss of DOC poses several threats. Most notably, its disappearance increases the exposure of littoral and shallow water organisms to higher UV radiation. Approximately 20% of the lakes in Ontario that have been surveyed by the Ontario Ministry of Environment (Neary et al., 1990) have natural DOC concentrations <3 mg/L, where even slight decreases in DOC will cause rapid increases in UV penetration because of the negative exponential nature of the relationship (Scully and Lean, 1994; Schindler et al., 1996b; Figure 8.6). A similar proportion of lake trout lakes have low DOC and are vulnerable to increasing UV if DOC decreases significantly (Figure 8.3).

Thermocline deepening and hypolimnetic heating occur in acid lakes as the result of declining DOC (Schindler et al., 1996a; Snucins and Gunn, 2000). Both will reduce summer habitat for lake trout and other cold stenotherms. As the result of clearer waters, there may be several other subtle changes to food chain relationships, as discussed above.

Climate warming

Climate warming and/or drought will also cause increased transparency of lakes because less allochthonous DOC is delivered to the lakes from their catchments (Schindler et al., 1996a, 1997; Gunn et al., 2001), and there is increased in-lake removal and bleaching as the result of longer residence times (Dillon and Molot, 1997; Schindler et al., 1997). The effects are similar to those mentioned above for acid lakes. At least in the early stages of climate change, effects are smaller than those caused by acidification (Schindler et al., 1996a,b). In general, climate and hydrology affect DOC in two ways. First, inputs of colored allochthonous DOC decline in direct proportion to streamflow decreases. The latter can result from less precipitation, increased evaporation at warmer temperatures, or a combination of the two (Schindler et al., 1996b). Second, as for acidification, DOC bleaching and removal in lakes increase as water and, hence, DOC residence times increase (Dillon and Molot, 1997; Schindler et al., 1997).

Exacerbating the effects of acid rain and climate warming on UV is the depletion of stratospheric ozone in the northern hemisphere. The three act in concert to affect the UV exposure of aquatic organisms, leading Gorham (1996) to declare that lakes are under a "three-pronged attack" (Figure 8.7).

In summary, DOC can have many important effects on lake trout habitat by altering physical, chemical, and biological properties. Maintaining the long-term integrity of lake trout habitat will require the management of activities in the catchments of lakes that affect DOC inputs to lakes. Climate change and acid rain also affect fish habitats by changing DOC concentrations in addition to their well-known direct effects.

Recommendations

At the local level, it is important to maintain DOC concentrations within the natural range expected for a given lake. To do this, greater care is needed when planning clearcuts, roads, dams, and other human activities.

Other influences on DOC are beyond local control. Strong national and international policies are needed to control emissions of acidifying substances, greenhouse gases, and stratospheric ozone, which alone and in combination can degrade the habitat required by lake trout and their prey. Such changes are particularly likely in small lakes, where even small declines in habitat can threaten fragile populations.

Additional research and monitoring are needed in several areas because the widespread importance of DOC was not recognized until recently. First, until recently, DOC

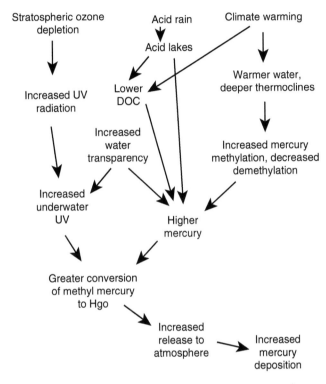

Hypothesized Interactions of Stratospheric Ozone
Depletion, Acid Rain, Climate Warning
and Mercury Biogeochemistry

Figure 8.7 A diagram showing how the "three-pronged attack" by stratospheric ozone depletion, climate warming, and acid rain on the aquatic UV environment is mediated by DOC. From Schindler (1999).

was not easy to analyze, particularly when <3 mg/L. It was also not widely recognized as important. This has changed, and long-term records for DOC will be important in assessing changes to habitats, UV exposures, and euphotic zones. Secondly, there are few data to show the impacts of landscape or drainage modification on DOC concentrations in lakes. Such changes are now easy to monitor, and DOC is clearly an important variable to measure in assessing the impacts of local activity.

Third, few studies have traced the effect of energy transfer from DOC to fisheries via heterotrophic bacteria and protozoans. Modern stable isotope methods make this possible, and such studies are critical to linking how changes in DOC concentration might affect the energy flow to fisheries in lake trout lakes. Finally, now that DOC has been shown to moderate the effects of climate warming, acid precipitation, stratospheric ozone depletion, and mercury cycling, its importance in linking these insults to aquatic effects needs to be explicitly studied.

Acknowledgments

We thank Dick Ryder, Norm Yan, and Rob Steedman for providing review comments. Data for the extensive survey of 2700 lake trout lakes were kindly provided by government agencies in Ontario, Quebec, Minnesota, New York, Michigan, and Wisconsin. Michael Malette assisted with data management and preparation of graphics.

References

Amyot, M., Mierle, G., Lean, D.R.S., and McQueen, D.J., 1997, Effects of solar radiation on the formation of dissolved gaseous mercury in temperate lakes, *Geochimica et Cosmochimica Acta* **61**: 975–987.

Bodaly, R.A. (Drew), St. Louis, V.L., Paterson, M.J., Fudge, R.J.P., Hall, B.D., Rosenberg, D.M., and Rudd, J.W.M., 1998, Bioaccumulation of mercury in the aquatic food chain in newly flooded areas, In *Mercury and Its Effects on Environment and Biology,* edited by H. Segel and A. Sigel, Marcel Dekker, New York, pp. 259–287.

Carignan, R., D'Arcy, P., and Lamontagne, S., 2000, Comparative impacts of fire and forest harvesting on water quality in boreal shield lakes, *Canadian Journal of Fisheries and Aquatic Sciences,* **57**(Suppl. 2): 105–117.

Carpenter, S.R., Cole, J.J., Kitchell, J.F., and Pace, M.L., 1998, Impact of dissolved organic carbon, phosphorus, and grazing on phytoplankton biomass and production in experimental lakes, *Limnology and Oceanography* **43**: 73–80.

Clark, C.W. and Levy, D.A., 1988, Diel vertical migrations by juvenile sockeye salmon and the antipredation window, *American Naturalist* **131**: 271–290.

Cooper, W.J., Chihwen, S., Lean, D.R.S., Gordon, A.S., and Scully, F.E., Jr., 1994, Factors affecting the distribution of H_2O_2 in surface waters, In *Environmental Chemistry of Lakes and Rivers,* edited by L.A. Baker, American Chemical Society, Washington, D.C., pp. 391–422.

Curtis, P.J., 1993, Effect of dissolved organic carbon on ^{59}Fe scavenging, *Limnology and Oceanography* **38**: 1554–1561.

Curtis, P.J. and Schindler, D.W., 1997, Hydrologic control of dissolved organic matter in low-order Precambrian Shield lakes, Northwestern Ontario, *Biogeochemistry* **36**: 125–138.

Dillon, P.J., Clark, B.J., and Molot, L.A., 2003, Predicting optimal habitat boundaries for lake trout (*Salvelinus namaycush* Walbaum) in Canadian Shield lakes, *Canadian Journal of Fisheries and Aquatic Sciences* (in press).

Dillon, P.J. and Molot, L.A., 1997, Dissolved organic and inorganic carbon mass balances in central Ontario lakes, *Biogeochemistry* **36**: 29–42.

Dodson, S.I., 1990, Predicting diel vertical migration of zooplankton, *Limnology and Oceanography* **35**: 1195–1200.

Donahue, W.F., Schindler, D.W., Page, S.J., and Stainton, M.P., 1998, Acid-induced changes in DOC quality in an experimental whole-lake manipulation, *Environmental Science and Technolgy* **32**: 2954–2960.

Driscoll, C.T., Blette, V., Yan, C., Schofield, C.L., Munson, R., and Holsapple, J., 1995, The role of dissolved organic carbon in the chemistry and bioavailability of mercury in remote Adirondack lakes, *Water, Air, and Soil Pollution* **80**: 499–508.

Elser, J.J., Chrzanowski, T.H., Sterner, R.W., and Mills, K.H., 1998, Stoichiometric constraints on food-web dynamics: a whole-lake experiment on the Canadian Shield, *Ecosystems* **1**: 120–136.

Engstrom, D.R., 1987, Influence of vegetation and hydrology on the humus budgets of Labrador lakes, *Canadian Journal of Fisheries and Aquatic Sciences* **4**: 1306–1314.

Fee, E.J., Hecky, R.E., Kasian, S.E.M., and Cruikshank, D.R., 1996, Effects of lake size, water clarity, and climatic variability on mixing depths in Canadian Shield lakes, *Limnology and Oceanography* **41**: 912–920.

Gorham, E., 1996, Lakes under a three-pronged attack, *Nature* **381**: 109–110.

Gunn, J.M., 2001, Impact of the 1998 El Nino event on a lake charr, *Salvelinus namaycush*, population recovering from acidification, *Environmental Biology of Fishes* **41**: 1–9.

Gunn, J.M., Snucins, E., Yan, N.D., and Arts, M.T., 2001, Use of water clarity to monitor the effects of climate change and other stressors on oligotrophic lakes, *Environmental Monitoring and Assessment* **67**: 69–88.

Gunn, J.M., Steedman, R.J., and Ryder, R.A. (Eds.), 2004, *Boreal Shield Watersheds: Lake Trout Ecosystems in a Changing Environment*. Lewis Publishers, Boca Raton.

Hassett, R.P., Cardinale, B., Stabler, L.B., and Elser, J.J., 1997, Ecological stoichiometry of N and P in pelagic ecosystems: comparisons of lakes and oceans with emphasis on zooplankton–phytoplankton interactions, *Limnology and Oceanography* **41**: 648–662.

Hecky, R.E., Ramsey, D.J., Bodaly, R.A., and Strange, N.E., 1992, Increased methylmercury contamination in fish in newly formed freshwater reservoirs, In *Advances in Mercury Toxicology*, edited by T. Clarkson and T. Suzuki, Plenum Press, New York, pp. 33–52.

Jackson, T.A. and Schindler, D.W., 1975, The biogeochemistry of phosphorus in an experimental lake environment: Evidence of humic-metal-phosphate complexes, *Internationale Vereinigung fur Theoretische und Angewandte Limnologie* **19**: 211–221.

Johnson, M.G., Leach, J.H., Minns, C.K., and Olver, C.H., 1977, Limnological characteristics of Ontario lakes in relation to associations of walleye (*Stizostedion vitreum vitreum*), northern pike (*Esox lucius*), lake trout (*Salvelinus namaycush*), and smallmouth bass (*Micropterus dolieui*), *Journal of the Fisheries Research Board of Canada* **34**: 1592–1601.

Jones, M.L., Marmorek, D.R., Reuber, B.S., McNamee, P.J., and Rattie, L.P., 1986, "Brown Waters": *Relative Importance of External and Internal Sources of Acidification on Catchment Biota*, ESSA Environmental and Social Systems Analysts, Toronto.

Keiber, R.J., Zhou, X., and Mopper, K.. 1990, Formation of carbonyl compounds from UV-induced photodegradation of humic substances in natural waters: fate of riverine carbon in the sea, *Limnology and Oceanography* **35**: 1503–1515.

Kelly, C.A., Rudd, J.W.M., Bodaly, R.A., Roulet, N.T., St. Louis, V.L., Heyes, A., Moore, T.R., Schiff, S., Aravena, R., Scott, K.J., Dyck, B., Harris, R., Warner, B., and Edwards, G., 1997, Increases in fluxes of greenhouse gases and methyl mercury following flooding of an experimental reservoir, *Environmental Science and Technology* **31**: 1334–1344.

Kerndorf, H. and Schnitzer, M., 1980, Sorption of metals on humic acid, *Geochimica et Cosmochimica Acta* **44**: 1701–1708.

Lazerte, B.D. and Dillon, P.J., 1984, Relative importance of anthropogenic versus natural sources of acidity in lakes and streams of central Ontario, *Canadian Journal of Fisheries and Aquatic Sciences* **41**: 1664–1677.

Marshall, T.R. and Ryan, P.A., 1987, Abundance patterns and community attributes of fishes relative to environmental gradients, *Canadian Journal of Fisheries and Aquatic Sciences* **44** (Suppl. 2): 198–215.

Martin, N.V. and Oliver, C.H., 1976, *The Distribution and Characteristics of Ontario Lake Trout Lakes*, Ontario Ministry of Natural Resources, Fisheries and Wildlife Research Branch, Research Report No. 97.

McMurty, M.J., Wales, D.L., Scheider, W.A., Beggs, G.L., and Dimond, P.E., 1989, Relationship of mercury concentrations in lake trout (*Salvelinus namaycush*) and smallmouth bass (*Micropterus dolomieu*) to the physical and chemical characteristics of Ontario lakes, *Canadian Journal of Fisheries and Aquatic Sciences* **46**: 426–434.

Meili, M., 1992, Sources, concentrations and characteristics of organic matter in softwater lakes and streams of the Swedish forest region, *Hydrobiologia* **229**: 23–41.

Miskimmin, B.M., Rudd, J.W.M., and Kelly, C.A., 1992, Influence of dissolved organic carbon, pH, and microbial respiration rates on mercury methylation and demethylation in lake water, *Canadian Journal of Fisheries and Aquatic Sciences* **49**: 17–22.

Molot, L. and Dillon, P.J., 1996, Storage of terrestrial carbon in boreal lake sediments and evasion to the atmosphere, *Global Biogeochemical Cycles* **10**: 483–492.

Mopper, K. and Zhou, X., 1990, Hydroxyl radical photoproduction in the sea and its potential impact on marine processes, *Science* **298**: 661–664.

Neary, B.P., Dillon, P.J., Munro, J.R., and Clark, B.J., 1990, *The Acidification of Ontario Lakes: An Assessment of Their Sensitivity and Current Status with Respect to Biological Damage*, Ontario Ministry of Environment, Dorset Research Centre, Dorset, Ontario.

O'Brien, W.J., 1987, Planktivory by freshwater fish: Thrust and parry in the pelagia, In *Predation: Direct and Indirect Impacts on Aquatic Communities*, edited by W.C. Kerfoot and A. Sih, University Press of New England, Hanover, pp. 3–16.

Oliver, B.G., Thurman, E.M., and Malcolm, R.L., 1983, The contribution of humic substances to the acidity of coloured natural waters, *Geochimica et Cosmochimica Acta* **47**: 2031–2035.

Pérez-Fuentetaja, A., Dillon, P.J., Yan, N.D., and McQueen, D.J., 1999, Significance of dissolved organic carbon in the prediction of thermocline depth in small Canadian shield lakes, *Aquatic Ecology* **33**: 127–133.

Rasmussen, J.B., Godbout, L., and Schallenberg, M., 1989, The humic content of lake water and its relationship to watershed and lake morphometry, *Limnology and Oceanography* **34**: 1336–1343.

Rosenberg, D.M., Bodaly, R.A., and Usher, P.J., 1995, Environmental and social impacts of large scale hydro-electric development: who is listening? *Global Environmental Change* **5**: 127–148.

St. Louis, V.L., Rudd, J.W.M., Kelly, C.A., Beaty, K.G., Bloom, N.S., and Flett, R.J., 1994, Importance of wetlands as sources of methyl mercury to boreal forest ecosystems, *Canadian Journal of Fisheries and Aquatic Sciences* **51**: 1065–1076.

St. Louis, V.L., Rudd, J.W.M., Kelly, C.A., Beaty, K.G., Flett, R.J., and Roulet, N.T., 1996, Production and loss of methylmercury and loss of total mercury from Boreal Forest catchments containing different types of wetlands, *Environmental Science and Technology* **30**: 2719–2729.

Sakamoto, M., 1971, Chemical factors involved in the control of phytoplankton production in the Experimental Lakes Area, northwestern Ontario, *Journal of the Fisheries Research Board of Canada* **28**: 203–213.

Salonen, K., Arvola, L., and Rask, M., 1984, Autumnal and vernal circulation of small forest lakes in southern Finland, *Internationale Vereinigung fur Theoretische und Angewandte Limnologie* **22**: 103–107.

Schiff, S.L., Aravena, R., Trumbore, S.E., Hinton, M.J., Elgood, R., and Dillon, P.J., 1997, Export of DOC from forested catchments on the Precambrian Shield of Central Ontario: clues from ^{13}C and ^{14}C. *Biogeochemistry* **36**: 43–65.

Schindler, D.W., 1971, Light, temperature and oxygen regimes of selected lakes in the Experimental Lakes Area (ELA), northwestern Ontario, *Journal of the Fisheries Research Board of Canada* **28**: 157–170.

Schindler, D.W., 1998, A dim future for boreal waters and landscapes: cumulative effects of climatic warming, stratospheric ozone depletion, acid precipitation and other human activities, *Bio-Science* **48**: 157–164.

Schindler, D.W., 1999, From acid rain to toxic snow, *Ambio* **28**: 350–355.

Schindler, D.W., Bayley, S.E., Curtis, P.J., Parker, B.R., Stainton, M.P., and Kelly, C.A., 1992, Natural and man-caused factors affecting the abundance and cycling of dissolved organic substances in Precambrian Shield lakes, *Hydrobiologia* **229**: 1–21.

Schindler, D.W., Bayley, S.E., Parker, B.R., Beaty, K.G., Cruikshank, D.R., Fee, E.J., Schindler, E.U., and Stainton, M.P., 1996a, The effects of climatic warming on the properties of boreal lakes and streams at the Experimental Lakes Area, Northwestern Ontario, *Limnology and Oceanography* **41**(5): 1004–1017.

Schindler, D.W., Curtis, P.J., and Parker, B., 1996b, Synergistic effects of climatic warming, acidification and stratospheric ozone depletion on boreal lakes, *Nature* **379**: 705–708.

Schindler, D.W., Curtis, P.J., Bayley, S.E., and Parker, B.R., 1997, DOC-mediated effects of climate change and acidification on boreal lakes, *Biogeochemistry* **36**: 9–28.

Scully, N.M. and Lean, D.R.S., 1994, The attenuation of ultraviolet radiation in temperate lakes, *Archiv für Hydrobiologie Beiheft. Ergebnisse der. Limnologie* **43**: 135–144.

Sellers P., Kelly, C.A., Rudd, J.W.M., and MacHutchon, A.R., 1996, Photodegradation of methylmercury in lakes, *Nature* **380**: 694–697.

Shao, C., Cooper, W.J., and Lean, D.R.S., 1994, Singlet oxygen formation in lake waters from mid-latitudes, In *Aquatic and Surface Photochemistry*, edited by G. R. Helz, R.G. Zepp, and D.G. Crosby, Lewis Publishers, Ann Arbor, pp. 215–221.

Snucins, E. and Gunn, J., 2000, Interannual variations in the thermal structure of clear and colored lakes, *Limnology and Oceanography* **45**(7): 1639–1646.

Sterner, R.W., Elser, J.J., Fee, E.J., Guildford, S.J., and Chrzanowski, T.H., 1998, The light:nutrient ratio in lakes: the balance of energy and materials affecting ecosystem structure and function, *American Naturalist* **150**: 663–684.

Urban, N.R., Bayley, S.E., and Eisenreich, S.J., 1989, Export of dissolved organic carbon and acidity from peatlands, *Water Resources Research* **25**: 1619–1628.

Weilenmann, U., O'Melia, C.R., and Stumm, W., 1989, Particle transport in lakes: Models and measurements, *Limnology and Oceanography* **34**: 1–18.

Wetzel, R.G., Hatcher, P.G., and Bianchi, T.S., 1995, Natural photolysis by ultraviolet irradiance of recalcitrant dissolved organic matter to simple substrates for rapid bacterial metabolism, *Limnology and Oceanography* **40**: 1369–1380.

Williamson, C., 1995, What role does UV-B radiation play in freshwater ecosystems? *Limnology and Oceanography* **40**: 386–392.

Xenopoulos, M.A., 1997, *Influence directe et indirecte du rayonnement ultraviolet-B sur la dynamique du phytoplancton,* M.Sc. Thesis, Université du Québec à Montréal, Montreal.

Xenopoulos, M.A. and Bird, D.F., 1997, Effect of acute exposure to hydrogen peroxide on the production of phytoplankton and bacterioplankton in a mesohumic lake, *Photochemistry and Photobiology* **66**: 471–478.

Xenopoulos, M.A., Prairie, Y.T., and Bird, D.F., 2000, Influence of ultraviolet-B radiation, stratospheric ozone variability, and thermal stratification on the phytoplankton biomass dynamics in a mesohumic lake, *Canadian Journal of Fisheries and Aquatic Sciences* **57**: 600–609.

Xenopoulos, M.A. and Schindler, D.W., 2001, The environmental control of near-surface thermoclines in boreal lakes, *Ecosystems* **4**: 699–707.

Yan, N.D., Keller, W., Scully, N.M., Lean, D.R.S., and Dillon, P.J., 1996, Increased UV-B penetration in a lake owing to drought-induced acidification, *Nature* **381**: 141–143.

Yan, N.D. and Miller, G.E., 1984, Effects of deposition of acids and metals on chemistry and biology of lakes near Sudbury, Ontario, In *Environmental Impacts of Smelters,* edited by J. Nriagu, John Wiley & Sons, New York, pp. 243–282.

chapter nine

Mercury contamination
of lake trout ecosystems

R.A. (Drew) Bodaly
Department of Fisheries and Oceans, Freshwater Institute
Karen A. Kidd
Department of Fisheries and Oceans, Freshwater Institute

Contents

Introduction

Mercury is a widespread contaminant in freshwater fish and is currently causing great concern because of its potential impact on the health of humans and wildlife. Most mercury (Hg) in fish flesh is present as methyl mercury (MeHg) (Bloom, 1992). This organic form of mercury is a powerful neurotoxin and in large doses causes motor, sensory, and developmental problems in humans and other vertebrate animals (Clarkson, 1992). Because of the concerns over human health impacts, Canadian provincial and federal agencies monitor Hg concentrations in fish from lakes with important fisheries and in commercial shipments. If total Hg (both methyl mercury and inorganic Hg) concentrations exceed the Canadian limit for commercial sale (0.5 µg g^{-1}), consumption advisories are issued and commercial sales are restricted. Though high concentrations of persistent pesticides have occasionally been the cause, mercury is by far the most common reason for fish consumption advisories in North American freshwaters (e.g., Quebec Ministère de l'Environnement et de la Faune, 1995; United States Environmental Protection Agency, 1998; Ontario Ministry of the Environment, 2003). For example, in Ontario 95% of fish consumption advisories in lakes were related to mercury, and consumption advisories for this contaminant applied to 1206 of the 1595 lakes tested (Ontario Ministry of the Environment, 2003). In the United States in 1997 mercury accounted for 78% of the fish consumption advisories in freshwaters (United States Environmental Protection Agency, 1998). Present-day exposure of humans to MeHg results almost wholly from the consumption of fish (Clarkson, 1992).

Lake trout (*Salvelinus namaycush*) frequently have high concentrations of mercury because of their position at the top of food chains and because the Boreal lakes that they inhabit often have conditions favorable for mercury bioaccumulation. As a result, most lake trout populations in Boreal lakes have fish consumption advisories. In this chapter we review current knowledge concerning Hg in lake trout populations in the southern Shield lakes of Ontario, Quebec, Minnesota, and New York.

Factors affecting Hg in freshwater fish are outlined, with discussion of the reasons why lake trout are frequently highly contaminated with Hg. Emphasis is on lakes that do not receive direct anthropogenic discharges of Hg but rather receive their Hg from atmospheric sources and local weathering of the earth's crust. Examples of Hg in lake trout populations are given, especially to demonstrate how Hg varies with size and trophic position of the fish and with the food-web structure of the lake. Approaches to managing mercury contamination in freshwater systems are outlined, including sampling needed to determine existing levels and advisory systems to advise the public of recommended consumption limits. Finally, some speculation is made about future trends of Hg in lake trout populations.

Mercury concentrations in lake trout populations in small Boreal lakes

Mercury concentrations in lake trout in Boreal lakes are frequently high, and most populations of lake trout have at least some fish with Hg concentrations greater than the Canadian marketing limit of 0.5 µg g^{-1}. For example, 74% of the lake trout lakes in Ontario (excluding the Great Lakes) contain lake trout with Hg concentrations greater than the Canadian limit, and therefore consumption advisories exist for this species (Ontario Ministry of the Environment, 2003). In Quebec, 80 of 105 lake trout populations (76%) that have been sampled had consumption advisories (Quebec Ministère de l'Environnement et de la Faune, 1995). Braune et al. (1999) noted that mercury in lake trout from northern Canadian lakes usually exceeds consumption guidelines. Fish consumption advisories exist for lake trout in all regions that have been sampled, demonstrating that mercury is a widespread problem for this species and its consumers.

Large differences in Hg concentrations are observed in fish from lakes in close proximity to one another. Within a region, mean Hg concentrations in predatory fish vary fivefold or more, even after standardization of these data for fish size and/or age. For example, in northern Quebec lakes standardized concentrations of Hg varied about fivefold in lake trout (Schetagne and Verdon, 1999), and predatory fish in six lakes in northwestern Ontario varied three- to fourfold (Bodaly et al., 1993). Similarly, mean Hg (standardized for fish size) in lake trout in almost 100 lakes from all regions of Ontario varied more than 20-fold, from 0.05 to more than 1 µg g^{-1} (McMurtry et al., 1989). Stafford and Haines (1997) also found mean Hg concentrations in lake trout from 120 randomly chosen lakes in Maine to vary more than eightfold, from 0.11 to 0.91 µg g^{-1}. These studies demonstrate that Hg concentrations in lake trout populations can vary considerably within a limited geographic area.

Despite the high variability in lake trout Hg concentrations within regions, some geographic trends remain evident. In southern Ontario, 74% of lakes have Hg in lake trout above 0.5 µg g^{-1} and 26% have Hg in lake trout exceeding 1.5 µg g^{-1}; in northern Ontario, only 58% of lakes have Hg in lake trout greater than 0.5 µg g^{-1} and only 8% have Hg in lake trout higher than 1.5 µg g^{-1} (Ontario Ministry of the Environment, 1997). In contrast to Ontario, mercury concentrations in lake trout from Quebec are lower in southern regions (Outaouais and Fleuve Saint-Laurent: 26 of 46 populations tested with recommended consumption limits less than four meals per month) than in the more remote northern areas (Lac Saint-Jean and Gaspésie–Côte-Nord: 22 of 26 lake trout populations with restricted consumption; La Grande Rivière and Grande Rivière de la Baleine: 32 of 33 populations; excluding reservoirs) (Quebec Ministère de l'Environnement et de la Faune, 1995). The underlying causes of these trends are not yet understood but may be related to differences in geologic or atmospheric sources of Hg.

Factors affecting mercury concentrations in lake trout

With the exception of lakes that have received direct discharges of Hg (e.g., the English-Wabigoon river system in northwestern Ontario; Parks and Hamilton, 1987), most Hg entering freshwater systems today is probably atmospheric in origin. This mercury originates from natural sources (such as geologic weathering, volcanic eruptions, and ocean degassing) and from anthropogenic sources (such as burning of coal, oil, and municipal wastes and industrial processes). Atmospheric concentrations of mercury in the northern hemisphere have increased since industrial times, and between one-half and three-fourths of the Hg in the atmosphere is anthropogenic in origin (Swain et al., 1992). This atmospherically derived Hg is mainly in inorganic forms and enters lakes directly and indirectly via watershed runoff. Atmospheric deposition of Hg to temperate and Arctic lakes is now about two to three times preindustrial rates (Lockhart et al., 1995). Whether increases in atmospheric deposition rates have caused increases in concentrations of Hg in freshwater fish, including lake trout, is unclear but some evidence suggests this (Kelly et al., 1975; Johnson, 1987; Swain and Helwig, 1989; Rolfhus and Fitzgerald, 1995).

Climate and rates of atmospheric deposition of Hg to lakes and their watersheds are similar within a given region and therefore cannot explain lake-to-lake differences in Hg concentrations in fish. The high variability in mercury levels in lake trout populations within regions must therefore be related to the physical, chemical, and biological characteristics of lakes and their watersheds. The lake-specific characteristics that are believed to affect Hg concentrations in freshwater fish include the rate of supply of inorganic Hg and methyl mercury to lakes and their watersheds, the trophic position and growth rates of different fish species, and the physical and chemical characteristics of lakes and their watersheds. The relatively high concentrations of Hg in lake trout in Boreal lakes are probably mainly the result of the trophic position of lake trout in freshwater systems, the relatively large size and age of many individual lake trout, and the chemical conditions of Boreal lakes that tend to promote high Hg concentrations in fish.

Almost all of the Hg in fish muscle is MeHg (Bloom, 1992; Lasorsa and Allen-Gil, 1995; Hammerschmidt et al., 1999), and this is the form of mercury that is accumulated in aquatic food webs. Lakes and their biota receive MeHg from three sources: precipitation, runoff from the surrounding watershed, and in-lake methylation of inorganic mercury (Rudd, 1995). Inputs of MeHg from precipitation are not sufficient to account for Hg in fish in the Boreal lakes of North America, and in-lake production of MeHg by methylation of inorganic mercury is thought to be a significant source of mercury to food chains and fish (Rudd, 1995).

MeHg production and its bioavailability are affected by chemical factors, and many of the conditions known to favor Hg methylation are observed in the Boreal lakes that support lake trout populations. For example, low pH and high dissolved organic carbon (DOC) concentrations are common in Boreal lakes, and these factors tend to be associated

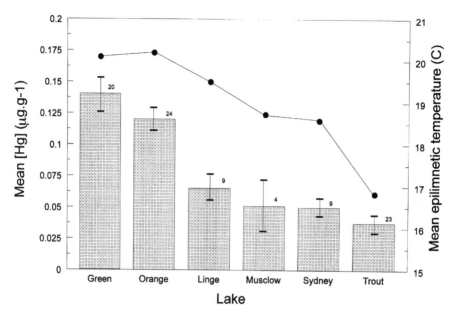

Figure 9.1 Mean mercury concentrations in axial muscle of yearling yellow perch from six lakes in northwestern Ontario. Vertical bars are mercury concentrations (with 95% confidence intervals and number of fish), and points are mean epilimnetic water temperatures (June–August, 1986–1989). Lakes are arranged in order of smallest (Green, 89 ha) to largest (Trout, 35,000 ha). From Bodaly et al., 1993, *Canadian Journal of Fisheries and Aquatic Sciences* **50**: 980–987.

with high Hg in fish in lakes (McMurtry et al., 1989; Wiener et al., 1990; Driscoll et al., 1994). The significant relationship between DOC and Hg in fish may be, at least in part, a result of the inputs of DOC-associated MeHg from wetlands in a lake's catchment (St. Louis et al., 1994).

The physical characteristics of lakes also affect Hg bioaccumulation in lake trout and other fish species. A study of six lakes in the Canadian Shield in northwestern Ontario that varied in their surface area from 89 to 35,000 ha but were similar in other chemical and physical characteristics revealed that lake size exerted a strong influence on Hg concentrations in fish (Bodaly et al., 1993). Concentrations of mercury in yearling yellow perch ranged from 0.04 $\mu g/g^{-1}$ in the largest lake studied to 0.14 $\mu g/g^{-1}$ (w/w) in the smallest lake that was studied and decreased in a regular pattern with lake size (Figure 9.1). These differences in mercury concentrations between the smaller and larger study lakes were also seen in predatory and planktivorous species. The methylation of mercury by microorganisms is a temperature-dependent process. The cooler epilimnetic temperatures in large lakes decrease mercury methylation rates and subsequent inputs of mercury to the food web (Figure 9.1) (Bodaly et al., 1993). In contrast, McMurtry et al. (1989) observed that Hg in lake trout was positively related to lake area in a study where the ratio of catchment to lake area was not kept constant as in Bodaly et al. (1993). Lakes with proportionately larger catchment areas will have greater inputs of mercury from drainage basin runoff and, most likely, higher mercury concentrations in the fish from these systems. Other studies have demonstrated that the watershed size in relation to the lake size is important in determining Hg concentrations in fish in Boreal lakes (Suns and Hitchin, 1990; Evans, 1986).

Fish obtain most of their mercury from their food and only a small proportion directly from the water via uptake across the gills (Hall et al., 1997; Rodgers, 1994; Harris and Snodgrass, 1993). Therefore, Hg in fish is influenced strongly by Hg concentrations in their diet (Borgmann and Whittle, 1992; Harris and Snodgrass, 1993; Harris and Bodaly, 1998). Top predators such as lake trout contain the highest concentrations of mercury in part because they tend to feed on prey with high mercury concentrations. The MeHg they are accumulating is successively concentrated from the base of the food web because it is much more efficiently absorbed and accumulated (Mason et al., 1996) and excreted more slowly (Trudel and Rasmussen, 1997) by organisms than the inorganic forms of Hg. Concentrations of MeHg increase from prey to predator, and high-trophic-level organisms tend to have the greatest concentrations of mercury in their tissues (Kidd et al., 1995; Cabana et al., 1994). As an example, in Lake Michigan mean dry weight concentrations of MeHg increase through the pelagic food web from 0.01 in zooplankton to 0.21 in the insectivorous bloater to 0.59 µg g^{-1} in lake trout (Mason and Sullivan, 1997), and similar increases are seen in Boreal lake food chains.

In recent studies of mercury accumulation through food webs, the trophic position of fish and invertebrates has been characterized using tissue ratios of stable nitrogen isotopes (^{15}N/^{14}N). The heavier isotope of nitrogen is enriched from primary producers to primary consumers, from primary consumers to secondary consumers, and so on up through the food web by an average of 3 to 5 parts per thousand (per mil; Peterson and Fry, 1989). This enrichment in the heavy isotope provides a continuous relative measure of an organism's trophic positioning within the food web and also reflects dietary habits over a period of months to years (Hesslein et al., 1993). Kidd et al. (1995) used stable nitrogen isotope ratios in fish muscle to quantify the trophic transfer of mercury through several food webs in northwestern Ontario. They found a highly significant relationship between muscle concentrations of Hg and the trophic position of fish (as quantified by stable nitrogen isotope analyses) in the six lakes examined (Figure 9.2). Similar relationships have been observed for mercury and other persistent pollutants in other freshwater and marine food webs (reviewed by Kidd, 1998). From the initial work done with this technique, it is evident from the differences in the slope of this relationship that the accumulation of Hg through food webs varies considerably from lake to lake; such variation may be related to varying efficiencies of carbon transfer in these systems.

The length of the underlying food chain also significantly affects the concentration of Hg in top predators such as lake trout. Such effects were unequivocally demonstrated by Cabana et al. (1994) and Cabana and Rasmussen (1994). They categorized temperate lakes into three classes based on the length of the pelagic food chain leading up to the top predator lake trout using the presence or absence of important prey species: Class 1 lakes had the shortest food chains with no mysids (a zooplanktivorous crustacean) or pelagic prey fishes (rainbow smelt [*Osmerus mordax*], lake cisco [*Coregonus artedi*], lake whitefish [*Coregonus clupeaformis*], alewife [*Alosa pseudoharengus*], and others); Class 2 lakes had an intermediate food chain length because of the presence of pelagic prey fishes; Class 3 lakes had the longest food chains because of the presence of both pelagic prey fishes and mysids. They found that the concentrations of mercury in lake trout increased 3.6-fold from the Class 1 to Class 3 lakes and were significantly related to the stable nitrogen isotope ratios in this species. It is likely that manipulations (both intended and accidental) of freshwater food webs will influence the concentrations of Hg in top predators. Vander Zanden and Rasmussen (1996) observed that Hg concentrations in lake trout were considerably higher in lakes that had introduced populations of smelt. They hypothesized that this was the result of increased food chain lengths in the systems that had introductions of this exotic species.

Concentrations of Hg in fish are also affected by the amount of its diet that it uses for growth relative to metabolism. The MeHg present in a fish's diet is efficiently absorbed and retained in its tissues; the excretion rates of MeHg are slow compared to that of Hg and are slower in older than in younger fish (Trudel and Rasmussen, 1997). Young, immature fish use a large proportion of their dietary carbon intake for growth and

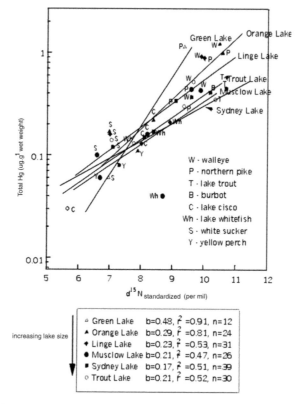

Figure 9.2 Relations between total mercury (μg.g^{-1} wet weight) and standardized $\delta^{15}N$ (per mil) of fish muscle from seven species from the Northwestern Ontario Lake Size Series (Kidd et al., 1995, *Water, Air, and Soil Pollution* **80**: 1011–1015.). [The $\delta^{15}N$ of the obligate benthivore, white sucker, generally increased with decreasing lake size, likely as a result of differences in in-lake cycling or sources of nitrogen. For this reason, $\delta^{15}N$ of all fish were standardized to the mean $\delta^{15}N$ of white sucker using the following formula: $\delta^{15}N_{fish} - \delta^{15}N_{white\ sucker} + 6.8$ (to account for the fact that suckers are secondary consumers and that $\delta^{15}N$ increases an average of 3.4 per mil with each trophic level)].

generally have low Hg concentrations in their tissues. This is often termed growth dilution, and although the MeHg is not "diluted," it results in concentrations in fast-growing, immature predators that can be similar to that of their prey. Larger, mature fish tend to be slow growing, and these fish use most of their ingested carbon for metabolism and reproduction (not growth) while retaining most of the ingested mercury. These fish therefore tend to have higher concentrations of this contaminant (Harris and Snodgrass, 1993; Rodgers, 1994; Harris and Bodaly, 1998), although Stafford and Haines (2001) did not find a relationship between growth rate and mercury in a lake trout population. As predatory fish grow, they tend to eat larger prey items with higher concentrations of contaminants, resulting in increasing concentrations of Hg with size (Figure 9.3). This relationship between Hg concentrations and fish size is commonly seen in temperate and Boreal lakes (e.g., Wiener et al., 1990).

Management of mercury exposure from consumption of lake trout

The primary objective of provincial and federal agencies in Canada for managing mercury in lake trout populations is to reduce health risks to humans. Fish are collected from lakes,

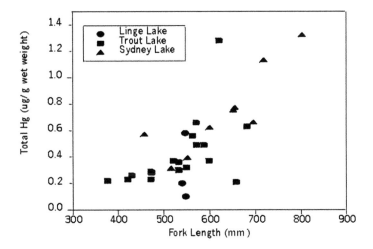

Figure 9.3 Relationship between Hg concentrations and fish length in lake trout from three lakes in northwestern Ontario. Data from Fudge et al., 1994 *Canadian Data Report of Fisheries and Aquatic Sciences 921.*

and the concentrations of mercury in axial muscle from individual fish are determined. Muscle tissue is typically analyzed for mercury because it is the main tissue consumed by people. The relation between mercury concentrations and fish size is then determined, and mercury concentrations are presented in relation to fish size in terms of the recommended human consumption limit for each species sampled (Ontario Ministry of the Environment, 1997; Quebec Ministère de l'Environnement et de la Faune, 1995). Advice on the sizes of fish fit for consumption is delivered to the public through booklets for anglers and signs posted on lake shores, and through the dissemination of information to communities. In aboriginal communities, fish are often an important and significant part of the diet. For this reason advice on the species, sizes, and quantities of fish that are safe to eat is based on consumption information specific to these communities (Health and Welfare Canada, 1984).

Safe consumption limits are based on recommendations from Health and Welfare Canada for maximum allowable intake of MeHg per day (0.47 µg/kg⁻¹ body weight/day; Health and Welfare Canada, 1984). Health Canada is currently recommending that Hg intake by children and by women of childbearing age not exceed 0.2 µg/kg⁻¹ body weight/day.

Sampling lakes to determine Hg concentrations in fish and its risk to human consumers is relatively straightforward. From each lake, at least 20 fish of each species likely to be caught and consumed by people should be obtained. Ideally, this sample should include a wide range of fish lengths and weights to ensure that analyses are conducted on all sizes of fish likely to be consumed. This eliminates the need for extrapolation of mercury concentrations to fish sizes not sampled and ensures a statistically reliable relationship between Hg and fish size for each species. To compare the concentrations of Hg in fishes across lakes, statistical methods are generally used to standardize Hg for differences in size and age. Several techniques have been used including regressions and transformations, alternative techniques such as multivariate analysis of univariate and bivariate statistics (Somers and Jackson, 1993), or polynomial regressions with indicator variables (Tremblay et al., 1998). As noted above, this information is often used by governmental agencies to advise sport fishers of recommended fish consumption limits for various water bodies (e.g., Ontario Ministry of the Environment, 1997).

Effects of mercury on fish

Although the main emphasis of research and management of mercury in freshwater fish populations has been placed on the human health implications, high MeHg concentrations may be affecting the fish themselves. There has been little research to date on the effects of Hg on fish at environmentally realistic concentrations or on possible effects on fish-eating wildlife (Wiener and Spry, 1996). However, there is some evidence that MeHg may impair reproduction in freshwater predators. For example, Friedmann et al. (1996) found that environmentally relevant MeHg concentrations of 0.1 and 1.0 μg g⁻¹ fed to juvenile walleyes affected their growth and gonadal development. Latif et al. (2001) examined the effect of MeHg in water, also at realistic concentrations, on walleye egg development and found significant reductions in egg survival at higher MeHg concentrations. Also, fathead minnows consuming food with elevated MeHg concentrations were found to show reduced spawning success (Hammerschmidt et al., 2002). The toxicologic significance of MeHg to fish is an important area for future research.

Possible future trends

Future trends in Hg levels in freshwater fish generally and lake trout specifically are difficult to forecast because many factors influence Hg concentrations in these organisms. There is some recent evidence that rates of atmospheric deposition of Hg in the northern hemisphere have recently begun to decrease (Engstrom and Swain, 1996). Some studies suggest that mercury concentrations in fish are related to rates of atmospheric deposition (Kelly et al., 1975; Swain and Helwig, 1989; Rolfhus and Fitzgerald, 1995). For this reason, reductions in atmospheric transport and deposition of Hg may lead to general decreases of this contaminant in lake trout populations.

Climate warming may have significant effects on Hg concentrations in fish in Boreal lakes by affecting rates of mercury methylation and the supply of MeHg to food chains. As noted above, Bodaly et al. (1993) found significant relationships between epilimnetic temperatures and Hg in fish in Boreal lakes. Climate warming may produce warmer and/or deeper epilimnia in the Boreal zone. Because mercury methylation probably takes place mainly in epilimnetic sediments and is known to be temperature dependent, climatic warming could increase rates of methylation. On the other hand, lower inputs of DOC to Boreal lakes with decreased precipitation and runoff (Schindler and Gunn, this volume) could in turn reduce the supply of MeHg to lakes from their watersheds.

The acidification of lakes by atmospheric deposition of pollutants may be increasing Hg concentrations in lake trout in small Boreal lakes. Experimental lake acidification was observed to increase Hg in fish (Wiener et al., 1990) and Hg in fish in lakes has often been observed to be negatively related to pH (e.g., McMurtry et al., 1989). Also, Hg methylation is stimulated by sulfate addition (Gilmour et al., 1992), and atmospheric sulfate deposition has increased concurrently with acidic deposition. Fortunately SO_2 emissions have declined substantially in recent years, with an approximate 40% decline in total North America emissions since 1980 (Jeffries et al., 2003).

The recent introduction and the spread of rainbow smelt into freshwater systems may also increase Hg concentrations in lake trout (Franzin et al., 1994; Cabana and Rasmussen, 1994; Futter, 1994). Because the presence of smelt is believed to increase the length of the food chain, lake trout from lakes with smelt tend to have higher Hg when compared to the same species from lakes without smelt (Vander Zanden and Rasmussen, 1996; Akielaszak and Haines, 1981).

Intensive fishing tends to decrease Hg in freshwater fish, at least temporarily (Verta, 1990). Exploitation will tend to decrease fish densities, increase growth rates, and decrease the mean age of the population, all of which will tend to decrease Hg in lake trout populations. Therefore, the presence of sport and commercial and subsistence fisheries on lake trout lakes may reduce Hg concentrations, at least on a size-adjusted basis.

Summary

Mercury is present in all freshwater fish. Lake trout, because of their piscivorous nature, are particularly susceptible to accumulating high concentrations of this contaminant. Most lake trout populations surveyed have mean concentrations of mercury greater than the 0.5 µg g^{-1} Canadian standard for human consumption, although there is a large amount of variation among lakes even within a given region. Mercury enters freshwater systems primarily from atmospheric deposition and is converted to MeHg, the form that is efficiently bioaccumulated through food webs. Factors affecting mercury concentrations in lake trout include the length of the food chain, the size and age of individual fish, and physical and chemical characteristics of lakes and their watersheds. It is difficult to predict future contaminant trends in predatory fish, including lake trout, in Boreal lakes. Declines in atmospheric deposition, increased fishing pressure, and a reduced supply of DOC to lakes from their watersheds may reduce mercury concentrations in lake trout. On the other hand, the spread of rainbow smelt populations into lakes and climate warming might increase the concentrations of mercury in lake trout and the risk to humans and fish-eating wildlife. At present, it is not possible to predict how these opposing factors will affect the concentrations of a widespread contaminant in freshwater fish.

References

Akielaszak, J.J. and Haines, T.A., 1981, Mercury in the muscle tissue of fish from three northern Maine lakes, *Bulletin of Environmental Contamination and Toxicology* 27: 201–208.

Bloom, N.S., 1992, On the chemical form of mercury in edible fish and marine invertebrate tissue, *Canadian Journal of Fisheries and Aquatic Sciences* 49: 1010–1017.

Bodaly, R.A., Rudd, J.W.M., Fudge, R.J.P., and Kelly, C.A., 1993, Mercury concentrations in fish related to size of remote Canadian shield lakes, *Canadian Journal of Fisheries and Aquatic Sciences* 50: 980–987.

Braune, B., Muir, D., de March, B., Gamberg, M., Poole, K., Currie, R., Dodd, M., Duschenko, W., Eamer, J., Elkin, B., Evans, M., Grundy, S., Hebert, C., Johnstone, R., Kidd, K., Koenig, B., Lockhart, L., Marshall, H., Reimer, K., Sanderson, J., and Shutt, L., 1999, Spatial and temporal trends of contaminants in Canadian Arctic freshwater and terrestrial ecosystems: a review, *Science of the Total Environment* 230: 145–207.

Borgmann, U. and Whittle, D.M., 1992, Bioenergetics and PCB, DDE, and mercury dynamics in Lake Ontario lake trout (*Salvelinus namaycush*), *Canadian Journal of Fisheries and Aquatic Sciences* 49: 1086–1096.

Cabana, G. and Rasmussen, J.B., 1994, Modelling food chain structure and contaminant bioaccumulation using stable nitrogen isotopes, *Nature* 372: 255–257.

Cabana, G., Tremblay, A., Kalff, J., and Rasmussen, J.B., 1994, Pelagic food chain structure in Ontario lakes: a determinant of mercury levels in lake trout (*Salvelinus namaycush*), *Canadian Journal of Fisheries and Aquatic Sciences* 51: 381–389.

Clarkson, T.W., 1992, Mercury: major issues in environmental health, *Environmental Health Perspectives* 100: 31–38.

Driscoll, C.T., Yan, C., Schofield, C.L., Munson, R., and Holsapple, J., 1994, The mercury cycle and fish in the Adirondack Lakes, *Environmental Science and Technology* 28: 136A-143A.

Engstrom, D.R. and Swain, E.B., 1996, Recent declines in atmospheric mercury deposition in the Upper Midwest, USA [Abstract], In: Fourth International Conference on Mercury as a Global Pollutant, Hamburg, Germany.

Evans, R.D., 1986, Sources of mercury contamination in the sediments of small headwater lakes in south-central Ontario, Canada, *Archives of Environmental Contamination and Toxicology* **15**: 505–512.

Franzin, W.G., Barton, B.A., Remnant, R.A., Wain, D.B., and Pagel, S.J., 1994, Range extension, present and potential distribution, and possible effects of rainbow smelt in Hudson Bay drainage waters of northwestern Ontario, Manitoba, and Minnesota, *North American Journal of Fisheries Management* **14**: 65–76.

Friedmann, A.S., Watzin, M.C., Brinck-Johnsen, T., and Leiter, J.C., 1996, Low levels of dietary methylmercury inhibit growth and gonadal development in juvenile walleye (*Stizostedion vitreum*), *Aquatic Toxicology* **35**: 265–278.

Fudge, R.J.P., Bodaly, R.A., and Strange, N.E., 1994, Lake variability and climate change study: fisheries investigations from the Northwestern Ontario Lake Size Series (NOLSS) lakes, 1987–1989, *Canadian Data Report of Fisheries and Aquatic Sciences 921*, Fisheries and Oceans Canada, Winnipeg, Manitoba.

Futter, M.N., 1994, Pelagic food-web structure influences probability of mercury contamination in lake trout (*Salvelinus namaycush*), *Science of the Total Environment* **145**: 7–12.

Gilmour, C.C., Henry, E.A., and Mitchell, R., 1992, Sulfate stimulation of mercury methylation in freshwater sediments, *Environmental Science and Technology* **26**: 2281–2287.

Hall, B.D., Bodaly, R.A., Fudge, R.J.P., Rudd, J.W.M., and Rosenberg, D.M., 1997, Food as the dominant pathway of methylmercury uptake by fish, *Water, Air and Soil Pollution* **100**: 13–24.

Hammerschmidt, C.R., Sandheinrich, M.B., Wiener, J.G., and Rada, R.G., 2002, Effects of dietary methylmercury on reproduction of fathead minnows, *Environmental Science and Technology* **36**: 877–883.

Hammerschmidt, C.R., Wiener, J.G., Frazier, B.E., and Rada, R.G., 1999, Methylmercury content of eggs in yellow perch related to maternal exposure in four Wisconsin lakes, *Environmental Science and Technology* **33**: 999–1003.

Harris, R.C. and Bodaly, R.A., 1998, Temperature, growth and dietary effects on fish mercury dynamics in two Ontario lakes, *Biogeochemistry* **40**: 175–187.

Harris, R.C. and Snodgrass, W.J., 1993, Bioenergetic simulations of mercury uptake and retention in walleye (*Stizostedion vitreum*) and yellow perch (*Perca flavescens*), *Water Pollution Research Journal of Canada* **28**: 217–236.

Health and Welfare Canada, 1984, *Methylmercury in Canada*, Volume 2, Health and Welfare Canada, Ottawa, Ontario.

Hesslein, R.H., Hallard, K.A., and Ramlal, P., 1993, Replacement of sulfur, carbon and nitrogen in tissues of growing broad whitefish (*Coregonus nasus*) in response to change in diet traced by δ^{34}S, δ^{13}C and δ^{15}N, *Canadian Journal of Fisheries and Aquatic Sciences* **50**: 2081–2076.

Jeffries, D.S., Clair, T.A., Couture, S., Dillon, P.J., Dupont, J., Keller, W., McNicol, D.K., Turner, M.A., Vet, R., and Weeber, P.J., 2003, Assessing the recovery of lakes in southeastern Canada from the effects of acid deposition, *Ambio* **32**(3): 176–182.

Johnson, M.G., 1987, Trace element loadings to sediments of fourteen Ontario lakes and correlations with concentrations in fish, *Canadian Journal of Fisheries and Aquatic Sciences* **44**: 3–13.

Kelly, T.M., Jones, J.D., and Smith, G.R., 1975, Historical changes in mercury contamination in Michigan walleyes (*Stizostedion vitreum vitreum*), *Journal of the Fisheries Research Board of Canada* **32**: 1745–1754.

Kidd, K.A., Hesslein, R.H., Fudge, R.J.P., and Hallard, K.A., 1995, The influence of trophic level as measured by δ^{15}N on mercury concentrations in freshwater organisms, *Water, Air, and Soil Pollution* **80**: 1011–1015.

Kidd, K.A., 1998, Use of stable isotope ratios in freshwater and marine biomagnification studies, In *Environmental Toxicology: Current Developments*, edited by J. Rose, Gordon and Breach Science Publishers, London, pp. 359–378.

Lasorsa, B. and Allen-Gil, S., 1995, The methylmercury to total mercury ratio in selected marine, freshwater, and terrestrial organisms, *Water, Air, and Soil Pollution* **80**: 905–913.

Latif, M.A., Bodaly, R.A., Johnston T.A., and Fudge, R.J.P., 2001, Effects of envrionmental and maternally derived methylmercury on the embryonic and larval stages of walleye (*Stizostedion vitreum*), *Environmental Pollution* 111: 139–148.

Lockhart, W.L., Wilkinson, P., Billeck, B.N., Hunt, R.V., and Wagemann, R., 1995, Current and historical inputs of mercury to high-latitude lakes in Canada and to Hudson Bay, *Water, Air, and Soil Pollution* 80: 603–610.

Mason, R.P., Reinfelder, J.R., and Morel, R.M.M., 1996, Uptake, toxicity and trophic transfer of mercury in a coastal diatom, *Environmental Science and Technology* 30: 1835–1845.

Mason, R.P. and Sullivan, K.A., 1997, Mercury in Lake Michigan, *Environmental Science and Technology* 31: 942–947.

McMurtry, M.J., Wales, D.L., Scheider, W.A., Beggs, G.L., and Dimond, P.E., 1989, Relationship of mercury concentrations in lake trout (*Salvelinus namaycush*) and smallmouth bass (*Micropterus dolomieu*) to the physical and chemical characteristics of Ontario lakes, *Canadian Journal of Fisheries and Aquatic Sciences* 46: 426–434.

Ontario Ministry of the Environment, 2003, *Guide to Eating Ontario Sport Fish 2003–2004*, Queen's Printer for Ontario, Toronto. Parks, J.W., and Hamilton, A.L., 1987, Accelerating recovery of the mercury-contaminated Wabigoon/English River system, *Hydrobiologia* 149: 2184–2202.

Peterson, B.J. and Fry, B., 1987, Stable isotopes in ecosystem studies, *Annual Review of Ecological Systems* 18: 293–320.

Québec Ministère de l'Environnement et de la Faune and Ministère de la Santé et des Services Sociaux, 1995, *Guide de Consommation du Poisson de Pêche Sportive en Eau Douce*, Québec.

Rodgers, D.W., 1994, You are what you eat and a little bit more: bioenergetics-based models of methylmercury accumulation in fish revisited, In *Mercury Pollution*, edited by C.J. Watras and J.W. Huckabee, Lewis Publishers, Boca Raton, pp. 427–439.

Rolfhus, K.R. and Fitzgerald, W.F., 1995, Linkages between atmospheric mercury deposition and the methylmercury content of marine fish, *Water Air and Soil Pollution* 80: 291–297.

Rudd, J.W.M., 1995, Sources of methyl mercury to freshwater ecosystems: a review, *Water, Air, and Soil Pollution* 80: 697–713.

St. Louis, V.L., Rudd, J.W.M., Kelly, C.A., Beaty, K.G., Bloom, N.S., and Flett, R.J., 1994, Importance of wetlands as sources of methyl mercury to boreal forest ecosystems, *Canadian Journal of Fisheries and Aquatic Sciences* 51: 1065–1076.

Schetagne, R. and Verdon, R., 1999, Mercury in fish of natural lakes of northern Quebec, In *Mercury in the Biogeochemical Cycle: Natural Environments and Hydroelectric Reservoirs of Northern Québec (Canada)*, edited by M. Lucotte, R. Schetagne, N. Thérien, C. Langlois, and A. Tremblay, Springer, Berlin, pp. 115–130.

Somers, K.M. and Jackson, D.A., 1993, Adjusting mercury concentrations for fish-size covariation: a multivariate alternative to bivariate regression, *Canadian Journal of Fisheries and Aquatic Sciences* 50: 2388–2396.

Stafford, C.P. and Haines, T.A., 1997, Mercury concentrations in Maine sport fishes, *Transactions of the American Fisheries Society* 126: 144–152.

Stafford, C.P. and Haines, T.A., 2001, Mercury contamination and growth rate in two piscivore populations, *Environmental Toxicology and Chemistry* 20: 2099–2101.

Suns, K. and Hitchin, G., 1990, Interrelationships between mercury levels in yearling yellow perch, fish condition and water quality, *Water, Air, and Soil Pollution* 650: 255–265.

Swain, E.B., Engstrom, D.R., Brigham, M.E., Henning, T.A., and Brezonik, P.L., 1992, Increasing rates of atmospheric mercury deposition in midcontinental North America, *Science* 257: 784–787.

Swain, E.B. and Helwig, D.D., 1989, Mercury in fish from northeastern Minnesota lakes: historical trends, environmental correlates, and potential sources, *Journal of the Minnesota Academy of Science* 55: 103–109.

Tremblay, G., Legendre, P., Doyon, J.-F., Verdon, R., and Schetagne, R., 1998, The use of polynomial regression analysis with indicator variables for interpretation of mercury in fish data, *Biogeochemistry* 40: 189–201.

Trudel, M. and Rasmussen, J.B., 1997, Modeling the elimination of mercury by fish, *Environmental Science and Technology* 31: 1716–1722.

United States Environmental Protection Agency, 1998, *Update: Listing of fish and wildlife advisories. Fact Sheet EPA-823-F-98–009*, Office of Water, Washington, D.C.

Vander Zanden, M.J. and Rasmussen, J.B., 1996, A trophic position model of pelagic food webs: impact on contaminant bioaccumulation in lake trout, *Ecological Monographs* **66**: 451–477.

Verta, M., 1990, Changes in fish mercury concentrations in an extensively fished lake, *Canadian Journal of Fisheries and Aquatic Sciences* **47**: 1888–1897.

Wiener, J.G., Fitzgerald, W.F., Watras, C.J., and Rada, R.G., 1990, Partitioning and bioavailability of mercury in an experimentally acidified Wisconsin lake, *Environmental Toxicology and Chemistry* **9**: 909–918.

Wiener, J.G. and Spry, D.J., 1996, Toxicological significance of mercury in freshwater fish, In *Environmental Contaminants in Wildlife — Interpreting Tissue Concentrations*, edited by W.N. Beyer, G.H. Heinz, and A.W. Redmon, Lewis Publishers, Boca Raton, pp. 299–343

Wiener, J.G., Martini, R.E., Sheffy, T.B., and Glass, G.E., 1990, Factors influencing mercury concentrations in walleyes in northern Wisconsin lakes, *Transactions of the American Fisheries Society* **119**: 862–870.

chapter ten

Acidic deposition in the northeastern United States: sources and inputs, ecosystem effects, and management strategies*

Charles T. Driscoll
Department of Civil and Environmental Engineering, Syracuse University

Gregory B. Lawrence
Water Resources, U.S. Geological Survey

Arthur J. Bulger
University of Virginia

Thomas J. Butler
Center for the Environment, Cornell University

Christopher S. Cronan
Department of Biological Sciences, University of Maine

Christopher Eagar
USDA Forest Service

Kathleen F. Lambert
Hubbard Brook Research Foundation

Gene E. Likens
Institute of Ecosystem Studies, Millbrook

John L. Stoddard
United States Environmental Protection Agency

Kathleen C. Weathers
Institute of Ecosystem Studies, Millbrook

* Modified from Driscoll et al., 2001. Acidic Deposition in the northeastern United States: sources and inputs, ecosystem effects, and management strategies. *BioScience* **51**(3): 180–198. Copyright, American Institute of Biological Sciences, with permission.

Contents

Introduction

Acidic deposition is the transfer of strong acids and acid-forming substances from the atmosphere to the surface of the earth. The composition of acidic deposition includes ions, gases, and particles derived from gaseous emissions of sulfur dioxide (SO_2), nitrogen oxides (NO_x), ammonia (NH_3), and particulate emissions of acidifying and neutralizing compounds. Over the past quarter century of study, acidic deposition has emerged as a critical environmental stress affecting forested landscapes and aquatic ecosystems in North America, Europe, and Asia. This complex problem is an example of a new class of environmental issues that are multiregional in scale and not amenable to simple resolution by policy makers. Acidic deposition can originate from transboundary air pollution and affects large geographic areas; is highly variable across space and time; links air pollution to diverse terrestrial and aquatic ecosystems; alters the interactions of many elements [e.g., sulfur (S), nitrogen (N), hydrogen ion (H^+), calcium (Ca^{2+}), magnesium (Mg^{2+}), aluminum (Al)]; and contributes directly and indirectly to biological stress and the degradation of ecosystems. Despite the complexity of the effects of acidic deposition, management actions in North America and Europe directed toward the recovery of damaged natural resources have resulted in recent decreases in both emissions and deposition of acidic S compounds.

Thus, acidic deposition is an instructive case study for coordination of science and policy efforts aimed at resolving large-scale environmental problems. Acidic deposition was first identified by R.A. Smith in England in the 19th century (Smith, 1872). Acidic deposition emerged as an ecologic issue in the late 1960s and early 1970s with reports of acidic precipitation and surface water acidification in Sweden and surrounding Scandinavia (Oden, 1968). The first report of acidic precipitation in North America was made at the Hubbard Brook Experimental Forest (HBEF) in the remote White Mountains of New Hampshire, based on collections beginning in the early 1960s (Likens et al., 1972). Controls on SO_2 emissions in the United States were first implemented following the 1970 Amendments to the Clean Air Act (CAAA). In 1990, Congress passed Title IV of the Acid Deposition Control Program of the CAAA to further decrease emissions of SO_2 and initiate controls on NO_x from electric utilities that contribute to acidic deposition. The Acid Deposition Control Program had two goals: (1) a 50% decrease or 9.1 million metric tons per year (or 10 million short tons per year) reduction of SO_2 utility emissions from 1980 levels that is expected to be fully implemented by 2010, and (2) an NO_x emission rate limitation (0.65 lb NO_x/m BTU in 1990 to 0.39 lb NO_x/m BTU in 1996) that will achieve a 1.8 million metric ton per year (2 million short tons per year as nitrogen dioxide) reduction in NO_x utility emissions from what would have occurred without emission rate controls. Both SO_2 and NO_x provisions are focused on large utilities. The legislation capped total utility emissions of SO_2 at 8.12 million metric tons per year (8.95 million short tons per year), whereas nonutility emissions of SO_2 were capped at 5.08 million metric tons per year (5.6 million short tons per year). Caps for NO_x emissions were not established in the legislation, and as a result, emissions may increase over time as the demand for electricity increases.

As we begin the 21st century, there is an opportunity to review the previous 10 to 30 years to assess the effects of the 1970 and 1990 Clean Air legislation on emission reductions, air pollution levels, trends and chemical impacts of acidic deposition, and ecosystem recovery. In this report, we focus on three critical questions to examine the ecologic effects of acidic deposition in the study region of New England and New York (Figure 10.1) and to explore the relationship between emission reductions and ecosystem recovery (see below). This analysis draws on research from the northeastern United States along with additional information from the mid-Atlantic and southeastern United States and eastern Canada. We rely heavily on data from the HBEF, a research site that provides the longest continuous records of precipitation and stream chemistry (Likens and Bormann, 1995). Because of its location in a region with bedrock that is resistant to chemical weathering and acidic soils, surface waters at the HBEF are representative of areas of the Northeast that are sensitive to acidic deposition. When stream chemistry from the biogeochemical reference watershed (watershed 6) at the HBEF was compared to results from the U.S. Environmental Protection Agency (EPA) synoptic survey of lakes in the Northeast collected through the Environmental Monitoring and Assessment Program (EMAP; Larsen et al., 1994; Stevens, 1994), only 4.9% of the lakes had lower concentrations of the sum of base cations (i.e., $Ca^{2+} + Mg^{2+} + Na^+ + K^+$), 67% had lower concentrations of SO_4^{2-}, and 5.7% had lower pH values. However, in comparision to populations of acid-sensitive EMAP lakes [acid-neutralizing capacity (ANC) < 50 μeq L^{-1}] 28, 77, and 32% of the lakes have lower concentrations of the sum of base cations, SO_4^{2-}, and pH, respectively, than stream water draining watershed 6 at the HBEF. Periodic review of knowledge gained from long-term monitoring, process-level research, and modeling is critical for assessing regulatory programs and solving complex environmental problems. The need to resolve the problem of acidic deposition is made more apparent as the many linkages between acidic deposition and other environmental issues are more clearly documented (Table 10.1). Much of the report that follows focuses on what has been learned since the

1990 CAAA concerning the effects of acidic deposition on forest vegetation, soils, and surface waters, and the influence of past and potential future emission reductions on ecosystem recovery in the northeastern United States.

Question 1: What are the spatial patterns and temporal trends for emissions, precipitation concentrations, and deposition of anthropogenic S, N, and acidity across the Northeastern United States?

Emissions

In the United States, there have been marked changes in emissions of SO_2 over the past 100 years. Total emissions of SO_2 increased from 9 million metric tons (9.9 million short tons) in 1900 to a peak of 28.8 million metric tons (31.7 million short tons) in 1973, of which 60% were from electric utilities (EPA, 2000). By 1998, total annual SO_2 emissions for the United States had declined to 17.8 million metric tons (19.6 million short tons). From 1970 to 1998, SO_2 emissions from electric utilities decreased by 24%, largely as a result of the 1970 and 1990 CAAAs. Emissions of NO_x have increased from about 2.4 million metric tons (2.6 million short tons) in 1900 to 21.8 million metric tons (24 million short tons) in 1990 and have remained fairly constant up to the present.

Emissions of SO_2 in the United States are highest in the Midwest. States clustered around the Ohio River Valley (Pennsylvania, Ohio, West Virginia, Indiana, Illinois, Kentucky, and Tennessee) comprised 7 of the 10 states with the highest SO_2 emissions in the nation during 1998 (Figure 10.1a). These 7 states accounted for 41% of the national SO_2 emissions during this period. Of these states, 5 (Pennsylvania, Ohio, Indiana, Illinois, and Tennessee) were also among the 10 states with highest total NO_x emissions for 1998 and comprise 20% of national emissions (Figure 10.1b). High emissions in this region are primarily from electric utilities and heavy manufacturing.

The 1990 CAAA required additional reductions in the emissions of SO_2 from electric utilities, starting in 1995 with Phase I of the Acid Deposition Control Program. This legislation helped to promote the continuing pattern of declining emissions between the periods of 1992–1994 and 1995–1997 for most states in the eastern United States (Figure 10.1a). For the United States, SO_2 emissions decreased 14% for the same period, whereas emissions decreased by 24% in the seven high-emission states in the Midwest. Decreases in emissions of NO_x between these periods, however, were only 2% nationally and 3% for the seven high-emission states in the Midwest (Figure 10.1b).

Atmospheric deposition of ammonium (NH_4^+) is derived from emissions of NH_3 and can contribute to the acidification of soil and water when these inputs are oxidized by soil microbes to nitrate (NO_3^-). The EPA has a national emissions inventory for NH_3, but little information is available on past emissions. Local and regional studies, however, have identified agricultural activities as the primary source of US emissions of NH_3 (Jordan and Weller, 1996). Livestock/poultry manure is generally considered the largest contributor; emissions from crop senescence may be as large but are difficult to measure accurately (Lawrence et al., 2000). Application of N fertilizer also contributes NH_3 to the atmosphere, but this source is less than 10% of emissions from manure handling in the Mississippi River Basin (Goolsby et al., 1999). Small sources of NH_3 emissions include automobiles and industrial processes (Fraser and Cass, 1998).

Patterns of precipitation and deposition of S and N

Acidic deposition can occur as wet deposition (as rain, snow, sleet, or hail); as dry deposition (as particles or vapor); and as cloud and fog deposition, more common at high

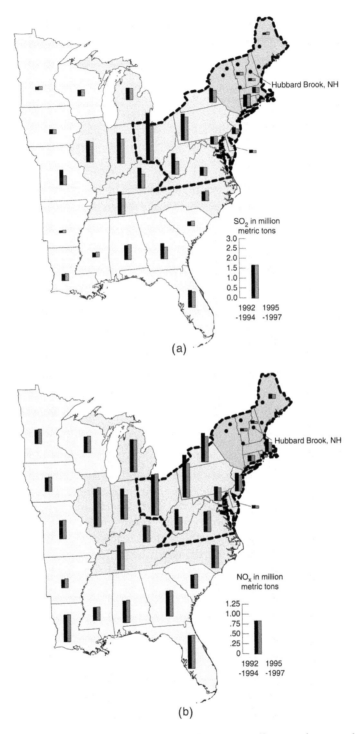

Figure 10.1 Study region for the analysis of acidic-deposition effects on forest and aquatic ecosystems is indicated by the shaded area; solid circles designate the location of the Hubbard Brook Experimental Forest (HBEF) and other National Atmospheric Deposition Program (NADP) sites in the study region; solid bars show state emissions of (a) SO_2 and (b) NO_x for the eastern United States for 1992–94, and open bars for 1995–97. The emissions source-area for the study region, based on 15-hour back trajectories, is indicated by bold dashed lines. The emissions source area, based on 21-hour back trajectories, is indicated by lighter shading (as calculated from Butler et al., 2001).

Table 10.1 Linkages Between Emissions of SO_2 and NO_x and Important Environmental Issues

Problem	Linkage to Acidic Deposition	Example/Reference
Coastal eutrophication	Atmospheric deposition is important in the supply of N to coastal waters	Jaworski et al., 1997
Mercury	Surface water acidification enhances mercury accumulation in fish	Driscoll et al., 1994a
Visibility	Sulfate aerosols are an important component of atmospheric particulates, decreasing visibility	Malm et al., 1994
Climate change	Sulfate aerosols increase atmospheric albedo, cooling the Earth and offsetting some of the warming potential of greenhouse gases. Tropospheric O_3 and N_2O act as greenhouse gases.	Moore et al., 1997
Tropospheric ozone	Emissions of NO_x contribute to the formation of ozone	Seinfeld, 1986

elevations and coastal areas. Wet deposition is monitored at over 200 U.S. sites by the interagency-supported National Atmospheric Deposition Program/National Trends Network (NADP/NTN), initiated in 1978. There are 20 NADP/NTN sites in the northeast study region. In addition, there are several independent sites where precipitation chemistry has been studied, in some cases for an even longer period (e.g., HBEF). Spatial patterns of wet deposition in the eastern half of the United States have been described by combining NADP/NTN deposition data with information on topography and precipitation (Grimm and Lynch, 1997).

Dry deposition is monitored by the EPA Clean Air Status and Trends Network (CAST-Net) at approximately 70 sites and by the National Oceanic and Atmospheric Administration AIRMON-dry Network at 13 sites. Most of the sites in these two networks are located east of the Mississippi River and began operation around 1988. There are seven CASTNet and five AIRMON-dry sites in the study region. An inferential approach is used in both CASTNet and AIRMON-dry to estimate dry deposition. This approach is dependent on detailed meteorologic measurements and vegetation characteristics, which can vary markedly over short distances in complex terrains (Clarke et al., 1997). As a result, the spatial patterns of dry deposition in the United States are poorly characterized.

Cloud and fog deposition in the northeastern United States have been monitored for limited periods at selected high-elevation (>1100 m) and coastal sites to support specific investigations (e.g., Weathers et al., 1988; Anderson et al., 1999). In recent years, the Mountain Acid Deposition Program (MADPro), as part of the EPA CASTNet Program, has involved the monitoring of cloud water chemistry at several sites in the eastern United States, including one site in the northeastern United States. Regional patterns and long-term trends are not well characterized, although cloud and fog deposition often contributes from 25 to over 50% of total deposition of S and N to high-elevation sites in the northeastern United States (Anderson et al., 1999).

Prevailing winds from west to east result in deposition of pollutants emitted in the Midwest that extend into New England and Canada. During atmospheric transport, some of the SO_2 and NO_x are converted to sulfuric and nitric acids; to ammonium sulfate and ammonium nitrate, which can be transported long distances; and nitric acid vapor, which has a shorter atmospheric residence time (Lovett, 1994).

Long-term data collected at the HBEF indicate that annual volume-weighted concentrations of SO_4^{2-} in bulk precipitation (precipitation sampled from an open collector) has declined (Figure 10.2) with national decreases in SO_2 emissions that followed the 1970

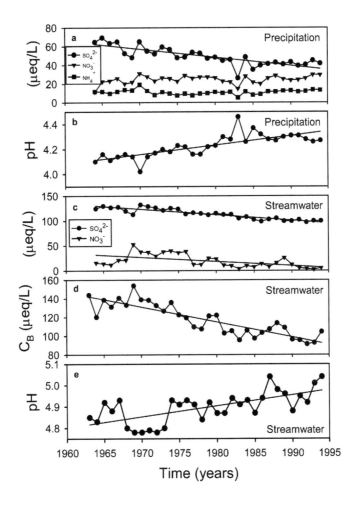

Figure 10.2 Long-term trends in volume-weighted annual mean concentrations of SO_4^{2-}, NO_3^-, NH_4^+, (a) and pH (b) in bulk precipitation, and SO_4^{2-}, NO_3^- (c), the sum of base cations (C_B; d), and pH (e) in stream water in watershed 6 of the Hubbard Brook Experimental Forest for 1963 to 1994.

CAAA (Likens et al., 2001). Using back trajectory analysis of air masses (Draxler and Hess, 1998), Butler et al. (2001) identified the approximate emissions source region for atmospheric deposition of S and N compounds to the study region in the northeastern United States (Figure 10.1). Annual mean concentrations of SO_4^{2-} in bulk precipitation at the HBEF were strongly correlated with annual SO_2 emissions based on both 15-hour ($r^2 = 0.74$; Figure 10.3) and 21-hour ($r^2 = 0.74$) back trajectories (Likens et al., 2001). Emissions from Ontario and Quebec appear to have contributed little (<10%) to the SO_4^{2-} deposition for the study region in the 1990s (Environment Canada, 1998; Butler et al., 2001). In contrast to SO_4^{2-}, there have been no long-term trends in annual volume-weighted concentrations of NO_3^- in bulk precipitation at the HBEF (Figure 10.2). This lack of a long-term pattern is consistent with the minimal changes in NO_X emissions over the last 30 years.

The beneficial influence of national clean air legislation is also reflected in the strong relationship between historical reductions in air emissions from the source region and decreased deposition of S throughout the northeastern United States, including the HBEF. As SO_2 emissions declined in the 1980s and 1990s in response to the CAAA, the geographic area exposed to elevated wet deposition of S in excess of 25 kg SO_4^{2-} ha^{-1}yr^{-1} decreased

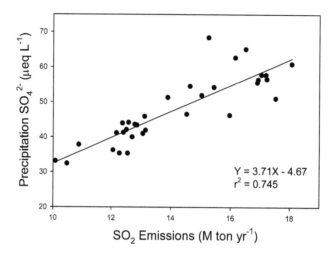

Figure 10.3 Volume-weighted annual concentrations of SO_4^{2-} in bulk precipitation at the Hubbard Brook Experimental Forest as a function of annual emissions of SO_2 for the source-area based on 15-hour back trajectories (see Figure 10.1; modified after Likens et al., 2001).

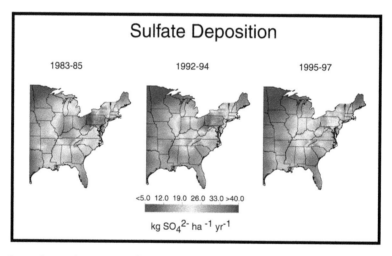

Figure 10.4 Annual wet deposition of SO_4^{2-} (in kg SO_4^{2-} ha^{-1} yr^{-1}) in the eastern United States for 1983–85, 1992–94, and 1995–97. Data were obtained for the NADP/NTN and the model of Grimm and Lynch (1997). See color figures following page 200.

(Figure 10.4). In 1995–1997, following implementation of Phase I of the Acid Deposition Control Program, emissions of SO_2 in the source area and concentrations of SO_4^{2-} in both bulk deposition at the HBEF (watershed 6) and wet-only deposition at NADP sites in the Northeast were about 20% lower than in the preceding 3 years, although not significantly different from the long-term trend (Likens et al., 2001). Nitrate and NH_4^+ concentrations decreased less than 10% during the same period. Year-to-year variations in precipitation across the region influenced the magnitude and spatial distribution of changes in S and N wet deposition between the periods of 1992–1994 and 1995–1997, which complicated the relationships between emissions and deposition (Lynch et al., 2000; Likens et al., 2001).

The Midwest is also a significant source of atmospheric NH_3. About half of the NH_3 emitted to the atmosphere is typically deposited within 50 km of its source (Ferm, 1998).

However, high concentrations of SO_2 and NO_x can greatly lengthen atmospheric transport of NH_3 through the formation of ammonium sulfate and ammonium nitrate aerosols; these submicron particles are transported distances similar to SO_2 (>500 km). Ammonium is an important component of atmospheric N deposition. For example, an average of 31% of dissolved inorganic N in annual bulk deposition at the HBEF occurs as NH_4^+.

Dry deposition contributes a considerable amount of S and N to the Northeast, although accurate measurements are difficult to obtain (see above). At 10 sites located throughout the United States, Lovett (1994) estimated that dry deposition of S was 9 to 59% of total deposition (wet + dry + cloud), dry deposition of NO_3^- was 25 to 70% of total NO_3^- deposition, and dry deposition of NH_4^+ was 2 to 33% of total NH_4^+ deposition. This variability is, in part, a result of proximity of sites to high-emission areas and of the relative contribution of cloud and fog deposition.

Question 2: What are the effects of acidic deposition on terrestrial and aquatic ecosystems in the northeastern United States, and how have these ecosystems responded to changes in emissions and deposition?

Terrestrial–aquatic linkages

Many of the impacts of acidic deposition depend on the rate at which acidifying compounds are deposited from the atmosphere compared to the rate at which acid-neutralizing capacity (ANC) is generated within the ecosystem. Acid-neutralizing capacity is a measure of the ability of water or soil to neutralize inputs of strong acid and is largely the result of terrestrial processes such as mineral weathering, cation exchange, and immobilization of SO_4^{2-} and N (Charles, 1991). Acid-neutralizing processes occur in the solution phase, and their rates are closely linked with the movement of water through terrestrial and aquatic ecosystems. The effects of acidic deposition on ecosystem processes must therefore be considered within the context of the hydrologic cycle, which is a primary mechanism through which materials are transported from the atmosphere to terrestrial ecosystems and eventually into surface waters.

The effects of acidic deposition on surface waters vary seasonally and with stream flow. Surface waters are often most acidic in spring following snowmelt and rain events. In some waters the ANC decreases below 0 µeq L^{-1} only for short periods (i.e., hours to weeks), when discharge is highest. This process is called episodic acidification. Other lakes and streams, referred to as chronically acidic, maintain ANC values less than 0 µeq L^{-1} throughout the year.

Precipitation (and/or snowmelt) can raise the water table from the subsoil into the upper soil horizons, where acid-neutralizing processes (e.g., mineral weathering, cation exchange) are generally less effective than in the subsoil. Water draining into surface waters during high-flow episodes is therefore more likely to be acidic (i.e., ANC < 0 µeq L^{-1}) than water that has discharged from the subsoil, which predominates during drier periods.

Both chronic and episodic acidification can occur either through strong inorganic acids derived from atmospheric deposition and/or by natural processes. Natural acidification processes include the production and transport of organic acids derived from decomposing plant material, or inorganic acids originating from the oxidation of naturally occurring S or N pools (i.e., pyrite, N_2-fixation followed by nitrification) from the soil to surface waters. Here we focus on atmospheric deposition of strong inorganic acids, which dominate the recent acidification of soil and surface waters in the northeastern United States.

Effects of acid deposition on soils

The observation of elevated concentrations of inorganic monomeric Al in surface waters provided strong evidence of soil interactions with acidic deposition (Driscoll et al., 1980; Cronan and Schofield, 1990). Recent studies have shown that acidic deposition has changed the chemical composition of soils by depleting the content of available plant nutrient cations (i.e., Ca^{2+}, Mg^{2+}, K^+), increasing the mobility of Al and increasing the S and N content.

Depletion of base cations and mobilization of aluminum in soils

Acidic deposition has increased the concentrations of protons (H^+) and strong acid anions (SO_4^{2-} and NO_3^-) in soils of the northeastern United States, which has led to increased rates of leaching of base cations and the associated acidification of soils. If the supply of base cations is sufficient, the acidity of the soil water will be effectively neutralized. However, if base saturation (exchangeable base cation concentration expressed as a percentage of total cation exchange capacity) is below 20%, atmospheric deposition of strong acids results in the mobilization and leaching of Al, and the neutralization of H^+ will be incomplete (Cronan and Schofield, 1990).

Mineral weathering is the primary source of base cations in most watersheds, although atmospheric deposition may provide important inputs to sites with very low rates of supply from mineral sources. In acid-sensitive areas, rates of base cation supply through chemical weathering are not adequate to keep pace with leaching rates accelerated by acidic deposition. Recent studies based on analysis of soil (Lawrence et al., 1999), long-term trends in stream water chemistry (Likens et al., 1996, 1998; Lawrence et al., 1999), and the use of strontium stable isotope ratios (Bailey et al., 1996) indicate that acidic deposition has enhanced the depletion of exchangeable nutrient cations in acid-sensitive areas of the Northeast. At the HBEF, Likens et al. (1996) reported a long-term net decline in soil pools of available Ca^{2+} during the last half of the 20th century as acidic deposition reached its highest levels. Loss of ecosystem Ca^{2+} peaked in the mid-1970s and abated over the next 15 to 20 years, as atmospheric deposition of SO_4^{2-} declined.

Without strong acid anions, cation leaching in forest soils of the Northeast is largely driven by naturally occurring organic acids derived from decomposition of organic matter, primarily in the forest floor. Once base saturation is reduced in the upper mineral soil, organic acids tend to mobilize Al through formation of organic Al complexes, most of which are deposited lower in the soil profile through adsorption to mineral surfaces. This process, termed podzolization, results in surface waters with low concentrations of Al that are primarily in a nontoxic, organic form (Driscoll et al., 1988). Acidic deposition has altered podzolization, however, by solubilizing Al with inputs of mobile inorganic anions, which facilitates transport of inorganic Al into surface waters. Input of acidic deposition to forest soils with base saturation values less than 20% increases Al mobilization and shifts chemical speciation of Al from organic to inorganic forms that are toxic to terrestrial and aquatic biota (Cronan and Schofield, 1990).

Accumulation of sulfur in soils

Watershed input–output budgets developed in the 1980s for northeastern forest ecosystems indicated that the quantity of S exported by surface waters (primarily as SO_4^{2-}) was essentially equivalent to inputs from atmospheric deposition (Rochelle and Church, 1987). These findings suggested that decreases in atmospheric S deposition, from controls on emissions, should result in equivalent decreases in the amount of SO_4^{2-} entering surface waters. Indeed, there have been long-term decreases in concentrations of SO_4^{2-} in surface waters throughout the Northeast following declines in atmospheric S deposition after the 1970 CAAA (Likens et al., 1990; Stoddard et al., 1999). However, recent watershed mass balance studies in the Northeast have shown that watershed loss of SO_4^{2-} exceeds atmospheric S deposition (Driscoll et al., 1998). This pattern suggests that decades of atmospheric S deposition have resulted in the accumulation of S in forest soils. With recent declines in atmospheric S deposition and a possible warming-induced enhancement of S mineralization from soil organic matter, previously retained S is gradually being released to surface waters (Driscoll et al., 1998).

Past accumulation of atmospherically deposited S is demonstrated by a strong positive relationship between wet deposition of SO_4^{2-} and concentrations of total S in the forest floors of red spruce stands in the Northeast (Figure 10.5a). It is now expected that the release of SO_4^{2-} that previously accumulated in watersheds from inputs of atmospheric S deposition will delay the recovery of surface waters in response to SO_2 emission controls (Driscoll et al., 1998). Imbalances in ecosystem S budgets may also be influenced by weathering of S-bearing minerals or by underestimation of dry deposition inputs of S. Further effort is needed to accurately quantify these processes.

Accumulation of nitrogen in soils

Nitrogen is generally considered the growth-limiting nutrient for temperate forest vegetation, and retention by forest ecosystems generally is high. As a result, concentrations of NO_3^- are often very low in surface waters draining forest landscapes. However, recent research indicates that atmospheric N deposition has accumulated in soils, and some forest ecosystems have exhibited diminished retention of N inputs. Total N concentration in the forest floor of red spruce forests is correlated with wet N deposition at both low (Figure 10.5b) and high elevations in the Northeast (McNulty et al., 1990). A record of stream chemistry in forest watersheds of the Catskill Mountains (New York) has shown increasing NO_3^- concentrations since 1920, apparently in response to increases in atmospheric N deposition (Charles, 1991). Increased stream NO_3^- concentrations have also been observed following experimental N additions to a small watershed in Maine (Norton et al., 1994). Nitrate behaves much like SO_4^{2-} facilitating the displacement of cations from the soil and acidifying surface waters.

Increased losses of NO_3^- to surface waters may be indicative of changes in the strength of plant and soil microbial N sinks in forest watersheds. Because microbial processes are highly temperature sensitive, fluctuations in microbial immobilization and mineralization in response to climate variability affect NO_3^- losses in drainage waters. Murdoch et al. (1998) found that annual mean NO_3^- concentrations in stream water were not related to annual wet N deposition but rather to mean annual air temperature; increases in temperature corresponded to increases in stream water concentrations. Mitchell et al. (1996) found that unusually low winter temperatures that led to soil freezing corresponded to increased loss of NO_3^- to surface waters. The sensitivity of NO_3^- release to climatic fluctuations tends to increase the magnitude and frequency of episodic acidification of surface waters.

Despite the linkage between atmospheric deposition of NH_4^+ and NO_3^- and loss of NO_3^- from forest ecosystems (Dise and Wright, 1995), future effects of atmospheric N deposition on forest N cycling and surface water acidification are likely to be controlled by climate, forest history, and forest type (Aber et al., 1997; Lovett et al., 2000). For example, forests regrowing after agricultural clearing or fire tend to have a higher capacity for accumulating N without release to surface waters compared to undisturbed forests (Aber et al., 1998; Hornbeck et al., 1997). The complexity of linkages of NO_3^- loss to climatic variation, land-use history, and vegetation type has slowed efforts to predict future

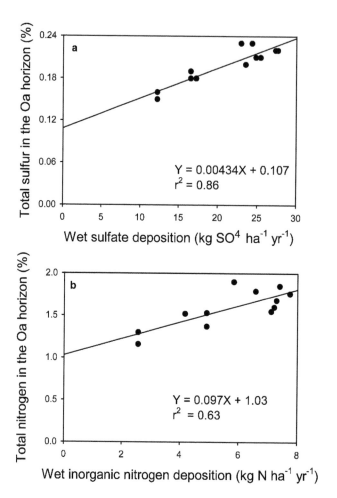

Figure 10.5 The concentration of total S in the soil Oa horizons as a function of wet SO_4^{2-} deposition (a) and total N in the soil Oa horizons as a function of total inorganic N (NO_3^- and NH_4^+) in wet deposition (b) in 12 red spruce stands located from the western Adirondacks in New York to eastern Maine (Lawrence, G., unpublished data).

responses of surface water ANC to anticipated changes in atmospheric N deposition associated with NO_x or NH_3 emission controls. Improved predictions will depend on continued progress in understanding how forest ecosystems retain N and in determining regional-scale information on land-use history. Despite this uncertainty, it is apparent that additional NH_4^+ and NO_3^- inputs to northeastern forests will increase the potential for increases in leaching losses of NO_3^-, whereas reductions in NO_x and NH_3 emissions and subsequent N deposition will contribute to long-term decreases in watershed acidification.

Effects of acidic deposition on trees

Observations of extensive dieback in stands of high-elevation red spruce *Picea rubens* beginning in the 1960s (Siccama et al., 1982) and in sugar maple *Acer saccharum* stands starting in the 1980s (Houston, 1999) led to investigations of effects of acidic deposition on trees. This research has focused on the direct effects of acidic precipitation and cloud-water on foliage and on indirect effects from changes in soils that alter nutrient uptake by roots. The mechanisms by which acidic deposition causes stress to trees are only partially understood but generally involve interference with Ca^{2+} nutrition and Ca-dependent cellular processes (DeHayes et al., 1999). The depletion of Ca^{2+} in forest soils, described earlier, raises concerns regarding the health and productivity of northeastern forests (McLaughlin and Wimmer, 1999; DeHayes et al., 1999). Progress on understanding the effects of acidic deposition on trees has been limited by the long response time of trees to environmental stresses, the difficulty in isolating possible effects of acidic deposition from other natural and anthropogenic stresses, and insufficient information on how acidic deposition has changed soils. To date, investigations of possible effects of acidic deposition on trees in the Northeast have focused primarily on red spruce and sugar maple.

Red spruce

There is strong evidence that acidic deposition causes dieback (reduced growth that leads to mortality) of red spruce by decreasing cold tolerance. Red spruce is common in Maine, where it is an important commercial species. It is also common at high elevations in mountainous regions throughout the Northeast, where it is valued for recreation, aesthetics, and as a habitat for unique and endangered species. Dieback has been most severe at high elevations in the Adirondack and Green Mountains, where over 50% of the canopy trees died in the 1970s and 1980s. In the White Mountains, about 25% of the canopy spruce died during that period (Craig and Friedland, 1991). Dieback of red spruce trees has also been observed in mixed hardwood–conifer stands at relatively low elevations in the western Adirondack Mountains that receive high inputs of acidic deposition (Shortle et al., 1997).

Results of controlled exposure studies show that acidic mist or acidic cloudwater reduces the cold tolerance of current-year red spruce needles by 3° to 10°C (DeHayes et al., 1999); this condition can be harmful because current-year needles are only marginally tolerant of minimum winter temperatures typical of upland regions in the Northeast. Hydrogen ion in acidic deposition leaches membrane-associated Ca^{2+} from needles, which increases their susceptibility to freezing. An increased frequency of winter injury in the Adirondack and Green Mountains since 1955 coincides with increased exposure of red spruce canopies to highly acidic cloudwater (Johnson et al., 1984). Recent episodes of winter injury (loss of current-year needles) have been observed throughout much of the range of red spruce in the Northeast (DeHayes et al., 1999).

Calcium depletion and Al mobilization may also affect red spruce in the Northeast. Low ratios of Ca^{2+} to Al in soil have been associated with dysfunction of fine roots, responsible for water and nutrient uptake (Shortle and Smith, 1988). Aluminum can block the uptake of Ca^{2+}, which can lead to reduced growth and increased susceptibility to stress. From an extensive review of these studies, Cronan and Grigal (1995) concluded that a Ca^{2+} to Al ratio of less than 1.0 in soil water indicated a greater than 50% probability of impaired growth in red spruce. They also cited examples of studies from the Northeast, where soil solutions in the field have been found to exhibit Ca/Al ionic ratios <1.0. These findings suggest that a Ca/Al ratio of 1.0 in soil waters of forest ecosystems may serve as a useful index for tracking the recovery of terrestrial ecosystems from the deleterious effects of acidic deposition.

To establish a stronger direct link between Ca/Al ionic ratios and red spruce dieback, several issues need to be addressed: (1) the uncertainty of extrapolating from controlled seedling experiments to responses of mature trees in the field, (2) the fact that declining forest stands may be exposed simultaneously to multiple stresses, and (3) the difficulty of quantifying the rhizosphere solution chemistry and Ca/Al ionic ratios of soil horizons containing roots of mature trees in the field. Other studies of historical changes in wood chemistry of red spruce have found a strong relationship between Ca concentrations in tree rings, trends in atmospheric deposition, and presumed changes in soil Ca^{2+} availability, suggesting that acidic deposition has altered the mineral nutrition of red spruce (Shortle et al., 1997). Although Ca concentrations in sapwood typically decrease steadily from older to younger wood, a consistent increase of Ca concentration in tree rings formed from about 1950 to 1970 has been documented in red spruce trees throughout the Northeast. Peak levels of acidic deposition during that period apparently caused elevated concentrations of Ca^{2+} in soil water and increased uptake of Ca^{2+} by roots (Shortle et al., 1997). Following that pulse of soil leaching, it is hypothesized that depletion of soil Ca resulted in decreased Ca^{2+} concentrations in soil water, decreased plant uptake of Ca^{2+}, and diminished Ca concentrations in subsequent tree rings. This scenario is illustrated by a trend in enrichment frequency of Ca concentrations in wood (the percentage of samples with a higher Ca concentration in 10 years of wood tissue than in the previous 10 years of wood tissue) that was relatively stable from 1910 to 1950, increased from 1950 to 1970, and then decreased to low levels in the period 1970 to 1990 (Shortle et al., 1997).

Sugar maple

Dieback of sugar maple has been observed at several locations in the Northeast since the 1950s but has recently been most evident in Pennsylvania, where crown dieback has led to extensive mortality in some forest stands (basal area of dead sugar maple ranging from 20 to 80% of all sugar maple trees; Drohan et al., 1999). High rates of tree mortality tend to be triggered by periodic stresses such as insect infestations and drought. Periodic dieback of sugar maple has been attributed to forest- and land-use practices that have encouraged the spread of this species to sites that are either drought-prone or have nutrient poor-soils. On these sites, the trees are less able to withstand stresses without experiencing growth impairment and mortality (Houston, 1999).

Acidic deposition may contribute to episodic dieback of sugar maple by causing depletion of nutrient cations from marginal soils. Long et al. (1997) found that liming ($CaCO_3$ addition) significantly increased sugar maple growth, improved crown vigor, and increased flower and seed crops of overstory sugar maple in stands that were experiencing dieback. Liming also increased exchangeable base cation concentrations in the soil and decreased concentrations of exchangeable Al.

Further evidence of a link between soil base cation status and periodic dieback of sugar maple has been reported by Horsley et al. (1999), who found that dieback at 19 sites in northwestern and north-central Pennsylvania and southwestern New York was correlated with combined stress from defoliation and deficiencies of Mg and Ca. Dieback occurred predominantly on ridgetops and upper slopes, where soil base availability was much lower than at mid- and low slopes of the landscape (Bailey et al., 1999). These studies suggest that depletion of nutrient base cations in soil by acidic deposition may have reduced the area favorable for the growth of sugar maple in the Northeast. Factors such as soil mineralogy and landscape position affect soil base status as well as acidic deposition, complicating assessments of the extent of sugar maple dieback attributable to acidic deposition.

Effects on surface waters

Inputs of acidic deposition to regions with base-poor soils has resulted in the acidification of soil waters, shallow ground waters, streams, and lakes in areas of the northeastern United States and elsewhere. In addition, perched seepage lakes, which derive water largely from direct precipitation inputs, are highly sensitive to acidic deposition (Charles, 1991). These processes usually result in decreases in pH and, for drainage lakes, increases in concentrations of inorganic monomeric Al. These changes in chemical conditions are toxic to fish and other aquatic animals.

Surface water chemistry

To evaluate the regional extent of lake acidification, data from a survey of lakes in the Northeast in 1991–1994 were used (EMAP; Larsen et al., 1994; Stevens, 1994). This probability-based survey allows inferences to be made about the entire population of lakes in the Northeast (10,381 lakes with surface area >1 ha in New York and New England). Other surveys conducted at different times, or with different criteria for minimum lake size, have shown somewhat different results (e.g., Kretser et al., 1989; Charles, 1991).

The Northeast EMAP survey was conducted during low-flow summer conditions, so the water chemistry likely represents the highest ANC values for the year. Lakes were subdivided into ANC classes. Lakes with ANC values below 0 µeq L^{-1} are considered to be chronically acidic; these lakes are acidic throughout the year. Lakes with ANC values between 0 and 50 µeq L^{-1} are considered susceptible to episodic acidification; ANC may decrease below 0 µeq L^{-1} during high-flow conditions in these lakes. Finally, lakes with ANC values greater than 50 µeq L^{-1} are considered relatively insensitive to inputs of acidic deposition.

Results from the EMAP survey indicate that in the Adirondack region of New York (1812 lakes) 41% of the lakes are chronically acidic or sensitive to episodic acidification (10% have ANC values <0 µeq L^{-1}; 31% have ANC 0–50 µeq L^{-1}). In New England and the eastern Catskill region of New York (6834 lakes), 5% of the lakes have ANC values <0 µeq L^{-1}, and 10% of the lakes have ANC values between 0 and 50 µeq L^{-1}. Most of the acidic and acid-sensitive surface waters in New York State are located in the Adirondack and Catskill regions. This regional variation in ANC is largely controlled by the supply of Ca^{2+} and Mg^{2+} to surface waters (ANC = -58 + 0.85 × (Ca^{2+} + Mg^{2+}); r^2 = 0.94; concentrations expressed in µeq L^{-1}).

To quantify the nature of the acid inputs, the distribution of anions was examined (i.e., SO_4^{2-}, NO_3^-, Cl^-, HCO_3^-, and organic anions) in acid-sensitive lakes of the Northeast (ANC <50 µeq L^{-1}; 1875 lakes). Naturally occurring organic anions were not measured directly but were estimated using the charge–balance approach (Driscoll et al., 1994b). Results of the analysis can be summarized as follows: 83% of the acid-sensitive lakes (ANC <50 µeq L^{-1}) were dominated by inorganic anions, with SO_4^{2-} constituting 82% of the total anionic charge; 17% of the acid-sensitive lakes were dominated by naturally occurring organic anions and were assumed to be naturally acidic lakes — organic anions accounted for an average of 71% of the total anions in that group of lakes. The acidity of organic-acid-dominated lakes was supplemented by sulfuric acid from atmospheric deposition, so that SO_4^{2-} contributed an average of 19% of the anionic charge in these naturally acidic lakes.

Seasonal and episodic acidification of surface waters

In the Northeast, the most severe acidification of surface water generally occurs during spring snowmelt (Charles, 1991); short-term acid episodes also occur during midwinter snowmelts and large precipitation events in summer or fall (Wigington et al., 1996).

Data from Buck Creek in the Adirondacks, part of the Episodic Response Project (ERP), illustrate the seasonal and episodic changes in water chemistry of acid-sensitive surface waters in the Northeast (Figure 10.6). In the ERP, acidic events and subsequent mortality of brook trout and blacknose dace were monitored in streams in the Adirondacks, Catskills, and Appalachian Plateau of Pennsylvania (Wigington et al., 1996). All streams had low ANC values and physical habitats judged suitable for fish survival and reproduction, and all had indigenous fish populations in at least part of the stream ecosystem (Baker et al., 1996).

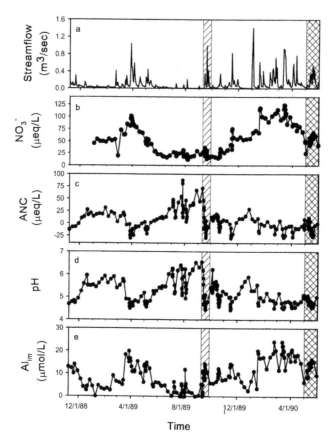

Figure 10.6 Seasonal changes in flow (a), nitrate (NO_3^-; b), acid-neutralizing capacity (ANC; c), pH (d), and inorganic monomeric aluminum (Al_{im}; e) at Buck Creek in the Adirondack region of New York. The cross-hatched area represents a significant event of episodic acidification. The double-cross-hatched area represents the period over which an *in situ* bioassay was conducted (see Figure 10.9).

Buck Creek exhibited both seasonal and event-driven changes in chemistry. The seasonal pattern in ANC corresponded to seasonal changes in NO_3^- ($r^2 = 0.44$). Stream NO_3^- concentrations were lowest in summer because of vegetation uptake of N, while ANC values were at the annual maximum. Stream NO_3^- increased and ANC decreased during fall, coinciding with increased flow and decreased plant activity. Nitrate concentrations increased and ANC values decreased during winter, with maximum NO_3^- concentrations and minimum ANC values occurring during spring snowmelt. Seasonal increases in NO_3^- were also associated with increases in inorganic monomeric Al concentrations ($r^2 = 0.93$). Superimposed on these seasonal patterns were event-driven changes in stream chemistry, such as occurred at Buck Creek on 15 September 1989 (see Figure 10.6). During this event, flow increased from 0.008 to 0.36 m^3 s^{-1}, which resulted in increases in NO_3^- concentrations (20 to 37 μeq L^{-1}), decreases in ANC (46 to –30 μeq L^{-1}) and pH (6.2 to 4.7), and increases in concentrations of inorganic monomeric Al (0.8 to 10 μmol L^{-1}).

Long-term changes in surface water chemistry

Unfortunately, there are limited data documenting the responses to atmospheric deposition since the time of the Industrial Revolution (Charles, 1991) and few tools to predict the future effects of atmospheric deposition. Acidification models have been used to estimate past and future acidification effects (Eary et al., 1989). The model PnET (PnET-CN; Aber and Federer, 1992; Aber et al., 1997; Aber and Driscoll, 1997) is a simple, generalized and well-validated model that provides estimates of forest net primary productivity, nutrient uptake by vegetation, and water balances. Recently, PnET was coupled with a soil model that simulates abiotic soil processes (e.g., cation exchange, weathering, adsorption, and solution speciation), resulting in a comprehensive forest-soil-water model, PnET-BGC, designed to simulate element cycling in forest and interconnected aquatic ecosystems (Kram et al., 1999; Gbondo-Tugbawa et al., 2001). The PnET models have been used extensively at the HBEF to investigate the effects of disturbance (e.g., cutting, climatic disturbance, air pollution) on forest and aquatic resources (Aber et al., 1997; Aber and Driscoll, 1997; Gbondo-Tugbawa et al., 2001).

From relationships between current emissions and deposition (e.g., Figure 10.3) and estimates of past emissions (USEPA, 2000), historical patterns of atmospheric deposition of S and N were reconstructed at the HBEF. In addition, we considered land disturbances to the watershed, including logging in 1918–1920 and hurricane damage in 1938. We calculated the response of vegetation, soil, and stream water to this deposition scenario with PnET-BGC (Figure 10.7). A detailed description of the application of PnET-BGC to the HBEF is available in Gbondo-Tugbawa et al. (2001). It is estimated that total atmospheric deposition (wet + dry) of S at the HBEF increased from 7 kg SO_4^{2-} ha^{-1} yr^{-1} in 1850 to a most recent peak of 68 kg SO_4^{2-} ha^{-1} yr^{-1} in 1973 and has decreased since that time. We also estimate that past soil base saturation (circa 1850) was ~20%, stream SO_4^{2-} concentration was approximately 10 μeq L^{-1}, stream ANC was about 40 μeq L^{-1}, stream pH was about 6.3, and stream Al concentration was below 2 μmol L^{-1}, 50% of which was in an organic form. Compared to model hindcast approximations, current conditions at the HBEF indicate that soil percentage base saturation has decreased to about 10% in response to acidic deposition and accumulation of nutrient cations by forest biomass. Further, acidic deposition has contributed to a nearly fourfold increase in stream SO_4^{2-}, a decrease in ANC from positive to negative values, a decrease in stream pH to below 5.0, and increases in stream Al, largely occurring as the toxic inorganic form (>10 μmol L^{-1}). Substantial deterioration in the acid-base status of soil and water at the HBEF is indicated over the 1850–1970 period. Model calculations suggest that strong acid inputs associated with mineralization of soil organic matter following forest cutting in the 1910s resulted in the short-term (i.e., 2 to 3 years) acidification of stream water.

Since 1964, stream water draining the HBEF reference watershed (watershed 6) has shown a significant decline in annual volume-weighted concentrations of SO_4^{2-} (–1.1 μeq L^{-1}yr^{-1}; Figure 10.2). This decrease in stream SO_4^{2-} corresponds to both decreases in atmospheric emissions of SO_2, and bulk precipitation concentrations of SO_4^{2-} (Likens et al., 2001). In addition, there has been a long-term decrease in stream concentrations of NO_3^- that is not correlated with a commensurate change in emissions of NO_X or in bulk deposition of NO_3^-. These long-term declines in stream concentrations of strong acid anions (SO_4^{2-} + NO_3^-; –1.9 μeq L^{-1} yr^{-1}) have resulted in small but significant increases in pH. The increase in stream pH has been limited by marked concurrent decreases in the sum of base cations (–1.6 μeq L^{-1} yr^{-1}).

A similar pattern is evident in the Adirondack and Catskill regions of New York and New England. Analysis of data from the EPA Long-Term Monitoring (LTM) Program, initiated in the early 1980s, showed significant declines in surface water SO_4^{2-} and in the sum of strong acid anions (SO_4^{2-} + NO_3^-) in the Adirondack/Catskill and New England subregions (Stoddard et al., 1999). Note that the rate of decline in SO_4^{2-} for Adirondack/Catskill surface waters (–1.9 μeq L^{-1} yr^{-1}) was somewhat greater than values observed for New England (–1.3 μeq L^{-1} yr^{-1}). In contrast to the patterns at the HBEF, regional sites

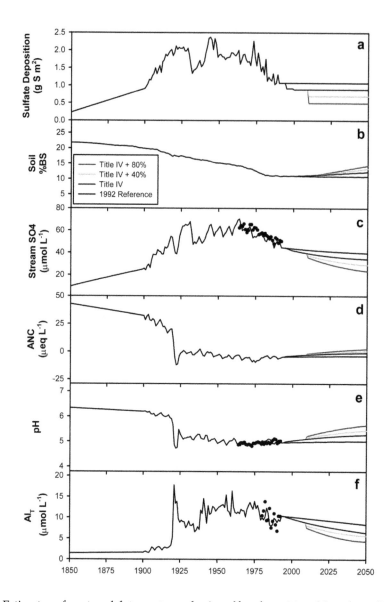

Figure 10.7 Estimates of past and future atmospheric sulfur deposition (a) and predictions of soil and stream water chemistry using the model PnET-BGC for watershed 6 of the Hubbard Brook Experimental Forest, New Hampshire. Model predictions include percentage soil base saturation (b), stream sulfate (SO_4^{2-}) concentrations (c), stream acid-neutralizing capacity (ANC; d), pH (e), and stream concentrations of aluminum (Al_T; f). Future predictions include four scenarios of atmospheric sulfur deposition: constant values from 1992, values anticipated following the 1990 Amendments to the Clean Air Act, 22% reduction in atmospheric sulfur deposition in 2010 beyond the 1990 Amendments to the Clean Air Act, and 44% reduction in atmospheric sulfur deposition in 2010 beyond the 1990 Amendments to the Clean Air Act. These latter two scenarios depict 40 to 80% reductions in utility emissions of sulfur dioxide. Actual annual volume-weighted concentrations of stream water are shown for comparison.

showed no significant trends in concentrations of NO_3^-. Surface waters in New England showed modest increases in ANC (+0.8 µeq L^{-1} yr^{-1}), but no increase in ANC was evident in the Adirondack/Catskill subregion. This difference was caused by the marked decrease in the sum of base cations in the Adirondack/Catskill subregion (–2.7 µeq L^{-1} yr^{-1})

compared to the New England subregion (-0.7 µeq L^{-1} yr^{-1}). These patterns suggest that the lack of recovery of Adirondack/Catskill surface waters in comparison to New England surface waters reflects the historically higher loading of acidic deposition in New York than in most of New England (Figure 10.4). This higher acid input has evidently resulted in greater depletion of exchangeable base cations in acid-sensitive watersheds in New York (Stoddard et al., 1999).

Effects on aquatic biota

Acidification has marked effects on the trophic structure of surface waters. Decreases in pH and increased Al concentrations contribute to declines in species richness and abundance of zooplankton, macroinvertebrates, and fish (Schindler et al., 1985; Keller and Gunn, 1995).

High concentrations of both H^+ (measured as low pH) and inorganic monomeric Al are directly toxic to fish (Baker and Schofield, 1982). Although Al is abundant in nature, it is relatively insoluble in the neutral pH range and thus unavailable biologically. Acid-neutralizing capacity largely controls pH and the bioavailability of Al (Driscoll and Schecher, 1990). Thus, surface waters with low ANC and pH and high concentrations of inorganic monomeric Al are less hospitable to fish. Calcium, however, directly ameliorates the toxic stress caused by H^+ and Al (Brown, 1983). Watershed supply of Ca^{2+} also contributes to ANC, and therefore lakes with higher Ca^{2+} are more hospitable to fish, as indicated by the synoptic survey conducted by the Adirondack Lakes Survey Corporation (ALSC; Gallagher and Baker, 1990). Of the 1469 lakes surveyed by the ALSC, one or more fish species were caught in 1123 lakes (76%), whereas no fish were caught in 346 lakes (24%). The 346 fishless lakes in the Adirondack region had significantly ($p < 0.05$) lower pH, Ca^{2+} concentration, and ANC as well as higher concentrations of inorganic monomeric Al, in comparison to lakes with fish (Gallagher and Baker, 1990).

Small, high-elevation lakes in the Adirondacks are more likely to be fishless than larger lakes at low elevation (Gallagher and Baker, 1990) because they may be susceptible to periodic winter kills, have poor access for fish immigration, have poor fish spawning substrate, or have low pH. Nevertheless, small, high-elevation Adirondack lakes with fish also had significantly higher pH compared to fishless lakes. Acidity therefore, is likely to play an important role in the absence of fish from such lakes.

Numerous studies have shown that fish species richness (the number of fish species in a water body) is positively correlated with pH and ANC (Rago and Wiener, 1986; Kretser et al., 1989; Figure 10.8). Decreases in pH result in decreases in species richness by eliminating acid-sensitive species (Schindler et al., 1985). Of the 53 species of fish recorded by the Adirondack Lakes Survey Corporation (ALSC; Kretser et al., 1989), about half (26 species) are absent from lakes with pH of less than 6.0. These 26 species include important recreational fishes such as Atlantic salmon *Salmo salar*, tiger trout *Salmo trutta* × *Salvelinus fontinalis*, redbreast sunfish *Lepomis auritus*, bluegill *Lepomis macrochirus*, tiger musky *Esox lucius* × *Esox masquinongy*, walleye *Stizostedion vitreum*, alewife *Alosa pseudoharengus*, and kokanee *Oncorhynchus nerka* (Kretser et al., 1989) plus ecologically important minnows that serve as forage for sport fishes. Significantly, the most common fish species caught by the ALSC (brown bullhead *Ameiurus nebulosus*, yellow perch *Perca flavescens*, golden shiner *Notemigonus crysoleucas*, brook trout *Salvelinus fontinalis*, and white sucker *Catostomus commersoni*) also show the greatest tolerance of acidic conditions, as evidenced by their occurrence in lakes with relatively low pH and high Al concentrations (Gallagher and Baker, 1990).

There is a clear link between acidic water resulting from atmospheric deposition of strong acids and fish mortality. *In situ* bioassays conducted during acidic events (pulses

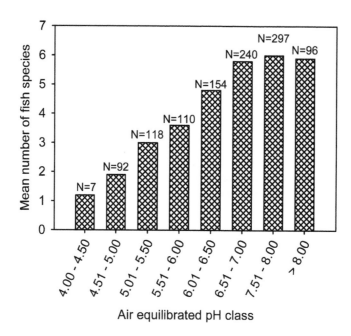

Figure 10.8 Distribution of the mean number of fish species for ranges of pH from 4.0 to 8.0 in lakes in the Adirondack region of New York. N represents the number of lakes in each pH category (modified from Kretser et al., *Adirondack Lakes Study. 1984–1987. An Evaluation of Fish Communities and Water Chemistry*, Adirondacks Lakes Survey Corporation, Ray Brook, New York. 1989).

of low-pH, Al-rich water following precipitation events or snowmelt) provide an opportunity to measure the direct, acute effects of stream chemistry on fish mortality; these experiments show that even acid-tolerant species, such as brook trout, are killed by acidic water in the Adirondacks (Figure 10.9; Baker et al., 1996; Van Sickle et al., 1996). Episodic acidification is particularly important in streams and rivers (compared to lakes) because these ecosystems experience large abrupt changes in water chemistry and provide limited refuge areas for fish. Baker et al. (1996) concluded that episodic acidification can have long-term negative effects on fish communities in small streams as a result of mortality, emigration, and reproductive failure.

The ERP study showed that streams with moderate to severe acid episodes had significantly higher fish mortality during bioassays than nonacidic streams (Van Sickle et al., 1996). The concentration of inorganic monomeric Al was the chemical variable most strongly related to mortality in the four test species (brook trout, mottled sculpin *Cottus bairdi*, slimy sculpin *Cottus cognatus*, and blacknose dace *Rhinichthys atratulus*). Because of their correlations with Al, variations in pH and Ca^{2+} concentrations were of secondary importance in accounting for mortality patterns. The ERP streams with high fish mortality during acid episodes also had lower brook trout density and biomass and lacked the more acid-sensitive species (blacknose dace and sculpins); radio-tagged brook trout in streams exhibiting episodic acidification emigrated downstream during episodes, whereas radio-tagged fish in nonacidic streams did not. In general, trout abundance was lower in ERP streams with median episode pH <5.0 and concentrations of inorganic monomeric Al >3.7 to 7.4 μmol L^{-1}. Acid-sensitive species were absent from streams with median episode pH <5.2 and concentrations of inorganic monomeric Al >3.7 μmol L^{-1}.

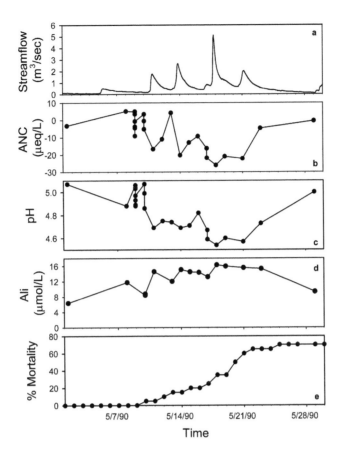

Figure 10.9 Results of an *in situ* bioassay in Buck Creek, Adirondacks, in spring 1990: discharge (a), acid-neutralizing capacity (b), pH (c), concentration of inorganic monomeric aluminum (d), and cumulative percentage mortality of brook trout over time (e). This figure represents the bioassay performed during the last shaded area in Figure 10.6.

Question 3: How do we expect emissions and deposition to change in the future, and how might ecosystems respond to these changes?

To date, major electric utilities in the United States have met or surpassed the Phase I SO_2 emission reduction target established by the Acid Deposition Control Program of the 1990 CAAA (see Figure 10.1; Lynch et al., 2000). Nevertheless, reports suggest that this emission target will not protect sensitive ecosystems (United States Environmental Protection Agency, 1995; Likens et al., 1996; Stoddard et al., 1999). This concern has spurred the introduction of several bills in Congress calling for deeper cuts in utility SO_2 and NO_x emissions. With acidic deposition resurfacing as a national environmental issue, decision makers need to determine if additional emissions reductions are needed and to what extent any further reductions will promote recovery from acidic deposition. To help address these issues, we present a conceptual framework for understanding ecosystem recovery from acidic deposition, suggest numerical indicators of chemical recovery, review current bills calling for utility emissions reductions, and use the model PnET-BGC to estimate changes in deposition corresponding to proposed emissions reductions and to predict ecosystems responses at the HBEF.

Ecosystem recovery

Acidic deposition disturbs forest and aquatic ecosystems by giving rise to harmful chemical conditions. Atmospheric S deposition to the northeastern United States has increased more than fivefold over the last 150 years (Charles, 1991; Figure 10.7), and most acid-sensitive ecosystems have been exposed to high inputs of strong acids for many decades. Since the 1970 CAAA there have been significant decreases in atmospheric S deposition, and as a result some aquatic ecosystems in the Northeast have been experiencing some chemical recovery (Stoddard et al., 1999). There are several critical chemical thresholds that appear to coincide with the onset of deleterious effects to biotic resources, including: a molar Ca/Al ratio of soil water <1 and soil percentage base saturation <20%, which indicates that forest vegetation is at risk with respect to soil acidification from acidic deposition (Cronan and Schofield, 1990; Cronan and Grigal, 1995), and surface water pH <6.0, ANC <50 μeq L^{-1}, and/or concentrations of inorganic monomeric Al >2 μmol L^{-1}, which indicate that aquatic biota are at risk from surface water acidification caused by acidic deposition (MacAvoy and Bulger, 1995). These values can also be used as indicators of chemical recovery (e.g., soil water Ca/Al >1, soil percentage base saturation >20%, surface water pH >6.0, ANC >50 μeq L^{-1}, inorganic monomeric Al <2 μmol L^{-1}), which are necessary for the restoration of ecosystem structure and function.

Although there is limited experience and understanding of acidification recovery, particularly at the ecosystem level, we envision that the process will involve two phases. Initially, decreases in acidic deposition following emission controls will facilitate a chemical recovery phase in forest and aquatic ecosystems. Recovery time for the first phase will vary widely across ecosystems and will be a function of (1) the magnitude of decreases in atmospheric deposition, (2) local depletion of exchangeable soil pools of base cations, (3) local rate of mineral weathering and atmospheric inputs of base cations, and (4) the extent and rate to which soil pools of S and N are released as SO_4^{2-} or NO_3^- to drainage waters (Galloway et al., 1983). In acidic soils with low base saturation, it is expected that reductions in concentrations of strong acid anions will result in little initial improvement in ANC of surface waters (Likens et al., 1996, 1998; Stoddard et al., 1999) and in soil Ca/Al ionic ratios. Only as the soil base status increases will these sites begin recovery of ANC. Other delays may occur in soils where atmospheric deposition has caused accumulation of S and N that will be released gradually through desorption or mineralization under conditions of lower atmospheric loading. In most cases it seems likely that chemical recovery will require decades, even with additional controls on emissions. Chemical recovery can be enhanced at specific sites (e.g., lakes, streams, watersheds of interest) by base addition (e.g., liming) (Driscoll et al., 1996).

The second phase in ecosystem recovery is biological recovery, which can occur only if chemical recovery is sufficient to allow survival and reproduction of plants and animals. The time period for biological recovery is uncertain (Yan et al., 2003). We know little about the mechanisms and time frame for recovery of terrestrial ecosystems following decreases in acidic deposition, but it is likely to be at least decades after soil chemistry is restored. Research suggests that stream macroinvertebrate populations may recover relatively rapidly (~3 years), whereas lake zooplankton populations are likely to recover more slowly (~10 years), in response to improved chemical conditions (Gunn and Mills, 1998). Some fish populations may recover in 5 to 10 years after the recovery of zooplankton populations. Recovery of fish populations could be accelerated by stocking. Although it is unlikely that aquatic ecosystems could be restored to the exact conditions that existed before acidification, improvement in the chemical environment is expected to allow for the recovery of ecosystem function that supports improved biotic diversity and productivity.

Proposed emission reductions

The rate and extent of ecosystem recovery is related to the timing and degree of emissions reductions. We reviewed SO_2 and NO_x emissions reductions associated with the 1990 CAAA and five prominent bills introduced in Congress aimed at controlling utility emissions (Table 10.2). The results of this review were used to define inputs to a model of acidic deposition and ecosystem effects at the HBEF.

The five bills analyzed here call for significant cuts in utility emissions of SO_2 and NO_x. We adopt the emission estimates for each of the bills as reported by the Congressional Research Service and assume equal levels of compliance under each bill (Parker, 2000). We rely on EPA estimates for emissions resulting from the 1990 CAAA (United States Environmental Protection Agency, 2000). Under these assumptions, the five bills would reduce utility SO_2 emissions by another 50 to 67% and decrease utility NO_x emissions another 56 to 72% beyond Phase II of the Acid Deposition Control Program of the 1990 CAAA. At present, Senate Bill 172 and House Bills 25 and 657 would modify CAAA standards least, while Senate Bill 1949 would reduce emissions most (Table 10.2).

Four of the bills reviewed set an implementation deadline of 2005, and Senate Bill 1949 set a deadline of 2010. All bills establish year-round requirements for utility emissions of SO_2 and NO_x. Senate Bill 172 includes additional NO_x cuts from May to September to achieve a higher level of protection during the ozone season. Several bills retain the cap and trade structure, expanding this approach to include NO_X. All bills pertain to the 50 U.S. states and the District of Columbia except Senate 172, which is limited to the 48 contiguous states and the District of Columbia. The size of the utilities affected varies somewhat among the bills. Most bills apply to units with a generating threshold of 15 megawatts or greater. Senate Bill 1949 applies to all electric utility generating units, and Senate Bill 172 sets a threshold of 25 megawatts or greater.

Modeling of emissions scenarios

We used information from the bills described above (Table 10.2) as input to the PnET-BGC model to predict ecosystem responses at the HBEF to a range of emission reductions (Figure 10.7). We compared (1) S deposition without implementation of the 1990 CAAA, (2) S deposition following implementation of the 1990 CAAA, and (3) S deposition following the 1990 CAAA with additional 40 and 80% cuts in utility SO_2 emissions in the year 2010. These latter percentages represent the full range of emissions reductions embodied in the five bills reviewed here and would be 22 and 44% of the total U.S. emissions of SO_2, respectively. We assume these decreases in SO_2 emissions will result in 22 and 44% decreases in total S deposition, respectively, in 2010. This 1-to-1 relationship between SO_2 emissions and SO_4^{2-} deposition is supported by recent observations by Butler et al. (2001). The proposed SO_2 emission reductions should have a marked effect on atmospheric S deposition in the Northeast. Therefore, we focused our analysis on S controls. We did not consider decreases in NO_3^- or NH_4^+ deposition. Controls on N emissions should also mitigate the effects of acidic deposition. If a condition of NO_3^- losses in surface waters equaling atmospheric N deposition develops in the Adirondacks within 50 years, the USEPA (1995) projects that the percentage of lakes with ANC <0 μeq L^{-1} will increase from 19 to 43%. It is unlikely that reductions in utility NO_X emissions alone will be sufficient to improve the N or acid-base status of sensitive forest ecosystems in the Northeast because utilities contribute less than one quarter of total NO_X emissions (Table 10.2). Indeed, these bills do not consider atmospheric N deposition originating from NH_3 or vehicle NO_X emissions, which are both important sources of N to the atmosphere.

Table 10.2 Summary of Estimated Utility Emissions Resulting from the 1990 Amendments of the Clean Air Act and Proposed Federal Legislation Aimed at Reducing Electric Utility Emissions That Contribute to Acidic Deposition and Ground-Level Ozone

Proposal	NO$_X$ Estimated utility emissions	NO$_X$ Percent of total emissions[1]	SO$_2$ Estimated utility emissions	SO$_2$ Percent of total emissions[2]	Timeframe for full implementation	Cap and trade structure
1990 Clean Air Act	5.16 (5.7)	24.8	8.07 (8.9)	54.6	2010	Yes - for SO$_2$
S. 172 Moynihan						
H.R. 657 Sweeney	2.14 (2.36)	12	4.04 (4.45)	37.6	2005	Yes
H.R. 25 Boehlert						
S. 1369 Jeffords						
H.R. 2645 Kucinich	1.5 (1.66)	8.8	3.24 (3.58)	32.6	2005	Yes
H.R. 2900 Waxman	1.63 (1.8)	9.4	2.8 (3.11)	29.6	2005	No
S. 1949 Leahy	1.27 (1.4)	7.5	2.63 (2.9)	28.2	2010	No
H.R. 2980 Allen	1.45 (1.6)	8.5	2.9 (3.2)	30.2	2005	No

Note: Emissions are in million metric tons, with values of million short tons indicated in parentheses. The percent of total U.S. emissions that utility emissions would contribute if each proposal were implemented is also shown.

[1] Assumes that total NO$_X$ emissions from other sources are constant and that total emissions decrease by the same amount as the reduction in utility emissions. NO$_X$ emission figures are based on 1997 levels for total (21.3 metric tons or 23.5 million short tons) and utility emissions (5.62 metric tons or 6.2 million short tons).

[2] Assumes that total SO$_2$ emissions from other sources remain constant and that total emission decrease by the same amount as the reduction in utility emissions. SO$_2$ figures are based on 1997 levels for total (18.5 metric tons or 20.4 million short tons) and utility emissions (11.8 metric tons or 13 million short tons).

As anticipated, model calculations show that decreases in atmospheric S deposition will result in beneficial changes in soil and surface water chemistry at the HBEF. The model calculations indicate that the Acid Deposition Control Program will result in modest improvements in average stream water chemistry at the HBEF for the period 1994 to 2005. Specifically, SO_4^{2-} will decrease by 12 µeq L^{-1}, ANC will increase by about 2 µeq L^{-1}, and pH will increase slightly, about 0.1 units. Additional controls on SO_2 emissions, such as those suggested in current proposals, should result in greater improvements in soil and water chemistry. The model predicts that a 22% decrease in atmospheric S deposition in 2010 beyond the levels anticipated from the Acid Deposition Control Program (40% decrease in utility SO_2 emissions) will decrease stream SO_4^{2-} concentrations by 8.1 µeq L^{-1} by 2025, compared to the condition expected if there were no controls beyond the 1990 CAAA. In contrast, a 44% decrease in S deposition (80% decrease in utility SO_2 emissions) would decrease stream SO_4^{2-} concentrations by about 15 µeq L^{-1} by 2025.

Despite marked reductions in atmospheric S deposition over the last 34 years (Likens et al., 2001), stream water ANC at the HBEF remains below 0 µeq L^{-1}. Because of the loss of available soil pools of nutrient cations as a result of atmospheric S deposition during the last century, the recovery of stream water ANC following decreases in strong acid loading has been delayed. For the condition of no controls beyond the 1990 CAAA, the rate of ANC increase predicted by the model for 2010 to 2025 is 0.06 µeq. Decreases of 22 and 44% in atmospheric S deposition in 2010 increase the predicted rate of ANC change to 0.09 and 0.15 µeq L^{-1} yr^{-1}, respectively. Model calculations suggest that a 44% reduction in atmospheric S deposition in 2010 beyond the 1990 CAAA will result in positive stream ANC values in 2023. In contrast, for a 22% decrease in atmospheric S deposition beyond the 1990 CAAA, stream ANC is predicted to reach positive values by 2038. Further, the model predictions also indicate that a 44% reduction in S deposition beyond the 1990 CAAA will result in stream pH values that will exceed 5.5 and concentrations of inorganic monomeric Al that will decrease to 2.7 µmol L^{-1} by 2050. Model calculations indicate that at the HBEF although marked improvements are predicted, full chemical and biological recovery may not be achievable by 2050, even with the most aggressive proposals for utility emission reductions.

Model calculations suggest that at the HBEF the greater the reduction in atmospheric S deposition, the greater the magnitude and rate of chemical recovery. Less aggressive proposals for controls on S emissions will result in chemical and biological recovery at a slower rate and in delays in the services of a fully functional ecosystem. Unfortunately, model calculations do not exist for the entire Northeast region. Because currently about 6% of the total lakes and 32% of the acid-sensitive lakes (ANC <50 µeq L^{-1}) are more acidic than watershed 6 of the HBEF, it seems likely that recovery of these surface waters would lag behind the values predicted for the HBEF. Finally, these calculations of future scenarios are made under the assumption that land disturbance (e.g., cutting, fire) and climate remain constant after the present. Note that model calculations are sensitive to these conditions; therefore, any land disturbance and/or climate change occurring in the future could significantly alter model predictions.

Summary

North America and Europe are in the midst of a large-scale experiment. Sulfuric and nitric acids have acidified soils, lakes, and streams, stressing or killing terrestrial and aquatic biota. It is therefore critical to measure and understand the recovery of complex ecosystems in response to decreases in acidic deposition. Fortunately, the NADP, CASTNet, and AIRMON-dry networks are in place to measure anticipated improvements in air quality and atmospheric deposition. Unfortunately, networks to measure changes in water quality are sparse, and networks to monitor soil, vegetation, and fish responses are even more limited. There is an acute need to assess the response of these resources to decreases in acid loading. It would be particularly valuable to assess the recovery of aquatic biota to changes in surface water chemistry because of their direct response to acid stress (Gunn and Mills, 1998, Yan et al., 2003).

Long-term research from the HBEF and other sites across the northeastern United States were used to synthesize data on the effects of acidic deposition and to assess ecosystem responses to reductions in emissions. Based on existing data, it is clear that in the northeastern United States:

- Reductions of SO_2 emissions since 1970 have resulted in statistically significant decreases in SO_4^{2-} in wet/bulk deposition and surface water.
- Emissions of NO_x and concentrations of NO_3^- in wet/bulk deposition and surface waters show no increase or decrease since the 1980s.
- There is considerable uncertainty in estimates of NH_3 emissions, although atmospheric deposition of NH_4^+ is important for forest management and stream NO_3^- loss.
- Acidic deposition has accelerated the leaching of base cations from soils, delaying the recovery of ANC in lakes and streams from decreased emissions of SO_2. At the HBEF, the available soil Ca pool appears to have declined 50% over the past 50 years.
- Sulfur and N from atmospheric deposition have accumulated in forest soils across the region. The slow release of these stored elements from soil has delayed the recovery of lakes and streams from emissions reductions.
- Acidic deposition has increased the concentration of toxic forms of Al in soil waters, lakes, and streams.
- Acidic deposition leaches cellular Ca from red spruce foliage, which makes trees susceptible to freezing injury, leading to over 50% mortality of canopy trees in some areas of the Northeast.
- Extensive mortality of sugar maple in Pennsylvania has resulted from deficiencies of Ca^{2+} and Mg^{2+}. Acidic deposition has contributed to the depletion of these cations from soil.
- Forty-one percent of lakes in the Adirondacks and 15% of lakes in New England exhibit chronic and/or episodic acidification; 83% of these impacted lakes are acidic as a result of atmospheric deposition.
- There have been only modest increases in the ANC of surface waters in New England and no significant improvement in the Adirondack and Catskill regions following decreases in atmospheric S deposition in recent decades.
- Acidification of surface waters results in a decrease in the survival, size, and density of fish and loss of fish and other aquatic biota from lakes and streams.
- Emissions of air pollutants have important linkages to other large-scale environmental problems including coastal eutrophication, mercury contamination, visibility impairment, climate change, and tropospheric ozone.

Further, it is anticipated that recovery from acidic deposition will be a complex, two-phase process in which chemical recovery precedes biological recovery. The time for biological recovery is better defined for aquatic than terrestrial ecosystems. For acid-impacted aquatic ecosystems, it is expected that stream macroinvertebrate and lake zoo-plankton populations would recover 3 to 10 years after favorable chemical conditions were reestablished, and fish populations would follow. For terrestrial ecosystems, trees would probably respond positively to favorable atmospheric and soil conditions over a period of decades.

Indicators of chemical recovery (soil percentage base saturation, soil Ca/Al ion ratios, and surface water ANC) were used to evaluate ecosystem response to proposed policy changes in SO_2 emissions. Projections using an acidification model (PnET-BGC) indicate that full implementation of the 1990 CAAA will not result in substantial chemical recovery at the HBEF and many similar acid-sensitive locations. Although uncertainties remain, our analysis indicates that current regulations will not adequately achieve the desired ecological outcomes of the 1990 CAAA. These desired outcomes include increases in the ANC of lakes and streams, improvements in the diversity and health of fish populations, decreases in the degradation of forest soil, and stress to trees (United States Environmental Protection Agency 1995). Model calculations indicate that the magnitude and rate of recovery from acidic deposition in the northeastern United States is directly proportional to the magnitude of emission reductions. Model evaluations of policy proposals calling for additional reductions in utility SO_2 and NO_x emissions, year-round emission controls, and early implementation (2005) indicate greater success in facilitating the recovery of sensitive ecosystems and accomplishing the goals of the Clean Air Act than current 1990 CAAA targets. Note that until transportation emissions of NO_X are curtailed, there will be increased potential for a condition in which improvements in acidic deposition from SO_2 controls by utilities will be offset somewhat by NO_X emissions. Specific emission reductions targets should be based on clear goals for the desired extent and schedule of recovery of sensitive aquatic and terrestrial ecosystems that are consistent with the goals of the Clean Air Act.

Acknowledgments

This paper is the result of a synthesis effort sponsored and organized by the Hubbard Brook Research Foundation with support from the Jessie B. Cox Charitable Trust, Davis Conservation Foundation, Geraldine R. Dodge Foundation, McCabe Environmental Fund, Merck Family Fund, John Merck Fund, Harold Whitworth Pierce Charitable Trust, The Sudbury Foundation, and the Switzer Environmental Leadership Fund of the New Hampshire Charitable Foundation.

We would also like to acknowledge the support of the National Science Foundation for the work of C. Driscoll and G. Likens, and the Andrew W. Mellon Foundation and Mary Flagler Cary Charitable Trust for the work of G. Likens. The U.S. Environmental Protection Agency, Clean Air Markets Division provided support to C. Driscoll to aid in the development of the model PnET-BGC.

We thank Dr. J. Lynch and J.M. Grimm, Pennsylvania State University, for their contribution of the wet sulfate deposition figure. We also thank S. Gbondo-Tugbawa for help with model calculations and K. Driscoll for her contribution to several figures in the paper.

We would like to acknowledge the following individuals for helpful comments on earlier versions of the manuscript: Ms. R. Birnbaum, US Environmental Protection Agency, Clean Air Markets Division; Dr. E. Cowling, North Carolina State University; Mr. R. Poirot, Vermont Department of Environmental Conservation, Air Resources Division; Dr. M. Uhart, National Acid Precipitation Assessment Program; and Dr. D. Burns, US Geological Survey.

Some data in this publication are part of the Hubbard Brook Ecosystem Study. The Hubbard Brook Experimental Forest is operated and maintained by the Northeastern Research Station, United States Department of Agriculture, Newton Square, Pennsylvania.

References

Aber, J.D. and Federer, C.A., 1992, A generalized, lumped-parameter model of photosynthesis, evapotranspiration and net primary production in temperate and boreal forest ecosystems, *Oecologia* **92**: 463–474.

Aber, J.D. and Driscoll, C.T., 1997, Effects of land use, climate variation, and N deposition on N cycling and C storage in northern hardwood forests, *Global Biogeochemical Cycles* **11**: 639–648.

Aber, J.D., Ollinger, S.V., and Driscoll, C.T., 1997, Modeling nitrogen saturation in forest ecosystems in response to land use and atmospheric deposition, *Ecological Modelling* **101**: 61–78.

Aber, J.D., McDowell, W., Nadelhoffer, K.J., Magill, A., Bernston, G., Kamakea, S.G., McNulty, W., Currie, W., Rustad, L., and Fernandez, I., 1998, Nitrogen saturation in temperature forest ecoystems, *BioScience* **48**: 921–934.

Anderson, J.B., Baumgardner, R.E., Mohnen, V.A., and Bowser, J.J., 1999, Cloud chemistry in the eastern United States, as sampled from three high-elevation sites along the Appalachian Mountains, *Atmospheric Environment* **33**: 5105–5114.

Bailey, S.W., Hornbeck, J.M., Driscoll, C.T., and Gaudette, H.E., 1996, Calcium inputs and transport in a base-poor forest ecosystem as interpreted by Sr isotopes, *Water Resources Research* **32**: 707–719.

Bailey, S.W., Horsley, S.B., Long, R.P., and Hallet, R.A., 1999, Influence of geologic and pedologic factors on health of sugar maple on the Allegheny Plateau, U.S., In *Sugar Maple Ecology and Health: Proceedings of an International Symposium*, edited by S.B. Horsley and R.P. Long, U.S. Department of Agriculture, Forest Service, Radnor, Pennsylvania, General Technical Report NE-261, pp. 63–65.

Baker, J.P. and Schofield, C.L., 1982, Aluminum toxicity to fish in acidic waters, *Water Air and Soil Pollution*, **18**: 289–309.

Baker, J.P., Van Sickle, J., Gagen, C.J., DeWalle, D.R., Jr., DeWalle, D.R., Sharpe, W.F., Carline, R.F., Baldigo, B.P., Murdoch, P.S., Bath, D.W., Kretser, W.A., Simonin, H.A., and Wigington, P.J., 1996, Episodic acidification of small streams in the northeastern United States: effects on fish populations, *Ecological Applications* **6**: 422–437.

Brown, D.J.A., 1983, Effect of calcium and aluminum concentrations on the survival of brown trout (*Salmo trutta*) at low pH, *Bulletin of Environmental Contamination and Toxicology* **30**: 582.

Butler, T.J., Likens, G.E., and Stunder, B.J., 2001, Regional-scale impacts of Phase I of the Clean Air Act Amendments: the relationship between emissions and concentrations, both wet and dry. *Atmospheric Environment* **35**: 1005–1028.

Charles, D.F., Ed., 1991, *Acidic Deposition and Aquatic Ecosystems. Regional Case Studies*, Springer-Verlag, New York.

Clarke, J.F., Edgerton, E.S., and Martin, B.E., 1997, Dry deposition calculations for the Clean Air Status and Trends Network, *Atmospheric Environment* **31**: 3667–3678.

Craig, B.W. and Friedland, A.J., 1991, Spatial patterns in forest composition and standing dead red spruce in montane forests of the Adirondacks and northern Appalachians, *Environmental Monitoring and Assessment* **18**: 129–140.

Cronan, C.S. and Grigal, D.F., 1995, Use of calcium/aluminum ratios as indicators of stress in forest ecosystems, *Journal of Environmental Quality* **24**: 209–226.

Cronan, C.S. and Schofield, C.L., 1990, Relationships between aqueous aluminum and acidic deposition in forested watersheds of North America and Northern Europe, *Environmental Science and Technology* **24**: 1100–1105.

DeHayes, D.H., Schaberg, P.G., Hawley, G.J., and Strimbeck, G.R., 1999, Acid rain impacts calcium nutrition and forest health, *BioScience* **49**: 789–800.

Dise, N.B. and Wright, R.F., 1995, Nitrogen leaching from European forests in relation to nitrogen deposition, *Forest Ecology Management* **71**: 153–162.

Draxler, R.R. and Hess, G.D., 1998, An overview of the Hysplit _4 modeling system for trajectories, dispersion, and deposition, *Austrian Meteorology Magazine* **47**: 295–308.

Driscoll, C.T. and Schecher, W.D., 1990, The chemistry of aluminum in the environment, *Environmental Geochemistry and Health* **12**: 28–49.

Driscoll, C.T., Baker, J.P., Bisogni, J.J., and Schofield, C.L., 1980, Effect of aluminum speciation of fish in dilute acidified waters, *Nature* **284**: 161–164.

Driscoll, C.T., Johnson, N.M., Likens, G.E., and Feller, M.C., 1988, The effects of acidic deposition on stream water chemistry: a comparison between Hubbard Brook, New Hampshire and Jamieson Creek, British Columbia, *Water Resources Research* **24**: 195–200.

Driscoll, C.T., Cirmo, C.P., Fahey, T.J., Blette, V.L., Bukaveckas, P.A., Burns, D.J., Gubala, C.P., Leopold, D.J., Newton, R.M., Raynal, D.J., Schofiled, C.L., Yavitt, J.B., and Porcell, D.B., 1996, The Experimental Watershed Liming Study (EWLS): comparison of lake and watershed neutralization strategies, *Biogeochemistry* **32**: 143–174.

Driscoll, C.T., Likens, G.E., and Church, M.R., 1998, Recovery of surface waters in the northeastern U.S. from decreases in atmospheric deposition of sulfur, *Water Air and Soil Pollution* **105**: 319–329.

Driscoll, C.T., Yan, C., Schofield, C.L., Munson, R., and Holsapple, J., 1994a, The mercury cycle and fish in the Adirondack lakes, *Environmental Science and Technology* **28**: 136A-143A.

Driscoll, C.T., Lehtinen, M.D., and Sullivan, T.J., 1994b, Modeling the acid-base chemistry of inorganic solutes in Adirondack, NY lakes, *Water Resources Research* **30**: 297–306.

Drohan, P.J., Stout, S.L., and Petersen, G.W., 1999, Spatial relationships between sugar maple (*Acer sacharum* Marsh.), sugar maple decline, slope, aspect, and topographic position in northern Pennsylvania, In *Sugar Maple Ecology and Health: Proceedings of an International Symposium*, edited by S.B. Horsley and R.P. Long, US Department of Agriculture, Forest Service, General Technical Report NE-261, Radnor, Pennsylvania, pp. 46–54.

Eary, L.E., Jenne, E.A., Eail, L.W., and Girvin, D.C., 1989, Numerical models for predicting watershed acidification, *Archives of Environmental Contamination and Toxicology* **18**: 29.

Environment Canada, 1998, *Annual Report on the Federal-Provincial Agreements for the Eastern Canada Acid Rain Program*, Minister of Public Works and Government Services Canada, Ottawa, Report nr 0–662–6300–3 (Cat. No. EN40–11/29–1997).

Ferm, M., 1998, Atmospheric ammonia and ammonium transport in Europe and critical loads: a review, *Nutrient Cycling in Agroecosystems* **51**: 5–17.

Fraser, M.P. and Cass, G.R., 1998, Detection of excess ammonia emissions from in-use vehicles and the implications for fine particle control, *Environmental Science and Technology* **32**: 1053–1057.

Gallagher, J. and Baker, J., 1990, Current status of fish communities in Adirondack Lakes, In *Adirondack Lakes Survey: An Interpretive Analysis of Fish Communities and Water Chemistry, 1984–1987*, Adirondacks Lakes Survey Corporation, Ray Brook, New York, pp. 3-11–3-48.

Galloway, J.N., Norton, S.N., and Church, M.R., 1983, Freshwater acidification from atmospheric deposition of sulfuric acid: a conceptual model, *Environmental Science and Technology* **17**: 541A-545A.

Gbondo-Tugbawa, S., Driscoll, C.T., Aber, J.D., and Likens, G.E., 2001, Validation of a new integrated biogeochemical model (PnET-BGC) at a northern hardwood forest ecosystem, *Water Resources Research* **37**: 1057–1070.

Goolsby, D.A., Battaglin, W.A., Lawrence, G.B., Artz, R.S., Aulenbach, B.T., Hooper, R.P., Keeney, D.R., and Stensland, G.J., 1999, Flux and sources of nutrients in the Mississippi-Atchafalaya River Basin, Report of Task Group 3 to the White House Committee on Environment and Natural Resources, Washington, D.C.

Grimm, J.W. and Lynch, J.A., 1997, Enhanced wet deposition estimates using modeled precipitation inputs, In *Final Report*, U.S. Forest Service, Northeast Forest Experiment Station Northern Global Change Research Program, Cooperative Agreement 23–721, Radnor, Pennsylvania.

Gunn, J.M. and Mills, K.H., 1998, The potential for restoration of acid-damaged lake trout lakes, *Restoration Ecology* **6**: 390–397.

Hornbeck, J.W., Bailey, S.W., Buso, D.C., and Shanley, J.B., 1997, Streamwater chemistry and nutrient budgets for forested watersheds in New England: variability and management implications, *Forest Ecology and Management* **93**: 73–89.

Horsley, S.B., Long, R.P., Bailey, S.W., Hallet, R.A., and Hall, T.J., 1999, Factors contributing to sugar maple decline along topographic gradients on the glaciated and unglaciated Alleghany Plateau, In *Sugar Maple Ecology and Health: Proceedings of an International Symposium, edited by S.B. Horsley and R.P. Long*, U.S. Department of Agriculture, Forest Service, General Technical Report NE-261, Radnor, Pennsylvania, pp. 60–62.

Houston, D.R., 1999, History of sugar maple decline, In *Sugar Maple Ecology and Health: Proceedings of an International Symposium*, edited by S.B. Horsley and R.P. Long, U.S. Department of Agriculture, Forest Service, General Technical Report NE-261, Radnor, Pennsylvania, pp. 9–26.

Jaworski, N.A., Howarth, R.W., and Hetling, L.J., 1997, Atmospheric deposition of nitrogen oxides onto the landscape contributes to coastal eutrophication in the Northeast United States, *Environmental Science and Technology* 31: 1995–2004.

Johnson, A.H., Friedland, A.J., and Dushoff, J.G., 1984, Recent and historic red spruce mortality: evident of climatic influence, *Water Air and Soil Pollution* 30: 319–330.

Jordan, T.E. and Weller, D.E., 1996, Human contributions to the terrestrial nitrogen flux, *BioScience* 46: 655–664.

Keller, W. and Gunn, J.M., 1995, Lake water quality improvements and recovering aquatic communities, In *Restoration and Recovery of an Industrial Region: Progress in Restoring the Smelter-damaged Landscape near Sudbury, Canada*, edited by J.M. Gunn, Springer-Verlag, New York, pp. 67–80.

Kram, P., Santore, R.C., Driscoll, C.T., Aber, J.D., and Hruska, J., 1999, Application of the forest-soil-water model (PnET-BGC/CHESS) to the Lysina catchment, Czech Republic, *Ecological Modelling* 120: 9–30.

Kretser, W., Gallagher, J., and Nicolette, J., 1989, *Adirondack Lakes Study. 1984–1987. An Evaluation of Fish Communities and Water Chemistry*, Adirondacks Lakes Survey Corporation, Ray Brook, New York.

Larsen, D.P., Thornton, K.W., Urquhart, N.S., and Paulsen, S.G.,1994, The role of sample surveys for monitoring the condition of the nation's lakes, *Environmental Monitoring and Assessment* 32: 101–134.

Lawrence, G.B., David, M.B., Lovett, G.M., Murdoch, P.S., Burns, D.A., Baldigo, B.P., Thompson, A.W., Porter, J.H., and Stoddard, J.L., 1999, Soil calcium status and the response of stream chemistry to changing acidic deposition rates in the Catskill Mountains of New York, *Ecological Applications* 9: 1059–1072.

Lawrence, G.B., Goolsby, D.A., Battaglin, W.A., and Stensland, G.J., 2000, Atmospheric nitrogen in the Mississippi River Basin — emissions, deposition and transport, *The Science of the Total Environment* 248: 87–100.

Likens, G.E., Bormann, F.H., and Johnson, N.M., 1972, Acid rain, *Environment* 14: 33–40.

Likens, G.E., Bormann, F.H., Hedin, L.O., Eaton, J.S., and Driscoll, C.T., 1990, Dry deposition of sulfur: a 23-year record for the Hubbard Brook Forest ecosystem, *Tellus* 42B: 319–329.

Likens, G.E. and Bormann, F.H., 1995, *Biogeochemistry of a Forested Ecosystem*, Second Edition, Springer-Verlag, New York.

Likens, G.E., Driscoll, C.T., and Buso, D.C., 1996, Long-term effects of acid rain: response and recovery of a forest ecosystem, *Science* 272: 244–246.

Likens, G.E., Driscoll, C.T., Buso, D.C., Siccama, T.G., Johnson, C.E., Lovett, G.M., Fahey, T.J., Reiners, W.A., Ryan, D.F., Martin, C.W., and Bailey, S.W., 1998, The biogeochemistry of calcium at Hubbard Brook, *Biogeochemistry* 41: 89–173.

Likens, G.E., Butler, T.J., and Buso, D.C., 2001, Long- and short-term changes in sulfate deposition: effects of the 1990 Clean Air Act Amendments, *Biogeochemistry* 52: 1–11.

Long, R.P., Horsley, S.B., Lilja, and P.R., 1997, Impact of forest liming on growth and crown vigor of sugar maple and associated hardwoods, *Canadian Journal of Forest Research* 27: 1560–1573.

Lovett, G.M., 1994, Atmospheric deposition of nutrients and pollutants in North America: an ecological perspective, *Ecological Applications* 4: 629–650.

Lovett, G.M., Weathers, K.C., and Sobczak, W.V., 2000, Nitrogen saturation and retention in forested watersheds of the Catskill Mountains, New York, *Ecological Applications* 10: 73–84.

Lynch, J.A., Bowersox, V.C., and Grimm, J.W., 2000, Changes in sulfate deposition in eastern USA following implementation of Phase I of Title IV of the Clean Air Act Ammendments of 1990, *Atmospheric Environment* 34: 1665–1680.

MacAvoy, S.E. and Bulger, A.J., 1995, Survival of brook trout (*Salvelinus fontinalis*) embryos and fry in streams of different acid sensitivity in Shenandoah National Park, USA, *Water Air and Soil Pollution* 85: 439–444.

Malm, W.C., Sisler, J.F., Huffman, D., Eldred, R.A., and Cahill, T.A., 1994, Spatial and seasonal trends in particle concentration and optical extinction in the United States, *Journal of Geophysical Research* **99**: 1347–1370.

McLaughlin, S.B. and Wimmer, R., 1999, Tansley Review No. 104, calcium physiology and terrestrial ecosystem processes, *New Phytologist* **142**: 373–417.

McNulty, S.G., Aber, J.D., McLellan, T.M., and Katt, S.M., 1990, Nitrogen cycling in high elevation forests of the northeastern U.S. in relation to nitrogen deposition, *Ambio* **19**: 38–40.

Mitchell, M.J., Driscoll, C.T., Kahl, J.S., Likens, G.E., Murdoch, P.S., and Pardo, L.H., 1996, Climatic control of nitrate loss from forested watersheds in the northeast United States, *Environmental Science and Technology* **30**: 2609–2612.

Moore, M.V., Pace, M.L., Mather, J.R., Murdoch, P.S., Howarth, R.W., Folt, C.L., Chen, C.Y., Hemond, H.F., Flebbe, P.A., and Driscoll, C.T., 1997, Potential effects of climate change on freshwater ecosystems of the New England/Mid-Atlantic region, *Hydrological Processes* **11**: 925–947.

Murdoch, P.S., Burns, D.A., and Lawrence, G.B., 1998, Relation of climate change to the acidification of surface waters by nitrogen deposition, *Environmental Science and Technology* **32**: 1642–1647.

National Acid Precipitation Assessment Program, 1998, *NAPAP Biennial Report to Congress. An Integrated Assessment*, National Acid Precipitation Program, Washington, D.C.

Norton, S.A., Kahl, J.S., Fernandez, I.J., Rustad, L.E., Schofield, J.P., and Haines, T.A., 1994, Response of the West Bear Brook Watershed, Maine, USA, to the addition of $(NH_4)_2SO_4$: 3-year results, *Forest and Ecology Management* **68**: 61–73.

Oden, S., 1968, The acidification of air precipitation and its consequences in the natural environment, *Bulletin of Ecological Research Communications NFR*, Translation Consultants Ltd., Arlington, VA.

Parker, L., 2000, *Electricity restructuring and air quality: comparison and proposed legislation.* Congressional Research Service. Report for Congress. Order Code RS20326, Washington, D.C.

Rago, P.J. and Wiener, J.G., 1986, Does pH affect fish species richness when lake area is considered? *Transactions of the American Fisheries Society* **11b**: 438–447.

Rochelle, B.P. and Church, M.R., 1987, Regional patterns of sulfur retention in watersheds of the eastern U.S., *Water Air and Soil Pollution* **36**: 61–73.

Schindler, D.W., Mills, K.H., Malley, D.F., Findlay, S., Shearer, J.A., Davies, I.J., Turner, M.A., Lindsey, G.A., and Cruikshank, D.R., 1985, Long-term ecosystem stress: effects of years of experimental acidification, *Canadian Journal of Fisheries and Aquatic Science* **37**: 342–354.

Seinfeld, J.H., 1986., *Atmospheric Chemistry and Physics of Air Pollution*, John Wiley & Sons, New York.

Shortle, W.C. and Smith, K.T., 1988, Aluminum-induced calcium deficiency syndrome in declining red spruce trees, *Science* **240**: 1017–1018.

Shortle, W.C., Smith, K.T., Minocha, R., Lawrence, G.B., and David, M.B., 1997, Acid deposition, cation mobilization, and stress in healthy red spruce trees, *Journal of Environmental Quality* **26**: 871–876.

Siccama, T.G., Bliss, M., and Vogelmann, H.W., 1982, Decline of red spruce in the Green Mountains of Vermont, *Bulletin of the Torrey Botanical Club* **109**: 162–168.

Smith, R.A., 1872, *Air and Rain*, Longmans, Green, London.

Stevens, D.L., 1994, Implementation of a national monitoring program, *Journal of Environmental Management* **42**: 1–29.

Stoddard, J.L., Jeffries, D.S., Lukewille, A., Clair, T.A., Dillon, P.J., Driscoll, P.J., Forsius, M., Johannessen, M., Kahl, J.S., Kellogg, J.H., Kemp, A., Mannio, J., Monteith, D.T., Murdoch, P.S., Patrick, S., Rebsdorg, A., Skjelkvale, B.L., Stainton, M.P., Traen, T., VanDam, H., Webster, K.E., Wieting, J., and Wilander, A., 1999. Regional trends in aquatic recovery from acidification in North America and Europe, *Nature* **401**: 575–578.

United States Environmental Protection Agency, 1995, *Acid Deposition Standard Feasibility Study Report to Congress*, Office of Air and Radiation, Acid Rain Division, Washington, D.C.

United States Environmental Protection Agency, 2000, *National Air Pollutant Emission Trends, 1900–1998*, U.S. Environmental Protection Agency, Report EPA - 454/R-00–002, Washington, D.C.

Van Sickle, J., Baker, J.P., Simonin, H.A., Baldigo, B.P., Kretser, W.A., and Sharpe, W.F., 1996, Episodic acidification of small streams in the northeastern United States: fish mortality in field bioassays, *Ecological Applications* **6**: 408–421.

Weathers, K.C., Likens, G.E., Bormann, F.H., Bicknell, S.H., Borman, B.T., Daube, B.C., Jr., Eaton, J.S., Galloway, J.N., Keene, W.C., Kimball, K.D., McDowell, W.H., Siccama, T.G., Smiley, D., and Tarrant, R., 1988, Cloud water chemistry from ten sites in North America, *Environmental Science and Technology* **22**: 1018–1026.

Wigington, P.J., Jr., Baker, J.P., DeWalle, D.R., Kretser, W.A., Murdoch, P.S., Simonin, H.A., Van Sickle, J., McDowell, M.K., Peck, D.V., and Barchet, W.R., 1996, Episodic acidification of small streams in the northeastern United States: Episodic Response Project, *Ecological Applications* **6**: 374–388.

Yan, N.D., Leng, B., Keller, W., Arnott, S.E., Gunn, J.M., and Raddum, G.G., 2003, Developing conceptual frameworks for the recovery of aquatic biota from acidification, *Ambio* **32**(3): 165–169.

section III

Biological effects and management reactions

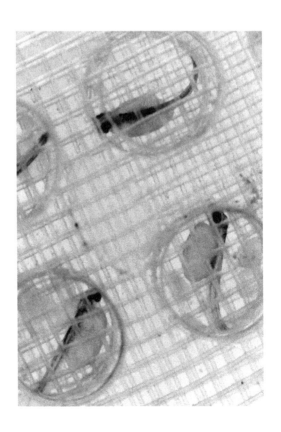

chapter eleven

The control of harvest in lake trout sport fisheries on Precambrian Shield lakes

Charles H. Olver
Ontario Ministry of Natural Resources (retired)
Daniel J. Nadeau
Société et de la faune et des parcs du Québec
Henri Fournier
Société et de la faune et des parcs du Québec

Contents

Introduction

Despite increasingly comprehensive and complex legislation intended to regulate exploitation, it is the most critical stress affecting the lake trout *Salvelinus namaycush* across the Precambrian Shield. The two main contributing factors are the lake trout's innate vulnerability to exploitation and the rapid growth of recreational fishing following the end of World War II. For instance, angling almost quadrupled in Ontario between 1950 and the early 1970s (Ontario Ministry of Natural Resources and Environment Canada, 1976). With the development and expansion of highways and forest access roads, the advent of the snow machine and the all-terrain vehicle, and the ready availability of aircraft, few lake trout lakes are inaccessible to anglers.

The sensitivity of lake trout to exploitation has been described by several authors (see, for example, Fry, 1949; Daley et al., 1965; Martin and Olver, 1980). Lake trout are easily caught by angling, so stocks can be depleted with only a moderate amount of fishing pressure (MacKay, 1956; Eschmeyer, 1964). Certain biological characteristics of the species (slow growth, late maturity, and low reproductive potential) and the unproductive nature of the waters it inhabits (low nutrient levels, cold temperatures) also contribute to the vulnerability of lake trout to exploitation. As a result of these limiting factors, lake trout occur in a small number of Precambrian Shield lakes, where they form sparse populations with low annual sustainable yields.

The main purpose of angling regulations is to ensure the perpetuation of fish stocks through regulation of the harvest. Simply stated, it is the kill of fish that needs to be controlled. Within this context, the intent of the regulations is to provide a continuing supply of fishing opportunities and to distribute the catch fairly among the participants. Regulations also prescribe the "rules of the game" and impart a sense of ethics by emphasizing the sporting aspects of angling.

Christie (1978) observed that short of closing lakes to fishing it is difficult to control angling effort given the open access, common property nature of resource management (e.g., Hardin, 1968). This is also true of lakes where stocks have been depressed because residual pressure may be sufficient to prevent their recovery. In open access fisheries, Christie also noted that traditional harvest control regulations (e.g., seasons, daily catch limits) work well until their effectiveness is undermined by large increases in the demand for recreational fishing. Anglers also often fail to appreciate the significance of their catch relative to the total harvest (Christie, 1978).

The growth of sport fishing for lake trout, and for many other species, has exceeded the ability of most agencies to be proactive. Attempts by managers to control exploitation pressures by introducing almost annual revisions to the angling regulations have resulted in a loss of credibility with anglers, who have become somewhat suspicious of managers' motives. Anglers usually see proposals for harvest reduction mainly in terms of a loss of fishing opportunities, whereas to the manager these proposals are seen as necessary to maintain the resource (Ontario Ministry of Natural Resources, 1978a).

Harvest is controlled by applying such methods as (1) limiting fishing opportunities (manipulating season length and regulating physical access); (2) using size-based regulations to exert biological control of the harvest; (3) applying daily catch limits to apportion the catch; (4) implementing quota controls to match actual harvest with maximum equilibrium yields; and (5) imposing gear restrictions to make anglers less efficient or to increase the survival of released fish. These measures are reviewed and directions for experimental management proposed.

Angling seasons

The establishment of a fishing season with fixed opening and closing dates is a common regulatory measure for most sport fish, including the lake trout. Ideally, the length of a fishing season indicates how intensively a species should be managed. It also reflects the social values and traditions that have developed for a particular sport species and its fishery. Seasons may be closed because a species may be especially vulnerable to exploitation at certain times of the year (such as the ice-out period for lake trout) or to protect mature fish during their spawning season.

Table 11.1 General Summary of Open Seasons for Lake Trout Angling in the Precambrian Shield Area of North America[a]

Jurisdiction	Seasons	Comments/exceptions
Saskatchewan	May 16–March 31	Central zone
	May 25–March 31	Northern zone
Manitoba	May 16–April 30	North-central and northwest divisions
	Generally all year	Northeast division
Ontario	January 1–September 30	Most common, but 12 different seasons, 4 split seasons (closed in late winter, reopened in spring) 5 divisionwide winter closures, approximately 50+ lakes elsewhere with specific winter closures
Quebec	April 25–September 1, 7, 14, or 30	Seasons may vary slightly in wildlife reserves and ZECs
Minnesota	May 9–September 30	Lakes both inside and outside and partly outside Boundary Waters Canoe Area adjacent to Ontario border have winter seasons
New York	April 1–October 15	All year in Finger Lakes, Lake Champlain Ice fishing permitted in designated trout waters

Note: ZECs, Zones d'Exploitation Contrôlée.

[a] Regulations pertaining to Great Lakes waters are not included. These regulations may have changed in recent years. Check with your local management agencies for updates.

In Manitoba and Saskatchewan, lake trout fishing is generally closed for only a short period of time in April or May (Table 11.1). Long seasons are also the norm in Ontario, where winter fishing is permitted in 30 of 35 fishing divisions. However, winter fishing for lake trout is not permitted in Quebec (with few exceptions), is confined to certain areas in Minnesota, and is confined to specified waters in New York. Lakes closed to late fall fishing are usually closed by or before September 30. Presumably the closing date selected is designed to protect lake trout from exploitation just prior to and during the spawning season.

In an attempt to reduce catch by about 30%, a number of measures were applied to free-access lake trout waters in southern Quebec in 1989. These are waters in which no access provisions apply, and no daily or annual fees other than a provincial fishing license are required to fish legally. These measures included closing the winter season, opening the summer season on June 1 (a loss of about 5 weeks of ice-free fishing), and lowering the daily catch limit to two fish. Season reductions were expected to have the most effect. A creel census was conducted on two lakes in 1991 to measure the changes in the fisheries. In Lac Matapédia there was no winter fishery, while in Lac des Trente et un Milles there were both winter and open-water fisheries. Changes in fishing effort (angler-hours), CUE (catch-per-unit-effort), and estimated harvest (number) were observed as follows:

Lac Matapédia	1985	1991	Difference (%)
Effort	3800	4883	+22
CUE	0.53	0.84	+59
Harvest	2016	4077	+102

Lac des Trente et un Milles	Average 1979–1987			1991 Summer	Difference (%) Summer	Difference (%) Total
	Winter	Summer	Total			
Effort	5,449	6,054	11,503	7,037	+10	−39
CUE	0.16	0.41	0.29	0.55	+34	+90
Harvest	874	2,472	3,346	3,885	+57	+16

In Lac Matapédia, despite the loss of 5 weeks of spring fishing, effort, fishing success, and harvest all increased substantially. In Lac des Trente et un Milles, although the winter fishery was closed and total fishing effort decreased substantially, a large increase in fishing success in summer resulted in an overall increase in total harvest. These examples show that anglers can quickly adapt to changes in season regulations, and that exploitation can still be quite high even when seasons are reduced.

A different approach to harvest control than in-year season changes has been the closure of lakes to fishing in alternate years. This system was instituted on a number of lake trout and brook trout *Salvelinus fontinalis* lakes in Algonquin Park, Ontario, in response to concerns that the large catches of mature fish in some of these lakes might adversely affect production and recruitment. Spawning escapement was not measurably enhanced because there was little difference in the size of year-classes produced in open and closed years (Martin and Baldwin, 1953). The quality of lake trout angling in lakes subject to alternate-year closure was also similar to that experienced in lakes open to fishing each year. A later study by Martin (1966) produced the same results. He suggested that a system of alternate-year closure might be most beneficial in intensively fished lakes that have fast growing fish and stable year-class production. He also noted that in lakes open for 2 consecutive years then closed for 1 year, the angling quality declined substantially in the second open year.

Winter fishing

One of the contentious issues of lake trout management is that of winter fishing. In the early 1950s, Fenderson (1953) noted that the effects of ice fishing on trout and salmon have been controversial for a long time in Maine. Concerns about the effect of ice fishing on lake trout stocks in Ontario arose shortly after winter open seasons for lake trout were established in inland lakes in 1957 (Armstrong, 1961). Hughson (1961) reported that in the Sudbury area, fishery managers were concerned with the large increase in winter fishing in recent years (primarily for lake trout and walleye, *Stizostedion vitreum vitreum*) and with the poor-quality angling occasionally experienced for these species during the summer. Ryder (1957), on the basis of a winter creel survey of 19 lakes in 1957, suggested that winter fishing did not overexploit lake trout populations in the Thunder Bay, Ontario, area as winter harvests were only about a tenth of those experienced during open water. However, some 15 years later the impact of winter fishing had changed dramatically. Ryder and Johnson (1972), citing just one example, estimated that 2 to 4 years of annual lake trout production was removed from one such Thunder Bay area lake by ice fishermen in 1 day.

The impact that winter angling can have on a small Precambrian Shield lake trout lake is shown in Figure 11.1. A similar exploitation pattern occurred in 1980 in Nelson Lake, a 308-ha rehabilitated lake trout lake also near Sudbury, Ontario (Gunn et al., 1988). Winter effort, harvest, and yield were estimated at 28,137 angler hour, 2132 fish, and 3.26 kg ha^{-1}, respectively. Catch rates (number/angler hour), mean fork length (FL), and mean weight of fish sampled generally declined monthly from January through March. The authorities responded by closing the lake to winter fishing. The closure is still in effect, and the lake is heavily fished in the spring. The rapid response of anglers and the resultant "high grading" that occurs when a closed lake is opened to uncontrolled fishing (by either new access or removal of sanctuary status) is a familiar pattern. It was also a common occurrence, at least in Ontario throughout the 1960s, because lake trout regulations were liberalized and the use of mechanized forms of winter travel allowed anglers access to previously inaccessible lakes.

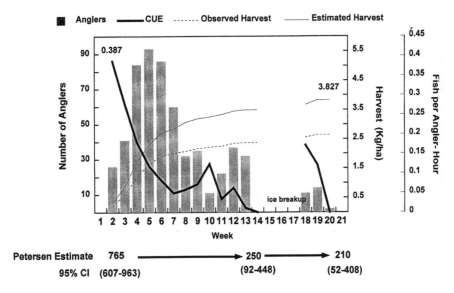

Figure 11.1 Impact of winter anglers on a remote lake trout lake near Sudbury, Ontario. Michaud Lake is a 148-ha lake accessible by a 26-km forestry road (12 km are used as a snowmobile trail in the winter). The lake was closed to angling for 7 years (1991–1997) to allow the lake trout population to be rehabilitated by hatchery stocking. The lake opened to fishing (without any public notice) on January 1, 1998, under the standard regulations for this area (unlimited access and entry, three fish per person daily catch [bag] limit). Catch rates were very high the first week of January (0.4 fish/angler-hour), but declined rapidly as more and more people (>90 people the 4th week) learned about the fishing in the lake. In spite of poor fishing late in the winter season, anglers continued to fish the lake until ice breakup occurred. They then returned for one last try in early spring. Catch rates were initially high in early spring (0.17 fish/angler-hour the first week of May), but declined again to zero by the third week of May. A creel survey was conducted every weekend day and on two random weekdays per week. A total of 384 harvested lake trout (2.6 kg/ha) were observed by the creel survey crew. By extrapolation, the estimated total harvest was 515 fish (3.8 kg/ha). Based on a marking program conducted in fall 1997 (220 fish marked and released), the adult population was reduced by about 72% (from 765 to 210) by anglers in the first few months that the lake was open. (Modified from Gunn and Sein, 2003; See Gunn and Sein, Chapter 14, this volume.)

One of the earliest studies comparing winter and open-water lake trout fisheries occurred on Algonquin Park lakes in the late 1940s and early 1950s. Martin (1954) found that in lakes where lake trout mature at a young age (5 years) and a small size (28 to 30.5-cm FL), as in Canisbay and Louisa Lakes, a summer fishery leaves a greater proportion of the mature fish to spawn each year than the winter fishery. In these lakes, harvests were generally greater in the winter. The catches sampled in the winter were made up of smaller trout and had a larger proportion of immature fish than in the summer fisheries. For example, in Canisbay and Louisa Lakes 60 and 50%, respectively, of the lake trout caught in winter had fork lengths less than 30 cm compared to 15 and 10%, respectively, in the summer fishery. These results in part contributed to the closure of Algonquin Park lakes to winter fishing after 1955.

Martin (1954) found that plankton feeding was generally at a low level during the winter, and with the absence of thermal barriers the smaller and more numerous plank-tivorous lake trout were able to feed on minnows more frequently. These fish are especially vulnerable to capture by anglers using bait fish. Lakes with an abundance of small lake trout (usually the smaller lakes) may attract more anglers and have greater effort on a per-unit-area basis than other lakes because the higher catch rates on such lakes provide the angler with "lots of action" and increase the likelihood of limit catches. The high harvests in some lakes during the winter may also be taken into account by the observation of Ryder and Johnson (1972) that "lake trout tend to form close schools in the winter, which, once located, can be harvested at will." This supports Martin's (1954) observation that the winter catch of lake trout in large lakes may be lower in part because of the difficulty in locating the fish.

An assessment of winter and summer lake trout fisheries was done by Schumacher (1961) on four lakes in Minnesota. In the lake with the longest time series (9 years) he found that both winter angling pressure and catchability were twice that of summer and resulted in the capture of five times the number of fish. Winter anglers took smaller trout, and nearly two-thirds of the fish caught were immature. Winter anglers were also the first to exploit each new year-class entering the fishery. Schumacher concluded that these lakes were overexploited.

Many other authors have observed that effort, fishing success, and catch tend to be higher in winter fisheries than in open-water fisheries and that winter catches contain higher proportions of small, young, and immature fish. Goddard et al. (1987) found that fishing effort is the most important factor affecting lake trout harvest in Ontario lakes, accounting for 73% of the annual variation in angler catch among lakes. Evans et al. (1991), in their review of Ontario lake trout sport fisheries, found that a disproportionate amount of annual effort occurred in winter because effort on lake trout was about 25% greater in winter than in spring and summer combined.

Lake trout are also very vulnerable to angling when the ice fishing season first opens, which is usually in January. A disproportionate amount of effort and harvest in winter lake trout fisheries has been observed in that month. In four Ontario lakes, for example, about one-half of the effort and one-half to two-thirds of the harvest occurred in the first month of the season (Vozeh, 1965; Samis, 1968; Purych, 1975; Bernier, 1977). Corresponding catch rates were from one-third to three times higher in January than in February or March in three of the four lakes. A similar pattern was observed in Michaud Lake (Figure 11.1). Catch rates and harvests may be lower in February and March simply because fewer fish are available for capture. If the season was not open until February or March, initial catch rates and harvests might also be as high as in January, although the cumulative harvest might be less because of a shorter season.

A high proportion of lake trout caught in winter angling fisheries are also immature. For example, Purych (1975) found that 80.7% of the lake trout sampled from the angler's catch at Summers Lake, Ontario were immature, and Walker (1978) reported that 77% of the lake trout caught at Lake Manitou, Ontario were immature. This could have a major effect on recruitment, as Martin and Fry (1973) have shown that in Lake Opeongo year-class strength is significantly related to spawning escapement, which is in turn largely governed by exploitation.

There is little doubt that winter fishing for lake trout has been harmful to many lake trout stocks. Evans et al. (1991) concluded that yields of lake trout in many Ontario lakes, especially for winter fisheries and small lakes, appear to be well above self-sustaining limits. As Schumacher (1961) pointed out, the winter angler is the first to catch each new year-class as it becomes vulnerable to fishing, and thus the winter fishery may have the dominating effect on the summer fishery as well as the total fishery.

Allocating the resource among winter and summer anglers is a social, not a biological, issue. As Christie (1978) noted, it makes little difference what time of year a fish is harvested as the same loss to the spawning population occurs whether it is caught many months before or just prior to the spawning season. DeRoche (1973) contended that winter fishermen have as much "right" to the resource as summer (open-water) anglers. In general, in winter most anglers tend to be local residents. Larger economic benefits may accrue in summer fisheries that attract a larger proportion of tourists. Social and economic interests may not be compatible with biological concerns, and compromises are inevitable. Managers must balance the demands of anglers with the demands of those who have an economic interest in the fisheries. However, more than 40 years ago, Weir and Martin (1961) noted that if the issues associated with winter fishing for lake trout are to be resolved, biological considerations should take precedence over nonbiological concerns.

There is justification for curtailing winter fishing in some manner because of the potential biological impact on lake trout stocks. This does not mean that winter fishing for lake trout should be eliminated across all lakes. Nevertheless, a reduction in the length of the winter fishing season in areas or on small lakes where exploitation is excessive, or even winter closures on individual lakes, may be an effective harvest control mechanism. In Ontario, for example, a review of lake trout management strategies (Olver et al., 1991) recommended that winter lake trout seasons be reduced to 1 month (February 15 to March 15) on all lake trout lakes less than 1000 ha, and that lake trout lakes less than 100 ha be closed to winter fishing. These recommendations have not been implemented.

One method to reduce harvest has been to delay the "traditional" opening date of January 1 to sometime in February or March and continue the season uninterrupted through to the fall. However, shortening the winter season from a few months to a few weeks might cause anglers to compress a similar amount of effort observed in a long season into a shortened season if opportunities to fish nearby open lake trout lakes are not available. Local conditions (sufficient ice cover, snow conditions, weather conditions) may also vary considerably from month to month, year to year, and lake to lake within the same fishing division, thereby making it difficult to establish regulations that are effective or suitable over broad areas.

Another option is to have a split season in which the season is closed for a period of time prior to the spring breakup of ice and then is reopened some time after the ice goes out. This approach deserves closer scrutiny, as the intent is to reduce harvest during a period when the lake trout appear to be especially vulnerable to angling. Regardless of whether season adjustments are applied across a fishing division or to individual lakes, the fine tuning of seasons by manipulating the opening and closing dates is a viable option to control harvest.

Access control

Access controls, in addition to fishing regulations, can be used to control harvest. These types of measures are usually designed to limit or to impede access to water bodies, restrict access of certain user groups, and limit development in remote areas or lakes. The traditional means of access control have been restrictions on travel by motorized road vehicles, road closures, and the siting of access roads away from relatively unexploited lakes. Access controls that may be used to lower fishing pressure include controls on the use of snow machines and power boats, the establishment of daily travel quotas at access points, the control of boats cached on public land, the allocation of fish in remote lakes or areas to specific outfitters, and the limitation of development or accommodation at outpost camps. Access controls are used in Quebec, but they are part of the quota management strategy used there and are discussed in that context.

Use of publicly owned access roads is usually permitted, although travel on such roads may be closed for various reasons. On private access roads, such as those owned by logging companies, public travel can be denied or negotiated between a private company and a government agency or potential local users. Road closures, however, often have not come into effect until after overexploitation has already occurred. Also, locating forest access roads away from lakes containing lake trout, as well as other species, is no longer an effective deterrent to exploitation in Ontario (Ontario Ministry of Natural

Resources, 1982) because of the ready availability of motorized vehicles that do not require roads for travel.

Regulations have also been enacted in some parks (such as Algonquin Park, Ontario), which prohibit the operation of snow machines and power boats or limit the horsepower of motor boats on certain lakes or to specified times of the year. The last restriction, for example, while primarily intended to separate canoe trippers and outboard motor users, also limits the ability of anglers to troll for lake trout on some of the larger lakes in the park. In Quetico Park, Ontario, which is much further from major population centers than Algonquin Park, access controls have also apparently had a strong influence on exploitation. Maher (1985) noted that in Quetico Park, the prohibition on power boats and snow machines reduced fishing pressure to a noncritical level. In Algonquin Park, quotas have also been established that limit the number of canoeing/camping parties entering the park interior from designated access points on any day at certain times of the year. The closure of Algonquin Park to winter fishing in 1955 and the use of access controls in more recent years are probably the major factors contributing to the retention of quality fishing in that park, as overexploitation of lake trout is largely confined to lakes adjacent to road access.

Limiting the numbers and locations of boats cached or left unattended on public lands is another form of access control that could also act as a harvest control mechanism on lakes where fishing pressure by anglers using such boats has contributed to the overexploitation of fishery resources. This idea has been used since the early 1980s in the northwestern part of Ontario (Ontario Ministry of Natural Resources, 1996), where all boats cached on public land must be authorized and identified by means of a validation decal. On some lakes boat caches may not be permitted, while on other lakes the number of boats and the conditions of use may be regulated.

Outpost camps on remote lakes may be allocated to specific tourist outfitters. The basic intent is to support the tourist industry by limiting competition and to prevent overexploitation. Agreements between government agencies and the tourist industry may also result in limits on the number of overnight accommodation units at these outpost camps. This in effect places a quota on the number of anglers who have access to these facilities. The intent is to maintain high-quality fishing by keeping projected harvests below annual production.

In summary, fishing pressure and ultimately harvest depend, in part, on accessibility. Hence, access controls to curb overexploitation offer a viable alternative to additional or more stringent angling regulations. They may also be used to complement existing angling regulations to effect the same result. Many types of access control have evolved in response to "local" problems. Consequently, they often lack a sense of perspective and can proliferate independently of one another and therefore have not been used to their full potential (Ontario Ministry of Natural Resources, 1982). A study of the role of these controls relative to the harvest of fishery resources would be both appropriate and timely.

Size limits

Size limit regulations have important management implications, because the size and age of recruitment and the entry of year-classes into a fishery can be determined and thus manipulated. Size-based regulations infer the release of live fish. Their use to achieve a particular social or biological goal is based on the assumption that the survival rate of released fish is sufficient to ensure sustainability of wild fish.

Although studies that test that assumption are not reviewed in detail, a few comments are presented to show the importance of hooking mortality to size-based regulations. Studies in large lakes during spring and summer have shown hooking mortality rates in the range of 7.0% (Great Bear and Great Slave Lakes, Northwest Territories; Falk et al.,

Color Plate 1
Lake 226 in the Experimental Lakes Area in northwestern Ontario during the famous experiment to demonstrate the effects of the addition of carbon, nitrogen, and phosphorus (far basin) vs. carbon and nitrogen only (near basin) on algal production. (Photo by I. Davies.)

Color Plate 2
Flooding of Lake 979 and associated wetland at the Experimental Lakes Area to study the effects of reservoir creation on Hg dynamics and greenhouse gas production. (Photo by K. Scott.)

Color Plate 3
Killarney Provincial Park in Ontario contains some of the clearest waters in the world, with dissolved organic carbon (DOC) concentrations <0.2 mg L^{-1} and Secchi depth transparency >30 m in some lakes. (Photo by E. Snucins.)

Color Plate 4
Meltwater bringing colored DOC into a Shield lake. DOC has widespread effects on the optical and thermal properties of lakes. (Photo by V. Liimatainen.)

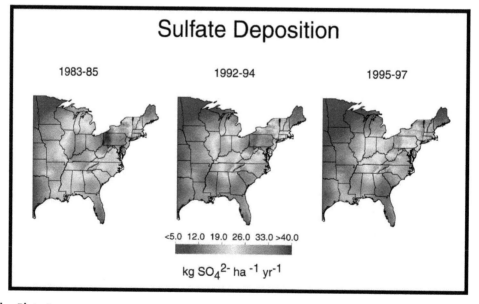

Color Plate 5
Declines in the annual wet deposition of SO_4^{2-} (in kg SO_4^{2-} ha^{-1} yr^{-1}) in the eastern United States based on data from 1983–1985, 1992–1994, and 1995–1997. Data were obtained for the NADP/NTN and the model of Grimm and Lynch (1997). (From Driscoll et al., 2001, *Bioscience* 51(3) 180–198.)

10-Year (1980-1989) Mean NO3 Wet Deposition (kg/ha/yr)

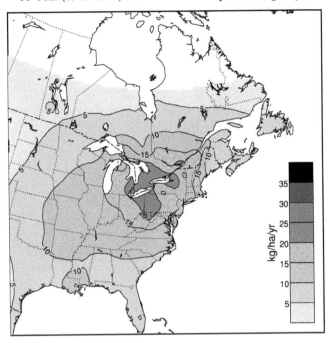

10-Year (1990-1999) Mean NO3 Wet Deposition (kg/ha/yr)

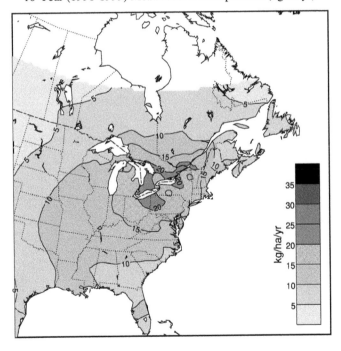

Color Plate 6
Mean nitrate deposition (kg ha^{-1} yr^{-1}) changed little between the periods 1980–1989 and 1990–1999, a reflection of minimal change in NO$_x$ emissions in North America. (From Ro, C.U. and Vet, R.J., 2003, Analyzed data fields from the National Atmospheric Chemistry Database (NatChem) and Analysis Facility. Air Quality Research Branch, Meteorological Service of Canada, Environment Canada, 4905 Dufferin St., Toronto, Ontario, Canada M3H 5T4.)

Color Plate 7
Groundwater seepage site on Pedro Lake, a thermal refuge area for lake trout in this warm, shallow (max. depth 11 m) lake. In a more typical deep thermally-stratified lake, the cool waters of the hypolimnion provide this refuge. (Photo by E. Snucins.)

Color Plate 8
Lake trout spawning over egg collectors (30 cm diameter) buried in the natural spawning substrate at Whitepine Lake. Most spawning sites in small Boreal Shield lakes are in depths of <1.5 m and are within 10 m of shore. (Photo by S. McAughey.)

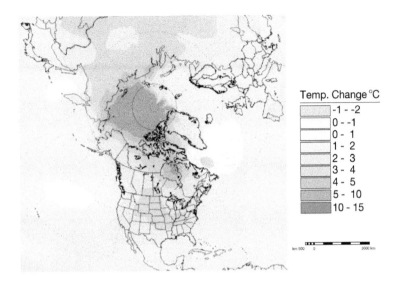

Color Plate 9
Changes in average annual air temperatures by 2050 as projected by the Canadian coupled climate model.
(From Meteorological Service of Canada, Environment Canada.)

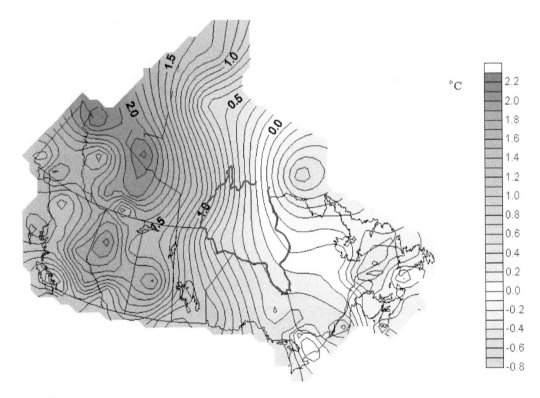

Color Plate 10
Observed changes in average annual air temperature between 1948 and 2000. (From Meteorological Service of Canada, Environment Canada.)

Color Plate 11
Forest fires are a natural disturbance in the Boreal Shield ecozone, with an average of over 2 million hectares burned each year. With present climate warming trends, larger and more frequent fires are expected in the future. (Photo by J. Shearer.)

Color Plate 12
Atmospheric emissions from nickel smelters in Sudbury, Ontario. Emissions of SO_2 have declined by 90% since the 1960s, when the Sudbury smelters were among the largest point sources of SO_2 in the world. Total emissions of SO_2 across North America are now 40% less than in 1980. (Photo by F. Prevost.)

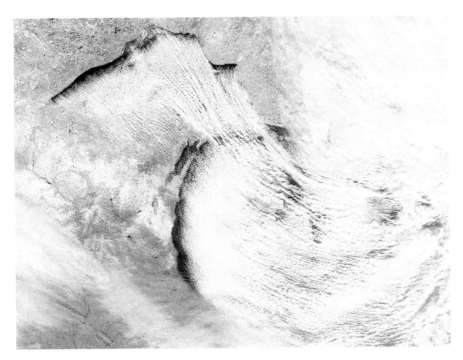

Color Plate 13
Winter storm on Lake Superior showing the "lake effect" of the Great Lakes on surrounding areas. (From http://visibleearth.nasa.gov/cgi-bin/viewrecord?6566.)

Color Plate 14
Teardrop Lake (3.4 ha surface area, 16.6 m max. depth), the smallest known lake trout lake in the Boreal Shield ecozone. Over 75% of the more than 3000 lake trout lakes in this area are less than 500 ha. In contrast, Lake Superior, the largest lake inhabited by lake trout, has a surface area of approximately 8,200,000 ha and maximum depth of nearly 400 m. (Photo by E. Snucins.)

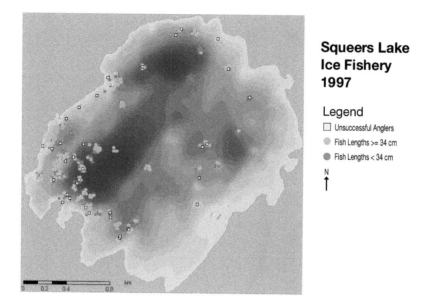

**Squeers Lake
Ice Fishery
1997**

Legend

☐ Unsuccessful Anglers

◉ Fish Lengths >= 34 cm

● Fish Lengths < 34 cm

N
↑

Color Plate 15
Angling effort and success varies widely among regions, lakes, and areas within lakes. Shown here is the distribution of angling locations and the size of captured fish on 384 ha (33.6 max. depth) Squeers Lake. (Digital bathymetry map by R. Kushneriuk, Ontario Ministry of Natural Resources.)

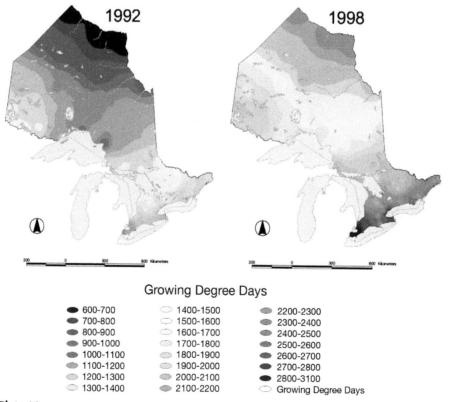

Growing Degree Days

● 600-700	○ 1400-1500	◉ 2200-2300
● 700-800	○ 1500-1600	◉ 2300-2400
● 800-900	○ 1600-1700	◉ 2400-2500
● 900-1000	○ 1700-1800	● 2500-2600
● 1000-1100	◉ 1800-1900	● 2600-2700
◉ 1100-1200	◉ 1900-2000	● 2700-2800
○ 1200-1300	◉ 2000-2100	● 2800-3100
○ 1300-1400	◉ 2100-2200	○ Growing Degree Days

Color Plate 16
An example of annual differences in air temperature across Ontario. Maps show the accumulated temperature (>5°C) for a cold (1992) and very warm year (1998). (Models by T. Marshall, Ontario Ministry of Natural Resources.)

1974) to 14.9% (Lakes Huron, Michigan, and Superior; Loftus et al., 1988). Unpublished Quebec and Ontario studies on smaller lakes of the Precambrian Shield have shown similar or slightly higher mortality rates. The higher mortality in some of the Ontario lakes was attributed to prolonged handling times, poor handling practices, and entanglement problems with tethering gear used to retain lake trout for assessment of delayed hooking mortality. In addition to hooking mortality, Lee and Bergersen (1996) noted that the release and subsequent high mortality of lake trout in late summer in lakes with inadequate thermal and dissolved oxygen refugia may nullify the intended benefit of size-based regulations.

Hooking mortality of lake trout during the winter may be higher than during the open-water period, especially when set-lining (i.e., still fishing) using a bait (usually cyprinids, dead rainbow smelt *Osmerus mordax*, or lake herring *Coregonus artedi*). In Gunflint Lake, Minnesota (Persons and Hirsch, 1994) the observed mortality rate using this method was 32.0%, and fish were still dying at the end of the 6-day holding period. In contrast, jigging caused a mortality rate of only 9%. Jigging is a more active method than still fishing, the line is closely tended, and the fish has less chance or time to deeply ingest the bait (as it must strike a moving target) before the hook is set. With still fishing, the fish generally swallows the bait before the hook is set, and the bait tends to be deeply ingested and may be difficult to remove without causing further injury or death. As noted, anglers tend to catch smaller lake trout in the winter than in open-water even in the same lake. The mortality rate of released small fish is generally considered greater than that for larger fish. Additional studies of winter lake trout hooking mortality are required. Hook placement (i.e., location) and hook size are important aspects of hooking mortality and need to be studied further.

It would seem prudent for anglers to leave deeply imbedded hooks in any fish they intend to release (at any time of the year) because removal is likely to induce bleeding or cause further injury or death. Lake trout that are bleeding should not be released unless required by a size-based regulation. The importance of effective techniques for hook removal, handling, and release of fish should be promoted through public education programs.

The general purpose of minimum length limits, in which all fish below some designated size must be released, is to permit fish to mature and spawn at least once before reaching legal size. In theory, some of the released sublegal fish will be harvested when they attain legal size or will increase catches by contributing to future spawning stocks. These length limits are generally applied in fisheries where growth is good and recruitment has been reduced because of high exploitation.

Maximum size limits, in which all fish above a specified size must be released, are used to protect brood stock or to counter a decline in age and size of spawning stock. They are applied to stocks for which growth potential is large, but high exploitation has resulted in a low density of mature fish and limited recruitment. Conversely, these limits may be used to maintain the population structure of unexploited stocks, especially in large lakes where large body size can be attained.

A combination of minimum and maximum size limits is the slot limit. A size range in which anglers may only keep fish below or above specified lengths is called a *protected slot*. All fish within the protected slot must be released. The intent is to protect the brood stock while still allowing anglers access to the more numerous, smaller fish and the less abundant large or trophy-size fish. The capture of fish below the slot should, in theory, reduce density, increase growth, and allow escapement into the protected slot size. Reproduction of adults in the slot should increase recruitment. A harvested slot in which only fish between a specified minimum and maximum size may be kept has not, to our knowledge, been applied to lake trout.

Table 11.2 Size Limit Regulations for Lake Trout in Small, Forested Precambrian Shield Lakes[a]

Jurisdiction	Minimum size (TL)	Comments/exceptions
Saskatchewan	None	Only one may exceed 65 cm (TL) provincewide
		In catch-and-release one lakes, only one of two may be larger than 65 cm
		In catch-and-release two and three lakes, none may be larger than 65 cm
Manitoba	None	Only one may be larger than 65 cm (TL) provincewide
		In northwest division, all lake trout larger than 65 cm must be released
		In northeast division, in high-quality lakes all lake trout larger than 65 cm (TL) must be released
Ontario	None	Slot limits (33–40 cm, 40–55 cm, 40–56 cm) in some lakes
		Maximum size 38 cm in 4 lakes
		Lake-specific one over certain size (40, 56, 65, 70 cm) in approximately 20 lakes
Quebec	None	Slot size 35–50 cm (FL) in Zones 1–15, 18; does not apply to parks, wildlife reserves, and ZECs
Minnesota	None	
New York	53 cm	Statewide
	38 cm	Finger Lakes, Lake Champlain
	46 cm	Specified waters

Note: TL, total length; ZECs, Zones d'Exploitation Contrôlée.

[a] Regulations pertaining to Great Lakes waters are not included. Regulations in some jurisdictions may have changed in recent years.

Minimum length limits have generally been avoided for lake trout throughout most of the Precambrian Shield except in New York, which has statewide and lake-specific minimum size limits (Table 11.2). Most authors reporting on the utility of size limits have indicated that the existing minimum length limits were too small to ensure sufficient spawning escapement.

Webster et al. (1959) reported that the 38-cm FL size limit in Cayuga Lake, New York, was ineffective as few trout smaller than this size were taken by anglers. Seamans (1960) noted that a 38-cm total length (TL) size limit was ineffective in controlling exploitation in New Hampshire lakes as few lake trout were mature below 46 cm long. He reported that the size limit of 38 cm in Newfound Lake, New Hampshire, was arbitrary and denoted a "desirable" keeper, and suggested that the limit be increased to 46 or 48 cm or abandoned entirely. It is still in effect on that lake today. AuClair (1976) reported that the size limit on lake trout in Moosehead Lake, Maine, was increased from 38 to 46 cm in 1972 to increase natural reproduction.

A minimum size limit has been in effect for lake trout for many years in Raquette Lake, New York. Overharvest was thought to be the major reason for the population decline (Shupp, 1973). He recommended an increase in the minimum size limit (TL) from 38 to 53 cm, as 90% of female trout were mature at lengths greater than 51 cm. Barnhart and Engstrom-Heg (1984) noted several changes in the lake trout population at Raquette Lake following an increase in the minimum size limit from 38 to 53 cm in 1973. There was a threefold increase in the spawning stock within 2 years, and the catch rate of fish retained increased from 0.03 lake trout h^{-1} in 1968 to 0.08 during 1973–1978. Natural reproduction increased substantially because about 75% of the juvenile lake trout by 1978 were wild fish, whereas at the time of the change in the size limit, they were almost entirely hatchery fish.

Genetic and environmentally induced variability in lake trout growth makes it imprac-
tical to set a single minimum legal length to protect immature fish over a broad geographic
area. Across the total range of the lake trout, first maturity has been reached at lengths
from 18 cm (FL) in stunted populations in national parks in western Canada (Donald and
Alger, 1986) to 66–76 cm (TL) in fast-growing lake trout in Seneca Lake, New York (Royce,
1951). Even in a relatively small area such as Algonquin Park, Ontario, size at first maturity
varied from 28 to 48 cm (FL) (Martin, 1966). The selection of a single minimum size limit
over any large area to allow part of the stock to spawn at least once before legal harvesting
would be difficult to determine and would result in lake trout that were almost entirely
unavailable to anglers in many lakes (Martin, 1966). This would be true of any region
where there is a large concentration of lake trout lakes, particularly with a mix of planktivorous and piscivorous stocks that vary considerably in their age, size, and maturity
structure. The practicality of minimum size limits is further complicated by the fact that
in the same lake, the winter fishery tends to catch smaller trout than the summer fishery
(Martin, 1954; Schumacher, 1961). Martin (1966) noted that summer fishing with wire line
and large spoons serves as a self-regulating control on the size of fish caught. MacCrimmon
and Skobe (1970) concluded that a length limit was unnecessary for lake trout in Lake
Simcoe because the selective trout baits usually catch the larger fish. This self-regulating
effect may explain why minimum size limits are not common in many areas.

Work by Trippel (1993) on low- (<40 µS/cm) and high-conductivity (>77 µS/cm) lakes
in northwestern Ontario suggested that managers may need to restrict the harvest of
females larger than 55 cm (FL) in low-conductivity lakes and of females larger than 60 cm
(FL) in high-conductivity lakes. Lake trout in the former lakes had slower growth, matured
at an older age, attained maturity at similar or smaller body sizes, and were less fecund
than lake trout in high-conductivity lakes. Simple possession limits are not sensitive to
interlake differences in lake trout population structure (growth, maturation, fecundity).
To regulate the harvest of large adults, Trippel proposed that depending on lake conductivity, the daily catch or possession limit be changed so that anglers are allowed to keep
only one fish larger than 55 or 60 cm. Alternatively, a maximum size in effect at these sizes
could be employed.

A maximum size limit requires anglers to release all fish above a specified size. It may
also be used in combination with a daily catch limit to control the harvest of large or
trophy-size fish. The daily catch limit may remain unchanged, but only one fish above a
large maximum size may be kept. In Manitoba, the general regulation is that only one
lake trout larger than 65 cm (TL) may be kept. However, a maximum size limit (65-cm
TL) is in effect across one fishing division and in designated high-quality lakes in another
fishing division (Table 11.2). In these lakes, all lake trout larger than 65 cm (TL) must be
released. Limiting the catch of large lake trout may be a useful regulatory device where
anglers target on large lake trout or where the establishment of a trophy fishery is a
management objective. Although a maximum size is usually imposed to limit the harvest
of large, trophy-size fish, a maximum size of 38 cm (TL) is in effect on four south-central
Ontario lakes as a means of protecting the unique genetic structure of these stocks.

A protected slot limit of 40 to 56 cm (TL) was placed on seven Highway 60 corridor
lakes in Algonquin Park, Ontario in 1989. Even though these lakes are closed to winter
angling, some were overexploited during the open-water fishery. On one lake studied
(Smoke Lake), initial results in 1992 were promising (Hicks, 1994). Effort continued to
decline, possibly in response to previous overharvest and because anglers tend to fish
elsewhere when more restrictive regulations are first imposed. Harvest declined by 38%,
catch rates improved, and fall spawning assessment studies showed that more fish in the
length range from 40 to 60 cm were appearing on the spawning shoals and that total
abundance of lake trout on the spawning shoals had increased.

Beginning in 1996, protected slot limits were also used to regulate harvest in other south-central Ontario lakes south of Algonquin Park. About 88 lakes now have a slot limit of 40–55 cm (TL), and a slot limit of 33–40 cm (TL) has been placed on 13 lakes with polyphagous lake trout stocks. Of the 101 lakes in both slot ranges, 20 have been closed to winter fishing and in the other 81 slot limit lakes where winter fishing is allowed, 70 lakes have had the number of lines allowed when fishing through the ice reduced from two to one. Anglers have been slow to accept these changes and have complained about what they perceive to be unnecessary restrictions and are concerned about the mortality of fish released within the slot ranges. Government response has been to reduce the number of lakes regulated by slot limits.

In 1993 Quebec imposed a 35- to 50-cm (FL) protected slot in southern areas of the province except in Zones d'Exploitation Controlée (ZECs; literally, a zone where exploitation is controlled), parks, and wildlife reserves, and in those areas where outfitting operations have exclusive fishing rights. Angling is managed by a quota system in these areas. The regulation was imposed after an evaluation of the status of lake trout stocks, the high survival of released fish, and the receptivity of anglers to the regulation (Ministère du Loisir de la Chasse et de la Pêche, 1991). To keep the system manageable, a single protected slot size was applied to more than 300 lakes in which many lake trout stocks have different growth rates and age and length at first maturity.

Lake trout lakes were grouped into three categories according to growth rate (slow, medium, fast). Size limit options (minimum, maximum, protected slot) were evaluated for each growth category using Ontario's Lake Trout Management Support System (LTMSS) age-structured population model (Korver, 1992). A single minimum size limit was not effective across such a broad range of biological variability in growth. Only protected slot size limits offered the needed reduction in catch across all growth categories. For the fastest growing fish the protected slot acts as a minimum size limit, while for fish with the slowest growth rates it acts as a maximum size limit.

Modeling showed that the 35- to 50-cm (FL) slot, by protecting a significant portion of the brood stock, should permit higher recruitment and higher population densities. In turn, release of part of the catch by anglers should translate into high catch rates (the total of released and kept fish) than historically observed in these fisheries. According to the simulation model, the slot regulation should be effective for an angling effort up to 12 angler hours/ha. In 1993, angling effort was on the order of 8 angler hours/ha.

Results of creel surveys conducted in 1993–1994 on two lakes to evaluate the effectiveness of the 35- to 50-cm (FL) protected slot are shown below (Legault, 1994; Nadeau, D., unpublished data). Results are in accordance with model simulations for harvest reduction and the proportion of fish released. The biological objectives of the regulation change were achieved.

	Lac des Trente et un Milles	Kipawa
Number of years after regulation change	1	2
Harvest reduction (%)	59	59
CUE variation (%)	−25	+27
Proportion released (%)	52	50

A phone survey of 318 lake trout anglers conducted 1 year after the regulation was introduced indicated that 61% of the anglers interviewed agreed with the slot size regulation, 30% disagreed, and 9% had no opinion (Montminy, 1994). The level of acceptance of the measure indicates that the social objectives were also met.

A slot limit may be combined with a daily catch limit to control the harvest of large fish whereby more fish may be kept as part of the daily catch limit below the protected slot than above. For example, if the daily catch limit for lake trout was three fish and a protected slot limit was in effect, the daily catch limit could be set at two lake trout below the protected slot lower size limit and one fish above the protected slot upper size limit.

Catch limits

Along with season manipulations, the daily catch limit is the most common form of regulation applied to the individual angler. As most anglers do not fish each day during the open season, the daily catch limit distributes the catch somewhat throughout the season. In theory, it provides for an equitable distribution of fish among anglers by limiting the daily catch of the more skillful anglers. However, without entry or effort restrictions directed at the individual angler the daily catch limit is not a very effective mechanism for controlling total harvest.

A daily catch limit for lake trout may be expressed as an individual limit for the species or as part of an aggregate limit (usually a combination of trout and salmon species) in which the lake trout limit is a specified number lower than the aggregate total. This type of species combination limit is usually employed in lakes or areas where there are a number of salmonid species available to anglers and where the individual catch limits for each species combined is quite high, or where hybrid crosses such as the splake or lake trout backcross (*S. namaycush* × *S. fontinalis*) may be difficult to distinguish from the parent species.

Daily catch limits for lake trout vary from one to four fish, but most often are two or three fish per day (Table 11.3). A daily limit of one lake trout may be in effect on individual water bodies. As noted, size limits are used in conjunction with daily catch limits in several jurisdictions, such as in Manitoba and Saskatchewan, to control the harvest of large fish (Table 11.3). To remain within the legal daily catch and possession limits, anglers often keep fish on a stringer in the water or in a live well, and release small fish as larger fish are caught. In an attempt to prevent this culling or "trading up" of fish to a larger size some jurisdictions, such as Maine and New Hampshire, require anglers to choose immediately to release a legal fish or to kill it and keep it as part of the daily catch limit. Stringers and live holding pens are prohibited in only one of Ontario's 35 fishing divisions. This approach might be considered elsewhere.

Table 11.3 Daily Catch Limits for Lake Trout in Small, Forested Precambrian Shield Lakes [a]

Jurisdiction	Number	Comments/exceptions
Saskatchewan	4	Only one may be larger than 65 cm (TL)
		In Lac la Ronge, two/day, four annually
	1–2	In catch-and-release lakes
Manitoba	2	One with conservation license
		One in high-quality management lakes
Ontario	2, 3	Two in 19 divisions, three in 16 divisions, 1 fish in fewer than 10 lakes in province
		One with conservation license
Quebec	2–4	Two in most zones
Minnesota	3	
New York	3	

Note: TL, total length.

[a] Regulations pertaining to Great Lakes waters are not included. Regulations in some jurisdictions may have changed in recent years.

Harvest may also be controlled by the use of a conservation license. In Manitoba and Ontario, the daily catch limit for most fish species is reduced if a resident or nonresident purchases a less expensive conservation license. In both provinces, the daily catch limit for the holder of such a license is one lake trout instead of two or three. This idea might also be tried in other jurisdictions.

Schumacher (1961) conducted the first study that evaluated the effect of daily catch limits on annual harvest of lake trout. He found that if angling pressure remained the same, a change in the daily catch limit from five to three on Minnesota lakes would reduce annual harvest by more than 16%. This measure would particularly affect winter anglers, because the highest proportion of anglers taking four and five lake trout occurred in January, February, and March. However, Schumacher concluded that the proposed reduction in the daily catch limit of lake trout in Minnesota lakes from five to three (now in effect) "should not be considered as a method of curing the problem of too much fishing" if total catches remained above sustainable limits.

Olver (1969), working on a group of Ontario lake trout lakes, showed that a reduction in the daily catch limit from five to three fish would have little effect on harvest in lakes with low yield and low catchability, whereas the harvest in high-yield, high-catchability lakes would be reduced considerably. However, at the then-present levels of fishing effort he concluded that a decrease in the daily catch limit alone from five to three would not likely be effective as a means of preventing overexploitation, and that additional and more stringent measures may be required to regulate harvest in high-yield lake trout lakes. This conclusion was confirmed by Purych (1975), who found that the reduction in the daily limit from five to three lake trout in 1970 did not significantly reduce the yield in the winter fishery at Summers Lake, Ontario. Similarly, Bernier (1977) found that the same reduction in the daily catch limit had little impact on the average annual yield of lake trout during the winter fishery at Bone Lake, Ontario. Daily catch limits are usually set too high to curtail annual harvests effectively. However, the real problem is the general, open-access nature of sport fishing, in which the total number of anglers is not regulated.

Martin (1966) noted the disparity in size between planktivorous and piscivorous lake trout in Algonquin Park lakes. He commented on the inequity of the daily catch limit that allowed an angler to catch a limit of three fish that would weigh about 1.2 kg in some lakes and about 4.5 kg in others. He suggested that a daily total or cumulative length or weight regulation should be considered. The efficacy of such an approach could be tested experimentally. Currently, the only weight limit applied to lake trout on a broad basis occurs in Maine, which has an aggregate limit of 3.4 kg for salmon, trout, and togue (lake trout) combined.

Quotas

Quota management is the ultimate harvest control method (Ontario Ministry of Natural Resources, 1982). Ideally, management of a fishery for any species should be based on a quota system in which the fishery for that species is closed for the rest of the season or year when a preset quota (equivalent to maximum equilibrium yield) is reached. Quotas can be applied in several ways. An individual angler may be permitted to keep a predetermined number of fish during the fishing season. This may be accomplished through regulations that require the angler to apply a tag to each fish the angler retains or notch or punch the angling license and record on the back of the license each fish that is kept. This type of regulation may be lake specific or apply across a broad area for the same species. Quotas may be applied without reference to an individual angler and be based on the cumulative catch of all anglers, either on a lake-specific basis or across a broad area.

Tag and quota systems may reduce the need for a plethora of local regulations. These systems represent an innovative change from the usual approach of implementing angling regulations on a piecemeal basis, which often results in acrimonious and incessant squabbling when the proposed regulation is presented to anglers. Angling opportunities are not reduced. Under a tag or quota system, anglers without tags or over quota are required to return all fish to the water even though some of these fish may be injured and subsequently die. Anglers with tags have the choice of releasing a fish or keeping and tagging it and applying it to their seasonal or annual quota.

An individual tag-and-quota system for lake trout has been in effect since 1991 in Clearwater Bay and adjacent Echo Bay and Cul de Sac, Lake of the Woods, Ontario, to curtail the harvest of trophy-size lake trout (Mosindy, 1998). Catch-and-release fishing for lake trout is in effect except for tag holders, who are selected by an annual draw. No more than one tag, which is nontransferable, is allotted per angler. Anglers who decide to harvest a fish must immediately affix a nonreusable tag (seal) to the harvested fish. The total annual quota is 110 fish. Other special regulations apply as well (e.g., barbless hooks, no fish or fish parts for bait, no winter fishing).

A quota may also be established without reference to the individual angler. An experimental quota fishery has been in effect for lake trout at Squeers Lake, Ontario (a fish sanctuary) since 1985. The fishery operates in mid-March, and the number of anglers is set through a lottery system to match a preselected harvest figure based on the projected angling success and effort of the successful lottery applicants.

The most intriguing use of a quota system approach to fisheries management occurs in the province of Québec, where it applies in many parks, wildlife reserves, areas where outfitters have exclusive fishing rights, and ZECs. All these areas or territories, as they are referred to, are described in regulation and encompass a wide spectrum of management options. Parks are managed by the government, while wildlife reserves are managed by a semipublic, profit-oriented corporation created and overseen by the government. Exclusive fishing rights are granted to some outfitters by lease for a specific territory. ZECs, however, are nonprofit organizations appointed by the government to manage angling in each individual ZEC (Ministère de l'Environnement et de la Faune, 1994).

The controlled zone concept was first introduced in the late 1970s when the exclusive fishing and hunting rights granted to many private clubs were abolished. There are now 63 hunting and fishing ZECs scattered throughout the province, and they encompass about 3% of the total area of the province. Each is managed by a board of directors, who are elected by the ZEC members. A written agreement exists for each ZEC between its board of directors and the government. The government sets the maximum fees that may be charged for fishing (and hunting) and carries out a financial audit each year. Regulatory powers are delegated to each ZEC, and when regulations are passed they have the force of provincial law. These regulations are often more stringent than those required by provincial law. Each ZEC usually hires a foreperson who runs the day-to-day operations of registering anglers, collecting fees and biological information, and overseeing road construction and maintenance.

Other areas of the province outside these four territories are called *free territories* or *free-access areas*. They are managed in a manner similar to public lands in other jurisdictions in that they are not regulated by fees other than a provincial fishing license, or on a quota basis.

Anglers need a provincial fishing license in all territories. In addition, parks and wildlife reserves charge a daily fee. Outfitters with exclusive rights have more flexibility in that they may allow only those anglers staying at one of their establishments to fish in lakes under their control, or they may charge nonguests a daily fee to fish in their territory.

In ZECs, anglers have the option of purchasing an annual license for each ZEC or paying a daily fee to fish.

Each territory (not including the free territories) has an exploitation plan in which catch quotas are set by government biologists. Initially, quotas are determined using standard yield equations. The quotas are expressed in biomass (kg or kg/ha), number of fish, or number of angler days per year. Quotas are reevaluated each year by government biologists and can be adjusted for individual lakes in response to specific management goals, such as maximum equilibrium yield, fishing success, or size of fish captured.

In parks, wildlife reserves, and territories where outfitters have exclusive rights, the managers decide on a daily basis how many anglers can fish each lake. The intent is to ensure the quality of the fishing experience by eliminating crowding. The quota is regulated by what is known as an access right that can be denied when the quota is reached. ZECs, however, do not have the authority to control the number of people fishing on any one lake on any one day. Fishing on a lake in a ZEC is closed by the government when the quota for the year is reached, even if this takes only a few days or a few weeks.

The manager of each territory may institute a number of fishing regulations (approved by the government) to meet specific needs. For example, to lower fishing success and maximize the number of angler days in some ZECs, fly fishing only is allowed. In Lac Maganasipi in ZEC Restigo it used to take only about 2 weeks to reach the quota using artificial lures. The lake is now open for the full regular season under a regulation for fly fishing only.

Delaying the opening of a season in early spring after ice-out, when fishing is generally considered to be excellent, can allow longer seasons at other times of the year. For example, Lac St-Patrice, which was open for only about 1 week in May before the quota was reached, is now open for about 4 weeks — from the last weekend in June to the last weekend in July.

In some territories with a common opening date, the quota is reached within a few weeks after the opening of the season in May. Progressive opening of different water bodies to angling, rather than one synchronized opening, however, can ensure sustained fishing opportunities throughout the summer. For example, in ZEC St-Patrice some lakes open at the start of the season in mid-May, others open in the first week of June, while the rest open in the last week of June.

Continuous monitoring is required to manage a quota system effectively. This necessitates an infrastructure to register anglers when they both enter and exit a territory. These checkpoints are costly to maintain. To reduce costs after anglers are registered on entry, they may be asked to declare their catch on a self-declaration form, which is to be deposited at boxes on exit.

Some information may be collected on exit if personnel on are duty. This is generally limited to the number of fish harvested and fishing effort. Sometimes fish are weighed and otoliths may be collected for age determination. This biological information is biased, however. Anglers tend to eat the smaller fish in their catch, and this can result in an overestimate of the average weight and age structure of the catch. Experience has shown that while these data are sufficient for management purposes, they are not adequate for scientific analyses.

Fisheries management in Quebec, exclusive of the free territories, has three attractive features:

1. It incorporates some form of a "user pays" basis.
2. In wildlife reserves and ZECs, the public is very much involved as a partner in fisheries management.
3. Lakes can be closed to angling when the potential harvest (i.e., quota) is reached.

The various forms of tag-and-quota controls (individual tag and quota, lake specific, across a fishing division, or across a state or province) each merit serious study. Such investigations are necessary not only from a biological perspective, but also as a means of understanding the social implications of this type of regulation.

Gear

Regulations that define the type of gear anglers may use to catch fish are intended to control harvest by reducing the efficiency of anglers, altering the size of fish caught, or reducing the mortality of released fish. In this review, gear includes the number of lines allowed while fishing, fishing techniques, bait restrictions, and the use of "high-technology" equipment.

Most jurisdictions allow the use of two lines while fishing through the ice and only one line at other times (Table 11.4). New York regulations are more liberal, and two lines can be used during both open-water and ice fishing seasons. In Minnesota only one line is permitted in the winter on designated trout lakes, while in Ontario a one-line restriction in the winter applies to only one division and several individual lakes across the province. This measure is directed at reducing harvest without curtailing fishing opportunities. In several areas anglers may not fish with an unattended line when fishing through the ice, and must be within a specified distance (see Table 11.5) from their fishing lines. The intent is to allow anglers to reach and retrieve hooked fish quickly to reduce hooking mortality.

Persons and Hirsch (1994) suggested that because of high mortality (32%) of lake trout observed using set lines in winter fisheries in Minnesota, the number of such lines allowed when using this technique should be curtailed, especially in lakes where release is desired or mandated. Instead of reducing the number of set lines, the state has disallowed their use altogether (Table 11.5).

The effect of the number of lines (or rods) fished on harvest is a complex issue. Not all anglers use two lines when fishing through the ice. Both Lester et al. (1991) and Evans et al. (1991) found that the average number of lines used per angler when ice fishing in Ontario was about 1.5. The reduction in total annual harvest would also depend on the

Table 11.4 Maximum Number of Lines Permitted while Angling for Lake Trout in Small, Forested Precambrian Shield Lakes[a]

Jurisdiction	Open water	Ice fishing	Comments/exceptions
Saskatchewan	1	2	
Manitoba	1	2	
Ontario	1	2	In line in winter in one division and specified lakes
Quebec	1	5	Generally no ice fishing for lake trout
Minnesota	1	2	Two through ice except on designated trout lakes and streams
New York	2		Two hand lines and five tip-ups may be used where ice fishing permitted
			Two hand lines and 15 tip-ups permitted while ice fishing on Lake Champlain

[a] Regulations pertaining to Great Lakes waters are not included. Regulations for some jurisdictions may have changed in recent years.

Table 11.5 Gear Restrictions for Lake Trout Angling in Small Forested Precambrian Shield Lakes[a]

Jurisdiction	Comments
Saskatchewan	Barbless hooks mandatory on catch-and-release waters
	Proposed that use of barbless hook be mandatory provincewide in 2000–2001 angling season
	Anglers must be within 25 m of fishing line
Manitoba	Barbless hooks provincewide
Ontario	Barbless hooks in two fishing divisions
	Ice fishers must remain within 60 m of any line while fishing
Quebec	Certain waters in some ZECs and wildlife reserves are reserved exclusively for fly-fishing
Minnesota	Angling with unattended line (>200 ft) or set line prohibited
New York	Anglers must be in immediate attendance when lines are in water

Note: TL, total length; ZECs, Zones d'Exploitation Contrôlée.

[a] Regulations pertaining to Great Lakes waters are not included. Regulations for some jurisdictions may have changed in recent years.

portion of the annual catch taken through the ice. Lester et al. (1991) found that the effects of using two lines instead of one were highly variable, but nonsignificant. The catch rate using two lines varied from 0.5 to more than 2.0 times the catch rate of a single line. Evans et al. (1991) observed that the effect of two lines was the greatest in small lakes (200 to 1000 ha). Party size also affects catch rate, and Lester et al. (1991) found the relationship between party size and rods per angler affects angler effectiveness. More studies should be conducted to determine the effectiveness of a one-line restriction in winter as a harvest control mechanism.

Barbless hooks are required throughout Manitoba and on all catch-and-release waters in Saskatchewan (Table 11.5). The use of barbless hooks is required in only one fishing division in Ontario and in a few designated lakes.

Most jurisdictions within the Precambrian Shield have not prohibited the use of live bait for lake trout fishing other than fish species (Table 11.6). Hence, regulations concerning bait generally refer to bait fish. Many jurisdictions regulate the use of bait fish by defining the fish species that may be used as bait; by regulating the conditions (alive or dead) under which they may be used, possessed, transported or imported, and by prohibiting their use in specified waters, sections of certain waters, or throughout fishing divisions. With the exception of Ontario, most Canadian jurisdictions in the Precambrian Shield do not allow the use of live fish as bait or live bait fish (as defined in their regulations). These restrictions seem reasonable as Mongillo (1984) concluded that using natural baits causes hooking mortality in critical areas about 50% of the time or 5 times greater than the use of artificial gear.

In Ontario, the use or possession of live bait fish (as defined in their regulations) is prohibited in three divisions. In one division (Algonquin Park), this restriction is primarily directed at retaining the integrity of the native fish fauna, particularly brook trout, and has little to do with control of exploitation of lake trout. A ban on live bait fish seems to be particularly applicable in areas or lakes where native salmonids such as lake trout and brook trout have coadapted.

Presumably, there is some correlation among hook size, bait fish size, and the size of lake trout caught by anglers. Anglers may use large single or treble hooks baited with such fish (where permitted) as rainbow smelt, lake herring or cisco, and suckers (*Catostomus* spp.) to catch large trout. Hickory (1973) noted that the use of cisco 15 to 25 cm in length as bait is size selective for mature lake trout brood stock and perhaps should be

Table 11.6 Restrictions on Use of Bait for Lake Trout Angling in Small, Forested Precambrian Shield Lakes[a]

Jurisdiction	Comments
Saskatchewan	Live fish may not be used as bait; only bait fish that have been commercially frozen or preserved may be used
Manitoba	Live bait fish not permitted in Precambrian Shield lakes (three of four divisions); frozen or preserved bait may be used in all divisions
Ontario	No live fish for bait permitted in three divisions
	No live or dead herring permitted in four divisions
	No live rainbow smelt maybe used as bait; no dead smelt permitted in six divisions
Quebec	Live or dead bait fish generally not permitted except where allowed in 4 zones; use of dead bait fish permitted in 11 zones (with exceptions)
Minnesota	No regulations noted in angling summary
New York	Use/possession of alewife, blueback herring prohibited in five counties

[a] Regulations pertaining to Great Lakes waters are not included. Regulations for some jurisdictions may have changed in recent years.

banned. The use of lake herring (alive or dead) as bait is prohibited in only four fishing divisions in Ontario. Live smelt may not be used as bait in Ontario, and the use of dead smelt is prohibited in six divisions in that province.

Some anglers impale large bait fish on gorge hooks, let the bait settle to the lake bottom, and allow a lake trout to swallow the bait deeply before setting the hook. This method is intended to catch the larger lake trout, which are usually retained by the anglers. Generally, these fish cannot be returned to the water alive because the baited hook or hooks are deeply ingested. A ban on the use of one or two species of large bait fish, such as lake herring or rainbow smelt, cannot be totally effective if the use of alternate bait fish species (alive or dead) of a similar size is permitted. To be effective, prohibitions on the use of bait fish should include both live and dead fish and be more inclusive than they tend to be. A ban on the use of all species of large bait fish might be considered in lakes when anglers use this method to target large lake trout and where the lake's reputation as a producer of trophy fish is considered to be endangered. Some jurisdictions have responded to this concern not by bait or gear restrictions (i.e., by banning the use of gorge hooks), but through a combination of size and daily catch limits on lake trout.

The use of sophisticated or so-called hi-tech fishing gear such as downriggers, fish finders, and trolling speed indicators has increased substantially in the last two decades. The use of these devices by anglers fishing for lake trout is now commonplace. The results of a study conducted on the lake trout trolling fishery at Lake Temagami, Ontario in the summer of 1981 (Duckworth, 1982) showed that anglers using a fish finder in combination with wire line had a success rate (0.202 fish man-h^{-1}) about twice that of other methods (0.099 to 0.108 fish man-h^{-1}). Hicks (1986) found that 70.5% of the 708 lake trout anglers interviewed in Lake Opeongo, Ontario in 1986 trolled with wire line, and 14.8% used a downrigger. Trolling with a wire line and a fish finder was the most successful method (0.18 fish man-h^{-1}) and was similar to the results reported earlier by Duckworth (1982).

Hicks (1986) also determined the effect that previous fishing experience on Lake Opeongo had on fishing success. The most experienced fishermen were the most successful, and anglers using a fish finder were generally more successful than anglers not using this device. He indicated that with some exceptions (which may be a function of sample size), catch rates increased with the level of prior experience on a lake, wire line trolling with or without a fish finder was generally more successful than a downrigger with or without that gear, and trolling with the aid of a fish finder was generally the most successful technique. The use of devices such as fish finders and downriggers can increase the catch of lake trout, especially when used by experienced fishermen.

Plosila (1977) suggested that if these types of fishing gear were contributing factors to overexploitation or juvenile (sublegal) mortality of lake trout, reduced daily catch limits or additional gear restrictions may be required on New York State lake trout waters. An evaluation of the use and impact of these gear types on lake trout stocks and fisheries is urgently needed.

Although it can be argued that the use of high-technology gear is not likely to be prohibited, this should not be considered an absolute. Further, even if such devices are not prohibited, their effectiveness should be evaluated so that measures to counteract their impact can be employed if necessary.

The insightful concluding remarks of Regier and Loftus (1972) in their article on the effects of exploitation on salmonid communities are still pertinent. They stated that "every technical innovation or improvement in gear or fishing techniques would require some compensatory modification in existing management regulations " to ensure the continuance of high-valued salmonid fisheries. The key is to have the minimum number of effective regulations to provide the necessary resource safeguards. The achievement of such a balance is a difficult task. If in doubt about the effectiveness of a regulation, managers should take a conservative or protectionist approach to controlling harvest.

Population models

Studies by Payne et al. (1990), Evans et al. (1991), and Shuter et al. (1998) have added greatly to our understanding of the population dynamics of lake trout and how its various components affect sustainable levels of harvest, fishing effort, and fishing mortality. Payne et al. (1990) observed an inverse relationship between lake surface area and yield of lake trout. Evans et al. (1991) developed a harvest equation for lake trout ($\log_{10} H = 0.60 + 0.72 \log_{10} A$) that takes into account this relationship. Shuter et al. (1998) constructed a population model that predicts sustainable levels of fishing mortality and fishing effort from lake area and total dissolved solids. Small lakes are more sensitive to overexploitation because the fishing mortality rate necessary to reach maximum equilibrium yield is much lower than it is for large lakes. Small lakes are less able to compensate for overfishing and are more likely to suffer stock collapse (Payne et al., 1990). Consequently, small lakes (<1000 ha; see Evans et al., 1991) may require more stringent regulations. This is particularly true for winter fisheries on small lakes, which usually have higher effort and catch rates in winter than during open-water seasons. As noted by Ryder et al. (1974), on a per-unit-area basis, smaller lakes are exploited more efficiently than larger lakes when both receive proportionally similar fishing effort. This has important management implications in that overharvest of lake trout is a more likely occurrence on the smaller lakes, suggesting, for example, that lake size should be a consideration when setting the length of fishing seasons. Managers may use the model as a reference to decide whether harvest, effort, and mortality values in a particular lake in their jurisdiction indicate the need for additional harvest controls.

The LTMSS and its current refinement, the Fisheries Management Support System (FMSS) (Korver and Kuc, 2002), also allow the user to specify and evaluate a wide range of management options, such as changes in effort, size limits, daily catch limits, and stocking rates. These models are particularly useful to managers because they allow testing of the potential effects of proposed harvest control regulations before they become law or determining what the most appropriate change might be.

Conclusions

In the Precambrian Shield, exploitation is the most critical stress affecting the lake trout. This is due to the lake trout's innate vulnerability to angling and the growth of sport fishing as a major recreational activity. The biological characteristics of the species, its exacting habitat requirements, and the unproductive nature of its environment also serve to increase its susceptibility to angling stress.

The open access, common property nature of resource management is an inherent constraint to effective harvest control. Without meaningful controls on effort, traditional angling regulations for lake trout are likely ineffective in preventing overexploitation. Harvest controls such as seasons and daily catch limits may provide managers and anglers alike with a false sense of confidence about their ability to control exploitation. An entire year's annual production or more can be removed in only a few days or weeks of fishing, particularly on small lakes, in an entirely lawful manner. Yet under such a system, angling is allowed to continue unabated until the season reaches its normal closing date. It seems prudent that some form of quota management in which biological control of the harvest supersedes social controls should be the focus of regulatory research and reforms in the near term.

Proposed regulations are seldom assessed in an experimental context, and those in effect are seldom evaluated after implementation. The result is that managers rarely know which regulations are effective, which are too liberal to control harvest, which are detrimental, and which are merely cosmetic. The need to apply a scientific approach to studying and implementing fisheries management regulations is long overdue.

One of the most difficult aspects of harvest control is adjusting or tailoring the regulations, particularly those that are size based, to account for the natural variability and diversity of lake trout populations. Differences in climate, growing degree days, human population densities, access, and effort across the Precambrian Shield also contribute to the difficulty of establishing regulations for this species that are effective over broad areas, let alone more specific or smaller areas. As Martin (1966) noted: "It is likely what is biologically ideal for each of the fisheries will always have to be tempered with what is administratively feasible for the fisheries as a whole." The challenge is to achieve a balance between broadly based but effective regulations and the need to develop harvest control mechanisms specific to individual lakes or sets of lakes.

Recommendations

The first set of recommendations suggests some areas of regulatory control that might be implemented or reviewed now:

- To determine whether new harvest control regulations need to be implemented, (1) use the harvest equation presented by Payne et al. (1990), which is $\log_{10}$ Harvest $= 0.500 + 0.830 \log_{10}$ Area and where Harvest is in kilograms/(hectares·year) and Area is in hectares, and (2) use the age-structured model developed by Shuter et al. (1998) to determine sustainable levels of harvest, fishing effort, and fishing mortality.
- Use the age-structured LTMSS (Korver, 1992) and FMSS (Korver and Kuc, 2002) models to test and evaluate effects of proposed changes in effort, size limits, or daily catch limits or to determine the most appropriate changes.
- In areas where exploitation during the winter or early spring is intense, adjust the length of the open season by either delaying the usual opening date of January 1 or creating a split season by closing it before the ice goes out and reopening it some time after the ice is out.

- Review the role of access restrictions as harvest control mechanisms.
- Review/consider regulations that prohibit culling or "trading up" of fish (may prohibit use of stringers or live wells).
- Review/consider regulations that reduce daily catch limits by means of a conservation license.
- Review/consider regulations that prohibit or restrict the use of set lines when fishing through the ice.

The second set of recommendations suggests some areas of regulatory control that should be initiated under a program of experimental management in which fisheries projects are treated as hypothesis-testing experiments. Casual observations drawn from creel surveys seldom provide the data necessary to make informed judgments about the effects of regulatory changes. It also takes many years to accumulate data. Experimental management, however, "with appropriate control lakes, can accelerate the process significantly " (Ontario Ministry of Natural Resources, 1978b). Lakes selected for study require preexperimental data on the fishery and the population to which the impacts of changes in regulations can be compared.

Experimental management projects should be limited, at least initially, to testing one variable at a time. This will avoid the problem of confounding variables by which two or more techniques are applied or changed at the same time such that the contribution of each cannot be determined. Tests using multiple tactics should only be conducted after the effects of each variable have been independently evaluated.

Experimental management programs should be designed to evaluate:

- Hooking mortality of lake trout caught during the period of ice cover
- Hook size as a possible mortality factor
- The practice of allowing only one fish of the daily catch limit above a specified size to be kept as a means of controlling the harvest of large fish
- Maximum size limits
- Slot size limits
- Total or cumulative daily length or weight limits in place of daily catch limits
- Seasonal or annual catch quotas for the individual angler, either lake specific or jurisdictionwide
- Seasonal or annual catch quotas for specific lakes or parts of lakes
- Reduced daily catch limits during the period of ice cover
- Effectiveness of a one-line restriction during the period of ice cover
- Effect of party size on angler catch rates
- Difference (if any) between catch per unit effort using live or dead bait fish
- A ban on the use of bait fish (alive and dead)
- A ban on large bait fish (alive and dead) in lakes where anglers use this method to target large lake trout
- Use and effectiveness of high-technology gear

Acknowledgments

We would like to thank Henk Rietveld and Mike Powell for their helpful comments on an early draft. Rob Korver, Nigel Lester, Tom Mosindy, and Neville Ward were generous with their time and advice. Mike Freutel provided constructive criticism of the first draft. Dick Ryder and John Gunn contributed both scientific advice and editorial assistance throughout the review process.

References

Armstrong, G.C., 1961, Winter sport fishing in Ontario, *Ontario Department of Lands and Forests Fish and Wildlife Review*, 1(3):2–7.

AuClair, R.P., 1976, Moosehead update, *Maine Fish and Wildlife*, 18:4–7.

Barnhart, G.A. and Engstrom-Heg, R., 1984, A synopsis of some New York experiences with catch and release management of wild salmonids, in *Wild Trout III Proceedings of the Symposium*, Richardson F. and Hamre R.H., Eds., pp. 91–101.

Bernier, M.-F., 1977, Bone Lake Creel Census, 1968–1977, Ontario Ministry of Natural Resources, Sault Ste. Marie District, Sault Ste. Marie, Ontario.

Christie, W.J., 1978, A Study of Freshwater Fishery Regulation Based on North American Experience, Food and Agricultural Organization Technical Paper FIRI/T180, Rome.

Daley, R., Hacker, V.A., and Wiegert, L., 1965, The Lake Trout, Its Life History, Ecology and Management, Wisconsin Conservation Department, Publication 233, Wisconsin.

DeRoche, S.E., 1973, Ice fishing versus open water fishing. A matter of rights and preferences, *Maine Fish and Game*, 15:16–18.

Donald, D.B. and Alger, D.J., 1986, Stunted lake trout (*Salvelinus namaycush*) from the Rocky Mountains, *Canadian Journal of Fisheries and Aquatic Sciences*, 43:608–612.

Duckworth, G.A., 1982, Lake Temagami and Cross Lake Summer Creel Census Report 1981 Including an Assessment of the Annual Lake Trout Fishery on Lake Temagami, Ontario Ministry of Natural Resources, Temagami District, Ontario.

Eschmeyer, P.H., 1964, The Lake Trout (*Salvelinus namaycush*), United States Department of the Interior, Fish and Wildlife Service, Fisheries Leaflet 555.

Evans, D.O., Casselman, J.M., and Willox, C.C., 1991, Effects of Exploitation, Loss of Nursery Habitat, and Stocking on the Dynamics and Productivity of Lake Trout Populations in Ontario Lakes, Lake Trout Synthesis Response to Stress Working Group, Ontario Ministry of Natural Resources, Toronto.

Falk, M.R., Gillman, D.V., and Dahlke, L.W., 1974, Comparison of Mortality Between Barbed and Barbless Hooked Lake Trout, Canada Department of the Environment, Fisheries and Marine Service Technical Report Series CEN/T-74-1.

Fenderson, C.N., 1953, An Investigation of the Branch Lake Fisheries with Emphasis on the Brown Trout (*Salmo trutta*), master of science thesis, University of Maine, Orono.

Fry, F.E.J., 1949, Statistics of a lake trout fishery, *Biometrics*, 5:27–67.

Goddard, C.I., Loftus, D.H., MacLean, J.A., Olver, C.H., and Shuter, B.J., 1987, An evaluation of the effects of fish community structure on the yield of lake trout, *Canadian Journal of Fisheries and Aquatic Sciences*, 44(Suppl. 2):239–248.

Gunn, J.M., McMurtry, M.J., Casselman, J.M., Keller, W., and Powell, M.J., 1988., Changes in the fish community of a limed lake near Sudbury, Ontario: effects of chemical neutralization or reduced atmospheric deposition of acids? *Water, Air, and Soil Pollution*, 41:113–136.

Gunn, J.M. and Sein, R., 2004, Effects of forestry roads on reproductive habitat and exploitation of lake trout, in *Boreal Shield Watersheds: Lake Trout Ecosystems in a Changing Environment*, chap. 14, Gunn, J.M., Steedman, R., and Ryder, R., Eds., Lewis Publishers, Boca Raton, FL.

Hardin, G., 1968, The tragedy of the commons, *Science*, 162:1243–1248.

Hickory, R., 1973, Dwindling Ontario lake trout populations, *Ontario Fisherman and Hunter*, 5(11):17.

Hicks, F., 1994, Early Response of Lake Trout to a Slot Size Limit, Ontario Ministry of Natural Resources, Algonquin Fisheries Assessment Unit, Fisheries Assessment Unit Report #94-3. Algonquin Park, Ontario.

Hicks, F.J., 1986, The Opeongo Creel Survey, 1985–86, Ontario Ministry of Natural Resources, Algonquin Fisheries Assessment Unit, unpublished report.

Hughson, D.R., 1961, Report on Penage Lake Angling Success, 1960, Ontario Department of Lands and Forests Resource Management Report 58:48–55, Sudbury District Report, Toronto.

Korver, R., 1992, Lake Trout Management Support System User's Guide Version 2.0, Ontario Ministry of Natural Resources, Fisheries Branch, Toronto.

Korver, R. and Kuc, M., 2002, Fisheries Management Support System Version 1.0.8, Ontario Ministry of Natural Resources, Fish and Wildlife Branch, Peterborough.

Lee, W.C. and Bergersen, E.P., 1996, Influence of thermal and oxygen stratification on lake trout hooking mortality, *North American Journal of Fisheries Management*, 16:175–181.

Legault, M., 1994, *Bilan de la Première Année d'Application de la Gamme de Taille Protégée pour le Touladi*, Ministère de l'Environnement et de la Faune, Québec.

Lester, N.P., Petzold, M.M., Dunlop, W.I., Monroe, B.P., Orsatti, S.D., Schaner, T., and Wood, D.R., 1991, *Sampling Ontario Lake Trout Stocks: Issues and Standards*, Lake Trout Synthesis Sampling Issues and Methodology Working Group, Ontario Ministry of Natural Resources, Toronto.

Loftus, A.J., Taylor, W.W., and Keller, M., 1988, An evaluation of lake trout (*Salvelinus namaycush*) hooking mortality in the upper Great Lakes, *Canadian Journal of Fisheries and Aquatic Sciences*, 45:1473–1479.

MacCrimmon, H.R. and Skobe, E., 1970, *The Fisheries of Lake Simcoe*, Ontario Department of Lands and Forests, Fish and Wildlife Branch, Toronto.

MacKay, H.H., 1956, The Lake Trout, Ontario Department of Lands and Forests, Sylva 12:25–28.

Maher, T., 1985, Management options for small lake trout lakes in the Atikokan District, in *1985 Lake Trout Seminar North Central Region*, Ontario Ministry of Natural Resources, Thunder Bay.

Martin, N.V., 1954, Catch and winter food of lake trout in certain Algonquin Park lakes, *Journal of the Fisheries Research Board of Canada*, 11:5–10.

Martin, N.V., 1966, The significance of food habits in the biology, exploitation and management of Algonquin Park, Ontario, lake trout, *Transactions of the American Fisheries Society*, 95:415–422.

Martin, N.V. and Baldwin, N.S., 1953, Effects of the alternate closure of Algonquin Park lakes, *Canadian Fish Culturist*, 14:26–38.

Martin, N.V. and Fry, F.E.J., 1973, Lake Opeongo: The Ecology of the Fish Community and of Man's Effects on It, Great Lakes Fishery Commission Technical Report 24, Ann Arbor, MI.

Martin, N.V. and Olver, C.H., 1980, The lake charr (*Salvelinus namaycush*), in *Charrs, Salmonid Fishes of the Genus* Salvelinus, Balon, E.K., Ed., Dr. W. Junk, Hague, The Netherlands, pp. 205–277.

Ministère de l'Environnement et de la Faune, 1994, *Controlled Zones (ZECs). Nature and Operation*, Ministère de l'Environnement et de la Faune, Service de la Gestion Déléguée, Québec.

Ministère du Loisir de la Chasse et de la Pêche, 1991, *Proposition Concernant l'Implantation d'une Gamme de Taille Protégée pour le Touladi*, Ministère du Loisir de la Chasse et de la Pêche, Québec.

Mongillo, P.E., 1984, A Summary of Salmonid Hooking Mortality, Washington Department of Game, Fish Management Division, unpublished report.

Montminy, L., 1994, *Enquête Auprès des Pechêurs de Touladi*, Ministère de l'Environnement et de la Faune, Québec.

Mosindy, T., 1998, The Licence Validation Tag and Fish Harvest Tag as Regulatory Tools, Ontario Ministry of Natural Resources, Lake of the Woods Fisheries Assessment Unit, Fisheries Assessment Unit Report #98-3, Kenora, Ontario.

Olver, C.H., 1969, Potential Effects of Reducing the Daily Catch Limit of Lake Trout from 5 to 3, Ontario Department of Lands and Forests, Sault Ste. Marie District, unpublished report.

Olver, C.H., DesJardine, R.L., Goddard, C.I., Powell, M.J., Rietveld, H.J., and Waring, P.D., 1991, *Lake Trout in Ontario: Management Strategies*, Lake Trout Synthesis Management Strategies Working Group, Ontario Ministry of Natural Resources, Toronto.

Ontario Ministry of Natural Resources, 1978a, An Allocation Policy for Ontario Fisheries, Report of SPOF Working Group Number 5, Ontario Ministry of Natural Resources, Toronto.

Ontario Ministry of Natural Resources, 1978b, Experimental Management, Report of SPOF Working Group Number 3, Ontario Ministry of Natural Resources, Toronto.

Ontario Ministry of Natural Resources, 1982, Control of Angler Exploitation, Report of SPOF Working Group Number 16, Ontario Ministry of Natural Resources, Toronto.

Ontario Ministry of Natural Resources, 1996, *The Northwest Region Boat Cache Program — Program Description and Procedures*, Ontario Ministry of Natural Resources, Kenora.

Ontario Ministry of Natural Resources and Environment Canada, 1976, *First Report. Federal–Provincial Strategic Planning for Ontario Fisheries. Preliminary Analysis of Goals and Issues*, Toronto.

Payne, N.R., Korver, R.M., MacLennan, D.S., Nepsy, S.J., Shuter, B.S., Stewart, T.J., and Thomas, E.R., 1990, The Harvest Potential and Population Dynamics of Lake Trout Populations in Ontario, Lake Trout Synthesis Population Dynamics Working Group, Ontario Ministry of Natural Resources, Toronto.

Persons, S.E. and Hirsch, S.A., 1994, Hooking mortality of lake trout angled through the ice by jigging and set-lining, *North American Journal of Fisheries Management*, 14:664–668.

Plosila, D.S., 1977, *A Lake Trout Management Program for New York State*, New York Department of Environmental Conservation, Division of Fish and Wildlife.

Purych, N.A., 1975, The 1968, 1969 and 1971 Winter Lake Trout Sport Fishery of Summers Lake, Ontario Ministry of Natural Resources, Blind River District, unpublished report.

Regier, H.A. and Loftus, K.H., 1972, Effects of fisheries exploitation on salmonid communities in oligotrophic lakes, *Journal of the Fisheries Research Board of Canada*, 29:959–968.

Royce, W.F., 1951, Breeding Habits of Lake Trout in New York, U.S. Department of the Interior, Fish and Wildlife Service Fishery Bulletin 59, 52:59–76.

Ryder, R.A., 1957, Winter Fishing Pressure on Lake Trout, Port Arthur District, 1957, Ontario Department of Lands and Forests, Port Arthur District, unpublished report.

Ryder, R.A. and Johnson, L. 1972, The future of salmonid communities in North American oligotrophic lakes, *Journal of the Fisheries Research Board of Canada*, 29:941–949.

Ryder, R.A., Kerr, S.R., Loftus, K.H., and Regier, H.A., 1974, The morphoedaphic index, a fish yield estimator — review and evaluation, *Journal of the Fisheries Research Board of Canada*, 31:663–688.

Samis, W.G.A., 1968, Wakomota Lake winter angling fishery. 1968, Ontario Department of Lands and Forests, Sault Ste. Marie District, unpublished report.

Schumacher, R.E., 1961, Some Effects of Increased Angling Pressure on Lake Trout Populations in Four Northeastern Minnesota Lakes, Minnesota Department of Conservation, Division of Game Fish, Series 3:20–42.

Seamans, R.G., 1960, Newfound Lake, Hebron–Groton–Bristol, New Hampshire. The Findings, Conclusions and Recommendations of a Fisheries Management Study Conducted from 1958 to 1960, New Hampshire Fish and Game Federal Aid Project F9R.

Shupp, B.D., 1973, Results of a Four Year Study of the Lake Trout (*Salvelinus namaycush*) in Raquette Lake, Hamilton County, NY, New York Department of Environmental Conservation, Division of Fish and Wildlife, Dingle Johnson Project F-22-R.45.

Shuter, B.J., Jones, M.L., Korver, R.M., and Lester, N.P., 1998, A general, life history based model for regional management of fish stocks: the inland lake trout fisheries of Ontario, *Canadian Journal of Fisheries and Aquatic Sciences*, 55:2161–2177.

Shuter, B.J. and Lester, N.P., 2004, Climate change and sustainable lake trout exploitation: predictions from a regional life history model, in *Boreal Shield Watersheds: Lake Trout Ecosystems in a Changing Environment*, Gunn, J.M., Steedman, R., and Ryder, R., Eds., Lewis Publishers, Boca Raton, FL, chap. 15.

Trippel, E.A., 1993, Relations of fecundity, maturation, and body size of lake trout, and implications for management in Northwestern Ontario lakes, *North American Journal of Fisheries Management*, 13:64–72.

Vozeh, G.E., 1965, A Winter Creel Census on Little Quirke Lake, Sault Ste. Marie District, Ontario Department of Lands and Forests, Sault Ste. Marie District, unpublished report.

Walker, V., 1978, Lake Manitou Winter Creel Census, 1975–1978, Ontario Ministry of Natural Resources, Espanola District, unpublished report.

Webster, D.A., Bentley, W.G., and Galligan, J.P., 1959, Management of the Lake Trout Fishery of Cayuga Lake, New York, with Special Reference to the Role of Hatchery Fish, Memoirs Cornell University Agricultural Experiment Station 357.

Weir, J.C. and Martin, N.V., 1961, Winter Fishing Committee Report, Ontario Department of Lands and Forests, unpublished report.

chapter twelve

Lake trout stocking in small lakes: factors affecting success

Michael J. Powell
Ontario Ministry of Natural Resources
Leon M. Carl
Great Lakes Science Center

Contents

Introduction

Stocking has traditionally been viewed as a primary solution to fisheries management problems. A common view among early fisheries workers was that natural reproduction was inefficient, and that artificial propagation and careful stocking of fry guaranteed healthy fish populations (Nevin, 1892; Fullerton, 1906). The bottleneck appeared to take place on the spawning grounds. Males and females arrived and ripened at different times; when spawning finally occurred the manner in which eggs and milt were released into the environment seemed to ensure that few eggs were ever fertilized. This prevailing view was affirmed by Nevin (1898, p. 20): "Nature's provisions for the survival and increase of the several species of fish are not adequate. To rectify this apparent error in nature's laws, we have resorted to artificial propagation with gratifying results."

As in the past, the decision to stock may seem eminently logical given that problems such as habitat loss, overexploitation, and species introductions share the common symptom of decreased fish. It is understandable, then, that "topping up" the population would be perceived as a positive action. However, the objectives of many stocking programs are often not well articulated and might best be described, rather vaguely, as increasing the population so more fish can be caught. With no clear objectives it is difficult to set quantifiable measures of success and therefore determine if the stocking program is achieving the desired results.

Although the value of fish stocking has traditionally gone unquestioned, the potential negative consequences of planting fish have now been well documented (Evans et al., 1991; Evans and Willox, 1991; Hindar et al., 1991; Krueger and May, 1991; Hilborn, 1992). There is better understanding of community ecology and the need to manage in a sustainable manner by preserving biodiversity and managing ecosystems rather than single species (Callicott, 1991; Olver et al., 1995; Winter and Hughes, 1995; Schramm and Hubert, 1996). As a result there has been a considerable amount of discussion about the acceptability of stocking fish under any circumstances (J. Martin et al., 1992; Daley, 1993; Utter, 1994; Rahel, 1997). We accept the debate as healthy and suspect that fish stocking will continue but will be scrutinized much more closely in the future. Our purpose here is not to enter into the debate but rather to review past lake trout *Salvelinus namaycush* stocking results and attempt to determine which factors have had an effect on survival, growth, and reproduction of the stocked fish.

Stocking objectives

A number of factors, many of which may be unknown, can converge to affect the success of any stocking event. It is therefore essential to have current information on the physical, chemical, and biological characteristics of the candidate lake and, using this information, set objectives for the stocking program.

Objectives should be measurable, have a designated time frame, and have a reasonable chance of success. Examples might be to provide a lake trout yield of $0.6 \text{ kg}^{-1} \text{ ha}^{-1}$ per year within 7 years or to establish a self-sustaining population of lake trout with a mortality rate less than 45% within 10 years. Objectives can also delineate benchmarks such as a target catch rate in standard sampling gear, a growth target, an angler catch rate, or a proportion of wild fish in the catch. Without periodic assessment a lake may continue to be stocked in the same manner indefinitely merely by default. Simple, easy-to-measure benchmarks that are set during the planning phase, and agreed to by local managers, will help ensure that monitoring is completed and original objectives are reviewed and modified if required.

Most lake trout stocking programs fall within the following four broad categories:

- **Put-grow-and-take programs** — Stocking in water bodies that do not support naturally reproducing populations of lake trout. Although other salmonids are sometimes stocked at a catchable size, lake trout are normally expected to use the lake's resources to grow into the fishery. The intent is to provide angling opportunities.
- **Introductions** — Stocking in water bodies where lake trout have never been present, including lakes both within and outside the natural geographic range of the species. The intent is usually to establish a self-sustaining population, but if regular stocking continues — because of poorly articulated objectives or lack of monitoring — this original plan often evolves into a put-grow-and-take fishery. The purpose of lake trout introduction may be to provide additional angling opportunities or a refuge for important lake trout stocks threatened in another lake.

- **Supplemental programs** — Stocking in water bodies that support naturally reproducing populations of the same species. This is sometimes called *enhancement* or
  maintenance stocking; the intent is to augment the native population so more lake
  trout can be taken by anglers.
- **Rehabilitation** — Stocking to rebuild extirpated lake trout stocks or populations
  no longer reproducing; the intent is to reestablish a self-sustaining population.
  Within this category we include conservation stocking, which has the shorter term
  aim of conserving threatened populations by stocking indigenous fish back into a
  lake until underlying problems are corrected to the point at which a self-sustaining
  population can be reestablished.

Few would argue about the desirability of stocking lake trout for rehabilitating
degraded fish communities. However, questions have been raised about some of the other
stocking categories listed above. The effects of supplemental stocking on existing native
lake trout stocks have been reviewed by Evans and Willox (1991); given these findings it
is difficult to support stocking hatchery lake trout for this purpose. Introductory stockings
have been extensive but can have serious effects on other species inhabiting the lake
(Kircheis, 1985; Donald and Alger, 1993; Crossman, 1995).

In addition, because lake trout "type" lakes that do not presently contain the species
are few, it may be prudent to reserve these as potential refugia for strains threatened in
their native lake or for controlled experiments that advance the understanding of ecosystem function. Put-grow-and-take stocking may be a valid practice in some situations but
may mask underlying problems in others. For example, although algal growth on spawning shoals may inhibit the survival of lake trout eggs, stocked lake trout may survive well
and give the impression of a healthy environment. If the underlying problem is nutrient
addition and this proceeds unabated, then hypolimnetic oxygen depletion may soon limit
survival of the stocked fish as well. Consideration of the purpose and objectives of any
lake trout stocking program will help minimize the future occurrence of adverse effects.

Environment

It is perhaps platitudinous to state that stocking objectives for any species will not be met
in the absence of basic environmental requirements. Although not described in detail here,
lake trout typically prefer an environment of clear, nutrient-poor, low-temperature waters
with high dissolved oxygen (N.V. Martin and Olver, 1980). MacLean et al. (1990) suggested
that "usable" lake trout habitat could be defined by an upper temperature limit of 15.5°C
and a lower oxygen limit of 4.0 mg/l while "optimum" habitat could be considered 10°C
and over 6.0 mg/l oxygen. On the Precambrian Shield such conditions normally occur
only in deep, clear lakes that stratify thermally during the summer months. Where oxygen
levels are too low or temperature is too high, all other niche requirements become moot.

Lakes that have the basic oxygen and temperature requirements for lake trout may
not contain all of the other habitat features required for self-sustaining populations.
Spawning areas, as described by N.V. Martin and Olver (1980) and MacLean et al. (1990),
or insufficient deep nursery habitat, as described by Evans et al. (1991), may limit or
prohibit natural reproduction. In addition, in many instances lake trout populations have
been eliminated or adversely affected by physical, chemical, or biological changes to their
environment (Beggs et al., 1985; Wilton, 1985; Legault et al., Chapter 5, this volume). These
stresses on lake trout environments can be expected to continue and perhaps increase in
the future. Shoreline development no doubt will continue to add oxygen-depleting nutrients to Shield lakes, especially those located close to urban areas. Particularly troubling
is what Gorham (1996) described as the three-pronged attack of acid precipitation, global

warming, and increased ultraviolet light. These three environmental problems can act alone or in harmony; effects will be most pronounced in the clear, cold Shield lakes in which lake trout are the primary sport fish (Schindler, 1998). One of the most obvious potential effects (other than the direct toxicity of depressed pH) is the reduction in lake trout thermal habitat volume, a parameter used to predict the harvest potential in lake trout lakes (Christie and Regier, 1988; Payne et al., 1990).

Although increasing temperature will ultimately affect lake trout populations, there is some evidence that the temperature requirements of lake trout may not be quite as stringent as previously thought. Sellers et al. (1998) described the pelagic distribution with respect to temperature, oxygen, and light of natural populations of lake trout in three small northwestern Ontario lakes and suggested that the species may be more tolerant of high temperatures and less tolerant of low oxygen concentrations than previously thought. They observed lake trout commonly inhabiting 20°C waters but avoiding waters with oxygen concentrations less than 5 mg/l. It was suggested that foraging was the prime reason for inhabiting warmer water, and that it was probable that individuals depended on refuge in cold water for some period of the day. Snucins and Gunn (1995) observed one introduced population surviving in a small lake that does not stratify during the summer months. Temperatures reach 20°C throughout the water column during the summer, and lake trout seek colder water by ascending to the surface at dusk to cluster around a small spring seepage. Situations such as this must be rare, and the population may be persisting only due to the absence of other stresses.

Sustainable yields are low in Shield lakes (Payne et al., 1990) and, coupled with the fact that lake trout are captured relatively easily with angling gear (Gunn et al., 1988; Gunn and Sein, 2000), make exploitation an added concern in all but the most remote lakes. As angler success rates decrease, lakes will appear to have been "fished out" even if the underlying cause of a population decline is environmental. The demand for stocking is likely to increase as lake trout populations decrease or are lost, and the complication of exploitation coupled with environmental change makes both problem identification and clear stocking objectives extremely important.

Stocking location and method

Much of the published work on stocking location was conducted either on the Great Lakes or on large inland lakes, probably because stocking location has not been considered a major problem on small inland lakes. Lake trout tend to disperse widely and move to deep water very quickly when stocked from shore on the Great Lakes (Pycha et al., 1965; Elrod, 1987) and on large inland lakes such as Cayuga Lake, New York (Webster et al., 1959), and Lake Simcoe, Ontario (MacLean et al., 1981). Lake trout stocked from barges over deep water in Lake Ontario have been observed by sonar to descend to the bottom (approximately 50 m) in less than 5 min (Elrod, 1997). Given these observations it would seem reasonable that lake trout released into small inland lakes would also disperse fairly quickly, and stocking from one point along a shoreline, scatter planting, or planting over deep water in the middle of the lake would have little effect on the ultimate results. Transitory predation by birds or warm water fish, which may be a concern on the Great Lakes (Elrod, 1997), should not pose a major problem on most small inland lakes.

General observations tend to support this suggestion. Of ten introductions in northeastern Ontario that resulted in establishment of naturally reproducing populations, seven were done by spreading the lake trout over deep water from a landed float plane, and three were done directly from shore (Hitchens and Samis, 1986). No attempt was made to plant the fish over suitable spawning substrate, yet the stocked trout established naturally sustaining populations in all of the lakes. Given the observations on large lakes, there is reason to believe that lake trout released from shore in small lakes will quickly move to deep water and that lake trout released over deep water will descend almost immediately to preferred habitat.

Lake trout are commonly moved from the hatchery to stocking sites on inland lakes by helicopters, float planes, or tanker trucks. Hatchery managers have been aware for some time that lake trout transported at a low loading density in cold, well-oxygenated water suffer the least stress. Although no comprehensive studies have been published on the effects of routine transportation stress on subsequent survival of stocked lake trout, Plosila (1977) found no correlation between time in the hatchery truck (3.3 to 5.4 h) and future survival. Studies have attempted to quantify stress levels and offer some insight into when stress occurs in lake trout. McDonald et al. (1993) investigated how routine transportation practices from a soft water hatchery affected blood levels of cortisol, glucose, Na^+, and Cl^- in trout. Results from transporting lake trout for 4.5 to 11 h at loading densities of 8.1 to 17 kg/100 l of water were compared to the effects of tightly confining a group of fish for 8 h at the hatchery. Water temperature was kept relatively constant, and oxygen levels normally increased during the transportation trials due to the use of compressed oxygen. No follow-up on the direct effects of transportation on survival in the wild were possible, but the following useful observations were made:

- Trip duration did not affect the stress indicator levels. Rather, stress increased rapidly during the initial handling, crowding, and confinement and either reached a steady state after about 4.5 h of transport or fell off as the fish adjusted to the new environment. The authors suggested that accumulation of nitrogenous wastes would be the limiting factor in trip duration rather than increased stress levels, and that under the study conditions trips of up to 39 h would be possible.
- Loading densities of up to 17.0 kg/100 l water had no significant effect on transport stress.
- Compressed oxygen resulted in supersaturation in the transportation tanks, which was stressful to the transported fish.
- Fish held in high-density confinement (>50 kg/100 l water) for 8 h at the hatchery had similar stress indicator levels as those transported to a stocking site. The stress levels in the fish held at the hatchery returned to normal within 24 h of release from confinement.

The above suggests that when proper procedures are used, transferring lake trout to a boat, helicopter, or second truck is probably more stressful than an extended trip. Proper handling procedures and stocking locations are common to all stocked fish whether the end goal is rehabilitation, introducing lake trout to a new lake, or establishing a put-grow-and-take fishery.

Variation among stocks

A prerequisite for the development of discrete stocks of fish is a high degree of reproductive segregation. Small, isolated inland lakes provide the opportunity for this to occur, and lake trout stocks that evolved separately may have developed adaptive traits important to the survival of the population. These specialized adaptations might be used to increase the chances of survival and reproduction if a particular strain could be matched to a specific environment. However, determining which survival strategies are present at the species level, and thus common to all lake trout, and which are at the strain level, indicating an adaptation, is a challenge.

A number of studies of inland lakes have demonstrated that lake trout native to an area experience greater poststocking survival than nonnative stocks. Plosila (1977) reported on returns from 5 years of matched plantings of lake trout in nine small Adirondack lakes. The source lakes were Seneca Lake, a large, deep, alkaline lake located in south-central New York, and Upper Saranac Lake, a lake located in the Adirondack region. Over 200,000 lake trout of each strain were planted during the study, but relative survival was 15.9:1 in favor of the native Adirondack strain. An earlier study by Haskell et al. (1952) compared survival of Seneca Lake and Raquette Lake trout stocked into Raquette Lake; the return ratio was 24.5:1 in favor of the native strain. In studies on three lakes in Algonquin Park, Ontario, MacLean et al. (1981) found that the native Opeongo Lake strain survived better than those from Lake Simcoe, a large, hard-water lake in southern Ontario. Native trout also survived better than two Lake Superior strains when stocked into four lakes in northern Minnesota (Siesennop, 1992). It is not clear exactly why lake trout native to an area often survive better than nonnative fish, but localized survival adaptations appear to be a reasonable hypothesis.

Horrall (1981) suggested that the mechanism for reproductive isolation in lakes may be spawning site imprinting and homing. If so, then lake trout introduced to new waters may have trouble locating spawning grounds. Although lake trout habitually return to the same spawning area in small lakes (N.V. Martin, 1960), the use of these sites is not obligatory. Individual lake trout have been observed to move regularly between spawning shoals in inland lakes, sometimes visiting several in a single night (DeRoche, 1969; MacLean et al., 1981; Monroe, 1995). Whether these roaming fish eventually choose the same shoal on which to spawn each year or spawn on different shoals, even within the same year, remains a question. Movements may be for the purpose of verifying a site or investigating groups of fish prior to spawning. However, some straying from traditional shoals must occur because native lake trout, in spawning condition, will populate artificial shoals whether the shoals were created for the purpose of enhancing reproductive habitat (N.V. Martin, 1960) or by accident (Prevost, 1957). Trout may move onto these areas relatively quickly or after a number of years, as was the case in Smoke Lake, Ontario, where a spawning area constructed in 1981 was not used by native lake trout until 1996 (B. Monroe, personal communication, 1998).

Native lake trout will also select new natural areas for spawning if old areas are destroyed. Wilton (1985) noted that lake trout in Bark Lake, Ontario moved to shallow-water spawning areas when a newly created dam covered traditional shoals with about 12 m of water. Likewise, McAughey and Gunn (1995) observed lake trout spawning at 20 new sites on Whitepine Lake, Ontario after about 35% of the traditional spawning areas had been rendered unusable. Hatchery-reared trout also seem to have little trouble seeking suitable substrate for spawning as evidenced by their successful colonization of many inland lakes (Hitchins and Samis, 1986; Evans and Olver, 1995).

Although lake trout native to a particular lake may populate new shoals, the overriding tendency is to spawn at the same location each year. This preference to spawn in a particular location appears to be retained by stocked fish as well because different lake trout strains stocked in the same lake may segregate at spawning time (MacLean et al., 1981; Perkins et al., 1995), an observation that is consistent with the development and maintenance of discrete stocks of fish. The cues that cause this separation are not understood, although Gunn (1995) suggested that individual lake trout may be able to recognize related lake trout and are thus able to seek spawning aggregations of the same genetic strain.

If different strains of lake trout preferentially spawn at sites with specific characteristics, then this attribute might be used to increase stocking success. For example, although no rigorous studies have been reported, there is some evidence that spawning depth may be an inherited trait. Hacker (1957) observed that lake trout introduced into Green Lake, Wisconsin, spawned at a depth of 18 to 30 m over substrate dominated by silt, hardpan clay, marl, and sand even though large areas of apparently suitable windswept rock rubble occurred around the shoreline of the lake. The stocked fish originated from Lake Michigan, where gametes were taken from fish caught in 55 to 110 m of water over a hard clay bottom. It is likely that the spawning fish were selecting for depth rather than substrate type because the fish readily spawned over rock rubble deposited over the deep spawning area. If deep-water spawning is a heritable trait, then self-sustaining lake trout populations might be reestablished in small reservoirs where water drawdowns have dewatered spawning shoals and caused reproductive failure.

Other genetic traits have the potential to affect stocking success, but few have been studied in detail. Ihssen and Tait (1974) found that Lake Simcoe lake trout were able to retain swimbladder gas better than Louisa Lake trout, and crosses between the two strains had intermediate gas retention ability. They observed that in Louisa Lake the summer thermocline developed at about 8 m, while in Lake Simcoe it was substantially deeper at about 20 m. Because of their superior ability to retain swimbladder gas, Lake Simcoe lake trout appeared better adapted to a deeper water existence. It might be surmised from this that a hatchery strain developed from a relatively small, shallow lake like Louisa would not survive or grow well when stocked into larger lakes with deep thermoclines.

Age at first maturity may also be genetically linked. Male lake trout from Louisa Lake were found to mature earlier, under controlled hatchery conditions, than Lake Opeongo lake trout, and crosses between the two stocks matured at an intermediate age (Krueger and Ihssen, 1995). Lake trout in Louisa Lake grow slowly and mature at about the size they enter the fishery, thus allowing most individuals the opportunity to spawn at least once prior to harvest. A trait such as this could be useful when attempting to rehabilitate small lake trout lakes readily accessible to anglers. However, early maturity can also be greatly influenced by environmental factors. For example, both male and female lake trout in Birch Lake, Minnesota, were estimated to be 100% mature by age 3, but the same strains of fish did not reach 100% maturity until age 6 in two other Minnesota lakes (Siesennop, 1992). The author felt that early maturity was probably related to the rapid growth of young fish when stocked into a lake with few competitors and an abundance of forage. Matched plantings of different lake trout strains would be required to separate the environmental from the genetic effects.

Where the purpose of stocking is to rehabilitate a previously self-sustaining population, particular attention should be given to the stock of fish that will be used. If remnant populations still exist, then chances of rehabilitation will be enhanced by collecting eggs and stocking the progeny because the native lake trout strain may have adaptations specific to the lake. When introducing lake trout to a new lake or, in the case of rehabilitation, if remnant populations do not exist, then introducing adults or yearlings obtained from a nearby lake with similar physical characteristics is advisable. It is recognized that this may not always be possible due to the costs and logistics of collecting and raising wild fish, so it is important that hatchery systems offer at least a modest selection of stocks to managers. The study by Ihssen and Tait (1974) suggests that even the simple choice of stocking small lakes with a lake trout strain that originated in a small lake is likely to improve stocking success.

Size at release

A number of inland studies have compared poststocking survival of lake trout fall fingerlings (FF) with spring yearlings (SY). Anderson (1962) compared angler returns from FF and SY plantings of lake trout in Little Ossipee Lake, Maine, a 228-ha lake that contained

Table 12.1 Relative Returns of Lake Trout Stocked as Fall Fingerlings (FF), Spring Yearlings (SY), and Fall Yearlings (FYs) in Small Inland Lakes

Lake	Stocking size (cm)	Stocking rate (Number·ha⁻¹ year⁻¹)	Survival ratio			Reference
			FF	SY	FY	
Little Ossipee, ME						
Fingerlings	5.1–10.2	43.8–219				
Yearlings	10.2–15.2	43.8–110	1	17	—	Anderson, 1962
Nine lakes, New York						
Fingerlings	6.9–9.6	20–32				
Yearlings	10.4–16.5	10.0–12.0	1	9.5	8.7	Plosila, 1977
Pallette, WI						
Fingerlings	13.5–14.0	59	1	2.9	—	Hoff and Newman, 1995
Yearlings	20.8–22.9	59				

no lake trout prior to stocking. The lake received an initial planting of 50,000 fingerlings (219/ha) in fall 1955, followed by 25,000 each of SY and FF in 1956 and 10,000 of each in 1957. Of 422 lake trout recovered in subsequent assessments 23 were from the first fingerling stocking, 119 were from the first yearling stocking, and 279 were from the second yearling stocking — a return ratio of 1:17 (FF:SY). No fingerlings from the second two stockings were observed. Plosila (1977) examined the relative returns of two strains of lake trout stocked as FF, SY, and fall yearlings (FYs) in nine lakes in upper New York State. The lakes held populations of lake trout prior to stocking and ranged in size from 78 to 879 ha. Lake trout were stocked annually over a period of 6 years at a rate of 20–32 fingerlings and 20–24 yearlings/hectare and were recovered by various means in subsequent years in the ratio 1:9.5:8.7 (FF:SY:FY). Hoff and Newman (1995) stocked Pallette Lake, Wisconsin, with FF (59/ha) and SY (59/ha) for 3 years. The 71-ha lake was devoid of lake trout prior to these stockings. Subsequent assessment captured lake trout in the ratio 1:2.3 (FF:SY). A summary of the results of the three studies on relative survival of yearling and fingerling lake trout stocked in small inland lakes is given in 12.1.

Gunn et al. (1987) stocked yearling and 2-year-old lake trout, originating from a single pooled egg collection, into six small lakes (24 to 108 ha) near Sudbury, Ontario. Acidic conditions prevented survival of lake trout in two of the lakes. Of the remaining lakes, one (Whitepine) was stocked in November 1980 with lake trout aged 25 months and again in May 1981 with the same lot of fish aged 31 months. The other three lakes (Whitefish, Beaver, and McCulloch) were stocked in May 1981 and October 1991 with fish aged 31 and 36 months, respectively. At each stocking (24/ha) the lake trout were divided into small-, medium-, and large-size groups and marked for later identification. It was found that in these lakes where there were no (or insignificant) existing populations of lake trout prior to stocking, the fish introduced first had a greater advantage, and that the stocking order was more important to survival than stocking size. However, when a group of lake trout of differing sizes was stocked at the same time the larger fish in the group survived considerably better than the smaller ones (Table 12.2).

The advantage of being first in a new environment is also evident in Little Ossipee Lake, where the only FF planting to show any return was the first (Anderson, 1962). Likewise, in Pallette Lake the initial planting of FF in 1982 survived substantially better than the larger SY planted in 1985 and 1986 (Hoff and Newman, 1995). The paired plantings in that study showed a return ratio of 1:2.0 (FF:SY) in the first year, 1:5.0 (FF:SY)

Table 12.2 Mean Weight and Returns of Lake Trout Stocked at Different Sizes and Ages in Four Study Lakes near Sudbury, Ontario

Lake	Stocking date	Age (months)	Weight (g) Small	Medium	Large
Whitepine	November 1980	25	25.3	41.6	63.4
	May 1981	31	43.1	64.2	93.7
Beaver	May 1981	31	42.6	64.1	94.9
	October 1981	36	61.0	125.6	197.7
Whitefish	May 1981	31	41.0	64.8	93.3
	October 1981	36	62.5	122	205.3
McCulloch	May 1981	31	42.0	65.1	94.5
	October 1981	36	74.4	105.8	145.6

			Number recaptured		
Whitepine			16	39	70
Beaver			11	12	34

Note: Age measured from time of fertilization.

Source: Modified from Gunn, J.M., McMurtry, M.J., Bowlby, J.N., Casselman, J.M., and Liimatainen, V.A., 1987, *Transactions of the American Fisheries Society*, 116:618–627.

in the next year, and no returns of fingerling lake trout in the last year. It appears that initial introductions of FF may survive well, but where stocking is to be continued for a number of years SY will provide better returns.

A number of studies, both inland and on the Great Lakes, have reported only marginal increases in survival for lake trout stocked when older than SY (Buettner, 1961; Plosila, 1977; MacLean et al., 1981; Gunn et al., 1987). However, when different size fish are stocked at the same age there is a positive correlation between size and survival (Pycha and King, 1967; Gunn et al., 1987).

When stocking lake trout for rehabilitation, introductions, or put-grow-and-take programs, planting large SY (>22 g) appears to be the best stocking strategy in most situations. Adult transfers are also a viable option for rehabilitation and introductory stocking.

Stocking rate

Recommendations regarding the number of lake trout that should be stocked are generally not found in the published literature, possibly due to the many variables that can influence the optimal stocking rate. The objective of a stocking program along with the size at stocking, growth rate, harvest rate, stocking history, lake size, and relative abundance and size of other species will vary from lake to lake and may change through time, thus confounding any efforts to "nail down" a specific stocking rate. Nevertheless, some insight is possible and some bounds may be placed on stocking rates based on theoretical knowledge and anecdotal evidence of what has worked in the past.

A simple simulation model produced on a spreadsheet can help establish a starting point for stocking rates. The example shown in Table 12.3 uses the average length-at-age and weight–length relationship for Ontario lake trout to estimate growth (Payne et al., 1990). Total mortality was set at 50%, and for simplicity natural and fishing mortality were assumed to act independently. Natural mortality was set at 0.22, the median value estimated by Payne et al. (1990), and fishing mortality was assumed to be 0.28 and to begin at age 4. Under these conditions, which emulate natural populations, it was calculated that an annual stocking of 1000 yearling lake trout would, over time, allow for a yearly harvest of approximately 151 kg of lake trout (0.151 kg of lake trout per yearling stocked).

Table 12.3   Potential Numbers and Weight of Lake Trout Harvested And Surviving Each Year
Based on Assumed Annual Stocking of 1000 Yearlings

Age	Length[a] (mm)	Weight[b] (g)	Natural mortality	Fishing mortality	Number harvested	Number surviving	Weight harvested (kg/year)
2	171	49	0.22	0	0	780	0
3	237	138	0.22	0	0	608	0
4	293	272	0.22	0.28	170	304	46
5	340	437	0.22	0.28	85	152	37
6	380	622	0.22	0.28	43	76	27
7	414	816	0.22	0.28	21	38	17
8	443	1009	0.22	0.28	11	19	11
9	467	1194	0.22	0.28	5	10	6
10	487	1368	0.22	0.28	3	5	4
11	505	1529	0.22	0.28	1	2	2
12	519	1674	0.22	0.28	1	1	1
					340	1995	151

Note: Ontario lake formulas used.

[a] Average fork length (cm) at age (years): $L_t = 59.9 (1 - e^{-0.168t})$.

[b] Average weight (g)–fork length (mm) relationship: $W = 3.88 \times 10^{-6} L^{3.18}$.

Source: From Payne, N.R. et al., 1990, *The Harvest Potential and Dynamics of Lake Trout Populations in Ontario,* Ontario Ministry of Natural Resources, Lake Trout Synthesis working group report, Toronto.

To allow estimation of a stocking rate (i.e., number to stock per hectare) it was assumed that lake trout lakes have an inherent productivity equal to the sustainable yields produced by natural populations. Payne et al. (1990) recommended using the July thermal habitat volume (THV) to estimate sustainable harvests from lake trout lakes but noted that due to the relationship between area and volume, lake trout harvest was also significantly correlated with area ($\log_{10}$ Harvest = 0.500 + 0.830 $\log_{10}$ Area, r^2 = .923). Using the formula based on lake area the potential harvest from lakes of 100, 1,000, and 10,000 ha would be 145, 977, and 6,607 kg/year, respectively. These potential harvests, when divided by the predicted annual harvest resulting from stocking each yearling (0.151 kg), allow an estimate of the total number of yearlings to stock (960, 6,470, and 43,755, respectively). The stocking rates of 10, 6, and 4 yearlings per hectare for the three sizes of lakes were obtained by simply dividing the numbers to stock by the lake area and rounding to the nearest whole number. The suggested lake trout stocking rate for lakes of different sizes is shown is Figure 12.1.

The variables in Table 12.3 come from an array of possible growth and mortality combinations, and the calculations are shown primarily to demonstrate an approach to calculating stocking rates for individual lakes. Variables such as the number and population size of other competing, predatory, and prey species can affect stocking success, and a simple model cannot deal with these complexities. However, the model input parameters (growth, mortality, sustainable harvest) are empirically derived and, in the absence of better lake-specific information, represent a reasonable starting point for yearling lake trout stocking rates. Where lakes are maintained totally by stocking it could be argued that there is no need to limit total mortality, and that stocking rates should be increased to meet whatever angling demand occurs. Increasing the number of yearlings stocked may be a valid approach to a point, but it is based on the assumption that lake trout harvests can be forced steadily upward in direct proportion to stocking rates with no limits to productive capacity. An increase in the number of stocked fish will, more likely, cause a concomitant increase in competition and predation, resulting in increased mortality

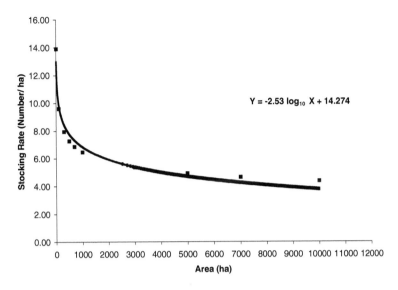

$$Y = -2.53 \log_{10} X + 14.274$$

Figure 12.1 Suggested annual yearling lake trout stocking rates for different lake sizes.

of stocked fish and diminished rates of return. Eventually, total returns may also decrease as too many stocked fish compete for limited resources. Attempting to emulate natural populations by keeping stocking rates low should result in increased survival and a population that will "grow into" the resource. If it is necessary to increase stocking rates, then it should be done slowly with a careful eye on growth and yield.

Where the objective is to establish, or reestablish, self-sustaining lake trout populations, annual stocking is not necessary. Plantings of small yearlings (8–13.6 g) at a rate of 19.2–72.1 yearlings/hectare established naturally reproducing populations in ten lakes in northeastern Ontario (Hitchens and Samis, 1986). Initial stockings in two of the lakes, at a rate of 19.2 and 24.1 yearlings/hectare, produced offspring at age 5. Four other lakes, stocked at a rate between 23.3 and 50.3 yearlings/hectare first produced young at age 6, and lake trout stocked in one lake at 72.1 yearlings/hectare did not spawn successfully until age 7. The data indicate that the higher stocking rates may have slowed growth and delayed maturity in these lakes, and it is probable that lower stocking rates would have established populations equally well. We recommend that to establish self-sustaining populations, yearling lake trout should be stocked every second year for a total of three stockings. Rates shown in Figure 12.1 will provide an adequate number of spawners.

In some situations, stocking larger lake trout may prove more efficient. Adult transfers from a donor lake that holds a particular strain of lake trout may be preferable to collecting spawn and raising the young for future stocking. Adult lake trout will spawn in new environments (Gunn et al., 1990), and a few fish are capable of producing an abundance of offspring under ideal conditions (Gunn and Keller, 1990). Sufficient numbers should be transferred to ensure that the newly established stock does indeed have the same genetic makeup as the donor stock. We recommend that any adult transfer program be continued for three consecutive years to help ensure genetic integrity and overcome any random event that might affect the transfer in any single year. We do not suggest minimal numbers that need to be collected and transferred because this will depend on factors such as logistics, the number that can be safely taken from the donor lake, and survival in the new lake. We do suggest that a genetic profile of the lake trout in the donor lake be taken for later comparison with their progeny in the new lake. This will verify that the desired strain has been successfully transferred.

Stocking larger lake trout may also be required in lakes where intense species inter-actions exist and a large predator is required to restore balance to the community.

Species interactions

Lakes that have simple communities with few predators or competing species but excellent lake trout habitat seem to be colonized readily by hatchery fish. There are many examples of successful lake trout introductions in inland lakes where size, strain, or method of stocking was given little or no attention (Olver and Lewis, 1977; Purych, 1980; Hitchens and Samis, 1986; Evans and Olver, 1995). Indeed some successful introductions, such as in Yellowstone Lake (Kaeding et al., 1996), have been unauthorized, and it is difficult to imagine that whoever made these initial introductions could have surreptitiously intro-duced large quantities of lake trout over a number of years. It is more likely that a small number were placed into a very hospitable environment. In contrast, hatchery lake trout stocked over robust native populations have shown insignificant returns in a large number of lakes (N.V. Martin and Fry, 1973; Purych, 1977; Powell et al., 1986).

Planting lake trout in lakes that have low native populations often results in high initial survival, but growth and survival of subsequent stockings can decrease markedly due to interactions with previously planted fish. Ihssen et al. (1982) reported that the first of three plants in four lakes near Espanola, Ontario had the highest survival. Returns from the second plant were severely suppressed through competition with the fish from the first plant. Survival of the third plant was intermediate, presumably due to the facts that the survivors from the first plant had shifted to larger food items and competition from the second plant was low due to their poor survival.

Returns of fingerlings and yearlings in Pallette Lake, Wisconsin showed a similar pattern, although the authors (Hoff and Newman, 1995) did not discuss it. Of the initial introduction of fingerlings in 1982, there were 74 captured during sampling in 1991 and 1992. Two ensuing plantings of fingerlings stocked at the same rate returned only two lake trout during the same sampling period. Yearlings, planted at the same rate in spring 1983, returned 149 lake trout while two additional yearling stockings in the spring of 1985 and 1986 returned a total of only 25 fish.

Gunn et al. (1987) found that the first stocking of lake trout into four small lakes near Sudbury, Ontario survived better than fish stocked later. Once established in the lakes the fish introduced first competed with subsequent stockings and exerted a controlling influ-ence, even though the lake trout that were stocked later were larger. The results of these studies indicate that the first cohort stocked into a new environment enjoys a considerable advantage, and that competition for resources limits survival of subsequent groups of stocked fish. Size differences between groups of stocked fish in these studies were insuf-ficient to consider cannibalism a major cause of mortality.

Reports on the significance of cannibalism as a source of mortality vary from lake to lake. In Algonquin Park, Ontario only one lake trout was observed in the stomachs of 725 lake trout taken from Redrock and Louisa Lakes (N.V. Martin, 1952), and only 11 were observed from 17,171 lake trout examined from Opeongo Lake over a period of 30 years (N.V. Martin, 1970). N.V. Martin and Olver (1980) concluded that while cannibalism had been observed in a number of lakes it was likely not a major source of mortality for young lake trout, with the possible exception of Arctic lakes which do not hold a diversity of prey species. However, other authors have suggested that cannibalism can be a significant source of mortality for stocked lake trout. Plosila (1977) reported that 17% of the large lake trout he examined from small Adirondack lakes contained lake trout remains, and that in one of the lakes 14 of 25 large lake trout examined contained lake trout. Powell et al. (1986) found 11 small lake trout in 54 adults in one study lake in northeastern Ontario.

Evans et al. (1991) reassessed the Opeongo data of N.V. Martin (1970) using digestion rates, the probability of sampling a juvenile lake trout from the stomach of adult lake trout, and the estimated population size. They estimated a possible annual mortality due to cannibalism in Lake Opeongo of 8.8 to 29.4%. This implied that cannibalism accounted for a large component of the estimated total annual mortality of juvenile lake trout (35% for fish aged 1 to 3 years) in Lake Opeongo.

It is probable that stocked lake trout suffer higher mortality through cannibalism than native trout of the same age. Culling of unwary young hatchery trout may occur quickly after placement in a lake with large predators. All cannibalism reported by Plosila (1977) was of stocked lake trout, and the lakes that showed the highest cannibalism levels were those that also had the highest native populations. Of 12 young lake trout examined by Powell et al. (1986), 4 were stocked fish and the remainder were too digested to determine origin. MacLean et al. (1981) noted that while only 11 lake trout (3 of which were stocked) were found in 17,171 lake trout from Lake Opeongo between 1936 and 1965, there was recovery of 8 newly stocked fish from 1,560 lake trout caught during the summer of 1973. These findings suggest that cannibalism may be a significant source of mortality for newly stocked fish attempting to adjust to a new environment.

Several authors have discussed interspecific interactions involving native lake trout that should also apply to stocked populations. Hackney (1973) and Day (1983) suggested that burbot *Lota lota* is a predator on lake trout. Lake herring *Coregonus artedi* have been described as competitors of juvenile lake trout in north-temperate lakes (N.V. Martin and Fry, 1973), and burbot and lake herring may interact (Carl, 1992), thus adding a confounding factor to the interplay among species. Day (1983) found a negative correlation between burbot and lake trout abundance, while Carl (1992) found no correlation between the two species. However, Carl et al. (1990) indicated that the relative abundance of lake trout in Ontario inland lakes was lower where burbot or lake whitefish *Coregonus clupeaformis* and lake herring were present compared to lakes in which they were absent. Carl (1997) found a strong relationship between the presence of lake whitefish or lake herring and poor recruitment by native lake trout populations in a 20-lake data set in the Algonquin Highlands, Ontario. There was an indication that smelt *Osmerus mordax* had a similar effect, and it was hypothesized that these species compete with young lake trout for food. Although the evidence was not conclusive, Evans and Loftus (1987) also suggested possible negative interactions between lake trout and smelt.

Evans and Olver (1995) suggested that species interactions and colonization history are important factors affecting reproduction of stocked lake trout in Ontario. In their review they found that lakes where reproduction of stocked lake trout occurred tended to have fewer species than those where reproduction was unsuccessful, indicating that species interactions were affecting the reproductive success of stocked lake trout. However, they also found that species richness in existing native lake trout lakes was similar to the stocked lakes in which reproduction had been unsuccessful. They attributed this apparent contradiction to the order in which colonization occurred. Lake trout are able to establish themselves more readily where competition and predation are not severe; once established they are able to control the population of species that colonize later. Thus, in these communities a feedback loop exists in which large lake trout prey on planktivores such as lake herring or smelt, and the planktivores limit the recruitment of lake trout. Feedback mechanisms such as this are normal in fish communities, but they are dependent on the maintenance of dynamic equilibrium among the populations. Any drastic reduction in the lake trout population (for instance, through changes to the environment or from excessive fishing) will allow a concomitant increase in the planktivore population, thus increasing competition with young native lake trout. The result may be a downward spiral of the lake trout population. In these situations young hatchery lake trout will also be suppressed by the dense planktivore populations, making reestablishment of a healthy lake trout population quite difficult.

Growth and survival of stocked lake trout may be expected to decrease in proportion to both the density of any existing lake trout population and the number of other species present in the lake (Powell et al., 1986; Evans and Olver, 1995). Therefore, if the intent is to reconstruct a fish population in a degraded lake that formerly contained lake trout it is important to stock the lake trout prior to reintroducing other species. If the objective of a stocking program is to establish a self-sustaining lake trout population — or even a put-grow-and-take fishery — in a lake that contains a large planktivore population, it may be necessary to reduce the size of the planktivore population prior to stocking yearling lake trout. One such experiment is being conducted in the Killarney area near Sudbury, Ontario, where acid precipitation extirpated native lake trout from a number of lakes (Snucins and Gunn, 2003). Although the pH has improved dramatically, reintroducing yearling lake trout has been unsuccessful in these lakes apparently due to a large lake herring population that developed in the absence of a predator population. Adult lake trout have been stocked in an attempt to reduce the lake herring populations and reestablish a dynamic equilibrium between predator and prey (E. Snucins, 1999, personal communication).

Management implications

Factors that affect the survival of stocked lake trout in small Precambrian Shield lakes may be roughly divided into ones that can be controlled prior to the fish entering the lake (strain, size, method, location, and rate) and those that exert an influence after the fish are stocked (angler harvest, physical and chemical environment, existing aquatic community). Of the second set of factors, only angler harvest may be readily controlled (although some may argue this point). Angler harvest can also be either an integral part of the stocking objective (when the objective is to provide a put-grow-and-take fishery) or a hindrance in reaching the stocking objective (when the objective is to establish a self-sustaining population). On small Shield lakes angling can drastically reduce a lake trout population in a very short time (Gunn et al., 1988; Gunn and Sein, 2000). Evans and Olver (1995), in their review of 183 lake trout introductions, found that natural populations were established more frequently in lakes where angling harvest was curtailed. We suggest that where the stocking objective is to develop a self-sustaining population the lake should be closed to lake trout angling for at least 10 years after the first introduction of yearling fish. This should allow the introduced lake trout to produce about three year-classes of progeny, but success should always be confirmed through assessment. Closure time may be shorter if older lake trout are introduced.

The physical and chemical environment and the existing fish community in a lake are the most important factors governing stocking success, and either can have an overriding influence on the growth, survival, and reproduction of stocked lake trout. The characteristics of small lake trout lakes (cold, clear, nutrient poor, oxygen rich) make them extremely sensitive to environmental changes such as acid precipitation, global warming, ultraviolet light, and nutrient additions (Gorham, 1996; Schindler, 1998; MacLean et al., 1990). How a candidate lake will be managed is largely dependent on these inherent lake conditions, so, prior to initiating any stocking program it is essential to ensure that the physical and chemical environment is favorable to lake trout.

An assessment of the existing fish community in a candidate lake is equally important. If it is found that the candidate lake already contains a viable, naturally reproducing lake trout population, then attempts to improve angling by supplementing the population with hatchery fish will be counterproductive (Evans and Willox, 1991). Improvements in catch and harvest in naturally reproducing lake trout lakes will be better attained through other management options, such as the harvest regulations described by Olver and coworkers (Chapter 11). Growth and survival of stocked lake trout can be expected to decrease in proportion to the number, density, and size of other major species (Powell et al., 1986; Evans and Olver, 1995), and dense populations of planktivores may limit stocking success. Under these conditions it may be difficult to develop a lake trout population without resorting to major interventions such as netting programs to remove competing species or stocking adult lake trout to control the prey species and, it is hoped, restore dynamic equilibrium in the fish community.

Lake trout are able to prey successfully on a number of organisms, and excellent yields are possible from lakes that have simple food webs and few fish species (N.V. Martin and Olver, 1980). Therefore, to establish a naturally reproducing population or to produce high yields of small trout no prey species should be introduced. If the objective is to restore a former fish community in a previously degraded lake, then stocking a suite of fish species may be required. To prevent the development of intense species interactions that might inhibit rehabilitation, other species should be introduced only after a lake trout population has become established. The lake should then be managed in a manner that maintains a healthy adult lake trout population capable of exerting control over prey populations that may compete with juvenile lake trout.

Although managers normally have little opportunity to manipulate conditions within a lake, they are able to control variables such as size, stocking rate, and strain. These considerations may determine the success of a stocking program, especially when conditions within the candidate lake are somewhat less than ideal. The common transportation methods (hatchery truck, boat, aircraft) and the location where lake trout are stocked are not likely to be a major concern on small inland lakes. Studies on the Great Lakes and on large inland lakes have shown that lake trout disperse quite quickly over a large area (Webster, 1959; Pycha et al., 1965; Elrod, 1987), so stocking from an access point, scatter planting along shorelines, or stocking offshore over deep water should not affect dispersal or survival in small lakes. Initial predation by predatory fish or birds will not normally be of concern in small inland lake trout lakes. However, where predation may be a concern, stocking at night or transporting the fish offshore may alleviate the problem. Stocking at night may not be as effective where populations of burbot, a nocturnal feeder, are high.

In the absence of competing species, stocked fingerlings will survive well initially, and fish of this size may suffice for introductory stocking or to rehabilitate a lake. However, studies of both small inland lakes and the Great Lakes have shown that with repeated stocking, large SY (>22 g) survive better than smaller fish (Pycha and King, 1967; Brown et al., 1981; Gunn et al., 1987). No advantage is evident when lake trout older than SY are stocked (Buettner, 1961; Plosila, 1977; Gunn et al., 1987). Where annual plantings of SY are prescribed we suggest that a stocking rate that decreases with lake size (Figure 12.1) should provide adequate growth and harvest. If stocking is to take place on an alternate-year basis, then interactions among stocked cohorts will be diminished and higher stocking rates may be possible. Excessive harvests are not normally a concern with put-grow-and-take fisheries; in these situations managers may be tempted to increase stocking rates to squeeze more production from a lake. Any increase should be carefully considered and closely monitored because increases in lake trout density will result in greater competition and predation on subsequent cohorts of stocked fish. In most situations more fish "in" will not necessarily equate to more fish "out."

Further research on community ecology and genetic adaptations should improve future stocking success. Studies of small Shield lakes have consistently demonstrated that stocked lake trout of native origin survive better than nonnative strains (Plosila, 1977; MacLean et al., 1981; Siesennop, 1992), although the specific cause-and-effect relationships have not been established. Further work is required to identify which adaptations are beneficial in which situations. For example, identification of a lake trout strain that consistently spawns in deep water would help resolve conflicts in a number of inland lakes where water drawdown is a problem. Similarly, identification of a strain or strains that mature early would help alleviate exploitation problems. Lake trout that mature closer to the age at which they enter a fishery would have greater opportunity to spawn at least once prior to being caught.

Specialized adaptations may be lost forever when discrete populations fall below some critical size. Genetic material may be preserved by establishing hatchery brood stocks or collecting spawn on the original lake and maintaining the population through an ongoing stocking program. However, both strategies require extensive use of hatchery space and are therefore restricted to the most important lakes. In addition, stocks maintained in a hatchery environment are not subjected to the same selective pressures as wild stocks, and genetic adaptations may be lost.

Another strategy has been to establish a population of a particular strain in a barren lake that has suitable lake trout habitat. Although this is preferable to maintaining a stock of wild fish in a hatchery, through time the fish may be subjected to selective pressures that differ from those of the original lake. The supply of barren lakes is now low, and the remainder should be identified and set aside for the preservation of lake trout stocks that face extirpation.

Although research on genetic adaptations offers promise to solve specific problems, it is not without risk. The temptation to stock genetically specialized strains as a "quick fix" on a broad scale will no doubt develop along with the science. This strategy would almost certainly be at the expense of native gene pools and should be vigorously resisted.

Acknowledgment

We wish to acknowledge and thank David Evans, Ed Snucins, and an anonymous reviewer whose valuable comments improved the original manuscript.

References

Anderson, R.B., 1962, A comparison of returns from fall and spring stocked hatchery reared lake trout in Maine, *Transactions of the American Fisheries Society*, 91:425–427.

Beggs, G.L., Gunn, J.M., and Olver, C.H., 1985, The Sensitivity of Ontario Lake Trout (*Salvelinus namaycush*) and Lake Trout Lakes to Acidification, Ontario Ministry of Natural Resources, Ontario Fisheries Technical Report Series 17, Toronto.

Brown, E.H., Jr., Eck, G.W., Foster, N.R., Horrall, R.M., and Coberly, C.E., 1981, Historical evidence for discrete stocks of lake trout (*Salvelinus namaycush*) in Lake Michigan, *Canadian Journal of Fisheries and Aquatic Sciences*, 38:1747–1758.

Buettner, H.J., 1961, Recoveries of tagged, hatchery-reared lake trout from Lake Superior, *Transactions of the American Fisheries Society*, 90:404–412.

Callicott, J.B., 1991, Conservation ethics and fishery management, *Fisheries*, 16(2):22–28.

Carl, L.M., 1992, The response of burbot, *Lota lota*, to change in lake trout, *Salvelinus namaycush*, abundance in Lake Opeongo, Ontario, *Hydrobiologia*, 243/244:229–235.

Carl, L.M., 1997, Lake Trout Recruitment Concerns in Inland Lakes, Ontario Ministry of Natural Resources, Management Brief, Peterborough, Ontario.

Carl, L.M., Bernier, M.-F., Christie, W., Deacon, L., Hulsman, P., Loftus, D., Maraldo, D., Marshall, T., and Ryan, P., 1990, *Fish Community and Environmental Effects on Lake Trout*, Ontario Ministry of Natural Resources, Lake Trout Synthesis Working Group Report, Toronto.

Christie, G.C. and Regier, H.A., 1988, Measures of optimal thermal habitat and their relationship to yields in four commercial fish species, *Canadian Journal of Fisheries and Aquatic Sciences*, 45:301–314.

Crossman, E.J., 1995, Introduction of the lake trout (*Salvelinus namaycush*) in areas outside its native distribution: a review, *Journal of Great Lakes Research*, 21(Suppl. 1):17–29.

Daley, W.J., 1993, The use of fish hatcheries: polarizing the issue, *Fisheries*, 18(3):4–5.

Day, A.C., 1983, Biological and Population Characteristics of, and Interactions between an Unexploited Burbot (*Lota lota*) Population and an Exploited Lake Trout (*Salvelinus namaycush*) Population from Lake Athapapuskow, masters of science thesis, University of Manitoba, Winnipeg.

Deroche, S.E., 1969, Observations on the spawning habits and early life history of lake trout, *Progressive Fish-Culturist*, 31:109–113.

Donald, D.B. and Alger, D.J., 1993, Geographic distribution, species displacement, and niche overlap for lake trout and bull trout in mountain lakes, *Canadian Journal of Zoology*, 71:238–247.

Elrod, J.H., 1987, Dispersal of three strains of hatchery-reared lake trout in Lake Ontario, *Journal of Great Lakes Research*, 13:157–167.

Elrod, J.H., 1997, Survival of hatchery-reared lake trout stocked near shore and off shore in Lake Ontario, *North American Journal of Fisheries Management*, 17:779–783.

Evans, D.O., Casselman, J.M., and Willox, C.C., 1991, *Effects of Exploitation, Loss of Nursery Habitat, and Stocking on the Dynamics and Productivity of Lake Trout Populations in Ontario Lakes*, Ontario Ministry of Natural Resources, Lake Trout Synthesis working group report, Toronto.

Evans, D.O. and Loftus, D.H., 1987, Colonization of inland lakes in the Great Lakes region by rainbow smelt, *Osmerus mordax*: their freshwater niche and effects on indigenous fishes, *Canadian Journal of Fisheries and Aquatic Sciences*, 44(Suppl. 2):249–266.

Evans, D.O. and Olver, C.H., 1995, Introduction of lake trout (*Salvelinus namaycush*) to inland lakes of Ontario, Canada: factors contributing to successful colonization, *Journal of Great Lakes Research*, 21(Suppl. 1):30–53.

Evans, D.O. and Willox, C.C., 1991, Loss of exploited, indigenous populations of lake trout, *Salvelinus namaycush*, by stocking of non-native stocks, *Canadian Journal of Fisheries and Aquatic Science*, 48(Suppl. 1):134–147.

Fullerton, S.F., 1906, Protection as an aid to propagation, *Transactions of the American Fisheries Society*, 35:59–64.

Gorham, E., 1996, Lakes under a three-pronged attack, *Nature*, 381:109–110.

Gunn, J.M., 1995, Spawning behavior of lake trout: effects on colonization ability, *Journal of Great Lakes Research*, 21(Suppl. 1):323–329.

Gunn, J.M., McMurtry, M.J., Bowlby, J.N., Casselman, J.M., and Liimatainen, V.A., 1987, Survival and growth of stocked lake trout in relation to body size, stocking season, lake acidity, and biomass of competitors, *Transactions of the American Fisheries Society*, 116:618–627.

Gunn, J.M., McMurtry, M.J., Casselman, J.M., Keller, W., and Powell, M.J., 1988, Changes in the fish community of a limed lake near Sudbury, Ontario: effects of chemical neutralization or reduced atmospheric deposition of acids? *Water, Air, and Soil Pollution*, 41:113–136.

Gunn, J.M. and Keller, W., 1990, Biological recovery of an acid lake after reductions in industrial emissions of sulphur, *Nature*, 345:431–433.

Gunn, J.M., Hamilton, J.G., Booth, G.M., Wren, C.D., Beggs, G.L., Rietueld, H.J., and Munro, J.R., 1990, Survival, growth, and reproduction of lake trout (*Salvelinus namaycush*) and yellow perch (*Perca flavescens*) after neutralization of an acidic lake near Sudbury, Ontario, *Canadian Journal of Fisheries and Aquatic Sciences*, 47:446–453.

Gunn, J.M. and Sein, R., 2000, Testing the effects of two potential impacts of forestry roads on lake trout populations: reproductive habitat loss, increased access and exploitation, *Canadian Journal of Fisheries and Aquatic Sciences*, 57(Suppl. 2):97–104.

Hacker, V.A., 1957, Biology and management of lake trout in Green Lake, Wisconsin, *Transactions of the American Fisheries Society*, 86:71–83.

Hackney, P.A., 1973, Ecology of the Burbot (*Lota lota*) with Special Reference to Its Role in the Lake Opeongo Fish Community, doctoral thesis, University of Toronto.

Haskell, D.G., Zillox, R.G., and Lawrence, W.M., 1952, Survival and growth of stocked lake trout yearlings from Seneca and Raquette Lake breeders, *Progressive Fish-Culturist*, 14:71–73.

Hilborn, R., 1992, Hatcheries and the future of salmon in the Northwest, *Fisheries*, 17(1):5–8.

Hindar, K., Ryman, N., and Utter, F., 1991, Genetic effects of cultured fish on natural fish populations, *Canadian Journal of Fisheries and Aquatic Sciences*, 48:945–957.

Hitchens, J.R. and Samis, W.G.A., 1986, Successful reproduction by introduced lake trout in 10 northeastern Ontario lakes, *North American Journal of Fisheries Management*, 6:372–375.

Hoff, M.H. and Newman, S.P., 1995, Comparisons of fingerling and yearling lake trout introductions for establishing an adult population in Pallette lake, Wisconsin, *North American Journal of Fisheries Management*, 15:871–873.

Horrall, R.M., 1981, Behavioral stock-isolating mechanisms in Great Lakes fishes with special reference to homing and site imprinting, *Canadian Journal of Fisheries and Aquatic Sciences*, 38:1481–1496.

Ihssen, P.E., Powell, M.J., and Miller, M., 1982, Survival and Growth of Matched Plantings of Lake Trout (*Salvelinus namaycush*), Brook Trout (*Salvelinus fontinalis*), and Lake X Brook F1 Splake Hybrids and Backcrosses in Northeastern Ontario Lakes, Ontario Ministry of Natural Resources, Ontario Fisheries Technical Report Series 6, Toronto.

Ihssen, P.E. and Tait, J.S., 1974, Genetic differences in retention of swimbladder gas between two populations of lake trout (*Salvelinus namaycush*), *Journal of the Fisheries Research Board of Canada*, 31:1351–1354.

Kaeding, L.R., Boltz, G.D., and Carty, D.G., 1996, Lake trout discovered in Yellowstone Lake threaten native cutthroat trout, *Fisheries*, 21(3):16–20.

Kircheis, F.W., 1985, Sport fishery management of a lake containing lake trout and an unusual dwarf arctic char, *Fisheries*, 10(6):6–8.

Krueger, C.C. and Ihssen, P.E., 1995, Review of genetics of lake trout in the great lakes: history, molecular genetics, physiology, strain comparisons and restoration management, *Journal of Great Lakes Research*, 21(Suppl. 1):348–363.

Krueger, C.C. and May, B., 1991, Ecological and genetic effects of salmonid introductions in North America, *Canadian Journal of Fisheries and Aquatic Sciences*, 48(Suppl. 1):66–77.

Legault, M., Benoît, J., and Bérubé, R., 2004, Impact of new reservoirs, in *Boreal Shield Watersheds: Lake Trout Ecosystems in a Changing Environment*, Gunn, J.M., Steedman, R., and Ryder, R., Eds., Lewis Publishers, Boca Raton, FL, chap. 5.

MacLean, J.A., Evans, D.O., Martin, N.V., and DesJardine, R.L., 1981, Survival, growth, spawning distribution, and movements of introduced and native lake trout (*Salvelinus namaycush*) in two inland Ontario lakes, *Canadian Journal of Fisheries and Aquatic Sciences*, 38:1685–1700.

MacLean, N.G., Gunn, J.M., Hicks, F.J., Ihssen, P.E., Malhoit, M., Mosindy, T.E., and Wilson, W., 1990, *Environmental and Genetic Factors Affecting the Physiology and Ecology of Lake Trout*, Ontario Ministry of Natural Resources, Lake Trout Synthesis working group report, Toronto.

Martin, J., Webster, J., and Edwards, G., 1992, Hatcheries and wild stocks: are they compatible? *Fisheries*, 17(1):4.

Martin, N.V., 1952, A study of the lake trout, *Salvelinus namaycush*, in two Algonquin Park, Ontario, lakes, *Transactions of the American Fisheries Society*, 81:111–137.

Martin, N.V., 1960, Homing behavior in spawning lake trout, *Canadian Fish Culturist*, 26:3–6.

Martin, N.V., 1970, Long term effects of diet on the biology of the lake trout and the fishery in Lake Opeongo, Ontario, *Journal of the Fisheries Research Board of Canada*, 27:125–146.

Martin, N.V. and Fry, F.E.J., 1973, Lake Opeongo: The Ecology of the Fish Community and of Man's Effects on It, Great Lakes Fishery Commission Technical Report No. 24.

Martin, N.V. and Olver, C.H., 1980, The lake charr, *Salvelinus namaycush*, in *Charrs: Salmonid Fishes of the Genus* Salvelinus, Balon, K.K., Ed., Dr. W. Junk, The Hague, The Netherlands.

McAughey, S.C. and Gunn, J.M., 1995, The behavioral response of lake trout to a loss of traditional spawning sites, *Journal of Great Lakes Research*, 21(Suppl. 1):375–383.

McDonald, D.G., Goldstein, M.D., and Mitton, C., 1993, Responses of hatchery-reared brook trout, lake trout and splake to transport stress, *Transactions of the American Fisheries Society*, 122:1127–1138.

Monroe, B., 1995, Homing Behaviour of Spawning Lake Trout and Its Effect on Estimating Population Abundance, Ontario Ministry of Natural Resources, Ontario Fisheries Assessment Unit Update 95-1, Peterborough.

Monroe, B., 1998, personal communication, Algonquin Fisheries Assessment Unit, Ontario Ministry of Natural Resources, Whitney, Ontario.

Nevin, J., 1898, Artificial propagation versus a closed season for the Great Lakes, *Proceedings of the American Fisheries Society,* 27:17–26.

Nevin, J., 1892, Planting fry versus planting fingerlings, *Transactions of the American Fisheries Society,* 21:81–86.

Olver, C.H. and Lewis, C.A., 1977, Reproduction of planted lake trout, *Salvelinus namaycush,* in Gamitagama, a small Precambrian lake in Ontario, *Journal of the Fisheries Research Board of Canada,* 34:1419–1422.

Olver, C.H., Nadeau, D., and Fournier, H., 2004, Control of harvest on Precambrian Shield lakes, in *Boreal Shield Watersheds: Lake Trout Ecosystems in a Changing Environment,* Gunn, J.M., Steedman, R., and Ryder, R., Eds., Lewis Publishers, Boca Raton, FL, chap. 11.

Olver, C.H., Shuter, B.J., and Minns, C.K., 1995, Toward a definition of conservation principles for fisheries management, *Canadian Journal of Fisheries and Aquatic Sciences,* 52:1584–1594.

Payne, N.R., Korver, R.M., MacLennan, D.S., Nepszy, S.J., Shutter, B.J., Stewart, T.J., and Thomas, E.R., , 1990, *The Harvest Potential and Dynamics of Lake Trout Populations in Ontario,* Ontario Ministry of Natural Resources, Lake Trout Synthesis working group report, Toronto.

Perkins, D.L., Fitzsimons, J.D., Marsden, J.E., Krueger, C.C., and May, B., 1995, Differences in reproduction among hatchery strains of lake trout in eight spawning areas in Lake Ontario: genetic evidence from mixed stock analysis, *Journal of Great Lakes Research,* 21(Suppl. 1):364–374.

Plosila, D.S., 1977, Relationship of strain and size at stocking to survival of lake trout in Adirondack lakes, *New York Fish and Game Journal,* 24:1–24.

Powell, M.J., Bernier, M.-F., Kerr, S.J., Leering, G., Miller, M., Samis, W., and Pellegrini, M., 1986, Returns of Hatchery-Reared Lake Trout from Eight Lakes in Northeastern Ontario, Ontario Ministry of Natural Resources, Ontario Fisheries Technical Report Series 22, Toronto.

Prevost, G., 1957, Use of artificial and natural spawning beds by lake trout, *Transactions of the American Fisheries Society,* 86:258–260.

Purych, P.R., 1977, Poor returns of hatchery-reared lake trout to the sport fishery of Flack Lake, Ontario, 1968–1974, *Progressive Fish-Culturist,* 39:185–186.

Purych, P.R., 1980, Successful reproduction of introduced lake trout in Horner Lake, Ontario, *Progressive Fish-Culturist,* 43:163–164.

Pycha, R.L., Dryer, W.R., and King, G.R., 1965, Movements of hatchery-reared lake trout in Lake superior, *Journal of the Fisheries Research Board of Canada,* 22:999–1024.

Pycha, R.L. and King, G.R., 1967, Returns of hatchery-reared lake trout in southern Lake Superior, 1955–62, *Journal of the Fisheries Research Board of Canada,* 24:281–298.

Rahel, F.J., 1997, From Johnny Appleseed to Dr. Frankenstein: changing values and the legacy of fisheries management, *Fisheries,* 22(8):8–9.

Schramm, H.L. and Hubert, W.A., 1996, Ecosystem management: implications for fisheries management, *Fisheries,* 21(12):6–11.

Sellers, T.J., Parker, B.R., Schindler, D.W., and Tonn, W.M., 1998, Pelagic distribution of lake trout (*Salvelinus namaycush*) in small Canadian Shield lakes with respect to temperature, dissolved oxygen, and light, *Canadian Journal of Fisheries and Aquatic Sciences,* 55:170–179.

Schindler, D.W., 1998, Sustaining aquatic ecosystems in boreal regions, *Conservation Ecology* [online], 2(2):18. Available at http://www.consecol.org/vol2/iss2/art18.

Siesennop, G.D., 1992, Survival, Growth, Sexual Maturation, and Angler Harvest of Three Lake Trout Strains in Four Northeastern Minnesota Lakes, Minnesota Department of Natural Resources Investigative Report 419, St. Paul.

Snucins, E.J., personal communication, 1999, Co-operative Freshwater Ecology Unit, Laurentian University, Sudbury, Ontario.

Snucins, E.J. and Gunn, J.M., 1995, Coping with a warm environment: behavioral thermoregulation by lake trout, *Transactions of the American Fisheries Society,* 124:118–123.

Snucins, E.J. and Gunn, J.M., 2003, Use of rehabilitation experiments to understand the recovery dynamics of acid-stressed fish population, *Ambio,* 32:240–243.

Utter, F.M., 1994, Detrimental aspects of put-and-take trout stocking, *Fisheries,* 19(8):8–9.

Webster, D.A., Bentley, W.G., and Galligan, J.P., 1959, Management of the Lake Trout Fishery of Cayuga Lake, New York, with Special Reference to the Role of Hatchery Fish, Cornell University Agricultural Experiment Station, Memoir 357, Ithaca, NY.

Wilton, M.L., 1985, Water Drawdown and Its Effects on Lake Trout (*Salvelinus namaycush*) Reproduction in Three South-Central Ontario Lakes, Ontario Ministry of Natural Resources, Ontario Fisheries Technical Report Series 20, Toronto.

Winter, B.D. and Hughes, R.M., 1995, AFS draft position statement on biodiversity, *Fisheries*, 20(4):20–26.

chapter thirteen

Species introductions and their impacts in North American Shield lakes

M. Jake Vander Zanden
University of Wisconsin
Karen A. Wilson
University of Wisconsin
John M. Casselman
Ontario Ministry of Natural Resources
Norman D. Yan
York University
Ontario Ministry of the Environment

Contents

Introduction

The aquatic biota of the world is rapidly being homogenized as a result of the introduction of species beyond their native range (Rahel, 2000; Ricciardi and MacIsaac, 2000). While the geographic range of species naturally changes in response to climate and other environmental factors, increased trade and human activities combined with current and past fisheries management practices have provided many aquatic species with the opportunity to colonize and survive in far-flung regions of the world that were never before accessible (Moyle, 1986; Claudi and Leach, 1999). For example, 176 exotic fish species (species originating from outside the continent) now occur within the United States (Claudi and Leach, 1999). Another 331 species native to the United States now occur outside their native range (Claudi and Leach, 1999). A variety of other aquatic invaders span a wide range of taxonomic groups, with amphibians, mollusks, plants, and crustaceans the taxa most well represented (for a listing, see Claudi and Leach, 1999). Invasive species are now widely recognized as a major threat to aquatic ecosystems and biodiversity (Sala et al., 2000; Coblentz, 1990; Soule, 1990; Wilcove and Bean, 1994; Naiman et al., 1995), and the rate of new invasions continues to increase (Mills et al., 1994). In addition, exotic species have caused tremendous economic impacts, estimated to exceed \$137 billion annually in the United States alone (Pimentel et al., 2000).

Despite the magnitude of the invasive species problem in freshwaters, perhaps the majority of species introductions have minor or no observable adverse impacts on native species and ecosystems. But for the smaller number of high-impact invaders, ecological effects can be severe and range from the extirpation of entire faunas (e.g., native cichlids by Nile perch *Lates niloticus* in Lake Victoria, native bivalves by zebra mussels *Dreissena polymorpha* in Lake St. Clair) to the complete restructuring of the ecosystem in which changes brought about by the invader cascade through the food web, producing a variety of unpredictable and often undesirable ecological alterations (Zaret and Paine, 1973; Spencer et al., 1991; Lodge, 1993; Strayer et al., 1999; Vander Zanden et al., 1999).

Throughout this chapter, we use terminology consistent with that of Lodge (1993). A "colonist" is a species that has arrived at a site outside its previous range. If a population establishes, it can be referred to as "introduced" or as an "invader." Species native to other continents are called "exotic," while species native to that continent but occurring outside their native range are "nonnative." Whether an invader has a measurable impact on the invaded ecosystem or native community is a separate consideration.

Another important distinction is the means by which a nonnative or exotic species arrives. Intentional introductions most often involve the stocking of game fish into previously unoccupied waters. In addition, nonnative fish and invertebrates have often been stocked to provide forage, usually for other nonnative species. A well-known example is the introduction of the freshwater shrimp *Mysis relicta* into lakes of western North America, Sweden, and Norway, which has dramatically altered the food web of these ecosystems (Richards et al., 1975; Goldman et al., 1979; Spencer et al., 1991). Exotics are also stocked for the purpose of biological control, such as the use of western mosquitofish *Gambusia affinis* to control biting insect populations.

In addition to these intentional introductions, many introductions are unintentional. The dumping of unused live bait has been identified as a particularly important vector of nonnative species dispersal (Litvak and Mandrak, 1993; Ludwig and Leitch, 1996; Litvak and Mandrak, 1999). Ballast water discharge of oceangoing ships has been most responsible for the introduction of exotic species, primarily of Eurasian origin, into the Laurentian Great Lakes (Ricciardi and MacIsaac, 2000). The Great Lakes, in turn, act as a source population from which these exotics disperse into smaller inland lakes.

While lakes of the Precambrian Shield have been invaded by a number of nonnative species, Shield lakes do not provide ideal habitat for many potential invasive species. Water temperatures are too cold for many fish of southerly (primarily U.S.) distribution. Furthermore, the low concentration of dissolved ions (typically Ca^{2+} <5 mg/l) will preclude potential invaders such as zebra mussels, which require dissolved calcium concentrations in the range of 15 to 30 mg L^{-1} (Mellina and Rasmussen, 1994; Ramcharan et al., 1992). For these reasons, Shield lakes are not likely to rival heavily invaded ecosystems such as the Laurentian Great Lakes, the Chesapeake Bay, and the San Francisco Bay estuary in terms of sheer numbers of invaders (Ricciardi and MacIsaac, 2000; Cohen and Carlton, 1998; Ruiz et al., 1999).

Yet despite the relatively small number of potential invaders, a developing literature indicates that Shield lake ecosystems and their biota can be highly sensitive to species invasions. While quantitative comparisons with other ecosystem types are not possible, dramatic impacts on native species and ecosystems in Shield lakes are well documented, perhaps more so than for many other ecosystem types. Because of the underlying ancient igneous bedrock, thin soils, a relatively recent (10,000 years) origin, and the lack of urban and agricultural development, Shield lakes are unproductive and support relatively few fish and invertebrate species. Barriers to fish and invertebrate dispersal during postglacial times also limited species distribution, further contributing to the low species richness. Compared to terrestrial and riverine ecosystems, lakes tend to be isolated from each other and can be considered islands of water in a sea of land (Magnuson, 1976).

The overall result is that Shield lakes have relatively simple, species-poor food webs that may be more vulnerable to perturbations than more productive, species-rich systems (McCann et al., 1998). In addition, these lakes are typically home to species such as lake trout *Salvelinus namaycush* and brook trout *Salvelinus fontinalis*, which are highly vulnerable to exploitation, habitat disturbance, and food web perturbations. So while relatively few nonnative species are presently invading Shield lakes, growing evidence indicates that they are having substantial impacts on Shield lake ecosystems (Evans and Loftus, 1987; Yan and Pawson, 1997; Vander Zanden et al., 1999; Yan et al., 2001).

Species invasions and introductions in Shield lakes must also be considered within the context of the predicted climate changes due to anthropogenic greenhouse gas emissions. Global circulation models (GCMs) that simulate a doubling of atmospheric CO_2 concentrations predict substantially warmer mean air temperatures as well as trends toward dryer conditions for much of the Canadian Shield (Magnuson et al., 1997). Climate warming will undoubtedly affect Shield lakes in a multitude of interconnected ways (reviewed in Magnuson et al., 1997), including a predicted increase in epilimnetic and hypolimnetic water temperatures (Destasio et al., 1996). Such warming will certainly have major implications for the thermal habitat of fish in lakes.

In addition, climate warming is predicted to increase the invasion rates of certain species (Jackson and Mandrak, 2002). The northern limit of smallmouth bass *Micropterus dolomieu* is effectively set by the short summer growing season of north-temperate lakes (Mandrak, 1989; Shuter and Post, 1990). Shuter and Post (1990) reported size-dependent over-winter starvation for smallmouth bass and yellow perch *Perca flavescens*. Population viability is thus contingent on their ability to complete a minimal amount of growth during their first summer (Shuter et al., 1980, 1989). Summer growth and over-winter survival of young-of-the-year (YOY) increase with water temperature and decrease as a function of latitude. Based on the Shuter model, the expected increases in water temperature would shift the zoogeographic boundaries for these cold-limited fish species (such as bass) northward by 500 to 600 km (Shuter and Post, 1990; Magnuson et al., 1997), which is likely to have important food web impacts (Vander Zanden et al. 1999).

The fundamental theme in this chapter is predicting, from easily measurable and readily available lake characteristics such as those presented in the appendices of this book, occurrences and impacts of invaders in individual Shield lakes. By focusing on predicting occurrences and impacts in individual lakes, lakes that are most vulnerable to invaders can be identified. This should be useful to lake managers for several reasons. For example, invader prevention efforts and education campaigns can target those lakes

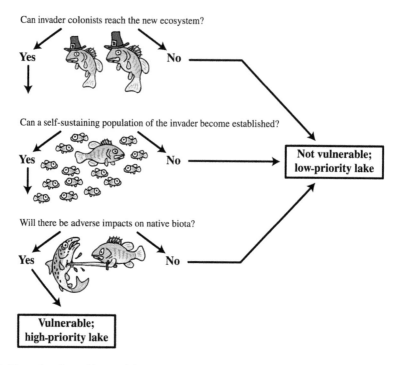

Can invader colonists reach the new ecosystem?

Yes   No

Can a self-sustaining population of the invader become established?

Yes   No

Will there be adverse impacts on native biota?

Yes   No

**Not vulnerable;
low-priority lake**

**Vulnerable;
high-priority lake**

Figure 13.1 Three levels or filters of the invasion process used to examine the vulnerability of Shield lakes to aquatic invaders.

identified as vulnerable, allowing optimal use of limited management resources. Furthermore, efforts to monitor invader distribution and impacts can target systems identified as most vulnerable (likely to be invaded).

In our examination of species invasions and impacts in Shield lakes, we deconstruct the invasion process into three sequential components or filters; each should be considered in an effort for ultimate prediction of the dynamics and impacts of a known invader for individual lakes (Figure 13.1). The three components can be assessed using semiqualitative criteria (for example, the presence or absence of public road access). Alternatively, quantitative techniques such as logistic regression, discriminant function analysis, or artificial neural networks (ANNs) can be used to predict species presence or absence (Ramcharan et al., 1992; MacIsaac et al., 2000; Olden and Jackson, 2001). In either case, assessment of the three filters requires some knowledge of the biology of the invader and its interactions with natural ecosystems. The information required to address these questions will often be available in public databases. It must be recognized that determining the vulnerability of an individual lake to a given invader is a probabilistic exercise, and that this approach represents a caricature of the highly complex and unpredictable dynamics of species invasions on the landscape. Still, the value of this approach is that it provides predictions of the specific location of species invasions before they occur (Vander Zanden et al., in press).

The first filter is whether colonists can reach an uninvaded ecosystem (Figure 13.1). This depends on the dispersal mechanisms and potential of the invader as well as interactions with both human and nonhuman dispersal vectors. Factors such as road access, the presence of boat launches, and urban and residential development may be important determinants, although natural dispersal through interconnected waterways must also be considered.

The second filter is whether the invader is capable of surviving, reproducing, and establishing a self-sustaining population in the novel ecosystem. In many cases invader colonists may reach a given ecosystem, but environmental or biotic conditions are not appropriate and a population cannot establish. It should be noted here that the failure of an invader to establish a population following introduction does not mean that conditions are not appropriate for establishment because stochastic factors play an important role in determining invader establishment (Pimm, 1991).

The third filter is whether an established invader has adverse impacts on the native ecosystem or biota. This will depend on the population size or density of the invader, the strength and nature of biotic interactions (predation and competition) between the invader and native species, whether the invader occupies an "empty niche," and whether the invader has ecosystem-altering potential in its new ecosystem. This third filter will most likely be the most difficult to address. An invader can only establish if the first two filters are satisfied (colonists reach the novel system, and the conditions are appropriate for the invader to establish). An invasion is of particular ecological concern if all three questions are answered affirmatively (Figure 13.1).

This chapter focuses on several animal invaders that may have already invaded Shield lakes, are likely to continue to spread, and have the potential for dramatic impacts on Shield lake ecosystems. For each invader we separately consider the filters of the invasion process. The invaders examined in this chapter are (1) smallmouth bass and rock bass *Ambloplites rupestris*, (2) rainbow smelt *Osmerus mordax*, (3) the spiny water flea *Bythotrephes*, (4) zebra mussel and quagga mussel *Dreissena bugensis*, (5) rusty crayfish *Orconectes rusticus*, and (6) *Daphnia lumholtzi*. In the final section, we briefly mention other potential invaders of Shield lakes. Recent efforts have been made to predict the identity of future invaders (Ricciardi and Rasmussen, 1998; Kolar and Lodge, 2001). It is hoped that efforts to predict the identity, occurrences, and impacts of future invaders will contribute to the development of management strategies that can limit the further spread of species with the greatest potential impacts on Shield lake ecosystems.

Invaders in Shield lakes

Smallmouth bass and rock bass in Ontario

Smallmouth bass and rock bass were historically confined to Mississippi and Great Lakes drainage systems (Scott and Crossman, 1973; Lee et al., 1980). During the past century, these and other species of the family Centrachidae have been widely introduced beyond their native range and now occur in much of western North America, many East Coast drainage systems, and northward into Shield lakes in regions of Ontario, Quebec, New Brunswick, Nova Scotia, and western Canada (MacCrimmon and Robbins, 1975; Lee et al., 1980; McNeill, 1995; Rahel, 2000). The northward range expansion of smallmouth bass and rock bass (hereafter referred to together as bass) into lakes of the Canadian Shield presently continues at a rapid pace. While resource management agencies no longer stock bass into new water bodies, bass continue to expand their range as a result of unauthorized introduction by anglers, accidental bait bucket transfers, and natural dispersal through drainage networks. Also, smallmouth bass and largemouth bass *Micropterus salmoides* have been introduced into dozens of countries on nearly every continent, although the ecological impacts of their introduction outside North America are virtually unknown (McDowall, 1968; Robbins and MacCrimmon, 1974; Welcomme, 1988).

Adult rock bass and smallmouth bass have broad, generalist diets and feed on a mix of prey fish, crayfish, and other zoobenthos with zooplankton, amphibians, songbirds, and small mammals in the diet on occasion (Hodgson and Kitchell, 1987; Hodgson et al.,

Table 13.1 Comparison of Central Ontario Lakes with and without Smallmouth Bass and Rock Bass

Type	Number of lakes	Prey fish species richness	Minnow catch rate[a]	Lake trout trophic position	Lake trout $\delta^{13}C$
Bass	5	2.4	6.6	3.28	−29.20
No bass	5	8.2[b]	35.8[c]	3.90[c]	−27.48

Note: Values are means across five lakes.

[a] Grams of fish/trap/day.

[b] $p < .001$ between lakes with and without bass (one-tailed t test).

[c] $p < .05$ between lakes with and without bass (one-tailed t test).

Source: Data from Vander Zanden et al. (1999).

1991; D.E. Schindler et al., 1997; Vander Zanden and Vadeboncoeur, 2002). Bass are efficient piscivores that can have substantial impacts on littoral prey fish diversity, abundance, and community structure in north-temperate lakes (Mittelbach et al., 1995; Chapleau et al., 1997; Vander Zanden et al., 1999; Whittier and Kincaid, 1999; Findlay et al., 2000). Considering the important top-down role of bass in structuring pelagic food webs and their range expansion during the last century, it is critical to examine the broader impacts of bass introductions on native species. Of particular concern is that reductions in forage fish following bass introductions into lakes could have adverse impacts on native top predators such as lake trout and brook trout, which rely on littoral prey fish (Olver et al., 1991; Vander Zanden et al., 1999).

Lakes of central and northern Ontario are rapidly being invaded by bass. We previously examined a series of nine Ontario lakes, five of which had been recently invaded, along with four uninvaded reference lakes (Vander Zanden et al., 1999). All of these lakes supported native, self-sustaining lake trout fisheries. Like most small headwater lakes in the region, these lakes lacked pelagic prey fish such as rainbow smelt, cisco, and lake whitefish, which are the preferred prey of lake trout. In the absence of these preferred prey fish, lake trout consume a mix of zooplankton, zoobenthos, and littoral prey fish such as minnows (family Cyprinidae) (Martin, 1970; Martin and Fry, 1972; Vander Zanden and Rasmussen, 1996). Among the nine lakes, littoral prey fish catch rates and species richness were significantly lower in lakes with bass relative to lakes without bass (Table 13.1). More compelling evidence comes from long-term (1981 to 1999) quantitative electrofishing monitoring of fish population abundance in seven lakes in the Haliburton Forest Preserve, Ontario. Abundance of cyprinids (expressed as number per square meter) is negatively correlated with centrarchid abundance (smallmouth bass and rock bass; Figure 13.2): log(cyprinid abundance) = −0.65*log(centrarchid abundance) + 0.70, $r^2 = .43$.

To address the broader food web consequences of bass introductions in central Ontario lakes, carbon and nitrogen stable isotopes were used to quantify differences in food web structure related to bass invasion (Vander Zanden et al., 1999). Corresponding with reduced littoral prey fish in invaded lakes, lake trout trophic position (based on $\delta^{15}N$ values) was reduced, indicating a diet consisting of invertebrates rather than fish. The $\delta^{13}C$ values indicated that lake trout relied primarily on littoral prey fish in lakes without bass and depended on zooplankton where they are sympatric with bass (Table 13.1, Figure 13.3).

In addition to this comparative analysis, long-term studies of two recently invaded lakes, MacDonald Lake and Clean Lake, revealed the food web consequences of bass impacts. In MacDonald Lake, littoral prey fish populations declined dramatically following bass establishment. Stable isotope analysis of freezer-archived muscle tissue samples collected throughout this period revealed a concurrent decline in lake trout trophic position (Figure 13.4). The invasion and establishment of bass into Clean Lake followed that

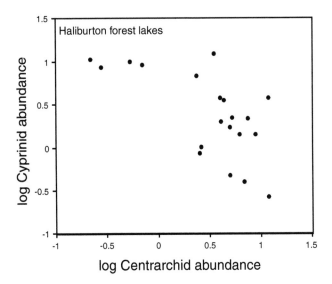

Figure 13.2 The relationship between centrarchid and cyprinid abundance (number of individuals) based on long-term (1981 to 1999) monitoring in seven lakes located in the Haliburton Forest Preserve, ON.

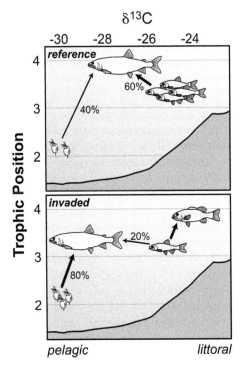

Figure 13.3 Food web structure based on carbon and nitrogen stable isotope studies of Shield lakes with and without smallmouth and rock bass. (Adapted from Vander Zanden et al., 1999.)

of MacDonald Lake, but some 6 years later, and the trophic position of Clean Lake lake trout did not show a marked change (Figure 13.4). The full impact of the bass invasions was not realized at that time, but has been subsequently. Ongoing monitoring of Clean Lake has chronicled a decline in prey fish, and Clean Lake has followed the same trajectory as MacDonald Lake (J.M. Casselman and D.M. Brown, unpublished data).

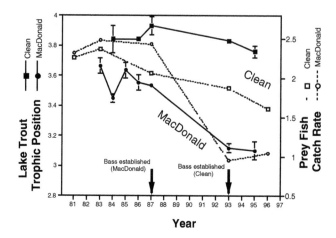

Figure 13.4 Long-term changes in minnow abundance as estimated by quantitative electrofishing and the corresponding shifts in lake trout trophic position. The arrows indicate the year both smallmouth bass and rock bass had become fully established. (Adapted from Vander Zanden et al., 1999.)

Invasion has affected angling success for lake trout. Although anglers initially saw increased catches, these catches quickly declined in response to change in the food web and lake trout predation activities and feeding. The more experienced anglers modified their fishing methods to simulate plankton and attract plankton-feeding lake trout. Subsequently, anglers have lost interest in this one-time spectacular recreational fishery. The loss of this resource has been far-reaching and insidious and has caused anglers to advocate stocking.

Competition between bass and lake trout has not been generally recognized, and it has been erroneously assumed that bass introductions have no effect on lake trout populations (Martin and Fry, 1972; Scott and Crossman, 1973; Olver et al., 1991). This interaction has been overlooked because bass inhabit inshore, littoral areas while lake trout inhabit offshore, pelagic areas. Despite these differences, bass and lake trout often share a common resource, and the introduction of bass has translated into the interruption of the trophic linkage of prey fish and lake trout. This change has directly affected lake trout growth rates, biomass, and productivity. Somatic growth and growth potential of lake trout were reduced 25 to 30% in MacDonald Lake following bass establishment. Even greater losses in reproductive growth were realized. This loss in lake trout growth and productivity, which was chronicled over time in MacDonald and Clean Lakes, has also been observed from point-in-time surveys in other lakes throughout the Haliburton Highlands of Ontario.

These invasions have been devastating to lake trout productivity. Invariably, anglers lose interest in these once-good lake trout fisheries and advocate the need for stocking, although such actions provide minimal benefit and could decrease the growth of existing lake trout because fish prey production has been diminished. The only advantage in stocking would be to provide potential prey for lake trout; this is an inefficient and unproductive way to try to bolster lake trout productivity and angling success.

Studies are under way to partition the relative importance of the different bass species in these invasions. This is not easy to separate given that smallmouth bass and rock bass often are coinvaders, and where one establishes it is not long until the other appears. There is, however, evidence that rock bass has the more important and devastating effect (J.M. Casselman and D.M. Brown, unpublished data).

Considering the tremendous number of Shield lakes (Olver et al., 1991), designing and implementing a management plan to minimize the adverse impacts of bass introductions is a daunting task. Using the framework of Figure 13.1, individual lakes in central Ontario that are vulnerable to bass invasion have been identified (Vander Zanden et al., in press). The analysis was performed using Geographic Information System (GIS) and included the central Ontario's more than 700 lakes containing a resident lake trout population. The study addressed the following questions:

1.   Which lakes are likely to receive bass colonists?
2.   Which lakes are likely to be able to support a bass population?
3.   Which lakes are likely to be adversely impacted if bass establish a population?

Each of these three filters was modeled separately, and the subset of lakes classified as positive for all three criteria is considered vulnerable. These individual lakes should be the focus of management efforts aimed at slowing or halting further bass impacts.

Which lakes are accessible to bass colonists? To be accessible, a lake either must have road access or must occur in a drainage system already invaded by bass. This is a reasonable set of assumptions because bass are rapidly expanding their range due to unauthorized introduction by anglers, accidental bait bucket transfers, and natural dispersal through drainage networks (M.J. Vander Zanden, personal observation). Because the vast majority of lakes in central Ontario have public road access, only a relatively small number of lakes located in provincial parks (notably Algonquin Provincial Park) are protected from bass colonists due to their remote location and roadless status.

Which lakes are capable of supporting bass populations? Models that predict bass presence or absence in Ontario lakes based on glacial history, local and regional environmental variables, and biotic variables have been developed (Vander Zanden et al., in press). Using ANN models, lakes were classified according to bass presence or absence with 77 to 90% accuracy. When the predictions of the neural network model were examined for the 771 central Ontario lakes containing lake trout, bass were predicted but not observed (i.e., false presence) in 59 of these lakes. Thus while bass do not presently occur in these 59 lakes, the model indicates that these lakes have the appropriate conditions for supporting self-sustaining bass populations. These lakes are likely to be capable of supporting bass populations (note that this observation is independent of whether colonists are able or likely to colonize these lakes).

In which lakes will bass have adverse impacts on the native biota? Food web studies using diet data and stable isotopes indicated that lake trout are linked to the pelagic food web in lakes containing pelagic prey fish such as rainbow smelt, lake herring *Coregonus artedi*, and lake whitefish *Coregonus clupeaformis* (Vander Zanden and Rasmussen, 1996; Vander Zanden and Rasmussen, 2002; Vander Zanden et al., in press). In lakes lacking pelagic prey fish, lake trout tend to be linked to the littoral food web through consumption of littoral prey fish (Vander Zanden and Rasmussen, 1996; Vander Zanden et al., 1999). Because the availability of littoral prey fish is a function of bass presence, competitive bass–trout interactions are predicted to occur only in lakes lacking pelagic forage fish. Thus, the presence of pelagic prey fish mediates the strength of bass–lake trout interactions. If pelagic prey fish are present, lake trout are buffered from impacts of bass on littoral prey fish populations (Figure 13.5) (Vander Zanden and Rasmussen, 2002; Vander Zanden et al., in press). With bass–lake trout interactions predictable from species composition, we can identify lake trout populations likely to be impacted by bass introductions. Of the 59 lake trout lakes classified as capable of supporting bass (Filter 2), 38 did not contain pelagic prey fish and are thus vulnerable to bass impacts based on food web considerations.

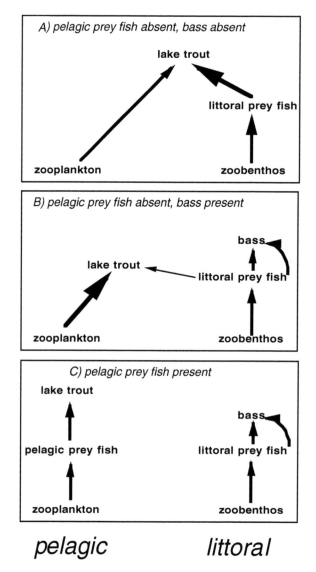

Figure 13.5 A summary of food web structure for three general food web types based on stable carbon and nitrogen isotopes: A) bass absent, pelagic prey fish absent; B) bass present, pelagic prey fish absent; C) pelagic prey fish present. (Based on Vander Zanden et al., 1999, in press; Vander Zanden and Rasmussen, 2002.)

The many thousands of Shield lakes that dot the north-temperate landscape provide a distinct management problem of how to apply limited resources to combat the spread of nonnative species and minimize potential adverse impacts. By separately considering the elements of the invasion process (Figure 13.1), lakes that are vulnerable to a particular invader were identified. In our study, roughly 5% of the lake trout lakes were classified as vulnerable to bass invasions, and these lakes should be the focus of efforts to prevent future invasion. While prevention of future introductions is the backbone of a successful invader management strategy, mitigating impacts where invaders have already established will require the development of techniques to reduce impacts. If historic levels of lake trout production are to be realized through natural reproduction and self-sustaining lake trout populations, then these bass invaders must be eliminated or at least substantially reduced. Yet to date, there are few examples of successful eradication of aquatic invaders: The extirpation of nutria from England during the 1980s and trout from small Sierra Nevada (California) lakes are among the few success stories. The limited potential for eradication of invaders underscores the central role of prevention as the most effective strategy for minimizing invader impacts.

Rainbow smelt

Rainbow smelt are an anadromous species native to coastal waters of Canada and the United States; they have a historical range that extends from coastal Labrador to New Jersey. In addition, there are a number of native landlocked freshwater populations of rainbow smelt along the Atlantic coast. Smelt were originally introduced into the Great Lakes drainage in 1912 into Crystal Lake, Michigan. Smelt spread to nearby Lake Michigan by 1923 and subsequently spread to the rest of the Great Lakes during the following decade (Dymond, 1944; Christie, 1974; Bergstedt, 1983).

Smelt have since dispersed beyond the Great Lakes into inland lake and river systems. Smelt were stocked into Lake Sakakawea, North Dakota, a reservoir on the Missouri River in 1971, and subsequently spread through much of the Missouri and Mississippi drainage systems (Mayden et al., 1987). Smelt have been stocked into other reservoirs of the western United States and have similarly expanded their range (Jones et al., 1994; Johnson and Goettl, 1999). This species now occurs in the Hudson Bay drainage waters of northwestern Ontario, Manitoba, and Minnesota (Franzin et al., 1994) and has recently reached Hudson Bay via the Nelson River (Remnant et al., 1997). Smelt continue to colonize Shield and non-Shield lakes within the Great Lakes drainage basin (Evans and Loftus, 1987; Hrabik and Magnuson, 1999). In the most comprehensive synthesis of smelt biology in inland lakes, Evans and Loftus (1987) reported the presence of smelt in 194 inland Ontario lakes, of which only 4 are thought to be native, relict populations. There are undoubtedly many more introduced smelt populations in lakes of the Canadian Shield, although little effort has been made to document their ever-expanding distribution.

In this section we examine the three filters of the invasion process (Figure 13.1) for rainbow smelt. Efforts to identify lakes that are likely to receive smelt colonists require an understanding of the mechanisms of smelt dispersal. Smelt can spread rapidly across the landscape once they have been introduced, as evidenced by their rapid downstream colonization of the Missouri/Mississippi and Hudson Bay drainages (Franzin et al., 1994; Remnant et al., 1997). Yet anthropogenic introductions, either intentional or accidental, are thought to be the primary vector of smelt introduction into new lakes.

This conclusion has been reached by numerous authors based on the close association of smelt with urban and cottage development and lake appearances that cannot be explained by dispersal from nearby lakes (Evans and Loftus, 1987; Hrabik and Magnuson, 1999). In northern Wisconsin, smelt have been deliberately introduced into lakes by anglers with the intention of increasing opportunities for netting smelt during their spring spawning runs (called *smelting*; T. Hrabik, July, 2002, personal communication). Another likely vector is the unintentional introduction of fertilized eggs into lakes while cleaning and processing smelt collected from other lakes. While perhaps discouraging, this also suggests that the spread of smelt is partially preventable and that educational efforts could reduce their spread into new waters. The available evidence indicates that lakes occurring in the same drainage as other smelt lakes as well as lakes with road access and cottage development should be considered open to rainbow smelt colonists. In addition, lakes with a large number of nearby smelt populations are far more likely to receive smelt colonists than lakes in regions lacking smelt populations.

Table 13.2 Characteristics of Ontario Lakes Containing Rainbow Smelt

	Mean	Minimum	Maximum
Lake area (km²)	52.3	0.1	4480
Mean depth (m)	11.6	2.0	38.7
Maximum depth (m)	35.7	4.0	213.5
TDS (mg/l)	49.5	5.5	231.4
Surface water pH	7.2	6.0	9.3
Secchi depth (m)	4.9	0.5	10.5
Fish species richness	11.3	3	63

Source: Data from Evans and Loftus (1987).

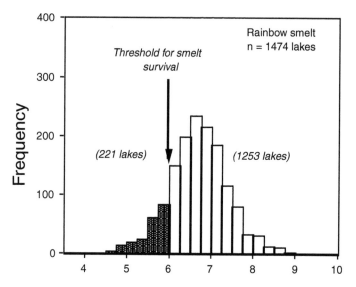

Figure 13.6 Frequency distribution of pH values for North American lake trout lakes (data from Appendix 2). The arrow indicates the threshold pH value of 6.0 for rainbow smelt *Osmerus mordax* occurrence as reported by Evans and Loftus (1987). Low pH is predicted to limit smelt occurrence in 221 of the 1253 Shield lakes for which data are available (<18% of lakes).

To identify lakes capable of supporting rainbow smelt populations Evans and Loftus (1987) summarized the morphometric and limnological parameters for Ontario lakes that contained smelt as of 1987 (reproduced in Table 13.2). While smelt typically inhabit lakes that are relatively deep, low in productivity, and with intermediate transparency, they occur in lakes that span a wide range of conditions, including lakes as small as a few hectares in size and as shallow as 4 m maximum depth (Evans and Loftus, 1987). One significant finding of Evans and Loftus (1987) was that smelt do not occur in lakes with pH less than 6.0, indicating a threshold pH value that may limit smelt occurrences. A frequency distribution of pH values for the available lake trout lakes in North America (n = 1474 lakes; data from Appendix 2) was plotted (Figure 13.6). The arrow indicates the pH 6.0 threshold for rainbow smelt; lakes with a pH to the left of the threshold (<18% of the Shield lakes; 221 of the 1474 lakes) are predicted not to support a smelt population. Interestingly, the pH of Shield lakes has been rising during the past two decades due to decreased SO_2 emissions (Stoddard et al., 1999). This result suggests that pH restriction of further smelt invasion might weaken as lakes recover from acidification.

Aside from effects of pH, rainbow smelt distribution does not appear to be severely limited by morphometric and limnological parameters, and rainbow smelt persist in a variety of lakes, reservoirs, and flowage ecosystems if minimal habitat and environmental conditions are met. Development of more sophisticated quantitative models that identify specific lakes likely to support rainbow smelt (Ramcharan et al., 1992; MacIsaac et al., 2000) will be a critical step toward characterizing lake vulnerability.

There has been little progress in understanding why smelt sometimes (but not always) have adverse impacts on native biota. A review of 35 individual cases of rainbow smelt introduction in inland lakes provided substantial evidence for adverse impacts of smelt on native biota, with the most frequent impact declines in lake whitefish and lake herring populations (Evans and Waring, 1987). Adverse impacts on lake trout, walleye *Stizostedion vitreum*, and burbot *Lota lota* populations were also noted. Lake whitefish recruitment failure in Twelve Mile Lake, ON, was attributed to intense rainbow smelt predation on YOY (Loftus and Hulsman, 1986). In Lake Simcoe, ON, declines in lake whitefish were also attributed to smelt introduction, although a positive correlation between yellow perch and smelt abundance was observed (Evans and Waring, 1987). Hrabik et al. (1998) reported different impacts of smelt introduction in two northern Wisconsin lakes. Smelt were a strong competitor of yellow perch in Crystal Lake, WI, but were a major predator of lake herring (cisco) in nearby Sparkling Lake, ultimately causing the extirpation of the lake herring population (Hrabik et al., 1998).

Explicitly viewing smelt from a predator–prey perspective serves as a basis for understanding their impacts as a predator and competitor of the native biota (Hrabik et al., 1998). The available case studies indicate that smelt are often a major player in aquatic food webs, acting as an important prey, predator, and competitor. Smelt often become the dominant fish species in the pelagic zone of invaded lakes. In addition, smelt are highly efficient predators and consume large prey items relative to other fish of similar body size. Consequently, smelt can dramatically reduce or even eliminate preferred prey. Smelt also have a broader diet than other forage fish, which commonly includes copepods, cladocerans, *Mysis*, zoobenthos, YOY smelt and other YOY fish (Vander Zanden and Rasmussen, 1996). This broad, generalist diet allows smelt to consume resources opportunistically as they become abundant and to switch to alternative prey when the abundance of preferred prey is reduced. Their generalist and omnivorous feeding behavior allows smelt to avoid food limitation, maintain a large population, and sustain strong predatory impacts on preferred prey (Courchamp et al., 2000).

Which factors might determine smelt impacts on native forage fish? One possibility is that in lakes with a greater diversity of alternative prey, smelt can become more abundant and have greater predatory impacts on preferred prey (i.e., YOY native fish). This suggests that smelt should have greater impacts on native fish in larger lakes in which there are more prey options to exploit. An alternative hypothesis is that the degree of spatial overlap between adult smelt and the vulnerable (YOY) life stage of native fish determines smelt impacts. One might predict that the biota in small, shallow lakes would be more vulnerable to smelt impacts because there is greater potential for spatial overlap between life stages of smelt and native fish. While the link between smelt introductions and declines of native pelagic fish is strong, smelt do not universally have adverse impacts, and the mechanism and magnitude of smelt impact vary widely from system to system. Predicting smelt impacts on native biota is an important area of future research.

A final impact to consider is the potential effect of smelt introduction on concentrations of contaminants in top predators. Smelt are a common prey item of game fish such as lake trout, walleye, and Atlantic salmon (Vander Zanden and Rasmussen, 1996; Vander Zanden et al., 1997). Of concern is the observation that smelt introduction corresponds with increased concentration of mercury and polychlorinated biphenyls (PCBs) in game fish

(Akielaszek and Haines, 1981; MacCrimmon et al., 1983; Mathers and Johansen, 1985; Vander Zanden and Rasmussen, 1996). The presumed explanation is that smelt have a higher trophic position than other forage fish due to their piscivorous diets, and fish that feed on smelt would also have an elevated trophic position and thus a greater scope for contaminant biomagnification. While other factors such as increased growth rates may counter the effects of longer food chains (Evans and Waring, 1987), the available evidence indicates that smelt introduction can lead to increased concentrations of contaminants in top predator fish.

Bythotrephes and Cercopagis

The spiny water flea (identified as either *Bythotrephes cederstromi* or *Bythotrephes longimanus*, hereafter referred to as *Bythotrephes*) is a predatory zooplankter native to northern Europe and Asia. This species was first discovered in North America in Lake Ontario in 1982 and was likely transported to North America via ship ballast water. From Lake Ontario it rapidly spread to the other Great Lakes, then to inland lakes in Ontario, Michigan, Minnesota, and Ohio (Yan et al., 1992). As of 1999, *Bythotrephes* had been identified in 50 North American lakes in the Great Lakes region (MacIsaac et al., 2000; Yan et al., 2002). Little is known about the vector of spread for *Bythotrephes*, although it is likely that they are unintentionally spread via bait buckets, bilge water, live wells, and the use and reuse of anchor ropes by recreational boaters. As with rainbow smelt and bass, any lake with road access and recreational boating traffic should be considered vulnerable to *Bythotrephes* colonists.

In Europe, *Bythotrephes* inhabit lakes spanning a wide range of physical, chemical, and biological conditions. MacIsaac et al. (2000) used discriminant function analysis (DFA) to predict *Bythotrephes* distributions and occurrences in Europe and then applied the resulting model to North American lakes. The significant variables in the model were water clarity, lake area, chlorophyll-a concentration, and maximum depth. The model correctly classified lakes according to *Bythotrephes* presence or absence in more than 90% of European lakes. When applied to North America, the model accurately predicted occurrences of *Bythotrephes* in lakes that presently contain the species (82%). More important, the model predicted the occurrence of *Bythotrephes* in many lakes that presently lack *Bythotrephes*. The high frequency of false negatives among North American lakes indicates that many are capable of supporting *Bythotrephes*, but have not yet been colonized. We plotted lake area versus Secchi depth for 1700 Ontario lakes (Figure 13.7) and superimposed the division for *Bythotrephes* presence or absence (estimated from MacIsaac et al., 2000). It is clear from this figure that conditions are favorable for *Bythotrephes* to inhabit the vast majority of Shield lakes, and that thousands of North American lakes appear to provide appropriate habitat for this invader.

Bythotrephes is a voracious predator on other zooplankton. Evidence from North American and European lakes indicates that the introduction of *Bythotrephes* often dramatically restructures the zooplankton community. While laboratory feeding studies indicate preference for large zooplankters (>2.0 mm) (Schulz and Yurista, 1998), field studies indicate that size-selective impacts are highly variable. Introduction of *Bythotrephes* reduced populations of large herbivorous zooplankton (*Daphnia*) in Lake Michigan (Lehman, 1988; Lehman and Caceres, 1993). However, in Harp Lake, Ontario, *Bythotrephes* reduced the abundance of smaller zooplankters such as *Bosmina*, *Chydorus*, and *Diaphanosoma*, while larger zooplankton species (*Holopedium*, *Daphnia*) actually increased in abundance (Yan and Pawson, 1997). While it is clear that *Bythotrephes* impact zooplankton communities through predation, the size-specific impacts on zooplankton communities have been generally unpredictable.

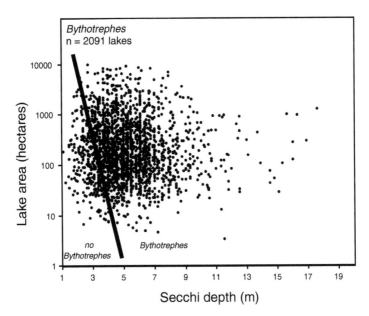

Figure 13.7 Lake area and Secchi depth as factors potentially limiting the distribution of *Bythotrephes* in North American Shield lakes (data from Appendix 2). The diagonal line indicates the predicted cutoff for *Bythotrephes* presence or absence based on the two most important predictor variables from MacIsaac et al. (2000). Lakes to the left of the diagonal line are predicted not to support *Bythotrephes*; lakes to the right are predicted potentially to support this invader.

Bythotrephes are large relative to other zooplankton species and are a preferred prey of zooplanktivorous fish (Mills et al., 1992; Coulas et al., 1998). This suggests that *Bythotrephes* abundances in inland lakes might be controlled by fish predation (Coulas et al., 1998; Yan and Pawson, 1998). While *Bythotrephes* are an important prey of adult planktivores, their large size and long spine likely inhibit consumption by YOY and juvenile fish. The replacement of edible zooplankters by inedible *Bythotrephes* may have negative impacts on growth rates of YOY fish and recruitment (Hoffman et al., 2001), although this question requires further examination in inland lakes.

A similar zooplankter, the fishhook water flea *Cercopagis pengoi*, native to the Ponto-Caspian region, was discovered in Lake Ontario in the summer of 1998 (MacIsaac et al., 1999). *Cercopagis* has since spread to Lake Michigan and the Finger Lakes in New York (Charlebois et al., 2001). This species probably has dispersal capabilities similar to that of *Bythotrephes* and is likely to spread to more inland lakes with recreational watercraft. *Cercopagis* have invaded aquatic ecosystems well beyond its native range in Europe (MacIsaac et al., 1999) and can thrive across a broad range of environmental conditions, indicating that this species may thrive in North American lakes (Ricciardi and Rasmussen, 1998). Projected impacts are similar to that for *Bythotrephes*.

Dreissenid mussels

Dreissenid mussel invaders to North America include zebra mussels and quagga mussels. Dreissenid mussels established in the Great Lakes in the 1980s and have since colonized large portions of North America. Much has been written about zebra mussels in North America, and a number of excellent reviews are available (Neary and Leach, 1992; MacIsaac, 1996; Strayer et al., 1999). Despite the rapid spread and widespread distribution of dreissenids in North America, most Shield lakes are not expected to support thriving

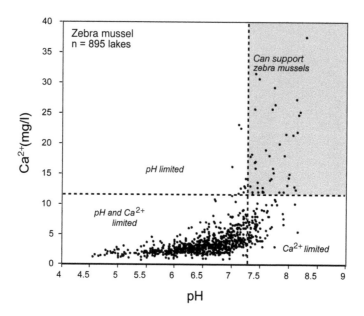

Figure 13.8 Thresholds potentially limiting the distribution of zebra mussels *Dreissena polymorpha* in North American lake trout lakes (data from Appendix 2). The vertical line indicates the pH threshold of 7.3, below which they do not survive (Ramcharan et al., 1992). The horizontal line indicates a Ca^{2+} threshold of 12 mg/l (Sprung, 1987), below which they are not likely to persist.

dreissenid populations. The presence of zebra mussels is limited primarily by low Ca^{2+} concentrations and low pH. In Europe, zebra mussels do not occur below a pH 7.3 threshold (Ramcharan et al., 1992). Several threshold values for Ca^{2+} have been reported: 12 mg/l (Sprung, 1987), 15 mg/l (Mellina and Rasmussen, 1994), and 28.3 mg/l (Ramcharan et al., 1992). Other potentially limiting factors such as oxygen concentrations, upper temperature, and salinity would not limit the distribution of dreissenids in Shield lakes (Karatayev et al., 1998). Using data from Appendix 2, we plotted Ca^{2+} versus pH for North American lake trout lakes (n = 895 lakes; Figure 13.8). The pH 7.3 threshold and the lowest of the three Ca^{2+} threshold values are indicated. The 43 lakes (4.8%) located in the upper right section of Figure 13.8 have the potential to support zebra mussels (i.e., both Ca^{2+} and pH threshold values are exceeded). Furthermore, Ca^{2+} concentrations in Shield lakes have been declining during the past two decades in correspondence with recovery from acidification (Keller et al., 2001). While this indicates that a small proportion of Shield lakes is capable of supporting dreissenids, this species typically has dramatic impacts on lake ecosystems where it occurs (MacIsaac, 1996), and the potential impacts of zebra mussels in select Shield lakes should not be neglected.

Rusty crayfish

Rusty crayfish *Orconectes rusticus* are native to the Ohio River Basin and since the 1960s have spread northward into Wisconsin, Michigan, Iowa, New York, Minnesota, Ontario, and all of the New England states. Although now banned for use as live bait in most states, rusty crayfish were once common live bait, and bait bucket releases are believed the major vector of rusty crayfish introductions. In some areas, rusty crayfish are thought to have been introduced to control nuisance aquatic vegetation (Magnuson et al., 1975). In northern Wisconsin lakes, Capelli and Magnuson (1983) found that, unlike the native

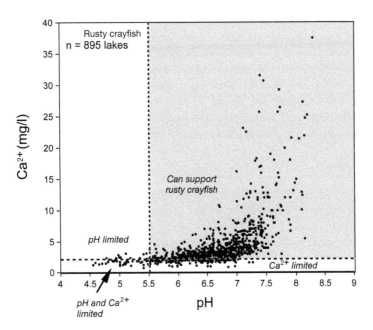

Figure 13.9 Thresholds potentially limiting the distribution of rusty crayfish *Orconectes rusticus* in North American lake trout lakes (data from Appendix 2). The vertical line is the pH threshold of 5.5, below which rusty crayfish tend not to occur in lakes (Berrill et al., 1985). The horizontal line indicates the Ca^{2+} threshold of 2.5 mg/l (Capelli and Magnuson, 1983), below which rusty crayfish are not observed.

crayfish *O. virilis* and *O. propinquus*, the distribution of rusty crayfish among suitable lakes was positively correlated with human activity and proximity to major roads. Once in a drainage system, rusty crayfish use natural waterways to spread along rivers and from lake to lake. However, natural dispersal is slow relative to other aquatic invaders because rusty crayfish do not have a pelagic larval stage. Colonization around the littoral zone of a single lake can take many years (Wilson, 2002; Wilson and Magnuson, in review).

Rusty crayfish distribution is likely limited by physiochemical factors, particularly dissolved calcium and pH. Berrill et al. (1985) rarely found rusty crayfish in lakes with a pH less than 5.5. In laboratory experiments, rusty crayfish stage III juveniles do not survive at pH 5.0 or lower, while adults do not survive at pH 4.7 or lower. Dissolved calcium is another limiting factor; lakes with dissolved Ca^{2+} concentrations less than 2.5 mg/l do not support rusty crayfish populations (Capelli and Magnuson, 1983). Based on these threshold values for pH and calcium, over 70% of Shield lakes are considered suitable for rusty crayfish colonization (Figure 13.9).

The availability of suitable habitat and the presence of predators, particularly fish, may also influence the establishment and abundance of rusty crayfish in lakes. Rusty crayfish prefer firm substrates (cobble or sand) over mucky substrates (Kershner and Lodge, 1995), and shelter in the form of logs or cobble is required for egg or young-carrying females and molting individuals. Rusty crayfish are consumed by several fish species, but quickly reach a size refuge from all but the largest predators (Stein and Magnuson, 1976). This species is also highly omnivorous, foraging opportunistically on aquatic plants, benthic invertebrates, fish eggs, detritus, and small fish (Momot, 1995), indicating a limited potential for resource limitation.

Once established in lakes and rivers, rusty crayfish tend to displace native crayfish (primarily *O. propinquus* and *O. virilis*) (e.g., Berrill, 1978; Capelli, 1982), sometimes reaching densities 20 times greater than that of the native crayfish (Wilson and Magnuson, in review). Rusty crayfish introduction is associated with declines in aquatic plant biomass and species richness, low densities of snails, and overall reductions in littoral fish abundance (Lodge and Lorman, 1987; Olsen et al., 1991; Lodge et al., 2000; Wilson, 2002). In summary, while rusty crayfish are no longer used as live bait and are a relatively slow natural disperser, the majority of Shield lakes are potentially vulnerable to rusty crayfish colonization. Furthermore, rusty crayfish tend to have dramatic, ecosystem-altering impacts and as such should be of high concern for lake managers.

Daphnia lumholtzi

This large, spiny cladorceran zooplankton species first appeared in North America in 1989 and has spread extremely rapidly through reservoirs in the southern and midwestern United States (Havel and Hebert, 1993), reaching Lake Erie by 1999 (Muzinic, 2000). The live wells, bait buckets, and bilge water of recreational anglers are thought to be the major transport vector for this exotic zooplankter (Dzialowski et al., 2000; Havel and Stelzleni-Schwent, 2000).

Feeding trials with juvenile bluegill sunfish indicated that *D. lumholtzi* were far less edible due to their spiny morphology than native *Daphnia* of the same size (Swaffar and O'Brien, 1996). The effect was particularly evident among the smallest bluegill (20 to 25 mm), indicating potential impacts on growth and survivorship of YOY fish. *Daphnia lumholtzi* prefers warm waters and is often abundant during middle-to-late summer in southern reservoirs (Havel et al., 1995; Kolar et al., 1997; East et al., 1999; Lennon et al., 2001). In laboratory studies, *D. lumholtzi* outperforms other *Daphnia* species in water between 25° and 30°C, although they can survive and reproduce at much cooler temperatures (Lennon et al., 2001) and are occasionally found in Missouri reservoirs during late fall (J. Havel, October, 2002, personal communication). These findings suggest that while some Shield lakes may support *D. lumholtzi* populations, they are not likely to thrive or dominate the zooplankton community in the cold waters of Shield lakes.

Other exotics

Another concern is that the Laurentian Great Lakes presently harbor a number of nonnative fish such as ruffe *Gymnocephalus cernuus*, round goby *Neogobius melanostomus*, and tubenose goby *Proterorhinus marmoratus* (Jude et al., 1992). While there is certainly potential for these fish species to be transported to inland lakes via angler's bait buckets (Litvak and Mandrak, 1993; Litvak and Mandrak, 1999) or even through natural dispersal, such occurrences are expected to be rare, and these species should spread at a slow rate. There are very few occurrences of these species in inland lakes, and we do not consider further the occurrences and impacts of these nonnative fish.

This chapter focuses on animal invaders in Shield lakes and as such does not consider invasive aquatic plants such as Eurasian watermilfoil *Myriophyllum spicatum* invasions. Invasive aquatic plants have the potential to alter the physical structure of the littoral zone and represent potentially large perturbations to lake ecosystems (Carpenter and Lodge, 1986). Eurasian watermilfoil has recently become problematic in acid lakes of eastern New England (Les and Mehrhoff, 1999; Crow and Hellquist, 2000) and central and northern Ontario and may continue to spread among Shield lakes. Eurasian watermilfoil can reproduce via fragments and winterbud — as well as seed and expansion — enabling easy transport via boats, boat trailers, and live wells.

Conclusions

Perhaps with the exception of *D. lumholtzi*, the invasive species highlighted in this chapter all have potential for dramatic impacts on Shield lake ecosystems. The first three invaders (bass, rainbow smelt, and *Bythotrephes*) are already a problem and are almost certain to continue to colonize new systems in the future, particularly in light of predicted global climate change scenarios. In addition to the handful of invaders considered here, there will undoubtedly be new waves of invaders that will establish in the future, and prediction of their identity is an important goal (Ricciardi and Rasmussen, 1998; Kolar and Lodge, 2001). For the exotic species it is critical to recognize that colonization of the Laurentian Great Lakes serves as a stepping stone for the eventual colonization of inland lakes. The Great Lakes can thus be viewed as source populations from which the species disperse outward to the smaller inland lakes. If the past is any predictor of the future, subsequent invaders of Shield lakes will continue to be a subset of the Great Lakes invader pool.

At timescales of decades to centuries, the identity and rates of nonnative invasions in Shield lakes will depend on a number of factors: the establishment of invaders in the Great Lakes, the trajectory of climate change, patterns of residential development and recreational boating (Padilla et al., 1996), educational and outreach efforts, and the development of legislation restricting the transport and introduction of nonnative species. In this chapter we have emphasized the importance of developing models to predict the occurrences and impacts of specific invaders in individual lakes. These predictions will ultimately derive from an understanding of the biology and physiological tolerances of the invader, attributes of the receiving community and ecosystem, and the often-subtle interactions between them. This approach can yield models to predict the location and impacts of invaders (Ramcharan et al., 1992; Koutnik and Padilla, 1994; Buchan and Padilla, 2000; MacIsaac et al., 2000; Vander Zanden et al., in press). By identifying vulnerable lakes and regions, management efforts aimed at preventing future invasions can be most efficiently focused. Indeed, prevention of nonnative introductions remains the most effective invasive management strategy because once an aquatic invader becomes established elimination of the invader is difficult, and further colonization of surrounding systems commonly follows (Hrabik and Magnuson, 1999). While prevention is imperative to stem the tide of invasions, we also note that serious research effort has not yet been invested into methods for controlling or eliminating aquatic invaders.

Shield lakes and their watersheds are subject to a diverse range of anthropogenic impacts, many of which are documented elsewhere in this volume. Introductions of nonnative species are clearly among the leading threats to lake ecosystems and in fact were recently noted as the leading threat to biodiversity in lakes (Sala et al., 2000). The diversity of threats and their potential interactions pose difficult management challenges that require adaptation to our increasing understanding and the ever-changing nature of the threats.

An understanding of some of the interactions between these diverse impacts is starting to emerge. Climate warming is predicted to promote further bass introductions, with potentially massive impacts on minnow diversity in Shield lakes (e.g., Jackson and Mandrak, 2002). Climate change, increased ultraviolet radiation, and acid deposition have cumulative and perhaps synergistic effects on boreal landscapes (D.W. Schindler, 1998). While a single species introduction may or may not be sufficient to cause an observable adverse impact on a Shield lake ecosystem, addition of multiple invaders, along with a variety of other simultaneous stressors, expands the scope for impacts dramatically. It becomes rapidly apparent that the greatest gap in our understanding is that of cumulative impacts and the potential interactions among multiple stressors. Addressing this issue is the greatest challenge for the future management of Shield lake ecosystems.

Acknowledgments

Norman Mercado-Silva, Helen Sarakinos, John Havel, and John Magnuson provided helpful comments on an early draft of this manuscript. Bill Feeny provided graphical assistance and artwork. The work of J.V.Z. was partially funded by a Nature Conservancy David H. Smith postdoctoral fellowship. This is Publication #DHS 2003-3 of the David H. Smith publication series.

References

Akielaszek, J.J. and Haines, T.A., 1981, Mercury in the muscle tissue of fish from three Northern Maine lakes, *Bulletin of Environmental Contamination and Toxicology,* **27**: 201–208.

Bergstedt, R.A., 1983, Notes on the origins of rainbow smelt in Lake Ontario, *Journal of Great Lakes Research,* **9**: 582–583.

Berrill, M., 1978, Distribution and ecology of crayfish in the Kawartha Lakes region of southern Ontario, *Canadian Journal of Zoology,* **56**: 166–177.

Berrill, M., Hollett, L., Margosian, A., and Hudson, J., 1985, Variation in tolerance to low environmental pH by the crayfish *Orconectes rusticus, O. propinquus,* and *Cambarus robustus, Canadian Journal of Zoology,* **63**: 2586–2589.

Buchan, L.A.J. and Padilla, D.K., 2000, Predicting the likelihood of Eurasian watermilfoil presence in lakes, a macrophyte monitoring tool, *Ecological Applications,* **10**: 1442–1455.

Capelli, G.M., 1982, Displacement of northern Wisconsin crayfish by *Orconectes rusticus* (Girard), *Limnology and Oceanography,* **27**: 741–745.

Capelli, G.M. and Magnuson, J.J., 1983, Morphoedaphic and biogeographical analysis of crayfish distribution in northern Wisconsin, *Journal of Crustacean Biology,* **3**: 548–564.

Carpenter, S.R. and Lodge, D.M., 1986, Effects of submersed macrophytes on ecosystem processes, *Aquatic Botany,* **26**: 341–470.

Chapleau, F., Findlay, C.S., and Szenasy, E., 1997, Impact of piscivorous fish introductions on fish species richness of small lakes in Gatineau Park, Quebec, *Ecoscience,* **4**: 259–268.

Charlebois, P.M., Raffenberg, M.J., and Dettmers, J.M., 2001, First occurrence of *Cercopagis pengoi* in Lake Michigan, *Journal of Great Lakes Research,* **27**: 258–261.

Christie, W.J., 1974, Changes in the fish species composition of the Great Lakes, *Journal of Fisheries Research Board of Canada,* **31**: 827–854.

Claudi, R. and Leach, J.H., 1999, *Nonindigenous Freshwater Organisms,* Lewis, Boca Raton, FL.

Coblentz, B.E., 1990, Exotic organisms: a dilemma for conservation biology, *Conservation Biology,* **4**: 261–265.

Cohen, A.N. and Carlton, J.T., 1998, Accelerating invasion rate in a highly invaded estuary, *Science,* **279**: 555–558.

Coulas, R.A., MacIsaac, H.J., and Dunlop, W., 1998, Selective predation on an introduced zooplankter (*Bythotrephes cederstroemi*) by lake herring (*Coregonus artedi*) in Harp Lake, Ontario, *Freshwater Biology,* **40**: 343–355.

Courchamp, F., Langlais, M., and Sugihara, G., 2000, Rabbits killing birds: modelling the hyperpredation process, *Journal of Animal Ecology,* **69**: 154–164.

Crow, G.E. and Hellquist, C.B., 2000, *Aquatic and Wetland Plants of Northeastern North America,* University of Wisconsin Press, Madison.

Destasio, B.T., Hill, D.K., Kleinhans, J.M., Nibblelink, N.P., and Magnuson, J.J., 1996, Potential effects of global climate change on small north temperate lakes: physics, fishes and plankton, *Limnology and Oceanography,* **41**: 1136–1149.

Dymond, J.R., 1944. Spread of the smelt, *Osmerus mordax,* in the Canadian waters of the Great Lakes, *Canadian Field-Naturalist,* **58**: 12–14.

Dzialowski, A.R., O'Brien, W.J., and Swaffar, S.M., 2000, Range expansion and potential dispersal mechanisms of the exotic cladoceran *Daphnia lumholtzi, Journal of Plankton Research,* **22**: 2205–2223.

East, T.L., Havens, K.E., Rodusky, A.J., and Brady, M.A., 1999, *Daphnia lumholtzi* and *Daphnia ambigua*: population comparisons of an exotic and a native cladoceran in Lake Okeechobee, Florida, *Journal of Plankton Research*, **21**: 1537–1551.

Evans, D.O. and Loftus, D.H., 1987, Colonization of inland lakes in the Great Lakes region by rainbow smelt, *Osmerus mordax*: their freshwater niche and effects on indigenous fishes, *Canadian Journal of Fisheries and Aquatic Sciences*, **44**(Suppl. 2): 249–266.

Evans, D.O. and Waring, P., 1987, Changes in the multispecies, winter angling fishery of Lake Simcoe, Ontario, 1961–83: invasion by rainbow smelt, *Osmerus mordax*, and the role of intra- and interspecific interactions, *Canadian Journal of Fisheries and Aquatic Sciences*, **44**(Suppl. 2): 182–197.

Findlay, C.S., Bert, D.G., and Zheng, L., 2000, Effect of introduced piscivores on native minnow communities in Adirondack lakes, *Canadian Journal of Fisheries and Aquatic Sciences*, **57**: 570–580.

Franzin, W.G., Barton, B.A., Remnant, R.A., Wain, D.B., and Pagel, S.J., 1994. Range extension, present and potential distribution, and possible effect of rainbow smelt on Hudson Bay drainage waters of Northwestern Ontario, Manitoba, and Minnesota, *North American Journal of Fisheries Management*, **14**: 65–76.

Goldman, C.R., Morgan, M.D., Threlkeld, S.T., and Angeli, N., 1979, A population dynamics analysis of the cladoceran disappearance from Lake Tahoe, California–Nevada, *Limnology and Oceanography*, **24**: 289–297.

Havel, J.E. and Hebert, P.D.N., 1993, *Daphnia lumholtzi* in North America — another exotic zooplankter, *Limnology and Oceanography*, **38**: 1823–1827.

Havel, J.E., Mabee, W.R., and Jones, J.R., 1995, Invasion of the exotic cladoceran *Daphnia lumholtzi* into North American reservoirs, *Canadian Journal of Fisheries and Aquatic Sciences*, **52**: 151–160.

Havel, J.E. and Stelzleni-Schwent, J., 2000, Zooplankton community structure: the role of dispersal, *Internationale Vereinigung fur Theoretische und Angewandte Limnologie*, **27**: 3264–3268.

Hodgson, J.R., Hodgson, C.J., and Brooks, S.M., 1991, Trophic interaction and competition between largemouth bass (*Micropterus salmoides*) and rainbow trout (*Oncorhynchus mykiss*) in a manipulated lake, *Canadian Journal of Fisheries and Aquatic Sciences*, **48**: 1704–1712.

Hodgson, J.R. and Kitchell, J.F., 1987, Opportunistic foraging by largemouth bass (*Micropterus salmoides*), *American Midland Naturalist*, **118**: 323–336.

Hoffman, J.C., Smith, M.E., and Lehman, J.T., 2001, Perch or plankton: top down control of *Daphnia* by yellow perch (*Perca flavescens*) or *Bythotrephes cederstroemi* in an inland lake? *Freshwater Biology*, **46**: 759–775.

Hrabik, T.R. and Magnuson, J.J., 1999, Simulated dispersal of exotic rainbow smelt (*Osmerus mordax*) in a northern Wisconsin lake district and implications for management, *Canadian Journal of Fisheries and Aquatic Sciences*, **56**: 35–42.

Hrabik, T.R., Magnuson, J.J., and McLain, A.S., 1998, Predicting the effects of rainbow smelt on native fishes in small lakes: evidence from long-term research on two lakes, *Canadian Journal of Fisheries and Aquatic Sciences*, **55**: 1364–1371.

Jackson, D.A. and Mandrak, N.E., 2002, Changing fish biodiversity: predicting the loss of cyprinid biodiversity due to global climate change, *American Fisheries Society Symposium*, **32**: 89–98.

Johnson, B.M. and Goettl, J.P., Jr., 1999, Food web changes over 14 years following introduction of rainbow smelt into a Colorado Reservoir, *North American Journal of Fisheries Management*, **19**: 629–642.

Jones, M.S., Goettl, J.P., Jr., and Flickinger, S.A., 1994, Changes in walleye food habits and growth following a rainbow smelt introduction, *North American Journal of Fisheries Management*, **14**: 409–414.

Jude, D.J., Reider, R.H., and Smith, G.R., 1992, Establishment of Gobiidae in the Great Lakes basin, *Canadian Journal of Fisheries and Aquatic Sciences*, **49**: 416–421.

Karatayev, A.Y., Burlakova, L.E., and Padilla, D.K., 1998, Physical factors that limit the distribution and abundance of *Dreissena polymorpha* (Pallas), *Journal of Shellfish Research*, **17**: 1219–1235.

Keller, W., Dixit, S.S., and Heneberry, J., 2001, Calcium declines in northeastern Ontario lakes, *Canadian Journal of Fisheries and Aquatic Sciences*, **58**: 2011–2020.

Kershner, M.W. and Lodge, D.M., 1995, Effects of littoral habitat and fish predation on the distribution of an exotic crayfish, *Orconectes rusticus, Journal of the North American Benthological Society,* **14**: 414–422.

Kolar, C.S., Boase, J.C., Clapp, D.F., and Wahl, D.H., 1997, Potential effect of invasion by an exotic zooplankter, *Daphnia lumholtzi, Journal of Freshwater Ecology,* **12**: 521–530.

Kolar, C.S. and Lodge, D.M., 2001, Progress in invasion biology: predicting invaders, *Trends in Ecology and Evolution,* **16**: 199–204.

Koutnik, M.A. and Padilla, D.K., 1994, Predicting the spatial distribution of *Dreissena polymorpha* (zebra mussel) among inland lakes of Wisconsin: modeling with a GIS, *Canadian Journal of Fisheries and Aquatic Sciences,* **51**: 1189–1196.

Lee, D.S., Gilbert, C.R., Hocutt, C.H., Jenkins, R.E., McAlister, D.E., and Stauffer, J.R., 1980, *Atlas of North American Freshwater Fishes,* North Carolina State Museum of Natural History, Raleigh.

Lehman, J.T., 1988, Algal biomass unaltered by food-web changes in Lake Michigan, *Nature,* **332**: 537–538.

Lehman, J.T. and Caceres, C.E., 1993, Food-web responses to species invasion by a predatory invertebrate: *Bythotrephes* in Lake Michigan, *Limnology and Oceanography,* **38**: 879–891.

Lennon, J.T., Smith, V.H., and Williams, K., 2001, Influence of temperature on exotic *Daphnia lumholtzi* and implications for invasion success, *Journal of Plankton Research,* **23**: 425–434.

Les, D.H. and Mehrhoff, L.J., 1999, Introduction of nonindigenous aquatic vascular plants in southern New England: a historical perspective, *Biological Invasions,* **1**: 281–300.

Litvak, M.K. and Mandrak, N.E., 1993, Ecology of freshwater baitfish use in Canada and the United States, *Fisheries,* **18**: 6–13.

Litvak, M.K. and Mandrak, N.E., 1999, Baitfish trade as a vector of aquatic introductions, in *Nonindigenous Freshwater Organisms: Vectors, Biology, and Impacts,* Claudi, R. and Leach, J.H., Eds., Lewis, Boca Raton, FL, pp. 163–180.

Lodge, D.M., 1993, Biological invasions: lessons for ecology, *Trends in Ecology and Evolution,* **8**: 133–137.

Lodge, D.M. and Lorman, J.G., 1987, Reductions in submersed macrophyte biomass and species richness by the crayfish *Oronectes rusticus, Canadian Journal of Fisheries and Aquatic Sciences,* **44**: 591–597.

Lodge, D.M., Taylor, C.A., Holdich, D.M., and Skurdal, J., 2000, Nonindigenous crayfishes threaten North American freshwater biodiversity: lessons from Europe, *Fisheries,* **25**: 7–19.

Loftus, D.H. and Hulsman, P.F., 1986, Predation on larval lake whitefish (*Coregonus clupeaformis*) and lake herring (*C. artedi*) by adult rainbow smelt (*Osmerus mordax*), *Canadian Journal of Fisheries and Aquatic Sciences,* **43**: 812–818.

Ludwig, H.R., Jr. and Leitch, J.A., 1996, Interbasin transfer of aquatic biota via angler's bait buckets, *Fisheries,* **21**: 14–18.

MacCrimmon, H.R. and Robbins, W.H., 1975, Distribution of the black basses in North America, in *Black Bass Biology and Management,* Stroud, R.H. and Klepper, H., Eds., Sport Fishing Institute, Washington, DC, pp. 56–66.

MacCrimmon, H.R., Wren, C.D., and Gots, B.L., 1983, Mercury uptake by lake trout, *Salvelinus namaycush,* relative to age, growth, and diet in Tadenac Lake with comparative data from other Precambrian Shield lakes, *Canadian Journal of Fisheries and Aquatic Sciences,* **40**: 114–120.

MacIsaac, H.J., 1996, Potential abiotic and biotic impacts of zebra mussels on the inland waters of North America, *American Zoologist,* **36**: 287–299.

MacIsaac, H.J., Grigorovich, I.A., Hoyle, J.A., Yan, N.D., and Panov, V.E., 1999, Invasion of Lake Ontario by the Ponto-Caspian predatory cladoceran *Cercopagis pengoi, Canadian Journal of Fisheries and Aquatic Sciences,* **56**: 1–5.

MacIsaac, H.J., Ketelaars, H.A.M., Grigorovich, I.A., Ramcharan, C.W., and Yan, N.D., 2000, Modeling *Bythotrephes longimanus* invasion in the Great Lakes basin based on its European distribution, *Archives of Hydrobiology,* **149**: 1–21.

Magnuson, J.J., 1976, Managing with exotics — a game of chance, *Transactions of the American Fisheries Society,* **105**: 1–9.

Magnuson, J.J., Capelli, G.M., Lorman, J.G., and Stein, R.A., 1975, Consideration of crayfish for macrophyte control, in *The Proceedings of a Symposium on Water Quality Management,* Brezonik, P.L. and Fox, J.L., Eds., University of Florida, Gainesville, pp. 66–74.

Magnuson, J.J., Webster, K.E., Assel, R.A., Bowser, C.J., Dillon, P.J., Eaton, J.G., Evans, H.E., Fee, E.J., Hall, R.I., Mortsch, L.R., Schindler, D.W., and Quinn, F.H., 1997, Potential effects of climate changes on aquatic ecosystems: Laurentian Great Lakes and Precambrian Shield region, in *Freshwater Ecosystems and Climate Change in North America: A Regional Assessment*, Cushing, C.E., Ed., John Wiley & Sons, New York, ch. 2, pp. 7–8.

Mandrak, N.E., 1989, Potential invasion of the Great Lakes by fish species associated with climatic warming, *Journal of Great Lakes Research*, **15**: 306–316.

Martin, N.V., 1970, Long-term effects of diet on the biology of the lake trout and the fishery in Lake Opeongo, Ontario, *Journal of the Fisheries Research Board of Canada*, **27**: 125–146.

Martin, N.V. and Fry, F.E.J., 1972, Lake Opeongo: effects of exploitation and introductions on the salmonid community, *Journal of the Fisheries Research Board of Canada*, **29**: 795–805.

Mathers, R.A. and Johansen, P.H., 1985, The effects of feeding ecology on mercury accumulation in walleye (*Stizostedion vitreum*) and pike (*Esox lucius*) in Lake Simcoe, *Canadian Journal of Zoology*, **63**: 2006–2012.

Mayden, R.L., Cross, F.B., and Gorman, O.T., 1987, Distributional history of the rainbow smelt, *Osmerus mordax* (Salmoniformes: Osmeridae), in the Mississippi River basin, *Copeia*, May 1987, **2**: 1051–1054.

McCann, K., Hastings, A., and Huxel, G.R., 1998, Weak trophic interactions and the balance of nature, *Nature*, **395**: 794–798.

McDowall, R.M., 1968, The proposed introduction of the largemouth bass *Micropterus salmoides* (Lacepede) into New Zealand, *New Zealand Journal of Marine and Freshwater Research*, **2**: 149–161.

McNeill, A.J., 1995, An overview of the smallmouth bass in Nova Scotia, *North American Journal of Fisheries Management*, **15**: 680–687.

Mellina, E. and Rasmussen, J.B., 1994, Patterns in the distribution and abundance of zebra mussel (*Dreissena polymorpha*) in rivers and lakes in relation to substrate and other physicochemical factors, *Canadian Journal of Fisheries and Aquatic Sciences*, **51**: 1024–1036.

Mills, E.L., Leach, J.H., Carlton, J.T., and Secor, C.L., 1994, Exotic species and the integrity of the Great Lakes, *BioScience*, **44**: 666–676.

Mills, E.L., O'Garman, R., DeGisi, J., Heberger, R.F., and House, R.A., 1992, Food of the alewife (*Alosa pseudoharengus*) in Lake Ontario before and after the establishment of *Bythotrephes cederstroemi*, *Canadian Journal of Fisheries and Aquatic Sciences*, **49**: 2009–2019.

Mittelbach, G.G., Turner, A.M., Hall, D.J., Rettig, J.E., and Osenberg, C.W., 1995, Perturbation and resilience: a long-term, whole-lake study of predator extinction and reintroduction, *Ecology*, **76**: 2347–2360.

Momot, W.T., 1995, Redefining the role of crayfish in aquatic systems, *Reviews in Fisheries Science*, **3**: 33–63.

Moyle, P.B., 1986, Fish introductions in North America: patterns and ecological impact, in *Ecology of Biological Invasions of North America and Hawaii*, Mooney, H.A. and Drake, J.A., Eds., Springer-Verlag, New York, pp. 27–43.

Muzinic, C.J., 2000, First record of *Daphnia lumholtzi* Sars in the Great Lakes, *Journal of Great Lakes Research*, **26**: 352–354.

Naiman, R.J., Magnuson, J.J., McKnight, D.M., and Stanford, J.A., 1995, *The Freshwater Imperative: A Research Agenda*, Island Press, Washington, DC.

Neary, B.P. and Leach, J.H., 1992, Mapping the potential spread of the zebra mussel (*Dreissena polymorpha*) in Ontario, *Canadian Journal of Fisheries and Aquatic Sciences*, **49**: 406–415.

Olden, J.D. and Jackson, D.A., 2001, Fish-habitat relationships in lakes: gaining predictive and explanatory insight by using artificial neural networks, *Transactions of the American Fisheries Society*, **130**: 878–897.

Olsen, T.M., Lodge, D.M., Capelli, G.M., and Houlihan, R.J., 1991, Mechanisms of impact of introduced crayfish (*Orconectes rusticus*) on littoral congeners, snails, and macrophytes, *Canadian Journal of Fisheries and Aquatic Sciences*, **48**: 1853–1861.

Olver, C.H., DesJardine, R.L., Goddard, C.I., Powell, M.J., Rietveld, H.J., and Waring, P.D., 1991, *Lake Trout in Ontario: Management Strategies*, Lake Trout Synthesis, Ontario Ministry of Natural Resources, Toronto.

Padilla, D.K., Chotkowski, M.A., and Buchan, L.A.J., 1996, Predicting the spread of zebra mussels (*Dreissena polymorpha*) to inland waters using boater movement patterns, *Global Ecology and Biogeography Letters*, **5**: 353–359.

Pimentel, D., Lach, L., Zuniga, R., and Morrison, D., 2000, Environmental and economic costs of nonindigenous species in the United States, *BioScience*, **50**: 53–64.

Pimm, S.L., 1991, *The Balance of Nature?* University of Chicago Press, IL.

Rahel, F.J., 2000, Homogenization of fish faunas across the United States, *Science*, **288**: 854–856.

Ramcharan, C.W., Padilla, D.K., and Dodson, S.I., 1992, Models to predict potential occurrence and density of zebra mussel, *Dreissena polymorpha*, *Canadian Journal of Fisheries and Aquatic Sciences*, **49**: 2611–2620.

Remnant, R.A., Graveline, P.G., and Bretecher, R.L., 1997, Range extension of the rainbow smelt, *Osmerus mordax*, in the Hudson Bay drainage of Manitoba, *Canadian Field Naturalist*, **111**: 660–662.

Ricciardi, A. and MacIsaac, H.J., 2000, Recent mass invasion of the North American Great Lakes by Ponto-Caspian species, *Trends in Ecology and Evolution*, **15**: 62–65.

Ricciardi, A. and Rasmussen, J.B., 1998, Predicting the identity and impact of future biological invaders: a priority for aquatic resource management, *Canadian Journal of Fisheries and Aquatic Sciences*, **55**: 1759–1765.

Richards, R.C., Goldman, C.R., Frantz, T.C., and Wickwire, R., 1975, Where have all the Daphnia gone? The decline of a major cladoceran in Lake Tahoe, California–Nevada, *Internationale Vereinigung fur Theoretische und Angewandte Limnologie*, **19**: 835–842.

Robbins, W.H. and MacCrimmon, H.R., 1974, *The Black Bass in America and Overseas*, Biomanagement and Research Enterprises, Sault Ste. Marie, Ontario.

Ruiz, G.M., Fofonoff, P., Hines, A.H., and Grosholz, E.D., 1999, Non-indigenous species as stressors in estuarine and marine communities: assessing invasion impacts and interactions, *Limnology and Oceanography*, **44**: 950–972.

Sala, O.E., Chapin, F.S., Armesto, J.J., Berlow, E., Bloomfield, J., Dirzo, R., Huber-Sanwald, E., Huenneke, L.F., Jackson, R.B., Kinzig, A., Leemans, R., Lodge, D.M., Mooney, H.A., Oesterheld, M., Poff, N.L., Sykes, M.T., Walker, B.H., Walker, M., and Wall, D.H., 2000, Biodiversity: global biodiversity scenarios for the year 2100, *Science*, **287**: 1770–1774.

Schindler, D.E., Hodgson, J.R., and Kitchell, J.F., 1997, Density-dependent changes in individual foraging specialization of largemouth bass, *Oecologia*, **110**: 592–600.

Schindler, D.W., 1998, A dim future for boreal waters and landscapes, *BioScience*, **48**: 157–164.

Schulz, K.L. and Yurista, P.M., 1998, Implications of an invertebrate predator's (*Bythotrephes cederstroemi*) atypical effects on a pelagic zooplankton community, *Hydrobiologia*, **380**: 179–193.

Scott, W.B. and Crossman, E.J., 1973, Freshwater fishes of Canada, *Bulletin of the Fisheries Research Board of Canada*, **184**: 1–996.

Shuter, B.J., Ihssen, P.E., Wales, D.L., and Snucins, E.J., 1989, The effect of temperature, pH and water hardness on winter starvation of young-of-the-year smallmouth bass (*Micropterus dolomieu* Lacepede), *Journal of Fish Biology*, **35**: 765–780.

Shuter, B.J., MacLean, J.A., Fry, F.E.J., and Regier, H.A., 1980, Stochastic simulation of temperature effects on first-year survival of smallmouth bass, *Transactions of the American Fisheries Society*, **109**: 1–34.

Shuter, B.J. and Post, J.R., 1990, Climate, population viability, and the zoogeography of temperate fishes, *Transactions of the American Fisheries Society*, **119**: 314–336.

Soule, M.E., 1990, The onslaught of alien species, and other challenges in the coming decades, *Conservation Biology*, **4**: 233–239.

Spencer, C.N., McClelland, B.R., and Stanford, J.A., 1991, Shrimp stocking, salmon collapse, and eagle displacement, *BioScience*, **41**: 14–21.

Sprung, M., 1987, Ecological requirements of developing *Dreissena polymorpha* eggs, *Archiv fur Hydrobiologie Supplement*, **79**: 69–86.

Stein, R.A. and Magnuson, J.J., 1976, Behavioral response of crayfish to a fish predator, *Ecology*, **57**: 751–761.

Stoddard, J.L., Jeffries, D.S., Lukewille, A., Clair, T.A., Dillon, P.J., Driscoll, C.T., Forsius, M., Johannessen, M., Kahl, J.S., Kellogg, J.H., Kemp, A., Mannio, J., Monteith, D.T., Murdoch, P.S., Patrick, S., Rebsdorf, A., Skjelkvale, B.L., Stainton, M.P., Traaen, T., van Dam, H., Webster, K.E., Wieting, J., and Wilander, A., 1999, Regional trends in aquatic recovery from acidification in North America and Europe, *Nature*, **401**: 575–578.

Strayer, D.L., Caraco, N.F., Cole, J.J., Findlay, S., and Pace, M.L., 1999, Transformation of freshwater ecosystems by bivalves, *BioScience*, **49**: 19–27.

Swaffar, S.M. and O'Brien, W.J., 1996, Spines of *Daphnia lumholtzi* create feeding difficulties for juvenile bluegill sunfish (*Lepomis macrochirus*), *Journal of Plankton Research*, **18**: 1055–1061.

Vander Zanden, M.J., Cabana, G., and Rasmussen, J.B., 1997, Comparing the trophic position of littoral fish estimated using stable nitrogen isotopes ($\delta^{15}N$) and dietary data, *Canadian Journal of Fisheries and Aquatic Sciences*, **54**: 1142–1158.

Vander Zanden, M.J., Casselman, J.M., and Rasmussen, J.B., 1999, Stable isotope evidence for the food web consequences of species invasions in lakes, *Nature*, **401**: 464–467.

Vander Zanden, M.J., Olden, J.D., Thorne, J.H., and Mandrak, N.E., in press, Predicting the occurrences and impacts of bass introductions on temperate lake food webs, *Ecological Applications*.

Vander Zanden, M.J. and Rasmussen, J.B., 1996, A trophic position model of pelagic food webs: impact on contaminant bioaccumulation in lake trout, *Ecological Monographs*, **66**: 451–477.

Vander Zanden, M.J. and Rasmussen, J.B., 2002, Food web perspectives on studies of bass populations in north-temperate lakes. In *Black Bass: Ecology, Conservation and Management*, Philipp, D.P. and Ridgway, M.S., Eds., American Fisheries Society Symposium, Bethesda, MD, **31**: 173–184.

Vander Zanden, M.J. and Vadeboncoeur, Y., 2002, Fishes as integrators of benthic and pelagic food webs in lakes, *Ecology.* **83**: 2152–2161.

Welcomme, R.L., 1988, *International Introduction of Inland Aquatic Species*, Food & Agriculture Organization of the United Nations, Rome Fisheries Technical Paper **294**: 1–318.

Whittier, T.R. and Kincaid, T.M., 1999. Introduced fish in Northeastern USA lakes: regional extent, dominance, and effects on native species richness, *Transactions of the American Fisheries Society*, **128**: 769–783.

Wilcove, D.S. and Bean, M.J., 1994, *The Big Kill: Declining Biodiversity in America's Lakes and Rivers*, Environmental Defense Fund, Washington, DC.

Wilson, K.A., 2002, Impacts of the Invasive Rusty Crayfish (*Orconectes rusticus*) in Northern Wisconsin Lakes, doctoral thesis, University of Wisconsin, Madison.

Wilson, K.A. and Magnuson, J.J., in review, Long-term transformation of a submersed aquatic macrophyte community by an invading crayfish (*Orconectes rusticus*): patterns of persistence.

Yan, N.D., Blukacz, A., Sprules, W.G., Kindy, P.K., Hackett, D., Girard, R.E., and Clark, B.J., 2001, Changes in zooplankton and the phenology of the spiny water flea, *Bythotrephes*, following its invasion of Harp Lake, Ontario, Canada, *Canadian Journal of Fisheries and Aquatic Sciences*, **58**: 2341–2350.

Yan, N.D., Dunlop, W.I., Pawson, T.W., and MacKay, L.E., 1992, *Bythotrephes cederstroemi* (Schoedler) in Muskoka lakes: first records of the European invader in inland lakes in Canada, *Canadian Journal of Fisheries and Aquatic Sciences*, **49**: 422–426.

Yan, N.D., Girard, R., and Bourdreau, S., 2002, An introduced invertebrate predator (*Bythotrephes*) reduces zooplankton species richness, *Ecology Letters*, **5**: 481–485.

Yan, N.D. and Pawson, T.W., 1997, Changes in the crustacean zooplankton community of Harp Lake, Canada, following invasion by *Bythotrephes cederstroemi*, *Freshwater Biology*, **37**: 409–425.

Yan, N.D. and Pawson, T.W., 1998, Seasonal variation in the size and abundance of the invading *Bythotrephes* in Harp Lake, Ontario, Canada, *Hydrobiology*, **361**: 157–168.

Zaret, T.M. and Paine, R.T., 1973, Species introduction in a tropical lake, *Science*, **182**: 449–455.

chapter fourteen

Effects of forestry roads on reproductive habitat and exploitation of lake trout*

John M. Gunn
Ontario Ministry of Natural Resources, Laurentian University
Rod Sein
Ontario Ministry of the Environment

Contents

Introduction

There are now very few parts of the Boreal Shield area of Ontario that are remote from access roads (Figure 14.1). This network of roads and associated human-directed impacts represents a largely unplanned legacy from forestry, mining, and other resource extraction industries. However, to date, there have been few attempts to measure the impacts of road-related effects on aquatic ecosystems. To do this, we adopted an experimental

* Modified from Gunn, J.M. and Sein, R., 2000, Effects of forestry roads on reproductive habitat and exploitation of lake trout (*Salvelinus namaycush*) in three experimental lakes, *Canadian Journal of Fisheries and Aquatic Sciences*, 57(Suppl. 2): 97–104.

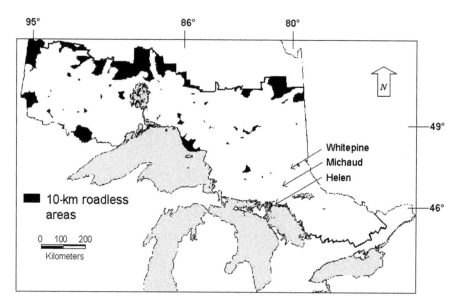

Figure 14.1 Areas within the outlined Boreal Shield landscape of Ontario, Canada, that are more than 10 km from roads. The locations of our study lakes are indicated. (Data provided by Ontario Ministry of Natural Resources.)

manipulation approach to test and compare two potential effects of forest access roads on fish populations: sedimentation and excessive exploitation.

We concentrated our studies on small lakes that supported naturally reproducing populations of lake trout *Salvelinus namaycush*, a species that receives special attention in forestry management plans, particularly in Ontario (Ontario Ministry of Natural Resources, 1988). Less than 1% of Ontario's lakes contain this prized sport fish species (Martin and Olver, 1976), and lake trout are considered highly sensitive to the disturbances that often accompany forestry operations. Under current guidelines in Ontario (Ontario Ministry of Natural Resources, 1988) all lake trout lakes must have terrestrial buffer strips (depending on slopes) of 30 to 90 m left around the entire shoreline, within which no road construction or tree harvesting is allowed. Road crossings of streams as potential point sources of sediment appear to represent more of a threat to nearshore spawning habitat of lake trout than silt inputs from clear-cut logging areas. Clear-cut areas in the low-relief landscape of much of the Boreal forest appear to have few, and probably quite temporary, effects on sediment transport to lakes (Blais et al., 1998; Steedman and France, 1999).

Spawning sites are identified as "critical fish habitat" in the guidelines and are considered essential areas that must be protected from sedimentation (Ontario Ministry of Natural Resources, 1988). The reproductive habitat is considered particularly vulnerable because the lake trout is a demersal spawner (Figure 14.2a and 14.2b) that broadcasts its eggs over clean, coarse substrates in very shallow (<2 m deep) nearshore (<10 m from shore) waters in small Shield lakes (Martin, 1957; DeRoche, 1969; McMurtry, 1986; Gunn, 1995). The developing embryos are deep in the interstitial spaces of the substrate for over 7 months (October to early May) and can readily be suffocated by sediments eroding from the catchment (Sly and Evans, 1996).

Ontario forestry guidelines and management plans usually give inadequate consideration to the very low productivity and high exploitation vulnerability typical of lake trout lakes (Ryder and Johnson, 1972; Shuter et al., 1998). Estimates of the sustainable harvest levels for lake trout populations are usually less than 1 kg·ha^{-1}·yr^{-1} (Healey, 1978; Martin and Olver, 1980). Shuter et al. (1998), using a simulation model they developed,

(a)

(b)

Figure 14.2 (a) Photo of spawning lake trout at Ox Narrows. (Photo by S. Skulason.) (b) Spawning lake trout on Whitepine Lake; note the clean coarse substrate where eggs are deposited. The egg collector is 30 cm in diameter. (Photo by S. McAugley.) See color figures following page 200.

predicted that small (100 ha) lakes on the Shield are more productive (maximum sustainable harvest of approximately 1.5 kg lake trout·ha^{-1}·yr^{-1}) than larger (1,000 to 10,000 ha) lakes, but are extremely vulnerable to overfishing. They estimated that small lakes might not be able to sustain more than approximately 7 h of fishing effort·ha^{-1}·yr^{-1} (Shuter et al., 1998).

We used three small lakes to test the effects of spawning habitat loss and exploitation on lake trout populations. In Whitepine Lake, we used opaque plastic sheeting and during 1992 to 1999 progressively covered available nearshore spawning habitat to simulate the effect of a sediment discharge to the lake (Figures 14.3 and 14.4). Helen Lake was added to the study in 1999 to verify the results of the habitat manipulation experiment using a lake with a more complex fish community. In Michaud Lake, we assessed the effects of exploitation on a remote lake when fishing resumed after a 7-year closure period, 1991 to 1997. Improved access to the lake was provided by the construction of a forest access road in 1994.

Methods

The three study lakes are all small (67- to 148-ha) headwater lakes located 40 to 90 km from Sudbury, Ontario (Figure 14.1), and are located within the Boreal Shield ecozone.

Figure 14.3 Aerial view of the habitat manipulation experiment. Opaque plastic sheeting was used to progressively cover lake trout spawning sites along the shore.

Figure 14.4 Installation of the plastic tarps to cover one of the largest of the traditional spawning sites on Whitepine Lake.

The two principal study lakes, Whitepine and Michaud, have very simple fish communities dominated by small-bodied lake trout, typical of populations with a mainly planktivorous or insectivorous diet. Other species present in the lakes include yellow perch *Perca flavescens*, Iowa darters *Etheostoma exile*, common white suckers *Catostomus commersoni*, and a few species of cyprinids. The lakes had previously been used in a variety of research projects related to acid rain (Gunn et al., 1987; Gunn and Keller, 1984, 1990) and had very similar histories. In the early 1980s they were both acidified (pH < 5.5), and their native lake trout populations were reduced to small remnant stocks of nonreproducing adults. Water quality improved throughout the 1980s because of reductions in emissions of SO_2 in Canada and the United States (Gunn and Keller, 1990).

Both principal study lakes were stocked with hatchery-reared lake trout to assist in rehabilitation. In Whitepine Lake, the hatchery stocking only occurred in 1980 and 1981 and then was stopped because the native population resumed reproduction in 1982. Reproduction occurred yearly thereafter, and a dense population of lake trout developed in Whitepine Lake by 1990. In Michaud Lake the native fish disappeared before natural recruitment resumed, and the hatchery stocking that began in 1984 was maintained periodically until 1992. Natural reproduction of the hatchery stocked fish in Michaud Lake began in 1990. A permanent sanctuary status was established for Whitepine Lake in 1980 to prevent angling exploitation. On Michaud Lake a temporary angling closure (1991 to 1997) was established.

Helen Lake (46°06′ N, 81°33′ E, 82 ha, 41.2 m maximum depth, 20.5 m mean depth) was a near-pristine lake located within the wilderness setting of Killarney Provincial Park and was used as an alternate lake to test the habitat manipulation technique. It has good water quality, a native reproducing lake trout population, and a complex fish community. In addition to lake trout it contains smallmouth bass *Micropterus dolomieu*, rock bass *Ambloplites rupestris*, cisco *Coregonus artedi*, brown bullhead *Ameriurus nebulosus*, bluegill *Lepomis macrochirus*, pumpkinseed *Lepomis gibbosus*, yellow perch, slimy sculpin *Cottus cognatus*, Iowa darter, and bluntnose minnow *Pimephales notatus*. Helen Lake is open to angling, but angling pressure is quite low because of its location and the fact that motorized access is prohibited.

Whitepine Lake: habitat loss experiment

Whitepine Lake was used as the principal experimental lake to test the hypothesis that spawning habitat reduction in shoreline areas would lead to recruitment failures in native lake trout populations. Previous publications from this experiment dealt with the behavioral response of fish to the habitat disturbances (McAughey and Gunn, 1995) and the accuracy of visual techniques of classifying spawning habitat (Gunn et al., 1996).

Whitepine Lake (47°17′ N, 80°50′ W) is a 67-ha lake (22 m maximum depth, 5.9 m mean depth) with a 4.7-km shoreline. It has a 328-ha terrestrial catchment area consisting of thin sandy soil over granitic bedrock and has a few small wetland areas. The area was logged in the past, and approximately 30% of the catchment was burned in 1975. At the start of this experiment in 1991 the forest cover was dominated by white pine *Pinus strobus* except for the burned area, where an early successional mixed cover existed. A band of trees and shrubs occurred along almost the entire shoreline, and there was no evidence of severe water-level fluctuations or excessive erosion or siltation. Few human disturbances existed in the area. Our small research cabin is the only building in the catchment area.

The habitat manipulation experiment began in October 1991 by mapping all of the traditional spawning sites used by lake trout. Spawning fish were located by cruising the entire shoreline each night of the spawning season (10 to 20 days in October) and spotting, with the aid of flashlights, spawning groups of lake trout over shallow (<2 m) nearshore areas of clean, coarse substrate. Associated studies have shown that lake trout in this lake select substrate that has a diameter of 2 to 10 cm (Gunn, 1995). Identified sites were later confirmed by examining the substrate for the presence of deposited eggs. In 1991 and 1992 egg deposition rates were measured using funnel collectors buried in the substrate (Gunn, 1995; Figure 14.2b). In subsequent years only the presence or absence of eggs was recorded. Frequent searches by boat and by diving eliminated the possibility of any undetected offshore sites in this study. This was also confirmed by tracking spawning fish tagged with ultrasonic transmitters.

The habitat manipulation began in 1992 by covering the spawning substrate with opaque plastic sheeting, which was left in place throughout the entire period of the experiment. The original spawning sites (seven sites, total surface area 40 m²) were removed as follows: 15% by area in 1992, 35% in 1993, 50% in 1994. Testing of the manipulation technique was done in 1995 by covering six previously used sites with sand as a more "natural" disturbance. Fish avoided sites covered with both plastic sheeting and sand. Therefore, only the plastic sheeting method was used to cover spawning sites in 1996, 1997, and 1998.

In addition to the annual assessment of spawning activity described above, we conducted detailed studies of annual abundance and size structure of the population. Population estimates were conducted in the spring (water temperature 6 to 15°C) by the continuous mark–recapture Schnabel method (Ricker, 1975). Fish were captured by angling or by using short-duration (30-min) sets of small mesh (38- to 51-mm stretched mesh) gill nets, and low handling mortality was confirmed through holding experiments (McAughey and Gunn, 1995). Accurate age assessment of the fish using otoliths was not possible because of the impact that lethal sampling for aging structures might cause. Instead, estimates of juvenile (<370 mm) abundance were made in terms of body size; these estimates were based on the assumption that the size-at-maturity relationships observed during the 1994 (Gunn et al., 1996) and 1997 spawning assessments applied throughout. To eliminate resampling the same fish, all captured-and-released juveniles were given a permanent adipose clip. To ensure detection of year-class failure, we attempted to capture and mark at least 100 juveniles each year.

The physical characteristics (depth, distance from shore, substrate size, and depth of interstitial spaces) of all new spawning sites were measured shortly after egg deposition ended. The location of the eggs was marked with a numbered brick, and a qualitative assessment of over-winter survival was then conducted by divers in late April to early May by excavating the sites and noting the presence or absence of live alevins.

Helen Lake: verification study

A test of the habitat manipulation technique was conducted in Helen Lake in 1999 to verify that the results in Whitepine Lake were not site specific or related to the absence of potential egg predators in the principal study site. Traditional spawning sites on Helen Lake were mapped in 1997 and 1998. In 1999 all five traditional spawning sites were covered with plastic sheeting, and the newly selected sites were located, marked, and assessed for alevin survival following the methods described above.

Michaud Lake: exploitation study

The exploitation experiment involved the assessment of the immediate impact of anglers on a remote lake (Michaud Lake) following construction of a forest access road and the lifting of the fishing ban. The final 12 km of a 26-km forest access road to Michaud Lake (46°49' N, 81°18' W, 148 ha, 24 m maximum depth, 7.0 m mean depth) was completed in 1994 providing access for snow machines and four-wheel-drive vehicles to within 100 m of the lake. In the summer of 1997 a netting survey was conducted to assess the relative abundance of lake trout in the lake. In the fall (October 1 to 30, 1997) a spawning assessment of adult lake trout in Michaud Lake was conducted, and 220 adults were fin marked and released.

On January 1, 1998, the fishing season opened under the standard regulations (unlimited entry and access, daily possession limit of three lake trout·angler⁻¹). No attempt was made to encourage fishing on the lake. There were no public notices in newspapers or elsewhere (the published regulations were actually in error and still indicated that the lake was closed to fishing), and there were no road signs to assist in locating the lake among the many other lakes and forest roads in the area. The newly constructed road was not plowed, so anglers had to use snowmobiles for the final 12 km to reach the lake. Once the fishing began, a random, stratified (by weekday or weekend day) creel survey was conducted to assess angling effort and harvest. The creel survey was maintained throughout the winter and early spring of 1998. Netting and spawning surveys were repeated in the summer and fall of 1998 to assess the effects of angling on the lake trout population.

Results and discussion

Habitat loss experiment: spawning site selection and quality

Lake trout proved to be highly adaptable to spawning habitat disturbances and repeatedly selected new sites in Whitepine Lake when previous spawning sites were covered. Similar results were found in Helen Lake, where nine new sites were selected after the traditional sites were covered (Table 14.1; Figure 14.5). In total, over 250 new spawning sites were selected in Whitepine Lake because of our experimental manipulations (Table 14.1; Figure 14.5). The accumulated impact of removing access to over 1600 m² of substrate during this 9-year experiment in Whitepine Lake did not prevent fish from spawning; however, it did appear to represent a severe enough impact that fish were forced to select what appeared to be marginal habitat. All the newly selected spawning sites in Whitepine Lake had at least a small patch of substrate within the preferred range (diameter of 2 to 10 cm), but many of the sites were very small (surface area <0.2 m²), had limited interstitial space beneath the substrate for eggs to settle (Figure 14.6), and were in shallow water (<0.4 m) where eggs appeared to be highly vulnerable to both predation and ice damage (Table 14.1). In the few relatively large sites, eggs were thinly dispersed and appeared to have drifted considerable distances before they became wedged within the substrate. The large number of widely dispersed sites (Figure 14.5) was further evidence of the severe disturbance imposed by the habitat removal. Lake trout usually spawn en masse (Figure 14.2a and 14.b) at relatively few sites (Martin, 1957; DeRoche, 1969; Gunn, 1995). It is rare to find an inland lake trout lake with more than ten traditional spawning sites, including lakes much larger than Whitepine (McMurtry, 1986).

In the early years of the study the trout exhibited strong fidelity to the seven traditional sites (Gunn, 1995; McAughey and Gunn, 1995). However, as the study proceeded it became evident that prior experience with a site proved unnecessary. Learned behaviors or chemosensory cues from previous use of a site (Foster, 1985; Hara, 1994) may assist a fish in returning to a site, but our study showed that the innate behavior of being able to readily identify usable habitat is very powerful in this species.

Evidence of recruitment failure

We were not able to accurately quantify egg deposition rate or survival rates of incubating embryos, but the new sites continued to produce alevins each year (Table 14.1), demonstrating that recruitment was not eliminated by the habitat disturbances. In the final year of the study, 21 of the 41 sites (51.2%) on Whitepine Lake contained live alevins on May 2, 2000. The presence of live alevins in the majority of the sites on Whitepine Lake occurred even though the amount of potential habitat lost during the experiment was more than

Table 14.1 Whitepine Lake and Helen Lake Spawning Sites

	Spawning sites used		Site characteristics				Spawning habitat covered		
Year	Number	Total area (m²)	Surface area (m²)	Water depth (m²)	Distance from shore (m)	Sites producing alevins (%)	Egg deposition area (m²)	Adjoining areas (m²)	Totals (m²)
Whitepine Lake									
1991	7	40.0	0.5–21.0	0.3–1.5	1.4–4.5	na	0	0	0
1992	15	74.8	0.5–21.0	0.2–2.0	na*	na	6.0	86.0	92.0
1993	18	64.9	0.5–21.0	0.1–1.2	na	44.4	13.0	86.0	99.0
1994	41	40.3	0.1–5.0	0.1–2.0	0.5–4.0	na	21.0	71.0	92.0
1995	44	82.7	0.2–10.0	0.3–0.8	0.3–0.9	na	6.0	0	6.0
1996	39	195.3	0.9–42.0	0.1–1.5	0.4–2.7	76.9	189.0	101.6	290.6
1997	52	126.3	0.1–15.0	0.1–1.4	0.4–6.0	44.3	195.3	28.4	223.7
1998	41	226.4	0.1–78.9	0.2–1.3	0.8–6.0	46.3	126.3	68.9	195.2
1999	41	98.1	0.1–13.2	0.2–1.5	0.2–15.0	51.2	226.4	378.6	605.0
Helen Lake									
1997	5	11.5	0.2–9.0	0.5–0.7	na	40.0	0	0	0
1999	9	44.2	0.2–15.0	0.3–0.8	na	55.5	11.5	138.5	150

Note: Spawning sites were defined as the area of egg deposition during the fall spawning period. Substrates were inspected immediately after ice off in April and early May to identify spawning sites that produced alevins. On Whitepine Lake the traditional spawning sites were gradually covered with plastic sheeting 1992 through 1994. During 1996 through 1999 (Whitepine Lake) and 1999 (Helen Lake) all previous spawning sites were covered each year.

* na, not available.

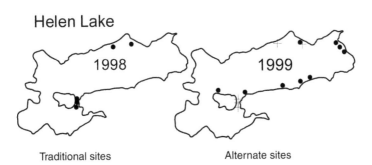

Figure 14.5 Locations of traditional spawning sites and the spawning sites used after the habitat manipulation experiment was completed in Whitepine Lake (1991, 1999) and Helen Lake (1998, 1999) (marked with •). The locations of spawning sites that were covered for this experiment are indicated (marked with +).

40 times that of the original area of substrate that supported the population in 1991. Similar results occurred in the 1-year manipulation of Helen Lake, where five of nine (55.5%) newly selected sites contained alevins on April 18, 2000.

The mark–recapture estimates of juvenile abundance on Whitepine Lake also failed to show any of the expected effects of the experimental treatment. We predicted that juvenile abundance would decline as a result of the habitat loss, expecting the decline to begin soon after all the traditional spawning sites were removed in 1994. The mean size of fish in the index gill nets was also expected to increase with time if abundance of juveniles, and resulting competition for food, declined.

There was no decline in the abundance of juveniles (fork length [FL] 260 to 370 mm) detected during the study (Table 14.2). The average size of fish in the population also did not increase (Figure 14.7), giving further evidence that recruitment was unaffected by the habitat loss. Age assessment data are not yet available for fish collected during recent years, but from age–size relationships obtained earlier (Gunn et al., 1996) it is clear that most of the abundant juveniles in Whitepine Lake are recruits from the new spawning sites. The difficulty in capturing very young fish (<3 years old) with our assessment methods means that year-class declines may still be detected in the future as a result of

Figure 14.6 One of the alternate sites on Whitepine Lake. This alternate site, like many others, appeared to have very marginal conditions (i.e., <10 cm deep, underlain by sand with little interstitial space for eggs).

Table 14.2 Estimated Abundance of Juvenile (260 to 370 mm) Lake Trout in Whitepine Lake from the Springtime Surveys

Year	Number Marked	Recaptured (%)	Estimated Total Number
1992	17	0	—
1993	63	22	180 (118–377)
1994	110	19	406 (284–709)
1995	77	10	507 (300–1653)
1996	45	16	169 (97–653)
1997	75	13	357 (220–938)
1998	89	6	955 (509–7738)
1999	139	9	1013 (647–2332)

Note: Abundance was estimated by Schnabel continuous mark-recapture of angled and gill netted fish. Estimated numbers with 95% confidence intervals (in parentheses) are presented.

possible reduction in the survival of the egg–alevin stages in the later years of the study. However, the continued production of alevins at the alternate sites throughout the study suggests that the habitat manipulations (i.e., continued removal) would have to be continued for many more years to detect a complete loss of recruitment.

Exploitation following improved road access

The effects of exploitation on Michaud Lake were not subtle. Fishermen were able to access the lake in midwinter by the new road. Anglers used snowmobiles and all-terrain four-wheel-drive vehicles to travel to the lake. Catch rates (Figure 14.8a) were very high when the fishery opened in the first week of January 1998 (0.4 fish·angler·h^{-1}), and most anglers were able to catch their allowable limit of three fish each day (angler-day). The number of angler-days increased steadily, reaching a maximum of 93 in the fourth week of January (Figure 14.8b).

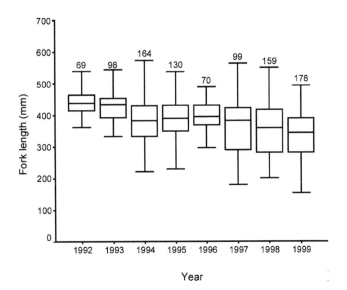

Figure 14.7 The size of lake trout captured in the index gill nets in Whitepine Lake, 1991 to 1999. Annual median, 25th percentile, 75th percentile, maximum, minimum, and sample size indicated.

The sustainable yield (kg·ha^{-1}) level for Michaud, a 148-ha lake, was estimated as 1.35 kg·ha^{-1} from the following equation of Payne et al. (1990) (Figure 14.8c):

$$Log_{10} (Harvest) = 0.50 + 0.83 \cdot log_{10} (Area)$$

The estimated maximum sustainable yield level was exceeded less than 3 weeks after anglers accessed the lake (Figure 14.8c). Fishing success rate declined through the winter, but anglers continued to harvest lake trout until ice melt forced an end to the fishery. Once the road was dry enough to permit truck travel, anglers brought boats to the lake and began fishing again on April 24. Fishing success rates improved for a brief period in the open-water period, but declined rapidly. Anglers largely abandoned the lake in early May.

In this brief fishery the estimated harvest reached 3.8 kg·ha^{-1} in fewer than 5 months. A total of 395 harvested fish were observed by the creel survey crews. Of the creel survey sample 22% were fin clipped, providing a mark–recapture estimate of 765 (95% confidence interval [CI], 607 to 963) for the preangling adult portion of the population. This population was reduced by approximately 72% (final population estimate 210; 95%CI, 52 to 408) in the winter and spring fisheries. The midsummer index netting survey provided consistent evidence of the population decline. Average catch rates of lake trout declined by 70% between 1997 and 1998 (Table 14.3).

Conclusions

The results of this study suggest that lake trout were not particularly vulnerable to damage from the physical alterations to spawning habitat that may occur through the construction of forest access roads in Boreal Shield areas of Ontario. First, there are management guidelines in place (e.g., Ontario Ministry of Natural Resources, 1988, 1990, 1991) for the construction of roads; if followed, the guidelines should limit siltation of streambeds and inlet areas of lakes. Unlike other *Salvelinus* species such as brook trout (Curry et al., 1997), lake trout appear rarely, if ever, to use streams or the area immediately adjoining inlet streams for spawning. In a survey of 95 lake trout lakes, McMurtry (1986) found that no

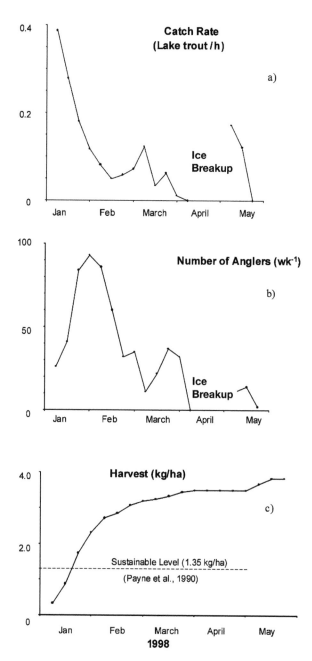

Figure 14.8 Weekly estimates of the number of a) fish captured per angler hour (lake trout/hour), b) anglers (wk^{-1}), and c) accumulated harvest (kg ha^{-1}) of lake trout in Michaud Lake during the winter and spring of 1998, the year the lake opened to fishing.

lake trout spawning sites were within 20 m of a stream inlet. Our study showed that if siltation of shorelines did actually occur, lake trout would seek spawning sites elsewhere and could maintain recruitment at these alternate sites. Siltation events would have to be extremely extensive and long lasting to eliminate reproduction completely in this long-lived species.

The fact that we have not yet detected a substantial effect of progressive spawning habitat loss on lake trout recruitment in two experimental lakes does not mean that there

Table 14.3 Effects of 1998 Winter Fishing on Abundance of Lake Trout in Michaud Lake

Year	Date	Number of Net Sets	Lake Trout Catch
1997	August 6–12	60	1.23 (1.33)
1998	August 12–19	60	0.37 (0.90)

Note: The population was assessed using multimesh (15-m single panels of 19-, 25-, 38-, 51-, and 64-mm stretched mesh with 5-m spacers between panels) gill nets set at random locations within the hypolimnion. The lake trout catch is the number of fish captured during a 2-h net set. Means with one standard deviation (in parentheses) are given.

is an unlimited supply of suitable spawning habitat on Shield lakes, or that lake managers should be any less diligent about protecting fish habitat from disturbances. Spawning habitat loss is surely one of a suite of stressors that have a cumulative effect on lake trout. However, our experimental results clearly suggest that resource managers need to question whether fish harvest controls and improved motor vehicle access to lake trout lakes have received sufficient attention in forest and fisheries management planning. We think that forest access roads and the increases in angling pressure they create have a far greater impact on Boreal Shield lake trout populations than spawning habitat loss due to sedimentation.

Acknowledgments

We appreciate the assistance of Robert Kirk, Lee Haslam, Scott McAughey, and David Gonder, who helped with the fieldwork, and Sylvia Donato for the graphics. Review comments were kindly provided by Greg Deyne, Warren Dunlop, Dave Evans, Steve Kerr, Rob Korver, Nigel Lester, Cheryl Lewis, Terry Marshall, Mike Powell, Bev Ritchie, Ed Snucins, Neville Ward, and two anonymous referees.

References

Anderson, B. and Potts, D.F., 1987, Suspended sediment and turbidity following road construction and logging in western Montana, *Water Resources Bulletin*, **23**: 681–690.

Blais, J.M., France, R.L., Kimpe, L.E., and Cornett, R.J., 1998, Climatic changes in northwestern Ontario have had a greater effect on erosion and sediment accumulation than logging and fire: evidence from 210 Pb chronology in lake sediments, *Biogeochemistry*, **43**: 235–252.

Curry, R.A., Brady, C., Noakes, D.L.G., and Danzman, R.G., 1997, Use of small streams by young brook trout spawned in a lake, *Transactions of the American Fisheries Society*, **126**: 77–83.

DeRoche, S.E., 1969, Observations on the spawning habits and early life of lake trout, *Progressive Fish Culturist*, **31**: 109–113.

Foster, N.R., 1985, Lake trout reproductive behavior: influence of chemosensory cues from young-of-the-year by-products, *Transactions of the American Fisheries Society*, **114**: 794–803.

Gunn, J.M., 1995, Spawning behavior of lake trout: effect on colonization ability, *Journal of Great Lakes Research*, **21**(Suppl. 1): 323–329.

Gunn, J.M., Conlon, M., Kirk, R.J., and McAughey, S.C., 1996, Can trained observers accurately identify lake trout spawning habitat? *Canadian Journal of Fisheries and Aquatic Sciences*, **53**(Suppl. 1): 327–331.

Gunn, J.M. and Keller, W., 1984, Spawning site water chemistry and lake trout (*Salvelinus namaycush*) sac fry survival during spring snowmelt, *Canadian Journal of Fisheries and Aquatic Sciences*, **41**: 319–329.

Gunn, J.M. and Keller, W., 1990, Biological recovery of an acid lake after reductions in industrial emissions of sulphur, *Nature (London)*, **345**: 431–433.

Gunn, J.M., McMurtry, M.J., Bowlby, J.N., Casselman, J.M., and Liimatainen, V.A., 1987, Survival and growth of stocked lake trout in relation to body size, stocking season, lake acidity and biomass of competitors, *Transactions of the American Fisheries Society*, **116**: 618–627.

Hara, T.J., 1994, The diversity of chemical stimulation in fish olfaction and gustation, *Reviews in Biology of Fishes*, **4**: 1–35.

Healey, M.C., 1978, Dynamics of exploited lake trout populations and implications for management, *Journal of Wildlife Management*, **42**: 307–328.

Lantz, R.L., 1971, *Guidelines for Stream Protection in Logging Operations*, Oregon State Game Commission, Portland, OR.

Martin, N.V., 1957, Reproduction of lake trout in Algonquin Park, Ontario, *Transactions of the American Fisheries Society*, **86**: 231–244.

Martin, N.V. and Olver, C.H., 1976, *The Distribution and Characteristics of Ontario Lake Trout Lakes*, Ontario Ministry of Natural Resources, Fish and Wildlife Research Branch Department, Report # 97, Toronto.

Martin, N.V. and Olver, C.H., 1980, The lake charr, *Salvelinus namaycush*, in *Charrs: Salmonid Fishes of the Genus* Salvelinus, Balon, E.K., Ed., Dr. W. Junk, Hague, The Netherlands, pp. 205–277.

McAughey, S.C. and Gunn, J.M., 1995, The behavioral response of lake trout to a loss of traditional spawning sites, *Journal of Great Lakes Research*, **21**(Suppl. 1): 375–383.

McMurtry, M.J., 1986, Susceptibility of Lake Trout (*Salvelinus namaycush*) Spawning Sites in Ontario to Acid Meltwater, Ontario Ministry of Natural Resources, Ontario Fisheries Acidification Technical Report Series 86-0001, Toronto.

Megahan, W.F. and Kidd, W.J., 1972, Effects of logging and logging roads on erosion and sediment deposition from steep terrain, *Journal of Forestry*, **80**: 136–141.

Ontario Ministry of Natural Resources, 1988, *Timber Management Guidelines for the Protection of Fish Habitat*, Ontario Ministry of Natural Resources Report, Toronto.

Ontario Ministry of Natural Resources, 1990, Environmental Guidelines for Access Roads and Water Crossings, Ontario Ministry of Natural Resources Report, Toronto.

Ontario Ministry of Natural Resources, 1991, Code of Practice for Timber Management Operations in Riparian Areas, Ontario Ministry of Natural Resources Report, Toronto.

Payne, N.R., Korver, R.M., MacLennan, D.S., Nepszy, S.J., Shuter, B.J., Stewart, T.J., and Thomas, E.R., 1990, The Harvest Potential and Dynamics of Lake Trout Populations in Ontario, Lake Trout Synthesis Population Dynamics Working Group Report, Ontario Ministry of Natural Resources, Toronto.

Plamondon, A.P., 1982, Increase in suspended sediments after logging and duration of effect, *Canadian Journal of Forest Research*, **12**(4): 883–892.

Ricker, W., 1975, Computation and interpretation of biological statistics of fish populations, *Bulletin of the Fisheries Research Board of Canada*, #191, Ottawa.

Ryder, R.A. and Johnson, L., 1972, The future of salmonid communities in North American oligotrophic lakes, *Journal of the Fisheries Research Board of Canada*, **29**: 941–949.

Schindler, D.W., 1998, A dim future for Boreal waters and landscapes: cumulative effects of climatic warming, stratospheric ozone depletion, acid precipitation, and other human activities, *BioScience*, **48**: 157–164.

Shuter, B.J., Jones, M.L., Korver, R.M., and Lester, N.P., 1998, A general life history based model for regional management of fish stocks: the inland lake trout (*Salvelinus namaycush*) fisheries in Ontario, *Canadian Journal of Fisheries and Aquatic Sciences*, **55**: 2161–2177.

Sly, P.G. and Evans, D.O., 1996, Suitability of habitat for spawning lake trout, *Journal of Aquatic Ecosystem Health*, **5**: 153–175.

Steedman, R.J. and France, R.L., 1999, Origin and transport of aeolian sediment from new clearcuts into boreal lakes, northwestern Ontario, Canada, *Water, Air, and Soil Pollution*, **122**: 139–152.

Models and issues associated with ecosystem management

chapter fifteen

Climate change and sustainable lake trout exploitation: predictions from a regional life history model

Brian J. Shuter
Ontario Ministry of Natural Resources
Nigel P. Lester
Ontario Ministry of Natural Resources

Contents

Introduction

Several authors have recently addressed the potential impacts of climate change on lake trout *Salvelinus namaycush* from quite different perspectives. Schindler (1998) and coworkers (1996) identified a variety of factors (e.g., increased wind exposure due to forest fires, increased water clarity due to DOC reductions) that could shrink hypolimnetic, summer refuge habitats for lake trout, particularly in smaller lakes, and thus negatively affect population sustainability.

Magnuson et al. (1990, 1997) and Magnuson and DeStasio (1997) looked at likely changes in the amount of habitat thermally suitable for lake trout and concluded that this factor alone would only lead to negative effects in shallow lakes, largely through increases in summer temperatures. Stefan et al. (2001) modeled changes in the annual temperature and oxygen regimes typical of small lakes in the continental United States and concluded that climate warming would produce significant contraction of usable habitat for many populations of coldwater fish (i.e., fish species with the low-temperature and high-oxygen requirements typical of lake trout; Ryan and Marshall, 1994).

As surface waters warm and longer stratification periods produce anoxia in deeper waters, usable habitat for lake trout will contract. Stefan et al. (2001) predicted that this effect would be severe, even along the northern border of the United States, with many populations expected to disappear entirely. Some empirical studies also suggest that negative impacts are likely. King et al. (1999) observed decreased lake trout growth in years with anomalously long stratification periods. Gunn (2002) observed that lake trout populations can persist in unstratified lakes at quite high summer temperatures (~22°C), but that persistence requires the presence of cool, groundwater seepage refuges and absence of competitors.

In many of these studies, population-level impacts (e.g., reduced abundance) are inferred directly from broad, qualitative shifts in environmental characteristics (e.g., longer stratification periods, deeper thermoclines). In this chapter, we use the lake trout life history model developed by Shuter et al. (1998a) to illustrate how to derive more specific predictions of the demographic consequences of climate-driven environmental changes. We also assess the implications of such changes for the management of exploitation.

A regional model for lake trout populations

Observed variation in lake trout life histories

The lake trout populations in the inland lakes of the Precambrian Shield exhibit considerable variation in their life histories, and this variation is associated with environmental differences among lakes (Shuter et al., 1998a). For lake trout, life history characteristics are closely linked with lifetime growth pattern, and thus variation in growth pattern can provide a reasonably good index of life history variation (Shuter et al., 1998a).

Lifetime growth pattern can be summarized by two parameters (Figure 15.1): growth rate early in life ω and maximum adult size L_∞. Using a life history database derived (by methods found in Shuter et al., 1998a) from 54 lake trout populations spread across the Precambrian Shield, we found that the variation in both of these parameters is just over twofold (Figure 15.2). In addition, variation in one parameter is essentially independent of variation in the other (ω vs. L_∞, $r^2 = .004$), so the range of possible growth patterns is very large, covering essentially all combinations of the parameter values, over their observed ranges (Figure 15.2).

Environmental variation, life history variation, and the parameters of a population harvest model

The lakes that support the 54 populations included in this life history database differ widely in both morphometry and water chemistry, with a range of variation that essentially covers the range exhibited by the full set of Precambrian Shield lake trout lakes (Table 15.1). While the range of variation in lake characteristics is large (a factor of ten or more) compared to the range of variation in growth history parameters (a factor of two), there are statistically significant links between them. Specifically, over 45% of the observed variation in L_∞ is associated with variation in lake area, over 22% of observed variation in ω is associated with variation in TDS, and over 90% of observed variation in harvest (kg/year) is associated with variation in lake area (ha). These associations form the

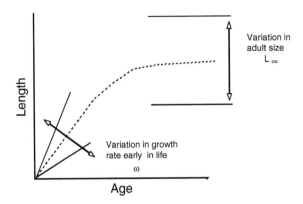

Figure 15.1 Parameters used to summarize the lifetime growth pattern typical of specific lake trout populations.

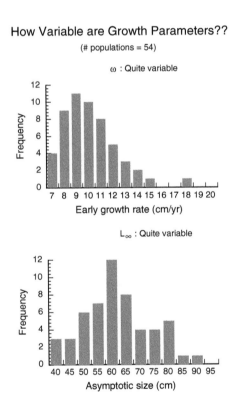

Figure 15.2 Interpopulation variation in the lifetime growth pattern of lake trout as reflected in interpopulation variation in early growth rate ω and maximum adult size L_∞. Histograms were derived from independent estimates for both growth parameters in 54 separate lake trout populations.

Table 15.1 Characteristics of Lakes Sampled for the Lake Trout Life History Database (Shuter et al., 1998) Compared to Precambrian Shield Lakes in General

Lake characteristics	Lake trout life history lakes (N = 54)	Precambrian Shield lakes in general
Surface area (ha)	661	192
	(25, 448,000)	(3, 9,800)
Mean depth (m)	16.5	11.3
	(6.8, 55)	(1.4, 54)
TDS (mg/l)	34	29
	(15, 180)	(2, 273)

Note: Median, maximum, and minimum values are given for each character.
        TDS = total dissolved solids

Table 15.2 Biological Interpretations of Statistical Linkages between Lake Characteristics and Lake Trout Population Parameters

Parameter	Environmental Correlate	Biological Interpretation of Correlation
ω	TDS	Variation in TDS reflects variation in basic lake productivity; high-productivity lakes provide more of the smaller prey items required by young lake trout and thus are able to support higher growth rates among young lake trout
L_∞	Lake area	Larger lakes support a more diverse set of potential prey items for adult lake trout and hence permit a more complete realization of lake trout growth potential throughout life
B_0	Lake area	Larger lakes provide more thermally suitable habitat for lake trout to live in; hence larger lakes will have the capacity to support more lake trout

Note: See Shuter et al. (1998a) for details.
        TDS = total dissolved solids.

biological basis (Table 15.2) for a set of predictive regression equations that use lake area and TDS values alone to develop a fully parameterized lake trout population harvest model (Figure 15.3) of the following form (see Shuter et al., 1998a, for details):

$$H = f\left(W_\infty, \omega, t_m, f_{max}, \alpha_{max}, M; \beta, B_0; t_c, F, q\right)$$

where H is lake trout harvest (kg·year^{-1}). In its original form (Shuter et al., 1998a), the model was defined in terms of yield (kg·ha^{-1}·yr^{-1}) rather than harvest, because it was being used in a comparative study of populations known to differ in yield. Since our focus in this paper is on the potential impact of climate change on the dynamics of an individual population we chose to use the harvest version of the model because it provides a clear description of the consequences for sustainable harvest of the changes in absolute amount of useable habitat that are likely to result from climate change.

The parameters of this model fall into three distinct groups: life history, habitat quantity, and fishery parameters.

1. Life history parameters define rates of growth, mortality, and reproduction and characteristics of specific habitats in the absence of intraspecific competition:
    W_∞ is the asymptotic mass of an adult fish (kg)

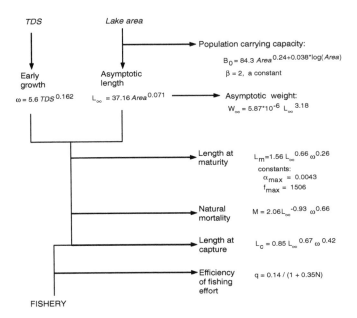

Figure 15.3 Schematic diagram illustrating how lake-specific estimates of surface area and total dissolved solids (TDS) are translated into a complete set of parameter estimates for the lake trout population model. The equations are empirical regression equations that reflect associations linking environmental characteristics and life history characters. N is population abundance (number of catchable fish per hectare); it is generated internally by the model and is used as input to the equation that defines the value of the catchability coefficient q.

ω is the rate of length growth (cm·yr^{-1}) in early life (Gallucci and Quinn, 1979) and is the product of the von Bertalanffy growth parameters K (yr^{-1}) and L_∞ (cm)

t_m is the age of maturation, with the knife-edge transition to maturity assumed

f_{max} is fecundity, the number of eggs per kilogram for a mature female at low population density

α_{max} is survival from egg to age 1 at low population density

M is the instantaneous natural mortality rate (year^{-1}) for fish aged 1 year and older

2. The habitat quantity parameters define both the number of individuals that can be supported by the available habitat and the rate at which intraspecific competition reduces early survival or fecundity as that number is approached:

B_0 is, a scaling parameter (kg) that reflects the amount of habitat available to the population and hence is directly related to both the carrying capacity of the population and its maximum sustainable harvest

β is a parameter that sets the rate at which early survival or fecundity declines as the population approaches its carrying capacity (i.e., a higher β value produces a more rapid decline)

3. Fishery parameters define the part of the population that is exploited and the intensity of exploitation:

t_c is the age of first capture by the fishery, with knife-edge transition to full vulnerability assumed

F is the instantaneous fishing mortality (year^{-1}) applied to all fish aged t_c or older

q is the catchability coefficient; it is the parameter that links fishing effort E (angler hours·year^{-1}) to fishing mortality ($F = q \cdot E$) and is a function ($q = a/[1 + b \cdot N]$) of the abundance of catchable fish (N)

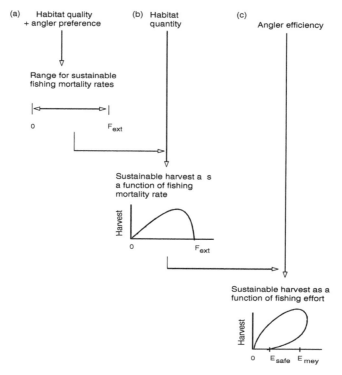

Figure 15.4 The roles of different kinds of information in determining the behavior of the lake trout population model. (a) Upper bound on sustainable fishing mortality rate F_{ext} is determined by the following aspects of habitat quality: productivity, prey availability, predator (i.e., angler) selectivity; (b) height of the harvest–fishing mortality curve is determined by the amount of suitable lake trout habitat available; (c) shape of the harvest–fishing effort curve is determined by the relationship linking angler catchability to population abundance. E_{safe} and E_{mey} delimit the range over which fishing effort is conditionally sustainable (i.e., sustainable provided population abundance is high enough; see Shuter et al., 1998b, for details); effort values equal to or less than E_{safe} are unconditionally sustainable, and effort values greater than E_{mey} are unsustainable.

Life history variation and variation in sustainable exploitation rates and sustainable harvests

The information in each of the model's parameter groups builds progressively (Figure 15.4) toward a complete picture of a population's capacity to provide sustained harvests under various levels of exploitation:

1. The characteristics of a lake that reflect the "quality" of lake trout habitat (i.e., prey diversity as reflected in lake area, TDS) are passed through a series of linked regressions (Figure 15.3) to define the age-specific fecundity and mortality schedules for the lake's population under conditions of zero exploitation rate; this set of population characteristics, coupled with information on the minimum size and age of fish preferred by anglers, then determines the range of fishing mortality rates that the population can sustain.
2. Lake characteristics that reflect the amount of lake trout habitat (i.e., thermal habitat volume, as reflected in lake area) determine the shape of the harvest–mortality curve over the range of sustainable fishing mortality rates.
3. Angler efficiency (i.e., catchability) determines both the range of fishing effort that is sustainable and the shape of the harvest–effort curve over that range.

Potential impacts of climate change on sustainable harvests and exploitation rates

Lakes will be affected by climate change in ways that will significantly impact resident lake trout populations (Table 15.3). Given the likely directions for these changes, we would expect many existing lake trout populations to respond to these changes by exhibiting significant contractions in the range of sustainable harvests, the range of sustainable levels of fishing mortality, and the range of sustainable levels of fishing effort (Figure 15.5). Climate-induced shifts in population behavior may also significantly increase angler efficiency, particularly in severely impacted populations, where many individual fish will be squeezed into reduced summer refuge habitats for extended periods of time. Since many lake trout populations are currently subjected to exploitation rates that approach or exceed sustainable levels (Evans et al., 1991), the overall effects of climate change on many populations will be to render current levels of use grossly unsustainable and to mandate levels of protection far more stringent than those currently in place.

While the directions of demographic impacts seem relatively easy to anticipate, specific forecasts of the extent and time frame for expected changes in specific lakes are much more difficult to derive. However, it is possible to identify at least two factors that will be important in influencing such forecasts: (1) lake location — many climate change scenarios (e.g., Taylor, 1996) suggest that warming and particularly drying effects will be most severe in the southwestern part of the Precambrian Shield and least severe in the northeastern part; (2) lake size — most of the anticipated negative effects on lakes (Table 15.3) will be most severe for small, shallow, dimictic lakes, and these negative effects include declines in production from reduced phosphorus input and loss of summer habitat due to shrinkage of hypolimnetic volume

Direct effects of climate change on aquatic habitats are likely (Shuter et al., 1998b) to be exacerbated by adaptation responses of the human population to these same changes in climate. As watersheds dry, increasing human demands will be put on what water remains: Pressure will build for additional water diversions to meet demands set by human consumption, agricultural irrigation, and hydroelectric power.

With these considerations in mind, it is possible to order lake trout populations according to their risk of suffering severe, deleterious impact (e.g., reductions in sustainable harvests, fishing mortality rates, and fishing effort levels): High-risk populations live in relatively small, shallow lakes in the southwest of the Shield, in watersheds that currently support significant human use; low-risk populations live in relatively large, deep lakes in the northeast of the Shield, in watersheds that do not currently support much human use.

Schindler (1998) pointed out that many of the anticipated impacts of climate change on boreal watersheds will reinforce existing impacts arising from acid rain and increased ultraviolet exposure. The approach to population-level impact assessment that is used in this chapter could be extended to provide a qualitative assessment of the interactive effects of all three of these factors. Given quantitative estimates of anticipated environmental changes (e.g., expected change in the annual temperature stratification cycle), semiquantitative estimates of impact could also be generated. Adaptations by human society to meet freshwater needs in the face of climate change will exacerbate direct impacts of climate change on Precambrian Shield lake trout populations. Procedures to ameliorate such effects should be included when developing comprehensive adaptation strategies human communities located on (or near) Precambrian Shield watersheds.

Table 15.3 Potential Impacts of a Drier, Warmer Climate on Lake Ecosystems and Likely Consequences for Lake Trout Populations

Physical/chemical change in lake systems	Impact on production of prey available to young lake trout = change in ω	Impact on diversity of prey available to adult lake trout = change in L_∞	Impact on volume of lake trout habitat = change in B_0
Drier watersheds lead to:			
— Overall reductions in lake volume	*Uncertain*	*Decrease:* overall species richness decreases with lake size (Mandrak, 1995)	*Decrease:* volume of lake trout habitat will decline with overall lake volume
		Uncertain	*Uncertain*
— Reductions in annual input of new phosphorus from watershed (Schindler, 1998)	*Decrease:* overall lake productivity declines with phosphorus input, leading to reduced prey production		
— Reductions in annual input of DOC from watersheds; for small dimictic lakes, this leads to increases in transparency sufficient to increase heating of deeper waters, generate deeper thermoclines, and in extreme cases, prevent stratification (Schindler, 1998)	*Decrease:* in lakes that no longer consistently stratify, summer exposure to suboptimal temperatures reduces potential growth, particularly in the presence of cool- or warm-water competitors	*Decrease:* in lakes that no longer consistently stratify, summer exposure to suboptimal temperatures reduces potential growth, particularly in the presence of cool- or warm-water competitors	*Decrease:* volume of habitat thermally suitable for lake trout during stratification period shrinks as thermocline deepens

Higher air temperatures and lower spring winds lead to:			
— Shallower thermoclines in large, dimictic lakes	*Uncertain*	*Uncertain*	*Increase:* shallower thermocline leads to increase in habitat thermally suitable for lake trout during stratification period. *Decrease:* in dimictic lakes, the period when lake trout are confined to cool hypolimnetic waters is extended; with this extension, comes a greater risk of further habitat restriction due to increased hypolimnetic oxygen depletion
— Longer ice-free period and, in dimictic lakes, longer stratification period	*Increase:* overall productivity of lower trophic levels will increase with longer ice-free periods in nutrient-rich systems (Shuter and Ing, 1997). *Decrease:* extension of slow-growth summer period	*Increase:* overall productivity of lower trophic levels will increase with longer ice-free periods in nutrient-rich systems (Shuter and Ing, 1997). *Decrease:* extension of slow-growth summer period	*Uncertain*
— Warmer surface water temperatures	*Increase:* overall productivity of lower trophic levels will increase with temperature in nutrient-rich systems (Shuter and Ing, 1997)	*Uncertain*	*Uncertain*

Note: See Shuter et al. (1998a) for details.

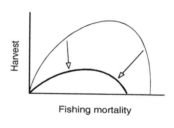

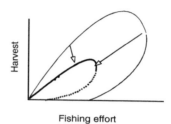

Figure 15.5 Potential negative impacts of climate change on the harvest curves for a Precambrian Shield lake trout population. The arrows indicate how the curves will shift as the climate changes. Populations at greatest risk of exhibiting such negative changes are those living in small, shallow lakes on the southwestern edge of the Shield.

References

Evans, D.O., Casselman, J.M., and Willox, C.C., 1991, Effects of Exploitation, Loss of Nursery Habitat and Stocking on the Dynamics and Productivity of Lake Trout Populations in Ontario Lakes, Lake Trout Synthesis Response to Stress Working Group Report, Ontario Ministry of Natural Resources, Peterborough.

Gallucci, V.F. and Quinn, T.J., II, 1979, Reparameterizing, fitting and testing a simple growth model, *Transactions of the American Fisheries Society*, 108:14–25.

Gunn, J., 2002, Impact of the 1998 El Nino event on a lake charr, *Salvelinus namaycush*, population recovering from acidification, *Environmental Biology of Fishes*, 64:343–351.

King, J.R., Shuter, B.J., and Zimmerman, A.P., 1999, Empirical links between thermal habitat, fish growth and climate change, *Transactions of the American Fisheries Society*, 128:656–665.

Magnuson, J.J., Meisner, J.D., and Hill, D.K., 1990, Potential changes in thermal habitat of Great Lakes fish after global climate warming, *Transactions of the American Fisheries Society*, 119:254–264

Magnuson, J.J. and DeStasio, B.T., 1997, Thermal niche of fishes and global warming, in *Global Warming: Implications for Freshwater and Marine Fish*, Wood, C.M. and MacDonald, D.G., Eds., Cambridge University Press, Cambridge, UK.

Magnuson, J.J., Webster K.E., Assel, R.A., Bowser, C.J., Dillon, P.J., Eaton, J.G., Evans, H.E., Fee, E.J., Hall, R.I., Mortsch, L.R., Schindler, D.W., and Quinn, N.F.H., 1997, Potential effects of climate changes on aquatic systems: Laurentian Great Lakes and Precambrian Shield region, *Hydrological Processes*, 11:828–871.

Mandrak, N.E., 1995, Biogeographic patterns of fish species richness in Ontario lakes in relation to historical and environmental factors, *Canadian Journal of Fisheries and Aquatic Sciences*, 52:1462–1474.

Ryan, P.A. and Marshall, T.R., 1994, A niche definition for lake trout (*Salvelinus namaycush*) and its use to identify populations at risk, *Canadian Journal of Fisheries and Aquatic Sciences*, 51:2513–2519.

Schindler, D.W., 1998, A dim future for boreal waters and landscapes, *BioScience*, 48:157–164.

Schindler, D.W., Bayley, S.E., Parker, B.R., Beaty, K.G., Cruikshank, D.R., Fee, E.J., Schindler, E.U., and Stainton, M.P., 1996, The effects of climatic warming on the properties of boreal lakes and streams at the Experimental Lakes Area, Northwestern Ontario, *Limnology and Ocean-ography*, 41:1004–1017.

Shuter, B.J. and Ing, K.K., 1997, Factors affecting the production of zooplankton in lakes, *Canadian Journal of Fisheries and Aquatic Sciences*, 54:359–377.

Shuter, B.J., Jones, M.L., Korver, R.M., and Lester, N.P., 1998a, A general, life history based model for regional management of fish stocks: the inland lake trout (*Salvelinus namaycush*) fisheries of Ontario, *Canadian Journal of Fisheries and Aquatic Sciences*, 55:2161–2177.

Shuter, B.J., Minns, C.K., and Regier, H.A., 1998b, Potential impacts of climate change on fisheries in Canada, in *Canada Country Study: Climate Impacts and Adaptation — National Sectoral Volume*, Koshida, G. and Avis, W., Eds., Environment Canada, Ottawa, chap. 6.

Stefan, H.F., Fang, X., and Eaton, J.F., 2001, Simulated fish habitat changes in North American lakes in response to projected climate warming, *Transactions of the American Fisheries Society*, 130:459–477.

Taylor, W., 1996, *Climate Change Scenarios for Canada: A User's Guide for Climate Impact Studies*, Environment Canada, Vancouver.

chapter sixteen

Monitoring the state of the lake trout resource: a landscape approach

Nigel P. Lester
Ontario Ministry of Natural Resources
Warren I. Dunlop
Ontario Ministry of Natural Resources

Contents

Introduction

> Oligotrophic lakes ... should be recognized for what they are, swimming pools carved out of granite, with low nutrient tributaries and a cold annual thermal regime. They are capable of having their environment and their communities severely altered — only too easily. Unless further increases of eutrophication and exploitation are brought to a halt, the lakes will be altered within the next three decades and their demise as producers of salmonid stocks may be irrevocable.

Ryder and Johnson, 1972, p. 948

Thus spoke the prophets 30 years ago. Were they correct? Have oligotrophic lakes been seriously eroded by human-induced stress? The anecdotal evidence is overwhelming, indicating that lake trout *Salvelinus namaycush* populations in many lakes have been seriously degraded (Evans and Willox, 1991; Evans et al., 1991a, 1991b, 1996). The extent of the damage, however, cannot be assessed objectively because the required data are not available. Much of our lake trout data come from case studies instigated in response to a perceived problem. Such data have enhanced our understanding of the stressors that impact lake trout, but they are not very useful in addressing questions about the overall condition of the lake trout resource. What is the current state of the resource? How many lakes have been degraded? What is the spatial extent of the damage? How much change has occurred? What is the rate of change? These questions cannot be answered simply by compiling the available historical data. A data collection approach that acknowledges the spatial and temporal scales of the questions is required.

Evaluating change in the state of a resource requires commitment to a long-term monitoring program that collects an unbiased sample of data. Such a program does not exist for the lake trout lakes of the southern Precambrian Shield. There are approximately 3000 lakes in this population. The Shield spans two Canadian provinces (Ontario and Quebec) and two American states (New York and Minnesota). In Ontario, there are approximately 2200 lake trout lakes; 20 of these are monitored by a network of Fisheries Assessment Units (FAUs). Data from these lakes are potentially useful, but the small sample size does not supply precise indicators of resource status at the landscape level. Other lakes in Ontario are sampled on an ad hoc basis by management offices. Because these inspections are often motivated by public concerns, the data cannot be used to monitor the overall condition of the lake trout resource. In Quebec, approximately 25 lakes are sampled on a regular basis (at least once every 5 years), but as in Ontario, a program that monitors lakes at the landscape level does not exist (D. Nadeau, 2000, personal communication).

One reason for the lack of a large-scale lake trout monitoring program is the apparent cost. Traditional monitoring programs involve intensive measurement of a large number of environmental variables repeated through time at a small number of fixed sites. Applying the same methods on a larger spatial scale (i.e., many sample sites) would result in an impossibly expensive monitoring program. This does not have to be a problem. The cost per lake can be managed in several ways to increase the number of lakes that can be sampled.

First, the program must be *focused*, monitoring only the variables that are needed to report on the condition of the resource. As Walters (1997) pointed out, scientists asked to develop a monitoring program will almost certainly identify a large set of variables to measure rather than focusing on key response variables. This strategy supplies data that may be needed to explain changes that occur, but it greatly increases the cost of the program. If the purpose of the program is to report change, not to explain it, then only key indicator variables should be monitored.

Second, the program must be *creative*, seeking new cost-effective methods of measuring indicator variables. Traditional methods of fisheries assessment call for intensive studies to monitor key variables within a lake. Such methods cannot be afforded when the survey domain is a population of lakes rather than a population of fish. The development of a rapid assessment technology is needed to sample at the landscape level. "Rapid" implies a cost-effective, yet scientifically valid, method of measurement (Hoenig et al., 1987; Oliver and Beattie, 1993; Jones and Stockwell, 1995).

Less-intensive sampling of individual lakes is one option that produces a more rapid assessment. Although this results in less-precise assessment of individual lakes, it allows more lakes to be assessed. Because the objective is to describe a population of lakes (not the state of individual lakes), a larger sample size in terms of lake number can compensate for a lack of precision in individual lakes estimates.

In this chapter, we describe an affordable approach to monitoring the health of the lake trout resource. First, we discuss how the state of lake trout lakes can be evaluated based on a model that describes the expected response to fishing. Second, we use the model to identify a set of indicators and criteria for evaluating state. Third, we describe cost-effective methods of collecting data to provide estimates of these indicators. Finally, we discuss how to sample a population of lakes to obtain precise estimates of resource status and monitor changes over time.

Evaluating the state of a lake trout population

Major stressors of lake trout populations in the 21st century are expected to be angling exploitation, changes in fish community, and habitat changes. Angling pressure is already a major stress in areas where lakes are located close to urban centers and are easily accessed (Olver, 1991; Olver et al., this volume, Chapter 11; Lewis et al., 1990). In other areas, many lakes have been protected because they are remote and difficult to access, but this protection will disappear as human populations expand and further development of roads facilitates access. Already the network of forest access roads created in Ontario has resulted in a small portion of the Boreal landscape farther than 10 km from the nearest road (Gunn and Sein, 2000).

Changes in the fish community and habitat are expected partly as a by-product of human encroachment on northern landscapes, but also because of global changes. Climate warming is expected to change the thermal and optical properties of Shield lakes (Schindler, 1998; Schindler and Gunn, Chapter 8, this volume), affecting the amount of habitat that supports production of lake trout and other species and driving changes in fish community structure. Changes in the fish community will be exacerbated by species introductions caused by humans (Vander Zanden et al., Chapter 13, this volume).

Changes in habitat also affect lake trout abundance, but their avenue of effect is more variable and difficult to observe than that of angling (Evans et al., 1991b). Habitat changes can result from many factors (eutrophication, silt loading, acidification, climate warming, water-level fluctuations). Some of these factors can have acute effects on lake trout survival (mainly during early life stages), but their effect is often indirect, affecting the amount of habitat that is suitable for lake trout. For example, eutrophication shrinks summer habitat due to oxygen depletion in the deep layer of the hypolimnion, which provides a thermal refuge for the cold-adapted lake trout. Similarly, climate warming is expected to shrink summer habitat in small lakes (i.e., <500 -ha) due to deepening of the thermocline (King et al., 1999a, 1999b; Clark et al., Chapter 6, this volume). The effect of these habitat changes is increased crowding and thus higher levels of competition and cannibalism. In short, they reduce the carrying capacity and potential abundance of lake trout.

Whereas the potential abundance of lake trout is limited by habitat and the biotic community, the observed abundance depends on angling pressure. Angling kills fish. Because anglers prefer large fish, sustained pressure reduces the number of adult fish and thus the production of eggs and new recruits. Although compensatory mechanisms (i.e., density-dependent growth and mortality) counteract these effects (e.g., Fabrizio et al., 2001; Negus, 1995; Rose et al., 2001), the degree of compensation is limited. The net result is that angling reduces the abundance of adult fish.

In prescribing a healthy level of lake trout abundance, the inherent capability of a lake to support a lake trout population and the effect of different levels of fishing on the population's ability to sustain itself must be considered. Also, one must be prepared to make compromises. Sustainability is maximized by eliminating the stress of fishing, but this option is generally not acceptable. Human exploitation of lake trout will persist, and the role of management is to choose a level of exploitation that will allow its persistence. This choice dictates which reference values will be used to evaluate the state of a population.

For much of the 20th century, fisheries management was dominated by the single objective of achieving the maximum sustained yield (MSY) from a stock. Larkin (1977), in his now-famous paper, "An Epitaph for the Concept of Maximum Sustainable Yield," argued that MSY was not attainable for single species and must be compromised to reduce the risk of collapse and to accommodate the interactions among species that comprise the aquatic community. Collapses of commercial fisheries worldwide have supplied ample evidence that an MSY-based approach makes stocks vulnerable to overfishing and collapse. These failures have fostered a more conservative "precautionary approach" (Food and Agriculture Organization, 1995, 1996; Mace, 2001) that defines a new role for MSY. MSY is now viewed as a threshold rather than a target. It defines a "limit reference point," specifying a level of exploitation to be avoided to safeguard the long-term productivity of a stock. This concept is embodied in several U.N. Food and Agriculture Organization agreements and guidelines. Here, we use the concept to supply a reference point for evaluating the state of a lake trout population.

We define the state of a lake trout population as healthy if lake trout are abundant and likely to remain so. Given this definition of a healthy state, diagnosis requires at least two things: a measure of abundance and a reference point for classifying abundance as high or low. If this test implies abundance is low, then diagnosis is complete: the population is not healthy. If the test says abundance is high, additional data and criteria are needed to evaluate stress. One must then decide whether current stress levels are likely to drive abundance down (below criterion). If the answer is yes, the population is at risk and is not deemed healthy.

A 1998 model of lake trout exploitation (Shuter et al., 1998) supplies a framework for setting these reference values. The model describes the expected equilibrium relationship between lake trout angling yield (kg·ha^{-1}) and fishing mortality rate (a direct measure of fishing stress). The relationship depends on lake characteristics that affect growth, reproduction, and natural mortality rates of lake trout and determine the carrying capacity of a lake. The example shown (Figure 16.1A) describes the expected relationship for one lake type (Area = 1000 ha, total dissolved solids [TDS] = 26 mg·l^{-1}). It demonstrates that angling yield Y has a dome-shaped relationship with fishing mortality rate F. When no fishing occurs ($F = 0$), yield is zero. As F increases, yield increases initially and attains a maximum level ($MSY = 1$ kg·ha^{-1}) when fishing mortality rate F_{msy} reaches 0.21 year^{-1}. At higher levels of F, yield decreases and reaches zero at a fishing mortality rate (0.32 year^{-1}) that drives the population to extinction.

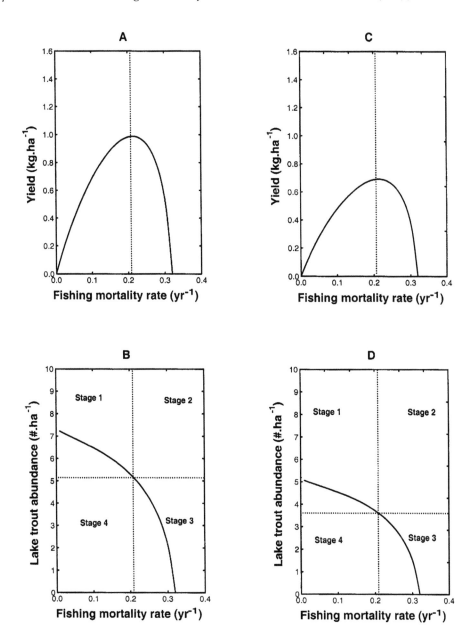

Figure 16.1 Predicted response to fishing stress for a 1000-ha lake with *TDS* = 26 mg·l^{-1} (based on Shuter et al., 1998): (A) how equilibrium yield changes as fishing mortality rate increases; (B) corresponding change in lake trout abundance; (C) and (D) effect of a change in habitat that reduces lake trout carrying capacity.

This dome-shaped relationship is due to the progressive reduction in lake trout abundance *N* as fishing mortality increases (Figure 16.1B). The abundance of fish (vulnerable to angling) is about 7.4 fish·ha^{-1} when no fishing occurs. It drops to 5 fish·ha^{-1} when the fishing mortality rate reaches F_{msy}, the value that maximizes yield. Then it decreases more rapidly as *F* approaches the extinction value.

The peak of the yield curve *MSY* supplies a useful reference point. The corresponding value of fishing mortality rate F_{msy} sets a prescribed upper limit on the level of exploitation

(Caddy and McGarvey, 1996). Thus, it supplies a criterion for judging health based on the level of stress. The corresponding value of abundance N_{msy} supplies an abundance criterion for judging health. Jointly, these criteria can be used to classify a lake into various stages of fishery development (Lester et al., 1991). These stages are identified as four quadrants in Figure 16.1B:

- Stage 1 (healthy): low fishing mortality and high abundance. These conditions are expected during the early stages of fishery development and in rehabilitated fisheries. They indicate the population is healthy. The population is not overexploited, and abundance is roughly at the correct point given the current level of fishing.
- Stage 2 (overexploited early): high fishing mortality and high abundance. These conditions are expected only during the early stages of overexploitation because stable combinations of fishing mortality rate and abundance do not exist in this quadrant. It represents a transient stage in an overexploited fishery and indicates a decline in abundance is expected if mortality remains high.
- Stage 3 (overexploited late): high fishing mortality and low abundance. This state indicates that the lake is overexploited, and the expected decline in fish abundance has occurred. A reduction in mortality rate is expected to increase sustainable harvest.
- Stage 4 (degraded): low fishing mortality and low abundance. This state indicates that the lake was probably overexploited in the past. It can result because management has imposed regulations to reduce fishing pressure on a Stage 3 fishery. It is also expected in the natural course of fishery development because anglers are likely to shift their effort to other lakes once their catch rates suffer due to the decline in abundance. The model predicts that stable combinations of abundance and mortality do not exist in this quadrant. If fishing mortality rate is kept low, a gradual transition to Stage 1, and the eventual reestablishment of stable, high-abundance levels, should occur. The fishery would then be described as rehabilitated. However, changes in the fish community resulting from the reduced abundance of the harvested species could slow this recovery or prevent it (Walters and Kitchell, 2001).

This schema suggests that measuring the state of the lake trout resource is a simple exercise. Estimates of abundance and fishing mortality for a sample of lakes could be compared to critical (*MSY*) levels predicted for each lake. The proportion of lakes judged as healthy (i.e., Stage 1) would supply an overall health index for a set of lakes. This index could be used to monitor changes in the condition of the resource if sampling was repeated at regular intervals.

One complication is that a lake's potential to produce lake trout may change, and this can affect criteria for judging its health (Shuter and Lester, Chapter 15, this volume). For example, if habitat changes caused by global warming reduce carrying capacity for lake trout, the yield–mortality curve will shrink (Figure 16.1C), the abundance–mortality curve will rotate downward (Figure 16.1D), and expected abundance at maximum sustainable yield N_{msy} will be lower. Use of this abundance criterion is more likely to result in a healthy prognosis than would the higher abundance criterion discussed above (Figure 16.1B). It is important, therefore, to identify which benchmark will be used in diagnosing health.

In this chapter we use a 20th century benchmark, basing reference levels on a current model that describes potential production of different types of lakes (Shuter et al., 1998). Because our objective is to describe ways of monitoring change in the condition of the resource, fixed criteria are needed. One consequence is that Stage 4 lakes may exist due to changes in habitat — historically high angling pressure does not have to be invoked as the cause. Either way, the lake would be classified as degraded, and this result would contribute negatively in reporting the health of the lake trout resource.

Criteria and indicators

Because lakes vary naturally in their ability to produce lake trout, appropriate criteria for evaluating state will differ among lakes. The expected yield curve depends on various factors, including the availability of suitable habitat, as well as the growth, maturation, and natural mortality of lake trout. Much of the variation in these factors is predicted by two easily measured lake parameters: surface area and total dissolved solids (*TDS*) (Payne et al., 1990; Shuter et al., 1998). *TDS*, an index of nutrient level (Ryder, 1964, 1965), has a positive effect on early growth rate (i.e., ω), measured as the slope of the growth curve at the origin (Figure 16.2A). Lake area is correlated with the asymptotic size of lake trout and maximum yields: Larger lakes tend to produce larger lake trout (Figure 16.2B) but support smaller maximum yields (Figure 16.2C). Size at maturity, size of first capture (Figure 16.2B), and natural mortality rate (Figure 16.2D) also vary with lake size and *TDS*. On larger lakes, fish mature and become vulnerable to angling at a larger size. Mortality rate in unexploited lake trout populations is positively correlated with early growth rate and negatively correlated with asymptotic length (Shuter et al., 1998), as predicted by Pauly's (1980) empirical formula. Consequently, natural mortality decreases with lake size and increases with *TDS* (Figure 16.2D).

The effect of these lake parameters on *MSY* levels of abundance and fishing mortality rate are described below for three lake sizes (100, 1,000, 10,000 ha) and three *TDS* values (13, 26, 92 mg·l^{-1}, median and 5 to 90% range). We also develop critical values of other indicators (angling effort and catch per unit effort [CUE]) that could be used to evaluate abundance or fishing stress. Appendix 16.1 supplies formulae for calculating critical values for any combination of lake size and *TDS*.

Lake trout abundance

Critical levels of N_{msy} (Figure 16.3A) refer to fish that are vulnerable to angling. Smaller lakes are expected to have a higher abundance of vulnerable fish when they are exploited at the *MSY* level. This result is somewhat misleading because the fish length criterion used to define the vulnerable population varies with lake size (Figure 16.2B). The lake size effect is less when critical abundance is calculated for a fixed minimum size of fish. A minimum size of 40 cm (the approximate size at first maturity) gives critical values that range from approximately 6 fish·ha^{-1} on small (100-ha) lakes to less than 2 fish·ha^{-1} on large (10,000-ha) lakes (Figure 16.3B). *TDS* has a positive effect on the critical abundance, but this effect is small compared to the lake size effect.

Angling CUE

The catch rate of anglers (*CUE*) is often used as an indicator of fish abundance and could be used to evaluate the state of lake trout populations. The *CUE* expected at *MSY* conditions can be calculated as

$$CUE_{msy} = q\,N_{msy} \qquad (16.1)$$

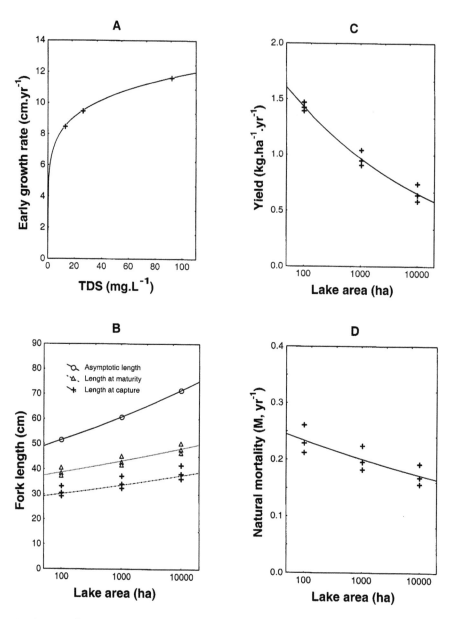

Figure 16.2 Expected growth, yield, and natural mortality rate in lake trout populations (based on Shuter et al., 1998): (A) relationship between initial growth rate ω and *TDS*; (B) relationship between lake area and size parameters asymptotic length L_∞, length at maturity, and initial length of capture by anglers; (C) relationship between lake area and *MSY* of lake trout; (D) predicted relationship between lake area and natural mortality rate *M*. In graphs B, C and D, results are shown for three levels of *TDS* corresponding to the points shown in graph A.

where *q* is the angling catchability coefficient, and N_{msy} is the abundance of lake trout vulnerable to angling. An estimate of *q* is available from Shuter et al. (1998). They found that *q* is not constant. It varies inversely with fish abundance:

$$q = \frac{0.14}{(1+0.35N)} \qquad (16.2)$$

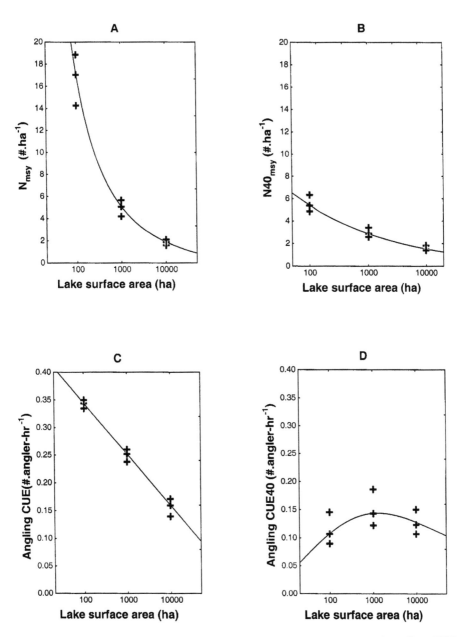

Figure 16.3 MSY levels of lake trout abundance and angling CUE: (A) N_{msy}, (B) angling CUE_{msy}, (C) $N40_{msy}$, (D) angling $CUE40_{msy}$.

Using this estimate of q, we calculated CUE_{msy} for different lake sizes and TDS levels (Figure 16.3C). We also calculated $CUE40_{msy}$, the catch rate of fish larger than 40 cm

$$CUE40_{msy} = CUE40 \frac{N40_{msy}}{N_{msy}}$$ (16.3)

to supply a large fish abundance criterion (Figure 16.3D).

These calculations imply that CUE_{msy} ranges from 0.15 to 0.34 fish·h^{-1}, depending on lake size. Small lakes are expected to support a higher CUE when exploited at MSY levels. The lake size effect disappears when only large fish are considered. At the median value of TDS (32 mg·l^{-1}), $CUE40_{msy}$ ranges from 0.12 to 0.14 fish·angler-hour^{-1}.

Mortality rate

Critical values of fishing mortality rate increase with lake size (Figure 16.4A) and TDS. As lake size increases from 100 to 10,000 ha, F_{msy} (at the median TDS value) increases from 0.14 year^{-1} to 0.29 year^{-1}. The lake size effect is partially due to differences in the natural mortality rates, which decrease with lake size (Figure 16.2D). Figure 16.4B shows that the range in critical values of total mortality rate ($Z_{msy} = F_{msy} + M$) is less than the range in F_{msy}. On small (100-ha) lakes, total mortality rate (at median TDS) is about 0.4 year^{-1} (33% annually) when sustained yield reaches a maximum. On large lakes (10,000 -ha), Z_{msy} is about 0.5 year^{-1} (40% annually).

Angling effort

In the same way that angling CUE is used as an indicator of abundance, angling effort can be used as an indicator of fishing mortality rate. Fishing mortality rate is related to angling effort E as

$$F = qE \qquad (16.4)$$

where q is the angling catchability coefficient discussed above. Thus, estimates of the critical effort intensity (angler-hour·ha^{-1}) can be calculated as

$$E_{msy} = \frac{F_{msy}}{q} \qquad (16.5)$$

where q is given by Equation (16.2).

Critical values of effort E_{msy} decrease with lake size (Figure 16.4C). As lake area increases from 100 to 10,000 ha, E_{msy} shrinks from around 6 angler-hour·ha^{-1} to values less than 4 angler-hour·ha^{-1}. This result seems counterintuitive given that F_{msy} shows the opposite trend, increasing with lake size. The difference is due to density-dependent catchability q. Because q increases as abundance decreases, strange things happen.

Rapid assessment methods

In this section we discuss various rapid assessment methods that could be used to measure lake trout abundance and fishing stress in a large set of lakes. Our intent is to identify a cost-effective suite of methods to allow assessment of multiple lakes. We discuss three methods of measuring lake trout abundance (mark–recapture, index fishing CUE, and angling CUE) and two measures of fishing stress (mortality rate, angling effort). We end with a brief discussion identifying which elements should be included in a landscape-level monitoring program.

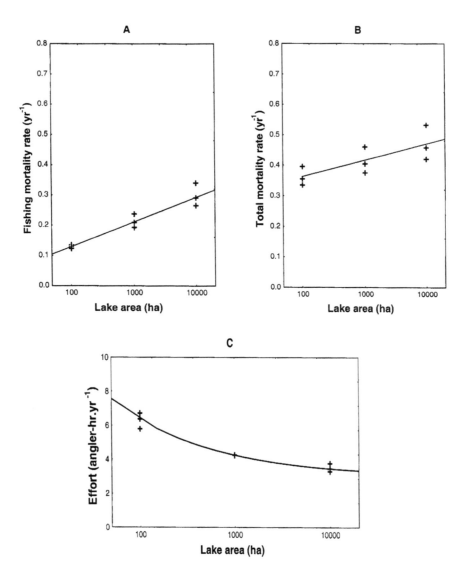

Figure 16.4 *MSY* levels of (A) fishing mortality rate, (B) total mortality rate, and (C) angling effort.

Mark–recapture estimates of lake trout abundance

Copious literature exists on methods of estimating animal abundance (Ricker, 1975; Seber, 1982; Pauly, 1984; Krebs, 1989; Skalski and Robson, 1992). A mark–recapture study is probably the best approach when a population estimate is needed for a relatively large area containing a relatively sparse population, such as lake trout. However, this method is expensive in terms of data requirements. It requires repetitive sampling of the population, first to mark a number of fish and then to recapture a suitable number of marked fish. In lake trout studies, marking is usually done in the fall when lake trout are clustered around spawning shoals. The recapture phase may be a spring index fishing survey or a survey of anglers' catches in the winter, spring, and summer following fall tagging. The method has been used successfully on several lake trout lakes in Ontario, but due to its high cost the method is not one that could be widely applied.

Index fishing CUE

A less costly alternative approach is index fishing. Index fishing is simply a standard method of fishing. It specifies fishing gear, a method of deploying the gear, a sampling time (e.g., season, time of day), and a design for selecting sample sites. Given a standard unit of effort, the mean catch per set C supplies an index of abundance that is expected to vary with the density D of fish in the sampled area of a lake,

$$C = qD \qquad (16.6)$$

where q is the catchability of the method. For example, Jones and Stockwell (1995) used the catch from a single electrofishing episode to predict the population estimate obtained from a more time-consuming multiple-pass removal method. They used this calibration as the basis of a rapid stream assessment protocol.

Ideally, index fishing should sample the entire lake so that survey results reflect lakewide abundance. To accomplish this, a lake could be stratified into a number of depth zones and within each depth zone into several pelagic layers. The combination of depth and pelagic layer identifies a number of strata. The mean catch that results from sampling in each stratum C_h would supply an index of its fish density N_h:

$$C_h = q N_h \qquad (16.7)$$

where q is the catchability coefficient of the sampling method. A lakewide index of abundance could be obtained by calculating a volume-weighted stratified mean:

$$C_{st} = \frac{1}{V} \sum (C_h V_h) \qquad (16.8)$$

where V_h is the volume of each stratum, and V is the lake volume. If catchability were known, this index could be converted to an estimate of fish density:

$$D = \frac{C_{st}}{q} \qquad (16.9)$$

This approach is difficult to implement. It calls for the use of fishing gear that samples both benthic and pelagic habitats. Also, its interpretation must either assume that catchability is the same in each stratum or estimate catchability in each stratum.

A shortcut method studied by the Ontario Ministry of Natural Resources is to focus sampling effort on a single habitat and calibrate an index of abundance for this habitat. This technique, known as Spring Littoral Index Netting (SLIN), seeks to obtain an index of the lake trout density in the littoral zone during the spring season (Lester et al., 1991; Hicks, 1999). It uses short-duration (90-min) sets of small-mesh (38-, 51-, and 64-mm) gill nets to sample lake trout during the day. Because lake trout are entangled, not wedged, by the gear, the method is not usually lethal and thus minimizes impacts of sampling. A spatially random design is used to select sites along the shore, and six panels of net (2.6 × 15.2 m) are set perpendicular to shore starting at a depth of 2.5 m. Given a total net length of 91 m, this setting procedure typically samples a depth range of about 3 to 15 m. The method is now widely used in Ontario to obtain an index of lake trout abundance, and studies designed to estimate its catchability are ongoing (Janoscik, 2001).

The Province of Quebec has also developed a standard method for indexing lake trout (D. Nadeau, 2000, personal communication). The Quebec method uses overnight sets of multimesh (25-, 38-, 51-, 64-, 76-, 102-, 127-, and 152-mm) gill nets. Sampling takes place in August before thermal destratification. Sampling intensity is determined by lake size, and the eight panels of net (1.8 × 7.6 m) are set randomly at a target depth of 15 to 40 m. Because fish are often wedged by this gear and nets are left in the water overnight, this sampling method is often lethal, but it offers advantages in terms of the data provided: Otoliths can be used to obtain better estimates of fish age, and sexual maturation can be determined.

The Scandinavian countries use a "Nordic standard" method for fish community assessment (Appelburg et al., 1995; Appelburg, 2000). The method uses a 30 m long multimesh (43-, 19.5-, 6.25-, 10-, 55-, 8-, 12.5-, 24-, 15.5-, 5-, 35-, 29-mm) monofilament gill net, composed of 12 seamlessly connected panels (1.5 × 2.5 m). The sampling takes place during summer thermal stratification and employs a depth-stratified design based on lake area and maximum depth. Nets are set for 12 h overnight and are deployed randomly around the lake; orientation to shore is also random. The technique was developed as a community sampling tool that allows use of a single gear to assess the entire fish community. Ontario is currently exploring the utility of this method to assess lake trout and other species in oligotrophic lakes. Initial results indicate it is an effective tool for characterizing the fish community and assessing lake trout abundance; however, there are concerns about sampling mortality of lake trout with this method.

For any of these methods to be used as an effective index of lake trout abundance, it is necessary to couple their use with research studies designed to estimate the catchability of the method. This necessitates intensive studies on lakes with known population size of lake trout. These studies are ongoing in Ontario.

Angling CUE

Angling *CUE* is another indicator that could be used to measure lake trout abundance. Its estimation requires that anglers submit information about the time spent fishing and the numbers (and sizes) of fish caught. *CUE* is then calculated as the total reported catch of lake trout divided by the total reported effort (angler-hours).

Active sampling methods, in which anglers are contacted shortly after (or during) a fishing trip, supply the best data for estimating angling *CUE*, but we do not recommend their use on a regular basis. Active sampling methods are expensive. We think the time spent by creel clerks contacting anglers would be better spent conducting an index fishing survey, especially considering that angling *CUE* supplies a less-reliable index of abundance. Factors affecting catch rates and size of fish captured are less controllable for angling data. For example, angling skill level and the use of specialized gear (e.g., fish-finders) may vary among lakes and over time. Such problems lead to a noisy interpretation of angling data and to questions of the value of investing heavily in its collection.

We suggest that angling data be collected on an opportunistic basis using passive methods of data collection. Anglers should be encouraged to keep diaries and record the numbers and sizes of fish caught. Angling associations should be encouraged to produce report cards based on the collection of data volunteered by their members. These data are a useful supplement to index fishing data. Angling *CUE* offers an alternative index of lake trout abundance and supplies fish size data that can be used in estimating mortality rate (see below). In addition, information about anglers' fishing patterns can be very useful in planning efficient surveys to estimate fishing effort (see below).

Angling records should include the following data about each fishing trip: date, start time, stop time, duration of any break periods (e.g., lunch), number of anglers, number of rods used, list of target fish species, number of fish caught and number kept (by species), length of each fish caught. These observations supply data for calculating an overall *CUE*, as well as *CUE* for a specified size range of fish.

While diary programs are cost-effective, some authors have identified drawbacks with using angler-reported data. Low participation rates, high turnover, and avidity bias have all been identified as shortcomings in angler diary programs. In a review of angler diary programs in Ontario, however, Cooke et al. (2000) found that diary programs with clearly defined program objectives, follow-up with program participants, and constantly recruited new participants to develop an angler base were successful in meeting the program objectives.

Mortality rate

Unbiased estimates of mortality rate may be derived from the age frequency of the catch if assumptions of constant year-class strength and mortality rate hold true and if all fish beyond some minimum age are equally vulnerable to the gear (Chapman and Robson, 1960). The method generally works well for lake trout, a slow-growing, long-lived fish that exhibits relatively little variation in year-class strength. An analysis of how recruitment variability can affect mortality rate estimates is reported in Appendix 16.2.

The fish captured by index fishing and angling supply data for estimating mortality. One problem is the determination of fish age. Although any bony structure (e.g., scales, fin rays, otoliths, etc.) can be used to age young fish, most structures underestimate the age of mature fish (Casselman, 1983; Sharp and Bernard, 1988). Otoliths offer the most reliable age estimates, but they necessitate killing of fish. Thus, good age data will not be available from the fish captured by an index fishing method that avoids killing fish. Although otoliths could be collected from fish killed by anglers, this option does not apply when most of the angling data are self-reported.

Length-based methods of estimating mortality rate (Beverton and Holt, 1956; Pauly and Morgan, 1987; Wang and Ellis, 1998) may offer an alternative that avoids the problems and costs of age determination. The approach is based on the principle that if growth did not vary among populations, mean length would be correlated with mean age. A small mean length would indicate high mortality (resulting in a low mean age). Similarly, a large mean length would indicate low mortality. The expected relationship is given by the formula (Beverton and Holt, 1956)

$$Z = k \frac{(L_\infty - \bar{L})}{(\bar{L} - L_c)} \tag{16.10}$$

where k and L_∞ are von Bertalanffy growth parameters, L_c is a length criterion, and $\bar{L}$ is mean length of fish longer than L_c. If the growth parameters (k and L_∞) are known, then the mean length of fish larger than some criterion (e.g., $L_c = 40$ cm) can be used to calculate Z.

The problem that arises is how to estimate growth parameters without age data. It cannot be done, but there are ways to reduce the amount of growth information needed. Equation (16.10) can be rearranged into the following regression form (Wetherall et al., 1987):

$$\bar{L} = a + bL_c \tag{16.11}$$

where

$$a = k \frac{L_\infty}{(Z+k)}$$

and

$$b = \frac{Z}{(Z+k)}$$

Parameters a and b can be estimated by regressing L on different L_c values (e.g., 40, 45, 50, etc.). Given these estimates of a and b, we can estimate Z as

$$Z = \frac{b}{a}\omega \qquad (16.12)$$

where $\omega = kL_\infty$. Omega ω is the slope of the growth curve at the origin (Gallucci and Quinn, 1979). It measures growth rate (cm·year^{-1}) early in life. Data from 51 lakes reported by Shuter et al. (1998) indicated that ω ranges from 7 to 20 cm·year^{-1}, but more than 50% of lakes fall within 9 to 12 cm·year^{-1}. That study also reported that some of this variation is explained by TDS (r^2 = .227, n = 51; Figure 16.2A):

$$\omega = 5.60\ TDS^{0.162} \qquad (16.13)$$

Substituting Equation (16.13) into Equation (12) gives

$$Z = \frac{b}{a}\ 5.60\ TDS^{0.162} \qquad (16.14)$$

This empirically based approach offers an entirely length-based method of estimating the mortality rate of lake trout. How well does it work? To answer this question we estimated mortality rate of 30 lake trout populations using length-based and age-based methods.

Length-based methods used 40 cm as the minimum length of full recruitment and obtained mortality estimates for fish longer than 40 cm (Figure 16.5A). Mean length above criterion was calculated at 5-cm intervals (e.g., 40, 45, 50, etc.) up to a maximum of 70 cm, and a self-weighted linear regression (mean length vs. the length criterion) was done to estimate A and B for each lake. These parameters and TDS of the lake were used in Equation (16.14) to estimate Z, the instantaneous mortality rate. Annual mortality rate A was then calculated as

$$A = 1 - e^{-Z} \qquad (16.15)$$

Age-based methods used a minimum age that matched the 40-cm minimum length chosen for length-based estimates (Figure 16.5B). Mortality rate was calculated from age frequency data using Chapman and Robson's (1960) estimator of survival S:

$$S = \frac{T}{(n+T-1)} \qquad (16.16)$$

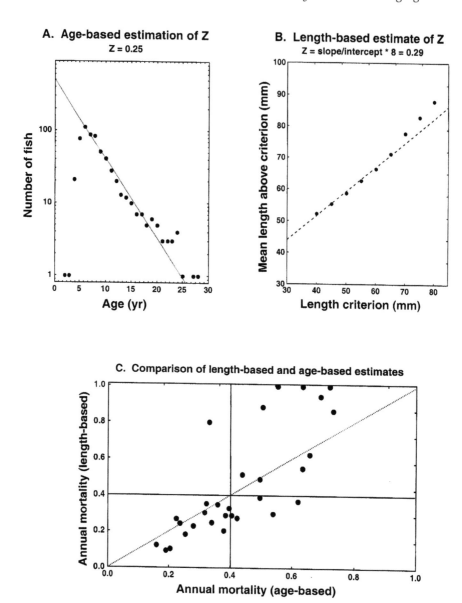

Figure 16.5 Methods of estimating mortality rate: (A) traditional age based and (B) length-based. Data in A and B are from one lake. (C) Age-based mortality estimates from a sample of lakes compared with length-based estimates.

where n is the number of fish older than or equal to the minimum age r, and T, total recoded age, is the sum of the number at each age multiplied by the recoded age (age − r). Annual mortality rate is then calculated as

$$A = 1 - S \tag{16.17}$$

The length-based method is quite effective (Figure 16.5C). A highly significant correlation exists between length-based and age-based estimates ($p < .001$). Length-based estimates closely match age-based estimates at low mortality rates (e.g., <40%), but they are much more variable when mortality rate is high. In spite of this variation, the length-based estimates do a good job of classifying lakes into categories of low and high mortality. Using a 40% mortality criterion to classify lakes (and assuming the age-based estimates are correct), we see that only 6 of the 30 lakes are misclassified, with 1 low-mortality lake classified as high and 5 high-mortality lakes classified as low. Overall, the length-based estimates would conclude that 63% of lakes are above the mortality criterion, agreeing fairly well with the age-based conclusion that 50% of lakes are above criterion. Thus we conclude that a length-based assessment could supply a reasonable estimate of the proportion of lakes overexploited.

Angling effort

Aerial surveys are appropriate for counting anglers and estimating effort over large areas (Pollock et al., 1994). A small airplane can cover 800 km in 4 h and sample a large number of lakes within a day. Compared to the cost of conducting ground surveys of each lake, aerial methods are cheap. The method has been used to survey lakes in various parts of Ontario (Kerr et al., 1992; Kerr and Cholmondeley, 1998; Kerr, 1999). The cost per lake is about $600 to obtain an annual (winter and summer) estimate of effort.

The principle of an aerial survey of fishing effort is the same as that of the roving creel survey (Robson, 1961; Pollock et al., 1994). "Instantaneous" counts of the number of anglers (or parties) fishing on a lake are taken at randomly selected times during the fishing day. Counts obtained from several days of sampling are averaged, and this mean count is multiplied by the duration of the fishing day to estimate daily effort. This estimate is then multiplied by the number of days in the fishing season to estimate total effort. If the survey counted anglers (not parties), the units of this estimate are angler-hours. If parties were counted, the units are party-hours, and the result must be multiplied by an estimate of mean party size (e.g., from angler diaries) to estimate effort in angler-hours.

The method calls for a sampling domain that spans the entire fishing day. Although fishing activity may be very low at certain times of day, counts at these times are still needed to obtain unbiased estimates of mean activity (and hence effort). If one has prior knowledge about the daily activity pattern, stratified sampling methods can be used to improve sampling efficiency. For example, the day can be divided into several periods (i.e., temporal strata). Allocating relatively more sampling to periods of high activity will supply a more precise estimate of effort. Alternatively, one could further reduce costs by opting for a biased estimate of effort. The survey could be designed to estimate effort for a specified portion of the fishing day (e.g., the expected period of high activity). This would provide a minimum estimate of fishing effort. If this estimate exceeds a critical effort level, further information is not needed to classify the lake. If not, some estimate of the bias would be needed to complete the diagnosis.

Based on ideas of Parker (1956) and Loftus (1984), Lester et al. (1991) proposed a technique that samples only during the middle of the day to estimate effort on lake trout lakes. That report supplied some initial estimates of the expansion factor needed to convert midday activity levels into an estimate of fishing effort. Kerr et al. (1992) reported that the proposed expansion factors did not work very well on a sample of lakes in southeastern Ontario and stressed the need for regionally based estimates of this factor. Here, we show how these expansion factors could be calculated using data obtained from an angler diary program.

Angler diaries can supply information about when anglers started and stopped fishing. These observations can be used to construct an angling activity profile that describes the relative number of anglers fishing at different times of the day. This profile can be used to estimate the bias that results when sampling only a portion of the fishing data. An example based on June fishing in Flack Lake demonstrates how this is done

A. Timing of fishing trips

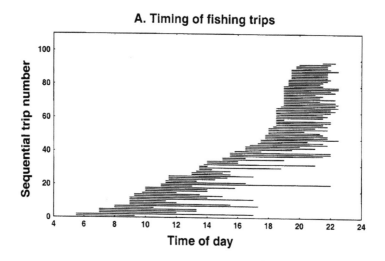

B. Estimated activity profile

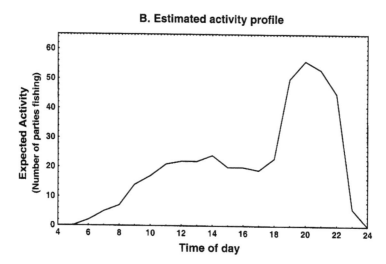

Figure 16.6 Demonstration of how complete trip information from angler diaries could be used to estimate expansion factors for estimating effort from "instantaneous" count data. See text for details.

(Figure 16.6). First, we produce a plot that describes when each fishing trip occurred (Figure 16.6A). Each horizontal line represents a trip, and the position of this line (along a time axis) indicates when the trip started and ended. Then, we calculate the number of trips in progress for each hour of the day (Figure 16.6B). In our example, the resulting profile indicates that fishing activity starts at 6:00 a.m., increases to an initial plateau by midday, remains fairly constant for several hours, increases to an evening peak around 8:00 p.m., and then rapidly declines after 10:00 p.m.

Knowledge of the daily pattern of angling allows estimation of fishing effort from counts of anglers taken at a specific time of day. Daily effort (expressed as party-hours of fishing) is estimated as Effort $= A_t D_t$, where A_t is the mean number of parties fishing at time t, and D_t is an expansion factor. An expansion factor for each hour of the day is calculated from an activity profile (e.g., Figure 16.6B) as the sum of all counts divided by the hourly counts. For example, if sampling was always done at noon, then daily fishing effort would be estimated as the mean midday activity level times 19.4, the midday expansion factor. If sampling was done at 8:00 p.m., a smaller expansion factor (7.6) would apply because a higher level of activity is expected at this time of day.

Activity profiles for different seasons and geographic areas could be constructed by pooling data from lakes where angler diary data were submitted. These profiles would supply valuable information for interpreting aerial survey data. Aerial surveys would have to be done in each season to obtain an annual estimate of effort. Activity profiles and estimates of mean party size could be constructed for each season of the year. This information would provide the background needed to optimize the design of aerial surveys and to estimate fishing effort from activity scores. Within each season, fishing effort E_i would be calculated as

$$E_i = A_i D_i S_i P_i \qquad (16.18)$$

where A_i is mean activity, D_i is a daily expansion factor (for activity recorded at a given time of day and season), S_i is the number of days in the season, and P_i is mean party size. Annual effort could then be calculated by summing across seasons.

Elements of a landscape-level monitoring program

The main program that we envisage includes:

- Index fishing to estimate lake trout abundance and mortality rate from fish length data
- Aerial surveys of angling activity to estimate fishing effort

Volunteer angler reports from representative lakes would be a useful supplement, providing valuable information to reduce the cost of other surveys. Daily and seasonal profiles of angling activity, obtained from angler diaries, support the optimal design of aerial surveys. Target species lists provide data for estimating targeted effort. Fish size data, submitted by anglers, increase the sample size for estimating mortality rate and thus its precision.

Selecting lakes

The objective of the proposed monitoring program is to describe the current and changing condition of the lake trout resource in a population of approximately 3000 lakes. Periodic assessment is needed to meet this objective. In each period, a sample of lakes must be surveyed to obtain a measure of the condition of the resource. How should this sample be chosen? How many lakes should be sampled in each period? Should the same lakes be sampled repeatedly, or should new lakes be picked in each period?

Decisions about how to sample a population are dictated largely by their effect on the precision of parameters estimated. We start, therefore, by showing how data from a sample of lakes could be used to evaluate the state of the lake trout resource and to describe the precision of this estimate. We then discuss strategies for selecting a sample of lakes to measure the condition of the resource at one period. We end by discussing alternative designs for sampling in multiple periods.

Measuring state of the resource from a sample of lakes

The sampling goal is to obtain a precise measure of the state of the lake trout resource from a small sample of lakes. The precision of the estimate depends on the among-lake variability in state, the number of lakes sampled, and the design used to select a sample. The precision obtained from a simple random sample (of fixed size) will decrease as among-lake variability in health increases, although a stratified design may be used to control this variability. For example, if the target population can be divided into strata that are more homogeneous internally than the population as a whole, then a stratified estimate will be more precise.

Among-lake variation in the condition of lakes is calculated on standardized indicators of abundance and fishing stress. Indicators are standardized for each lake by dividing each by a critical value predicted from the lake trout exploitation model described above. For example, an estimate of abundance N is divided by the predicted MSY level N_{msy}. The logarithm of this ratio (e.g., $\log N/N_{msy}$) supplies a standardized measure of abundance that can be used to measure among-lake variability in condition (based on abundance). Similarly, $\log F/F_{msy}$ or $\log E/E_{msy}$ supply standardized measures of fishing stress. Knowledge of among-lake variability in these indicators is of interest in planning a survey because precise estimates of both (i.e., abundance and stress) are sought.

How these types of measures could be used to estimate the condition of the lake trout resource and the precision of this estimate is best illustrated by an example (Figure 16.7). The example uses angling CUE as an index of abundance and fishing effort as a measure

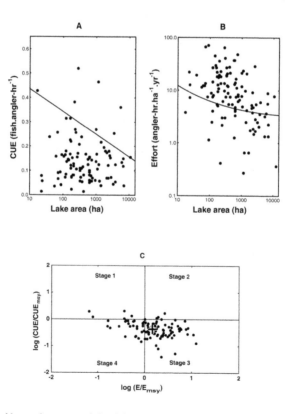

Figure 16.7 Example of how the state of the lake trout resource could be evaluated using indicators of lake trout abundance and fishing stress from a sample of lakes. (A) Estimates of angling CUE (fish·angler-hour^{-1}) for 98 lakes are plotted against lake area; predicted values of CUE_{msy} are shown by the solid line. (B) Estimates of lake trout angling effort (angler-hour·ha·year^{-1}) for the same lakes are plotted against lake area; predicted values of E_{msy} are shown by the solid line. (C) The CUE ratio (i.e., CUE/CUE_{msy}) of each lake is plotted against its effort ratio E/E_{msy} to classify lakes into various stages of development (i.e., four quadrants that correspond to those shown in Fig. 16.2B).

of fishing stress. Angling *CUE*, rather than index fishing estimates of abundance, is used because the index estimates are not yet available for a large sample of lakes. Estimates of angling *CUE* and fishing effort from 89 Ontario lakes are plotted against lake area in Figure 16.7. Critical values of each variable (CUE_{msy}, E_{msy}) were calculated based on lake area and *TDS* (see equations in Appendix 16.1) for each lake. Observed values were then divided by critical values, and the *CUE* ratio (log CUE/CUE_{msy}) was plotted against the effort ratio (log E/E_{msy}). The distribution of points in this graph (Figure 16.7C) supplies a picture that portrays the state of the resource.

These results suggest the lake trout resource is in bad shape. Only 4 of 89 (4.5%) of points lie in the healthy quadrant (Stage 1), indicating that abundance is high and fishing stress is low. Of the lakes, 73% are overexploited (Stages 2 and 3). The remainder (14%) are degraded (Stage 4, low abundance and low stress). The mean position of points and its precision can be described using the means and standard errors of log E/E_{msy} (mean = 0.099, SE = 0.05) and log CUE/CUE_{msy} (mean = −0.37, SE = 0.04). The mean state is depicted in Figure 16.7C by a cross. The size of this cross approximates the 95% confidence limits of each indicator. Because the cross lies entirely in the third quadrant, one can say with a fair degree of certainty that the average condition of these lakes is low abundance and high fishing stress.

It is important to note that these data supply a very biased view of the status of the lake trout resource. The data set is a collection of available data, not a statistical sample of lake trout lakes. In many cases, lakes were surveyed in response to anglers' concerns about the status of their lake trout populations. The sample is heavily biased in favor of lakes in the southern, highly populated portion of Ontario where the quality of fishing has been in decline for some time (Evans et al., 1991b). It is hardly surprising, therefore, that this sample of lakes gets a poor report card.

Unbiased selection of lakes is needed to obtain a fair assessment of the condition of the resource and to monitor changes that occur over time. Although our example used a biased sample of lakes, it demonstrates how data collected from an unbiased sample of lakes could be used to provide a snapshot of the resource status. If the exercise were repeated at regular intervals (i.e., each decade), a series of snapshots would be produced. These snapshots could be analyzed in various ways to describe changes in the state of the resource. For example, the proportion of lakes classified as Stage 1 is one simple method of tracking changes. This approach treats all lakes as equal. Because lake size may vary by several orders of magnitude, it may not result in a fair appraisal. Larger lakes should carry more weight when judging the state of the resource because they contribute more in terms of fishing opportunities. Thus, an area-weighted measurement, such as the proportion of total area classified as Stage 1, is probably a better indicator of change in the state of the lake trout resource.

The precision of this index will depend on the number of lakes sampled and the variability of lakes. If estimates of the abundance log-ratio and effort log-ratio are similar across lakes, a precise estimate of health can be obtained from relatively few lakes. If these measures are highly variable, more lakes will be needed to achieve the same level of precision.

Selecting a sample of lakes

Watershed units supply a suitable framework to select a sample of lakes that could be used to estimate the condition of the resource. The Shield landscape of Ontario is comprised of 76 watershed areas with an average of 23 lake trout lakes per area (Figure 16.8). In a statistical sense, these areas would be treated as primary sampling units and the lakes within each area as secondary sampling units (Cochran, 1977). The process of selecting a

Figure 16.8 Tertiary watershed map of Ontario. The number shown in each watershed is the number of known lake trout lakes.

sample is known as multistage sampling. First, a sample of watersheds is selected, then a sample of lakes within each watershed is selected.

Multistage sampling based on watershed units offers several advantages. First, watersheds define clusters of lakes that are likely to share the same types and levels of stress. Because fishing stress depends mainly on proximity to populated areas, the location of a watershed largely dictates the fishing pressure placed on its lakes. Exceptions exist because some lakes are more accessible, but overall it can be expected that variability in fishing stress is higher among watersheds than it is among lakes within watersheds. The same could be said for other stressors such as introduced species, habitat manipulations, and long-range transport of atmospheric pollutants. Stratification of watersheds may therefore be an effective way of reducing variation and obtaining a more precise estimate of resource condition. Stratification of watersheds may also be useful for administering a survey and for meeting specific information needs. For example, watersheds could be stratified along state/provincial boundaries (and regional boundaries within these) so that each jurisdiction could administer its own survey and produce its own report, yet still contribute to a grand survey of lake trout lakes.

Another advantage of watershed units is that they organize lakes into groups that can be sampled at one time. Logistically, an aerial survey of fishing effort is cost-effective when it samples a group of lakes that are found in the same area. It would not be efficient in surveying a random sample of lakes chosen from the entire population because these lakes would be widely dispersed on the landscape. The same principles would apply in selecting a sample of lakes for index fishing.

When implementing a multistage design one must choose a sample size at each stage. The usual recommendation is to sample the same fraction of elements (e.g., lakes) in each primary unit (e.g., watershed) as this will achieve a near-optimal estimate of the mean (Cochran, 1977, p. 323). The optimal fraction depends on the relative variances of the two levels (among watersheds, among lakes within watersheds). For example, fishing stress may be similar on lakes within a watershed, but highly variable among watersheds. If this is the case, relatively few lakes per watershed and more watersheds should be sampled. A useful rule of thumb (Krebs, 1989) is to sample an average of m lakes per watershed, where

$$m = \sqrt{\frac{V_2}{V_1}} \qquad (16.19)$$

V_2 is the variance among lakes within watersheds, and V_1 is the variance among watersheds.

The recommended sampling fraction is $f = m/M$, where M is the average number of lakes in each watershed. Once the second-stage sample size has been fixed, one can calculate how many watersheds n must be sampled to obtain a desired standard error of mean SE:

$$n = \frac{1}{SE^2}\left(V_1 + V_s\left(\frac{1}{m} - \frac{1}{M}\right)\right) \qquad (16.20)$$

Because diagnosis involves two types of indicators (abundance and fishing stress), these calculations should be done twice. Optimal sampling requirements for each indicator may differ, and some compromise would be needed to choose a strategy that works for both indicators.

Sampling through time

Measuring changes over time calls for periodic assessment. In each period, a sample of lakes must be selected to describe the health of the resource. Investigations of this type have three common designs (Schwarz, 1998), which can be labeled as (1) fixed sites, (2) variable sites, and (3) mixed (fixed and variable) sites. The merits of these designs can be evaluated in terms of their ability to meet the objectives of a monitoring program. These objectives include estimates of resource state and good statistical power to detect a change in state.

Design 1 (fixed sites) is often referred to as a permanent plot design: Lakes are selected in the first survey, and the same lakes are remeasured in each successive period. An advantage of this design is that comparisons over time are free of the additional variability that would be introduced if new lakes were selected at each point. According to Urquhart et al. (1998), this is the design most ecologists seem to favor for trend detection (see Skalski, 1990), although it has drawbacks in terms of measuring the overall condition of lakes.

Also, a potential problem is that effects of sampling (e.g., killing of fish) may impact lakes. Although the effect from each sampling may be small, cumulative effects could affect the condition of sampled lakes.

The other extreme is Design 2 (variable sites). A new sample of lakes is selected in each period. Data collected from one sample of lakes are used to describe the condition of the resource at one point, and a new sample is selected to describe the condition at each subsequent point. Comparison of results from two independent samples assesses change. One pitfall of this design is that the precision of the estimated change in condition may be poor because of the additional variability introduced by sampling different units each time.

Design 3 (mixed sites) is intermediate to the previous two designs. It involves repeat sampling of some lakes and selection of new lakes each time.

Urquhart et al. (1998) evaluated these types of sampling designs for measuring the current and changing condition of ecological resources and concluded that a design that balances the number of lakes visited with revisiting of individual lakes offers a sound compromise between competing concerns of status reporting and trend detection. Design 2 (variable sites) performs best, over the long term, in supplying precise estimates of resource status (due to the large sample of lakes), but it has less power to detect changes. Design 1 (fixed sites), which has the smallest sample size, has poor precision, but is far superior to Design 2 in detecting changes. The mixed design was as good as Design 1 in terms of trend detection and almost as good as Design 2 in terms of precision.

Discussion

Ecological sustainability is widely touted as a primary goal of natural resource management, yet few agencies have implemented long-term monitoring programs to measure their success in achieving this goal. "Monitoring programs have been, for the most part, piecemeal, intermittent, and short term" (Bricker and Ruggiero, 1998, p. 326).

Ontario made a good start 20 years ago when it created a network of FAUs to monitor a sample of lakes (Ontario Ministry of Natural Resources, 1978). The original design of the program included 74 lake trout lakes broadly representative of Ontario's watersheds. The program was expected to provide an understanding of how fish populations respond to stress and to devise indicators for evaluating conditions on other lakes (Ontario Ministry of Natural Resources, 1984a, 1984b, 1990). In addition, it supplied a fixed sample of lakes (i.e., Design 1) that could make a valuable contribution in monitoring changes in the status of Ontario's aquatic resources.

Unfortunately, due to funding constraints, the FAU program was never fully implemented, and monitoring has been severely curtailed in recent years. Although 74 lake trout lakes were initially identified for monitoring, only 20 lakes are currently monitored. In spite of these constraints, the program has lead to better understanding of how lake trout populations respond to stress. Its data have been used to develop a lake trout exploitation model (Shuter et al., 1998) that supplies guidelines for management and reference points for judging the health of lake trout lakes. In this regard, the program has been very successful. It would be a mistake, however, to rely solely on data from the 20 lake trout lakes sampled by this program to monitor the status of lake trout at the landscape level. Clearly, a larger sample of lakes is needed to represent a population with 2200 lakes in Ontario.

The design proposed for monitoring the lake trout resource calls for a mixture of fixed and variable sites. *Fixed sites* refer to a sample of lakes that is selected at startup and monitored through time. The FAU program supplies a set of lakes that could serve this purpose, but because this set is restricted to Ontario more fixed sites would be needed to span the domain of Boreal Shield ecosystems. *Variable sites* refer to a new sample of lakes selected each time. These sites increase the sample size for evaluating the condition of the resource. Their inclusion results in a more precise estimate of resource status and offers protection against errors that could result from relying on relatively few fixed sites.

Large scale does not have to imply extremely high cost. We have described a number of ways costs can be controlled to implement a broad-scale sampling program. Most important, the program must be focused. The luxury of measuring everything, in case it is important, does not exist when sampling a large number of lakes. Because the objective is to describe change, not to understand change, only key indicators should be measured. According to Angermeirer and Karr (1994), indicators should be responses to stress, easily measured and evaluated, distinguishable from natural variation, and societally significant. We chose lake trout abundance as the most important variable for evaluating the condition of the lake trout resource and suggested various indices of abundance that could be used. We also identified fishing stress as an important variable that must be measured to interpret abundance information.

Cost-effective assessment of many lakes is possible given a program that is focused on only two key indicators (i.e., lake trout abundance and fishing stress). Further cost savings can result from the development of an indirect rapid methods approach in which easily measured statistics substitute for parameters of ultimate concern. Examples include the use of index fishing to measure lake trout abundance, mean angling activity to measure angling effort, and mean fish length to measure mortality rate. The validation and calibration of these methods call for more detailed study of a sample of lakes. In Ontario, these needs are met by FAU lakes, which are intensively monitored by a variety of methods. In general, the monitoring program's fixed sites provide a sample of lakes that could fulfill this purpose.

Another tactic for managing cost is patience. The budget dictates the time frame for detecting change in resource status. Earlier detection results from sampling more lakes or sampling each lake more intensively. A smaller budget implies less sampling would be done on an annual basis; therefore more years of sampling would be needed to detect changes. The important question about budget is not how much money is needed, but rather what could be accomplished within a given time frame by a certain sum of money. McGuiness et al. (2000) supplied estimates of how the power to detect changes in the lake trout resource is affected by sampling intensity and design. These estimates, however, are based on limited knowledge of variability in the state of lake trout lakes. Better estimates will be available when the proposed monitoring program has been completed for one time step.

The reference conditions (i.e., criteria) proposed for evaluating the state of lake trout lakes are based on a 20th century model of lake trout production (i.e., Shuter et al., 1998). An important question is whether these criteria will be appropriate for evaluating conditions during the 21st century. Because lake size is a major determinant of potential production, the model assumes that the relationship between lake area and amount of habitat suitable for lake trout will not change. The model also assumes no change in fish community structure. These assumptions may not hold. Factors such as acid rain, climate warming, and various human activities (e.g., logging, road building) may affect light penetration and thermal properties of lakes and thus alter the volume of cold-water habitat suitable for lake trout (Schindler, 1998; Schindler and Gunn, Chapter 7, this volume; Shuter and Lester, this volume, Chapter 15). In addition, fish community structure may change because of habitat change or due to species invasions (Vander Zanden et al., this volume, Chapter 13).

These changes imply that a 20th century model of lake trout production may not supply a good reference for evaluating the state of the lake trout resource in the future.

Changes in lake trout habitat or fish community structure would affect lake trout carrying capacity and critical abundance (Figure 16.3). If these changes were detrimental to lake trout, 20th century criteria would paint a dimmer picture because fewer lakes would have lake trout populations with an abundance that exceeded the critical level. The 20th century criteria would be useful in describing change in the condition of the resource, but they would not be useful in evaluating wise use of the resource. Some lakes would be classified as low in abundance because the 20th century criteria place unrealistic expectations on abundance. Lower abundance criteria would be needed to determine whether lake trout abundance is above the level expected when exploited at the *MSY* level. Although the 20th century criteria would paint a dim picture of lake trout status relative to a "new" lower potential, it would still indicate how much resource loss has occurred due to the combination of exploitation and changes in habitat and community structure. A sampling program capable of separating exploitation effects from other stresses affecting lake trout abundance would require some additions to the program described above. These additions include the measurement of physical variables affecting the supply of lake trout habitat and biological variables that describe the biotic community.

Acknowledgments

We thank R. Ryder, J. Gunn, and R. Steedman for helpful advice and D. Evans for supplying yield and effort data. This is contribution 2000-09 of the Aquatic Research Development Section, Ontario Ministry of Natural Resources.

References

Angermeirer, P.G. and Karr, J.R., 1994, Biological integrity versus biological diversity as policy directives, *BioScience*, 44: 690–697.

Appelberg, M., 2000, Swedish Standard Methods for Sampling Freshwater Fish with Multi-mesh Gillnets, Fiskeriverket Information 2000:1, Goteborg, Sweden.

Appelberg, M., Berger, H.M., Hesthagen, T., Kleiven, E., Kurkilahti, M., Raitaniemi, J., and Rask, M., 1995, Development and intercalibration of methods in Nordic freshwater fish monitoring, *Water, Air and Soil Pollution*, 85: 401–406.

Bricker, O.P. and Ruggiero, M.A., 1998, Toward a national program for monitoring environmental resources, *Ecological Applications*, 8: 326–329.

Beverton, R.J.H. and Holt, S.J., 1956, A review of methods for estimating mortality rates in fish populations with special reference to sources of bias in catch sampling, Rapports st proces-verbaux des reunions, *International Council for the Exploration of the Sea*, 140: 67–83.

Caddy, J.F. and McGarvey, R., 1996, Targets or limits for management of fisheries? *North American Journal of Fisheries Management*, 16: 479–487.

Casselman, J.M., 1983, Age and Growth Assessment of Fish and Their Calcified Structures — Techniques and Tools, U.S. Department of Commerce National Oceanic Atmospheric Administrative Technical Report, National Marine Fisheries Service, 8: 1–17.

Chapman, D.G. and Robson, D.S., 1960, The analysis of a catch curve, *Biometrics*, 16: 354–368.

Clark, B.J., Dillon, P.J., and Molot, L.A., 2004, Lake trout (*Salvelinus namaycush*) habitat volumes and boundaries in Canadian Shield lakes, in *Boreal Shield Watersheds: Lake Trout Ecosystems in a Changing Environment*, Gunn, J.M., Steedman, R., and Ryder, R., Eds., CRC/Lewis, Boca Raton, FL, Chap. 6.

Cochran, W.G., 1977, *Sampling Techniques*, John Wiley & Sons, New York.

Cooke, S.J., Dunlop,W.I., MacLennan, D., and Power, G., 2000, Applications and characteristics of angler diary programmes in Ontario, Canada, *Fisheries Management and Ecology*, 7: 473–487.

Evans, D.O., Brisbane, J., Casselman, J.M., Coleman, K.E., Lewis, C.A., Sly, P.G., Wales, D.L., and Willox, C.C., 1991a, Anthropogenic Stressors and Diagnosis of Their Effects on Lake Trout Populations in Ontario Lakes, Ontario Ministry of Natural Resources, Lake Trout Synthesis, Toronto.

Evans, D.O., Casselman, J.M., and Willox, C.C., 1991b, Effects of Exploitation, Loss of Nursery Habitat, and Stocking on the Dynamics and Productivity of Lake Trout Populations in Ontario Lakes, Ontario Ministry of Natural Resources, Lake Trout Synthesis, Toronto.

Evans, D.O., Nicholls, K.H., Allen, Y.C., and McMurtry, M.J., 1996, Historical land use, phosphorus loading, and loss of fish habitat in Lake Simcoe, Canada, *Canadian Journal of Fisheries and Aquatic Sciences*, 53(Suppl. 1): 194–218.

Evans, D.O. and Willox, C.C., 1991, Loss of exploited, indigenous populations of lake trout, *Salvelinus namaycush*, by stocking of non-native stocks, *Canadian Journal of Fisheries and Aquatic Sciences*, 48(Suppl. 1): 134–147.

Fabrizio, M.C., Dorazio, R.M., and Schram, S.T., 2001, Dynamics of individual growth in a recovering population of lake trout (*Salvelinus namaycush*), *Canadian Journal of Fisheries and Aquatic Sciences*, 58: 262–272.

Food and Agriculture Organization, 1995, Precautionary Approach to Fisheries. Part 1: Guidelines on the Precautionary Approach to Capture Fisheries and Species Introductions, Food and Agriculture Organization, Fisheries Technical Paper 350/1.

Food and Agriculture Organization, 1996, Precautionary Approach to Fisheries. Part 2: Scientific Papers, Food and Agriculture Organization, Fisheries Technical Paper 350/2.

Gallucci, V.F. and Quinn, T.J., 1979, Reparameterizing, fitting, and testing a simple growth model, *Transactions of the American Fisheries Society*, 108: 14–25.

Gunn, J.M. and Sein, R., 2000, Effects of forestry roads on reproductive habitat and exploitation of lake trout (*Salvelinus namaycush*) in three experimental lakes, *Canadian Journal of Fisheries and Aquatic Sciences*, 57(Suppl. 2): 97–104.

Hicks, F., 1999, Manual of Instructions, Spring Littoral Index Netting, Ontario Ministry of Natural Resources, Peterborough.

Hoenig, J.M., Heisey, D.M., Lawing, W.D., and Schupp, D.H., 1987, An indirect rapid methods approach to assessment, *Canadian Journal of Fisheries and Aquatic Sciences*, 44(Suppl. 2): 324–338.

Janoscik, T., 2001, Monitoring the Abundance of Lake Trout (*Salvelinus namaycush*) with Index Netting, master of science thesis, University of Toronto.

Jones, M.L. and Stockwell, J.D., 1995, A rapid assessment procedure for the enumeration of salmonine populations in streams, *North American Journal of Fisheries Management*, 15: 551–562.

Kerr, S.J., 1999, A Survey of 12 Winter Fisheries in Lanark County during the Winter of 1998–99, Southcentral Sciences Section, Ontario Ministry of Natural Resources, Kemptville.

Kerr, S.J. and Cholmondeley, R.F., 1998, A Survey of Angling Activity on a Set of Inland Lakes in Southeastern Ontario during the Winter of 1997–98, Southcentral Sciences Section, Ontario Ministry of Natural Resources, Kemptville.

Kerr, S.J., Hoyle, J.A., and Grant, R.E., 1992, Results of an Aerial Creel Survey of Lake Trout and Splake Fisheries in Divisions 9, 10, and 29, Winter, 1992, Eastern Region, Ontario Ministry of Natural Resources, Kemptville.

King, J.R., Shuter, B.J., and Zimmerman, A.P., 1999a. Empirical links between thermal habitat, fish growth, and climate change, *Transactions of the American Fisheries Society*, 128: 656–665.

King, J.R., Shuter, B.J., and Zimmerman, A.P., 1999b, Signals of climate trends and extreme events in the thermal stratification pattern of multibasin Lake Opeongo, Ontario, *Canadian Journal of Fisheries and Aquatic Sciences*, 56: 847–852.

Krebs, C.J., 1989, *Ecological Methodology*, Harper and Row, New York.

Larkin, P.A., 1977, An epitaph for the concept of maximum sustainable yield, *Transactions of the American Fisheries Society*, 106: 1–11

Lester, N.P., Petzold, M.M., Dunlop, W.I., Monroe, B.P., Orsatti, S.D., Schaner, T., and Wood, D.R., 1991, Sampling Ontario Lake Trout Stocks: Issues and Standards, Ontario Ministry of Natural Resources, Lake Trout Synthesis, Toronto.

Lewis, C.A., Cunningham, G.L., and Chen, T., 1990, Analysis of Questionnaire on Stresses Acting on Lake Trout Lakes, Ontario Ministry of Natural Resources, Lake Trout Synthesis, Toronto.

Loftus, D.H., 1984, Sample Size and the Relative Precision of Estimates Obtained from Creel Survey Data, Ontario Ministry of Natural Resources, Haliburton–Hastings Fisheries Assessment Unit manuscript report.

Mace, P.M., 2001, A new role for *MSY* in single-species and ecosystem approaches to fisheries stock assessment and management, *Fish and Fisheries*, 2: 2–32.

McGuiness, F., Lester, N., Fruetel, M., Jackson, D., Powell, M., Dunlop, W., and Marshall, T., 2000, Monitoring the State of the Lake Trout Resource: Program Design and Costs, Ontario Ministry of Natural Resources, Peterborough.

Negus, M.T., 1995, Bioenergetics modeling as a salmonine management tool applied to Minnesota waters of Lake Superior, *North American Journal of Fisheries Management*, 15: 60–78.

Oliver, I. and Beattie, A.J., 1993, A possible method for the rapid assessment of biodiversity, *Conservation Biology*, 7: 562–568.

Olver, C.H., Desjardine, R.L., Goddard, C.I., Powell, M.J., Reitveld, H.J., and Waring, P.D., 1991, Lake Trout in Ontario: Management Strategies, Ontario Ministry of Natural Resources, Lake Trout Synthesis, Toronto.

Olver, C.H., Nadeau, D., and Fournier, H., 2004, The control of harvest on Precambrian Shield lakes, in *Boreal Shield Watersheds: Lake Trout Ecosystems in a Changing Environment*, Gunn, J.M., Steedman, R., and Ryder, R., Eds., CRC/Lewis, Boca Raton, FL, Chap. 11.

Ontario Ministry of Natural Resources, 1978, Designation of Assessment Units, Report of SPOF Working Group Number 1, Ontario Ministry of Natural Resources, Toronto.

Ontario Ministry of Natural Resources, 1984a, Fisheries Assessment Units in Ontario, Report of SPOF Working Group Number 13, Ontario Ministry of Natural Resources, Toronto.

Ontario Ministry of Natural Resources, 1984b, The Transfer, Exchange and Application of Fisheries Assessment Unit Intelligence to District Fisheries Management, Report of SPOF Working Group Number 18, Ontario Ministry of Natural Resources, Toronto.

Ontario Ministry of Natural Resources, 1990, Fisheries Assessment Unit Core Data Program, Ontario Ministry of Natural Resources, Fisheries Branch manuscript report, Toronto.

Parker, R.A., 1956, Discussion, in *Symposium on Sampling Problems in Creel Census, March 19, 1956*, Carlander, K., Ed., Iowa Cooperative Fisheries Research Unit, Iowa State College, Ames, pp. 59–62.

Pauly, D., 1980, On the interrelationships between natural mortality, growth parameters, and mean environmental temperature in 175 fish stocks, *ICES Journal of Marine Science*, 39: 175–192.

Pauly, D., 1984, *Fish Population Dynamics in Tropical Waters: A Manual for Use with Programmable Calculators*, International Center for Living Aquatic Resources Management, Manila, Philippines.

Pauly, D. and Morgan, G.R., 1987, *Length-Based Methods in Fisheries Research, ICLARM Conference Proceedings No. 13*, International Center for Living Aquatic Resources Management, Manila, Philippines, and Kuwait Institute for Scientific Research, Safat, Kuwait.

Payne, N.R., Korver, R.M., MacLennan, D.S., Nepszy, S.J., Stewart, T.J., and Thomas, E.R., 1990, The Harvest Potential and Dynamics of Lake Trout Populations in Ontario, Ontario Ministry of Natural Resources, Lake Trout Synthesis, Toronto.

Pollock, K.H., Jones, C.M., and Brown, T.L., 1994, *Angler Survey Methods and Their Applications in Fisheries Management*, American Fisheries Society, Bethesda, MD.

Ricker, W.E., 1975, Computation and Interpretation of Biological Statistics of Fish Populations, Bulletin of the Fisheries Research Board of Canada 191.

Robson, D.S., 1961, On the statistical theory of a roving creel census of fishermen, *Biometrics*, 17: 415–437.

Rose, K.A., Cowan, J.H., Winemiller, K.O., Myers, R.A., and Hilborn, R., 2001, Compensatory density dependence in fish populations: importance, controversy, understanding and prognosis, *Fish and Fisheries*, 2: 293–327.

Ryder, R.A., 1964, Chemical characteristics of Ontario lakes as related to glacial history, *Transactions of the American Fisheries Society*, 93: 260–268.

Ryder, R.A., 1965, A method for estimating the potential fish production of north-temperate lakes, *Transactions of the American Fisheries Society*, 94: 214–218.

Ryder, R.A. and Johnson, L., 1972, The future of salmonid communities in North American olig-otrophic lakes, *Journal of the Fisheries Research Board of Canada*, 29: 941–949.

Schindler, D.W., 1998, A dim future for Boreal waters and landscapes: cumulative effects of climatic warming, stratospheric ozone depletion, acid precipitation, and other human activities, *Bioscience*, 48: 157–164.

Schindler, D.W. and Gunn, J.M., 2004, Dissolved organic carbon as a controlling variable in Boreal Shield lakes, in *Boreal Shield Watersheds: Lake Trout Ecosystems in a Changing Environment*, Gunn, J.M., Steedman, R., and Ryder, R., Eds., CRC/Lewis, Boca Raton, FL, Chap. 8.

Schwarz, C.J., 1998, Studies of uncontrolled events, in *Statistical Methods for Adaptive Management Studies*, Land Management Handbook No. 42, Sit, V. and Taylor, B., Eds., British Columbia Ministry of Forests, Research Branch, Victoria, BC, pp. 19–40.

Seber, G.A.F., 1982, *The Estimation of Animal Abundance and Related Parameters*, Griffon, London.

Sharp, D. and Bernard, D.R., 1988, Precision of estimated ages of lake trout from five calcified structures, *North American Journal of Fisheries Management*, 8: 367–372.

Shuter, B.J., Jones, M.L., Korver, R.M., and Lester, N.P., 1998, A general, life history based model for regional management of fish stocks: the inland lake trout (*Salvelinus namaycush*) fisheries of Ontario, *Canadian Journal of Fisheries and Aquatic Sciences*, 55: 2161–2177.

Shuter, B.J. and Lester, N.P., 2004, Climate change and sustainable lake trout exploitation: predictions from a regional life history model, in *Boreal Shield Watersheds: Lake Trout Ecosystems in a Changing Environment*, Gunn, J.M., Steedman, R., and Ryder, R., Eds., CRC/Lewis, Boca Raton, FL, Chap. 15.

Skalski, J.R., 1990, A design for long-term monitoring, *Journal of Environmental Management*, 30: 139–144.

Skalski, J.R. and Robson, D.S., 1992, *Techniques for Wildlife Investigations: Design and Analysis of Capture Data*, Academic Press, San Diego, CA.

Urquhart, N.S., Paulsen, S.G., and Larsen, D.P., 1998, Monitoring for policy-relevant regional trends over time, *Ecological Applications*, 8: 246–257.

Vander Zanden, M.J., Wilson, K.A., Casselman, J.M., and Yan, N.D., 2004, Impacts of invasive species on food web dynamics, in *Boreal Shield Watersheds: Lake Trout Ecosystems in a Changing Environment*, Gunn, J.M., Steedman, R., and Ryder, R., Eds., CRC/Lewis, Boca Raton, FL, Chap. 13.

Walters, C., 1997, Adaptive policy design: thinking at different spatial scales, in *Wildlife and Landscape Ecology: Effects of Pattern and Scale*, Bissonette, A., Ed., Springer-Verlag, New York, pp. 386–395.

Walters, C. and Kitchell, J.F., 2001, Cultivation/depensation effects on juvenile survival and recruit-ment: implications for the theory of fishing, *Canadian Journal of Fisheries and Aquatic Sciences*, 58: 39–50.

Wang, Y.-G. and Ellis, N., 1998, Effect of individual variability on estimation of population param-eters from length-frequency data, *Canadian Journal of Fisheries and Aquatic Sciences*, 55: 2393–2401.

Wetherall, J.A., Polovina, J.J., and Ralston, S., 1987, Estimating growth and mortality in a steady-state fish stocks from length frequency data, in *Length-Based Methods in Fisheries Research*, ICLARM Conference Proceedings No. 13, International Center for Living Aquatic Resources Management, Manila, Philippines, and Kuwait Institute for Scientific Research, Safat, Kuwait, pp. 53–75.ß

Appendix 16.1

Calculation of criteria based on lake area and TDS

$$M_{msy} = 0.2200 \frac{TDS^{0.1061}}{Area^{0.06658}}$$

$$F_{msy} = 0.054 + 0.028 \log_{10} Area - 0.063 \log_{10} TDS + 0.038 \log_{10} Area * \log_{10} TDS$$

$$\log_{10} MSY = 0.594 - 0.239 \log_{10} Area - 0.066 \log_{10} TDS + 0.046 \log_{10} Area * \log_{10} TDS$$

$$\log_{10} E_{msy} = 0.0054 + \frac{1.892}{Area^{.16}} - 0.222 \log_{10} TDS + 0.073 \log_{10} Area * \log_{10} TDS$$

$$N_{msy} = \frac{CUE_{msy} E_{msy}}{F_{msy}}$$

$$N40_{msy} = 12.218 * \frac{TDS^{0.134}}{Area^{0.274}}$$

$$CUE_{msy} = 0.529 - 0.0796 \log_{10} Area - 0.00903 \log_{10} Area * \log_{10} TDS$$

$$CUE40_{msy} = \frac{0.14 \ N40_{msy}}{1 + 0.35 \ N40_{msy}}$$

Appendix 16.2

The effect of recruitment variability on estimating survival rates*

Mike Fruetel[†]
Ministry of Natural Resources

Introduction

Estimates of total annual survival or mortality rates are often useful in diagnosing the status of a fish population. In lake trout populations for example, sustainable total mortality rates at maximum equilibrium yield range from 29 to 44% depending on lake size and productivity (Shuter et al., 1998). Valid estimates of survival or mortality require that a few basic assumptions are met. These include constant mortality for all age classes and constant recruitment. The latter assumption can be troublesome, particularly for walleye populations that demonstrate highly variable year-class strengths.

Chapman and Robson (1960) provide a simple method to estimate annual survival rates based on the age at full recruitment and mean age of fully recruited fish. This issue of the *Tackle Box* will use an age structured simulation model to explore the effect of recruitment variability on estimates of survival using the method of Chapman and Robson (1960).

Methods

Age distributions were generated using an age structured population model that randomly selects the number of 1-year-old fish recruiting to the population according to a user specified recruitment probability curve and a predetermined "true" survival rate. Annual survival estimates were calculated using the method of Chapman and Robson (1960), Equation (16A2.1). In the model, the recruitment probability curve follows a negative binomial distribution which can produce "normally" distributed recruitment when the mean and variance are similar, or skewed recruitment when the variance is much larger than the mean (Figure 16A2.1). The model is strictly used to produce age distributions

Figure 16A2.1 Number of 1-year-old recruits generated from recruitment probability curves with a mean recruitment of 31 per year, but with different variances (low variance = 90, high variance = 675; N = 500 runs).

with variable year-class strength, and does not consider stock-recruitment dynamics. The model was run using two separate recruitment probability curves representing high and low recruitment variability, and three levels of annual survival (50, 60, and 70%). Each simulation was run for a 500 year period using ages one to 20 to estimate annual survival rates.

S = annual survival rate (%)
T = mean age of fully recruited fish
t' = age of fully recruited fish
n = number of fish in the sample

$$S = \frac{T - t'}{1 + T - t' - (1/n)} \qquad (16A2.1)$$

Data from Escanaba Lake (Serns, 1986) were used to set recruitment parameters for the high-variability runs. Serns estimated that the abundance of 1-year-old walleye in Escanaba Lake varied from 0.96 to 91.5 per hectare (mean = 31, variance = 675) between 1956 and 1974. These recruitment parameters produced age distributions similar to those observed from walleye index fishing data from northwestern Ontario. Recruitment parameters for the low-variability runs were set at a mean of 31 and variance of 90, which produced 1-year-old recruits ranging from 10 to 60.

Results and discussion

Estimates of survival became more variable as recruitment variability increased and actual survival levels decreased. When true survival was high, and recruitment variability low, survival estimates did not vary appreciably (Figure 16A2.2). Conversely, high recruitment variability and 50% survival produced survival estimates that ranged from 35 to 68% (5th to 95th percentile; Table 16A2.1). These simulations suggest that for species with relatively high recruitment variability, we need to be cautious when utilizing survival estimates derived from a single sample. In species that demonstrate fairly consistent recruitment, we may expect that estimates of survival will be fairly stable from year to year. From 72 to 82% of the variation in survival estimates was explained by the number of 1-year-olds entering the model population. A strong year-class of 1-year-olds effectively lowered the mean age of the sample and produced a low estimate of survival.

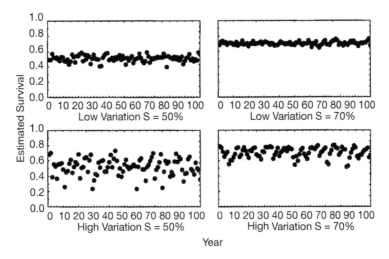

Figure 16A2.2 Effect of recruitment variability on estimates of survival (only runs 1 to 100 shown).

Table 16A2.1 Summary of Results of Survival Estimate Simulations Run over 500 Years with All Values Reported as Mean (5th to 95th Percentiles)

Recruitment variability	True S	Estimated S	Number of fish	Number of ages	Mean age
Low	50	51.0 (44–58)	60.4 (44–78)	6.4 (6–7)	2.0 (1.8–2.3)
	60	60.8 (55–66)	76.6 (58–96)	8.5 (8–9)	2.5 (2.3–2.9)
	70	70.4 (67–74)	102.4 (84–122)	11.9 (11–13)	3.4 (3.0–3.8)
High	50	52.0 (35–68)	57.9 (23–105)	6.0 (5–7)	2.2 (1.5–3.1)
	60	61.2 (47–72)	73.9 (37–123)	7.8 (6–9)	2.7 (1.9–3.6)
	70	70.7 (60–79)	98.2 (52–161)	10.9 (9–13)	3.5 (2.5–4.7)

The number of age classes present in the sample appeared to be a reasonable surrogate of survival rates. In both low and high recruitment variability simulations, fewer age classes were present at low survival rates. When true survival was 70%, 11 or 12 age classes were present in samples of 100 fish. At a true survival rate of 50%, the mean number of age classes present was about six in samples of 60 fish. The relationship between sample size, mortality, and the number of age classes must be considered when interpreting survival estimates. In low recruitment variability runs, there was little overlap in the number of ages present (Table 16A2.1). However, the high recruitment variability runs reduced the minimum number of ages observed in the model population. If recruitment did not vary at all, one might expect to see seven fully recruited age classes in a sample of 100 fish when survival was 50%. When survival is 70%, 12 fully recruited age classes would be present (Figure 16A2.3).

As expected, recruitment variability can cause large changes in abundance. The number of fish in the model population became more variable with increasing recruitment variability. Under the low variability runs, the 5th and 95th percentile of the total number of fish varied by a factor of 1.4 to 1.8 compared with 3.0 to 4.6 fold variations in total number in high variability runs (Table 16A2.1). These values appear to be reasonable for walleye populations. Serns (1982) reported a fourfold variation in the adult abundance of walleye in Escanaba Lake over a 20-year period where S was estimated at about 60%. If the recruitment parameters utilized in the high variability runs are realistic, then index netting CUE might also be expected to vary in a similar fashion.

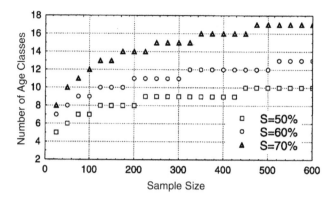

Figure 16A2.3 Effect of sample size on the expected number of age classes present for survival rates of 50, 60, and 70%. These data were generated with no recruitment variability.

Summary

The results from this modeling exercise demonstrate that recruitment variability can have a dramatic effect on the precision of mortality and abundance estimates. Biologists should also consider the effect of sampling variability. Biased samples and variation in vulnerability can also affect the precision of mortality estimates. In addition, survival estimates in one year are not independent of those in previous years. Estimates of mortality might be expected to change in a non-random fashion as exceptionally strong year-classes move through a fishery.

The Chapman–Robson equation provides a simple method to estimate survival rates based on the mean age of the catch. Survival estimates become less reliable as recruitment variability increases. The number of age classes present in a sample can be a useful indicator of survival.

Suggested Reading

Chapman, D.G. and Robson, D.S., 1960, The analysis of a catch curve, *Biometrics*, 16:354–368.

Serns, S.L., 1982, Influence of various factors on the density and growth of age-0 walleyes in Escanaba Lake, Wisconsin, 1958–1980, *Transactions of the American Fisheries Society*, 111:299–306.

Serns, S.L., 1986, Cohort analysis as an indication of walleye year-class strength in Escanaba Lake Wisconsin, 1956–1974, *Transactions of the American Fisheries Society*, 115:849–852.

Shuter, B.J., Jones, M.L., Korver, R.M., and Lester, N.P., 1998, A general life history based model for regional management of fish stocks: the inland lake trout (*Salvelinus namaycush*) fisheries in Ontario, *Canadian Journal of Fisheries and Aquatic Sciences*, 55:2161–2177.

section V

Synthesis

chapter seventeen

Boreal Shield waters: models and management challenges*

Robert J. Steedman
Ontario Ministry of Natural Resources
John M. Gunn
Ontario Ministry of Natural Resources, Laurentian University
Richard A. Ryder
RAR and Associates

Contents

This book was designed to address two important questions related to the waters of the Boreal Shield:* (1) Can we effectively manage human interactions with Boreal Shield waters and aquatic biota, at local-to-global spatial scales, now and in the future? and (2) Can lessons from Boreal Shield waters and watersheds serve as useful models for other regions and ecosystems?

Although human behavior has had great influence on Boreal Shield waters, we have limited ability to constructively affect these exquisitely complex and dynamic systems. Our influences, often harmful, have generally been through gross structural changes involving harvest or exploitation of ecological products and services such as fish, trees, hydraulic energy, and waste disposal. Our development practices also cause the insidious continued devastation of the shoreline ecotone, which serves as a center of attraction and

---

* We used the Boreal Shield ecozone, the largest of the 15 ecozones in Canada, as the geographic focus of this book. *Boreal* or *northern* forest refers to the mainly coniferous forest that covers most of the northern portion of the ecozone. *Shield* refers to the exposed Precambrian Shield bedrock that extends across the entire ecozone (Ecological Stratification Working Group, 1995).

production for many species of plants and wildlife. A sustainable relationship between humans and Boreal Shield waters therefore requires that we control and improve our environmental interactions and adopt a more cautious and protective role (i.e., the "pre-cautionary principle"). By adopting this principle, we recognize that we will often have to make choices from an array of potentially harmful actions, attempting to select the least harmful one.

Glacial legacies

In earlier chapters we introduced the idea that *Salvelinus namaycush* is a northern species pushed to southern latitudes by the glaciers, then stranded there in freshly gouged lake basins when the glaciers retreated, water levels declined, and the land began to rebound. We may infer that over time natural forces subsequently eliminated lake trout populations from some of those new habitats, particularly shallow, polymictic lakes; lakes with eroding shorelines and turbid waters; and lakes with high nutrient inputs, where competitors or predators such as walleye, northern pike, and bass became dominant.

Now, 8,000 to 12,000 years later, we see three general types of lake trout lakes on the Shield. The first is a southern group of very large lakes, including a large portion of the Laurentian Great Lakes. In these massive, deep lakes, natural lake trout populations were vertically isolated from most competitors and predators, and were extremely abundant in all of these lakes until about 60 to 90 years ago. Availability of deep-water habitats in the Laurentian Great Lakes may also have favored the development of distinctive sympatric stocks or morphotypes.

The second type is a group of intermediate-latitude lakes (i.e., boundary waters between Minnesota and Ontario, central and northern Ontario and southern Quebec) consisting of relatively shallow dimictic lakes (10 to 100 m) that generally provide well-oxygenated hypolimnia in the summer. The largest and deepest of these lakes often support diverse fish communities. In the shallowest of these lakes, lake trout populations are generally small and in constant jeopardy of exclusion by competing species and overharvesting by humans.

The third type is a group of lakes north of the Boreal Shield ecozone (i.e., from northern Quebec and Labrador to Alaska) that includes several very large lakes, but also many small and shallow polymictic lakes where summer water temperatures are cool enough to support lake trout. Generally, fewer species exist in the most northerly lakes.

A recent example of colonization

Although derived from a relatively small geographic area, the recent findings from Sud-bury, Ontario serve to recapitulate and illustrate some of the postglacial lake trout history outlined above.

Acid deposition from the Sudbury nickel smelters, the largest point source of SO_2 in the world in 1960, exterminated lake trout and many associated biota from nearly 100 lakes near Sudbury during the 1960s and early 1970s. Fortunately SO_2 emissions have been reduced by about 90% since 1960, resulting in water quality improvements in many former lake trout lakes (Gunn and Keller, 1990). Many lakes are still seriously damaged and further SO_2 reductions are required, but fishery rehabilitation projects have begun; preliminary results are both encouraging and illuminating (Gunn and Mills, 1998). For example, reestablishment of reproducing lake trout populations has proven to be very difficult in lakes with abundant competitors or predators (bass, walleye, whitefish, etc.), but almost routine in lakes with relatively simple fish communities (Hitchins and Samis, 1986; Gunn et al., 1987; Evans and Olver, 1995). The availability of spawning sites does not appear to be a limiting factor in these recovering lakes or for that matter in most other Shield lakes; newly established populations appear to quickly find enough suitable substrate sites for egg deposition (Gunn, 1995). When lake trout are restocked in very warm shallow lakes, the fish die in years with particularly warm summers (Gunn, 2002). When the watersheds of former lake trout lakes are heavily urbanized and the lakes are polluted by nutrient-rich stormwater runoff, oxygen levels decline in deep water habitats, spawning shoals are fouled with attached algae, reproduction is lost, and recolonization fails. When angling harvest is uncontrolled and excessive, lake trout populations are jeopardized as well (Gunn and Sein, 2000).

Modern threats

These histories and our recent observations suggest that lake trout face four major threats in the 21st century (Loftus and Regier, 1972; Evans et al., 1991; Ryder and Orendorff, 1999):

1. Overexploitation (lakes are relatively unproductive; fish are easy to catch)
2. Cultural eutrophication (loss of suitable hypolimnetic and reproductive habitat)
3. Introduction of invasive species (often via bait buckets)
4. Climate warming (lake trout in small, shallow lakes in northern areas are vulnerable to increased summer water temperatures; warm-water competitors increase in abundance)

The challenge of integrative ecosystem indicators

Lake trout become large, long-lived, and both ecologically and commercially valuable top predators in healthy environments. As such they have attracted great interest as integrative indicators of the health of Shield catchments and atmospheric conditions (Maitland et al., 1981; Ryder and Edwards, 1985; Marshall et al., 1987). The existence of robust, high-quality native lake trout populations clearly tells us that something is right about the biosphere in general and Boreal Shield waters in particular. For example, where lake trout thrive we may infer that the various physical, chemical, and cultural stressors identified in Table 17.1 and emphasized throughout this book are inactive, active at low levels, or active but recent. Where lake trout are declining or threatened, we may conclude that chronic or unsustainable stresses are active.

Human activity influences forests and waters directly and indirectly over large and small spatial scales (Table 17.1). There is considerable variation in the quality of scientific understanding about these influences, and as might be expected, large-scale impacts tend to be associated with greater uncertainty. Complex, multiscale threats to lake trout waters are difficult to identify, quantify, predict, and mitigate. For this reason, managers and researchers have explored integrative surrogate indicators, such as lake trout, capable of providing diagnostic, quantitative, advance warning of impending degradation of Shield ecosystems. This evolution was spurred in part by regulatory guidelines that specified protection of "biotic integrity" as required by the 1972 U.S. Federal Water Pollution Control ("Clean Water") Act, the 1978 Great Lakes Water Quality Agreement, and the Canadian National Parks Act.

Although a wide range of physical, chemical, and biotic indicators have been proposed for ecosystems, few have been implemented by management agencies. Unfortunately, the problems associated with finding practical, affordable, and technically unambiguous indices still appear rather intractable. Nonetheless, the scientific studies and debates associated with these various initiatives have greatly increased our knowledge and our recognition of the complexity of these ecosystems (Ryder and Orendorff, 1999).

Table 17.1 Spatial Scale of Human Impacts on Boreal Shield Waters

Stressor	Spatial Scale			Examples of Possible Impacts	Strength of Available Evidence	Notes
	Lake	Catchment	Regional to Global			
1. Climate change			X	Loss of cold-water habitats, expansion of cool- and warm-water species	Weak to moderate	Refers to recent climate impacts caused by human activity
2. Long-range atmospheric transport and deposition of pollutants			X	Acidification of water and soil, local extinction of aquatic biota, bioaccumulation of mercury and persistent organic contaminants	Good	
3. Ozone depletion			X	Cellular, genetic damage from UV-B radiation, biotic impairment	Weak	Strong potential interaction with Stressors 1 and 4 via dissolved organic carbon and water transparency
4. Non-point-source land use and forest disturbance		X		Mild to severe nutrient enrichment and hypolimnetic oxygen depletion, contamination, biotic impairment	Good	Relative impacts: urbanization > agriculture > forestry, wildfire
5. Introduced species	X	X		Behavioral and niche shifts, decreased production	Good	Introduction, e.g., of centrarchid fishes
6. Point-source discharges	X	X		Mild to severe nutrient enrichment and hypolimnetic oxygen depletion, contamination, behavioral changes, biotic impairment, local extinction	Good	Effluent from, e.g., pulp mills, mine tailings ponds, municipal sewage treatment plants; includes discharge of heated water from thermal power plants
7. Impoundment, dewatering, or diversion	X	X		Seasonal dewatering of spawning habitat, impaired reproduction, local mercury methylation	Good	Depends on water regulation regime
8. Shoreline or basin modification	X			Loss of habitat, impaired reproduction, nutrient loading, burial of lake or stream	Good	For instance, road construction, mines or tailings ponds, cottage development; includes impacts described in Stressor 4
9. Angling	X			Direct mortality, demographic change, local extinction	Good	Depends on harvest management

Source: After Regier (1979) and various chapters in this book.

For the last 50 to 80 years researchers have used Shield lakes as models in their studies of aquatic ecosystem response to disruption by human activities. As a result, many human effects on Shield waters are well documented and are at least partially predictable in an empirical or qualitative sense (Table 17.2). Many of these studies highlight the importance of catchment morphology and disturbance regimes and confirm the importance of long-term monitoring, comparative studies, and ecosystem experiments. A major challenge that remains is the need to develop useful diagnostic information when confronted with multiple stressor interactions. We now recognize that ecosystems are typically influenced simultaneously by multiple stressors, each potentially associated with quite similar responses from lake trout populations (Rapport et al., 1985). Lake trout may for instance be simultaneously subjected to harvest, habitat disruption, persistent contaminants, and introduction of exotic aquatic species. All of these stresses may contribute in part to a reduction in lake trout reproductive success and ultimate population size (Evans et al., 1991).

The science summarized in Table 17.2 spans spatial scales from local to biospheric (global). Model outputs (e.g., descriptions such as those that deal with thermal structure, water chemistry, composition of biotic assemblages, and lake trout demographics are but four examples) may be useful in some contexts as direct indicators of lake trophic status or as surrogates of large-scale (e.g., ecozone or landscape) phenomena. The most quantitative models tend to be regional in scope and relevant primarily to water yield and water quality rather than to habitat and biota (Carignan and Steedman, 2000). Some of these models (e.g., Ryder, 1965; Dillon and Rigler, 1975) have been used at various times as formal regulatory or assessment tools.

Some recent findings suggest that lake trout lakes may be less responsive to certain types of watershed disturbances than previously thought. Water renewal times for deep Shield lakes, where lake volume is large relative to catchment area, may range from a decade or so for small headwater lakes, to a century or more in the case of Lake Superior. This combination of lake morphology and drainage position creates significant hydrologic, thermal, and chemical inertia that may protect lake trout lakes to some degree. For example, temporary catchment disturbances that alter runoff hydrology or chemistry typically exert only small annual influences on lake trout lakes, and these effects may dissipate before the lake responds significantly (Schindler et al., 1980; Carignan et al., 2000; Steedman, 2000). In contrast, the serious consequences of chronic watershed disturbance have been repeatedly and thoroughly documented (in this volume, Legault et al. [Chapter 5], Driscoll et al. [Chapter 10], Krueger [Chapter 10], Steedman et al. [Chapter 4]). Slow water renewal rates may also delay the recovery of lakes from contaminant spills (also airborne contaminants) and can also increase the exposure and breakdown of DOC (dissolved organic carbon), leading to increased clarity and deeper penetration of solar radiation, including ultraviolet (UV) radiation (Schindler et al., 1997)

Boreal Shield ecosystems are among the best-studied natural ecosystems on earth, especially from a hydrological and geochemical perspective. However, there are still many challenging research questions to pursue, particularly when we try to understand the links between the physical and the biotic components. For example, Boreal Shield ecosystems are effective at collecting persistent organic pollutants (POPs) on the waxy surfaces of coniferous trees (Wania and McLachlan, 2001). However, the chronic effects of these trace contaminants on the biota are poorly understood. So, too, and perhaps more surprising, is the lack of quantitative information on the role of the littoral zone in the productivity and energy dynamics on Boreal Shield lakes. In fact, ecologists are just beginning to describe the community composition of some of the dominant species in the littoral zone, such as the species-rich microcrustaceans (Walseng et al., 2003). Important ecological events such as the annual ice melt (Figure 17.1), which may trigger and structure much

Table 17.2 Science Relevant to Conceptual and Quantitative Models of Boreal Shield Aquatic Ecosystem Response to Disturbance

Spatial Scale	Spatial Integration (km²)	Environmental Context	Model Name (if Applicable)	Inputs (Landscape or Environment Drivers)	Outputs (Lake-Scale Attributes or Indicators)	Primary References
Global to regional	1,000–10,000	Anthropogenic climate change, long-range transport of atmospheric pollutants, ozone depletion		Average annual temperature, land cover, precipitation	Various physical and biological attributes of lakes (ice cover duration, water renewal times, dissolved organic carbon, airborne contaminants in biota)	Magnuson et al. (1990); Schindler et al. (1990, 1996); Shuter and Meisner (1992); Yan et al. (1996); Snucins and Gunn (1995); Schindler (1998, this volume)
	1–1,000	Biogeography and biodiversity: distribution of fish species		Lake area, latitude, biogeography	Fish species richness and community composition	Matuszek and Beggs (1988); Matuszek et al. (1990); Minns (1989)
				Lake morphometry, productivity, pH, alkalinity, conductivity	Lake trout presence	Conlon et al. (1992); Gunn and Keller (1990); Ryan and Marshall (1994); Driscoll et al. (Chapter 10, this volume); Mills et al. (2000)
Watershed	1–1,000	Cumulative hydrologic impacts		Catchment forest disturbance, terrain model	Water yield, extreme flows	Buttle and Metcalfe 2000; Buttle et al. (2000)
		Material exports from land (water, carbon, nutrients, forest litter, sediment)	Ontario Trophic Status and refinements	Lake P budget (from catchment geology and land use, aerial deposition, sedimentation)	Water clarity (chlorophyll and Secchi depth)	Dillon and Rigler (1975); Hutchinson et al. (1991); Dillon et al. (1991, Chapter 7, this volume); Beaty (1994); Bayley et al. (1992); Snucins and Gunn (2000)

System	Scale	Management issue	Model	Model inputs	Model output	References
Lake	1–100	Water quality in lakes with burned and logged catchments		Catchment disturbance, lake morphology	Concentration of dissolved nutrients, carbon, cations	Carignan et al. 2000; Steedman (2000); Knapp et al. (2003);
				Lake morphometry, catchment morphometry, and drainage patterns	Concentration of dissolved organic carbon	Rasmussen et al. (1989); Schindler et al. (1997); Molot and Dillon (1997)
			Ontario end-of-summer oxygen profile	Lake morphometry, total phosphorus, dissolved oxygen at spring turnover	Late summer dissolved oxygen profile	Molot et al. (1992); Clark et al. (Chapter 6, this volume)
		Mercury accumulation in aquatic biota		Watershed slope, forest disturbance, reservoir age, lake and watershed morphology, fish species and size, sediment characteristics	Mercury concentration in zooplankton and fish	Garcia and Carignan (1999, 2000); Bodaly et al. (1984), Chapter 9, this volume; Jackson (1991); McMurtry et al. (1989); Legault et al. (this volume)
		Benchmark expectations for fish production	Morphoedaphic index (MEI) and refinements	Mean lake depth, total dissolved solids, thermal habitat volume	Long-term commercial fishery harvest (large lakes)	Ryder (1965, 1982); Christie and Regier (1988); Shuter et al. (1998)
		Effects of angling, introductions of exotic species		Lake morphometry, temperature profiles, angler effort and harvest, age-structured mortality and growth rates	Maximum sustained yield, allowable yield; production, structure, and dynamics of lake trout populations	Payne et al. (1990); Shuter et al. (1998); in this volume, Lester and Dunlop, Chapter 16, Vander Zanden et al.; Chapter 13, Gunn and Sein (2000)
		Shoreline disturbance		Shoreline disturbance by logging and wildfire	Whole-lake and littoral water temperature	Steedman and Kushneriuk (2000); Steedman et al. (1998, 2001)
					Littoral sedimentation Littoral fish populations	Steedman and France (2000) Steedman (2003)

Figure 17.1 Spring thaw. Climate models suggest that the ice-free season will be much longer in the future. (Photo by V. Liimatainen.)

of what happens biologically for the rest of the season, have not been thoroughly studied. The role of UV radiation and PAR (photosynthetically active radiation) in habitat use, the use of thermal refuge areas, the impact of invasive species, the effect of climate warming on lake productivity ... the research challenges, both old and new, remain.

The challenge of ecosystem sustainability

Given the high frequency of fire and insect outbreaks in the boreal forests and the glacial history of this region, Shield ecosystems may seem quite resilient. It is not known what these ecosystems were like in the interglacial periods, but it is probably safe to assume that each time the "slate was wiped clean" due to glacial action, functional ecosystems reestablished themselves. The individual Shield ecosystems seen today are therefore only one part of a temporal series of ecosystems that existed at this site, and the present conditions are in fact quite young in geological time, from a few decades or centuries to at most 12,000 years. The soils are also young, an interesting feature that Wright (2001) considered important in the high resilience of glaciated areas from the impacts of air pollutants (SO_2). In catchments lacking this glacial history, soils are older, and sulfur is strongly absorbed to iron and aluminum sesquioxides, making these nonglaciated systems slow to recover. The terrestrial flora and fauna of Boreal Shield ecosystems have many well-known adaptations to disturbances such as fire, but the aquatic biota also exhibit many specialized adaptations to changing conditions. For example, many zooplankton have resting stages that can remain dormant for decades or centuries until conditions improve (Hairston et al., 1995). Fish migration also occurs among connected lakes, often with surprising ease for some species (Jackson et al., 2001).

Boreal Shield ecosystems may therefore prove to be more adaptable to changing conditions than we might have originally expected, but one aspect of their identity appears unchangeable: they contain relatively low-productivity waters. Attempts to increase lake trout production by modifying habitat features (such as creating or cleaning spawning sites) are therefore destined to fail. Unfortunately, hundreds of these so-called enhancement projects have been conducted in lakes where the real management problem is excessive lake trout harvest, cultural eutrophication, or the impact of introduced species. Some people may argue that such habitat enhancement projects are still useful because they encourage public involvement in fisheries management and conservation. However, ineffective habitat enhancement projects more likely simply delay development of science and policy addressing the real problems and discourage the well-intentioned volunteers when they see that nothing comes of their efforts.

The inherently limited productivity of Shield waters also constrains the usefulness of other management actions such as hatchery stocking. Stocking may be necessary for rehabilitation purposes when a particular species has been extirpated from a lake, but when used in an attempt to supplement depressed natural populations it can often do more harm than good (Evans and Willox, 1991). For example, it can create highly unrealistic public expectations and thus increase fishing pressure to the point that irreplaceable remnant stocks of native fish are lost along with the introduced fish. Genetic introgression and disease transmission may be additional undesirable side effects of inappropriate stocking efforts (Powell and Carl, Chapter 12, this volume).

Rather than focus on how more lake trout can be produced, our desire should be to raise the value of the lake trout we have (Figure 17.2). One way to do this is to celebrate their role as environmental sentinels. What we are suggesting here is that the lake trout can be the "miner's canary" of Boreal Shield lakes, a species with narrow environmental tolerances (stenoecious species) that can serve as an early-warning signal for the ecosystem. As the largest and longest lived of the salmonid fishes native to the Shield, the lake trout also provides a longer term record because it carries within its body a physical and chemical history of the Boreal Shield environment (Figure 17.3). One of the most compelling of these stored signals is mercury body burden (Figure 17.4), which in recent years has been recognized as significantly affected by long-range atmospheric transport and deposition of fossil fuel emissions that originate far beyond Boreal Shield watersheds. Lake trout in many lakes exceed mercury consumption guidelines, even in the absence of local watershed disturbance. Due to its preference for deep, clear lakes, the lake trout is not always the most contaminated fish species (i.e., see walleye in Figure 17.4) in Boreal Shield waters. However, this does not mitigate the fact that distant human activity has polluted hundreds of lake trout lakes and other Boreal Shield lakes via this mechanism.

If we look ahead 100 or perhaps 500 years into the future, there is no doubt that lake trout ecosystems of the Boreal Shield will still be highly valued by humans but perhaps for different reasons. In future centuries urbanization and other demographic changes (e.g., depopulation of many northern towns) will likely continue, and participation levels in fishing and hunting may decline, but humans will no doubt still passionately value this landscape. The value of Boreal Shield ecosystems as sources of nutritious food will likely remain, and the importance of these ecosystems for clean drinking water, energy, and fiber will likely increase enormously, as will their value for recreation, art, and escape from the hectic urban life. The fact that we cannot imagine what this future will be simply reinforces the need to implement effective monitoring and conservation programs now to help protect this landscape.

Finally, it needs to be recognized that the often-repeated comment that the "fishing is not as good as it used to be" is not just a memory lapse. Overfishing and habitat degradation have occurred (Post et al., 2002), invading species (including humans) have arrived from all over the world, and climate changes are occurring with unknown effects (Schindler, 1998). Boreal Shield ecosystems are not static and in many ways may not be considered particularly fragile, but they need proactive and adaptive protection now. We cannot expect to "manage" these ecosystems in the same way or with the same control that we might try to manage a business. However, new information about the most important, tractable problems can be used to develop a healthier and more sustainable relationship

Figure 17.2 First lake trout — memories for a lifetime. (Photo by C. Cahill.)

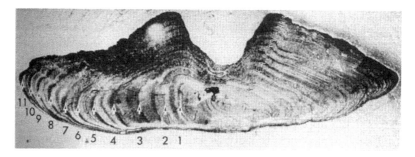

Figure 17.3 Age interpretation from an acetate replicate from a transverse section of an otolith from an age 11+ lake trout (65.9 cm total length, 2600 g male). In addition to detailed information on age and growth, new advances in chemical probe analyses are revealing much about the past environment of the fish from the records stored in otoliths and other calcified tissues. (Photo from J. Casselman.)

with Shield ecosystems. The lake trout was an influential spiritual and scientific icon during many years of public outcry to reduce acid rain. We hope that *Salvelinus namaycush* will continue to rally new science and recovery efforts addressing overharvest, pollution, harmful shoreline development, and other human activities that threaten the integrity of Boreal Shield waters (Figure 17.5).

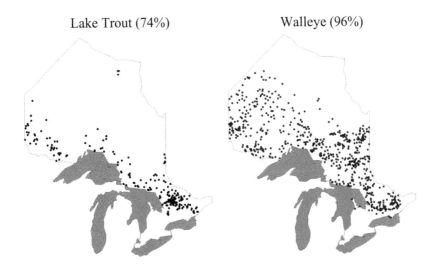

Lake Trout (74%)   Walleye (96%)

Figure 17.4 Inland lakes in Ontario where some of the larger or older lake trout and walleye exceed the Canadian marketing limit of 0.5 μg/g for mercury. The percentage of sampled lakes with mercury exceedances is indicated. (Data from the Ontario Ministry of the Environment.)

Figure 17.5 Reproducing lake trout, an icon for ecosystem sustainability. (Photo by S. Skulason.)

Acknowledgments

We thank Jake Vander Wal, Bill Keller, Ed Snucins, Nigel Lester, and Terry Marshall for many thoughtful discussions of the topics discussed in this chapter. Al Hayton of the Ontario Ministry of the Environment kindly provided the mercury data for Figure 17.4. Carissa Brown and Christine Brereton assisted with all aspects of the book production.

We are especially indebted to Judi Orendorff, who first proposed this project and provided funding and other support.

References

Bayley, S.E., Schindler, D.W., Beaty, K.G., Parker, B.R., and Stainton, M.P., 1992, Effect of multiple fires on nutrient yields from streams draining boreal forest and fen watersheds: nitrogen and phosphorus, *Canadian Journal of Fisheries and Aquatic Sciences* 49: 584–596.

Beaty, K.G., 1994, Sediment transport in a small stream following two successive forest fires, *Canadian Journal of Fisheries and Aquatic Sciences*, 51: 2723–2733.

Bodaly, R.A., Hecky, R.E., and Fudge, R.J.P., 1984, Increases in fish mercury levels in lakes flooded by the Churchill River diversion, northern Manitoba, *Canadian Journal of Fisheries and Aquatic Sciences*, 41: 682–691.

Bodaly, R.A. and Kidd, K.A., 2003, Mercury contamination of lake trout ecosystems, in *Boreal Shield Watersheds: Lake Trout Ecosystems in a Changing Environment*, Gunn, J.M., Steedman, R.J., and Ryder, R.A., Eds., Lewis/CRC, Boca Raton, FL, Chap. 9.

Buttle, J.M., Creed, I.F., and Pomeroy, J.W., 2000, Advances in Canadian forest hydrology, *Hydrological Processes*, 14: 1551–1578.

Buttle, J.M. and Metcalfe, R.A., 2000, Boreal forest disturbance and streamflow response, northeastern Ontario, *Canadian Journal of Fisheries and Aquatic Sciences*, 57(Suppl. 2): 5–18.

Carignan, R., D'Arcy, P., and Lamontagne, S., 2000, Comparative impacts of fire and forest harvesting on water quality in Boreal Shield lakes, *Canadian Journal of Fisheries and Aquatic Sciences*, 57(Suppl. 2): 105–117.

Carignan, R. and Steedman, R.J., 2000, Impacts of major watershed perturbations on aquatic ecosystems. Introduction to *Canadian Journal of Fisheries and Aquatic Sciences* Volume 57, Supplement S2, *Canadian Journal of Fisheries and Aquatic Sciences*, 57(Suppl. 2): 1–4.

Christie, G.C. and Regier, H.A., 1988, Measures of optimal thermal habitat and their relationship to yields for four commercial fish species, *Canadian Journal of Fisheries and Aquatic Sciences*, 45: 301–314.

Clark, B.J., Dillon, P.J., and Molot, L.A., 2003, Lake trout (*Salvelinus namaycush*) habitat volumes and boundaries in Canadian Shield lakes, in *Boreal Shield Watersheds: Lake Trout Ecosystems in a Changing Environment*, Gunn, J.M., Steedman, R.J., and Ryder, R.A., Eds., Lewis Publishers, Boca Raton, FL, Chap. 6.

Conlon, M., Gunn, J.M., and Morris, J.R., 1992, Prediction of lake trout (*Salvelinus namaycush*) presence in low-alkalinity lakes near Sudbury, Ontario, *Canadian Journal of Fisheries and Aquatic Sciences*, 49(Suppl. 1): 95–101.

Dillon, P.J., Clark, B.J., and Evans, H.E., 2004, Effects of phosphorus and nitrogen on lake trout (*Salvelinus namaycush*) production and habitat, in *Boreal Shield Watersheds: Lake Trout Ecosystems in a Changing Environment*, Gunn, J.M., Steedman, R.J., and Ryder, R.A., Eds., Lewis/CRC, Boca Raton, FL, Chap. 7.

Dillon, P.J., Molot, L.A., and Scheider, W.A., 1991, Phosphorus and nitrogen export from forested stream catchments in central Ontario, *Journal of Environmental Quality*, 20: 857–864.

Dillon, P.J. and Rigler, F.H., 1975, A simple method for predicting the capacity of a lake for development based on lake trophic status, *Journal of the Fisheries Research Board of Canada*, 32: 1519–1531.

Driscoll, C.T., Lawrence, G.B., Bulger, A.J., Butler, T.J., Cronan, C.S., Eagar, C., Lambert, K.F., Likens, G.E., Stoddard, J.L., and Weathers, K.C., 2004, Acidic deposition in the northeastern United States: sources and inputs, ecosystem effects, and management strategies, in *Boreal Shield Watersheds: Lake Trout Ecosystems in a Changing Environment*, Gunn, J.M. Steedman, R.J., and Ryder, R.A., Lewis Publishers, Boca Raton, FL, Chap. 10.

Ecological Stratification Working Group, 1995, A National Ecological Framework for Canada, Agriculture Canada, Agri-Food Canada, Center for Land and Biological Resources Research, and Environment Canada, Ottawa/Hull.

Evans, D.O., Brisbane, J., Casselman, J.M., Coleman, K.E., Lewis, C.A., Sly, P.G., Wales, D.L., and Willox, C.C., 1991, Anthropogenic Stressors and Diagnosis of Their Effects on Lake Trout Populations in Ontario lakes, Lake Trout Synthesis, Response to Stress Working Group, Ontario Ministry of Natural Resources, manuscript report, Toronto.

Evans, D.O. and Olver, C.H., 1995, Introduction of lake trout, *Salvelinus namaycush*, to inland lakes of Ontario, *Journal of Great Lakes Research*, 21(Suppl. 1): 313–322.

Evans, D.O. and Willox, C.C., 1991, Loss of exploited indigenous populations of lake trout, *Salvelinus namaycush*, by stocking of non-native stocks, *Canadian Journal of Fisheries and Aquatic Sciences*, 48(Suppl. 1): 134–147.

Garcia, E. and Carignan, R., 1999, Impact of wildfire and clear-cutting in the boreal forest on methyl mercury in zooplankton, *Canadian Journal of Fisheries and Aquatic Sciences*, 56: 339–345.

Garcia, E. and Carignan, R., 2000, Mercury concentrations in northern pike (*Esox lucius*) from boreal lakes with logged, burned, or undisturbed catchments, *Canadian Journal of Fisheries and Aquatic Sciences*, 57(Suppl. 2): 129–135.

Gunn, J.M., 1995, Spawning behavior of lake trout: effects on colonization ability, *Journal of Great Lakes Research*, 21(Suppl. 1): 323–329.

Gunn, J.M., 2002, Impact of the 1998 El Nino event on a lake charr, *Salvelinus namaycush*, population recovering from acidification, *Environmental Biology of Fishes*, 64: 343–351.

Gunn, J.M. and Keller, W., 1990, Biological recovery of an acid lake after reductions in industrial emission of sulphur, *Nature*, 345: 431–433.

Gunn, J.M., McMurtry, M.J., Bowlby, J.N., Casselman, J.M., and Liimatainen, V.A., 1987, Survival and growth of stocked lake trout in relation to body size, stocking season, lake acidity, and the biomass of competitors, *Transactions of the American Fisheries Society*, 116: 618–627.

Gunn, J.M. and Mills, K.H., 1998, The potential for restoration of acid-damaged lake trout lakes, *Restoration Ecology*, 6: 390–397.

Gunn, J.M. and Sein, R., 2000, Effects of forestry roads on reproductive habitat and exploitation of lake trout (*Salvelinus namaycush*) in three experimental lakes, *Canadian Journal of Fisheries and Aquatic Sciences*, 57(Suppl. 2): 97–104.

Gunn, J.M. and Sein, R., 2004, Effects of forestry roads on reproductive habitat and exploitation of lake trout, in *Boreal Shield Watersheds: Lake Trout Ecosystems in a Changing Environment*, Gunn, J.M., Steedman, R.J., and Ryder, R.A., Eds., Lewis Publishers, Boca Raton, FL, Chap. 14.

Hairston, N.G., Jr., Van Brunt, R.A., Kearns, C.M., and Engstrom, D.R., 1995, Age and survivorship of diapausing eggs in a sediment egg bank, *Ecology*, 76: 1706–1711.

Hitchins, J.R. and Samis, W.G.A., 1986, Successful reproduction of introduced lake trout in 10 northeastern Ontario lakes, *North American Journal of Fisheries Management*, 6: 372–375.

Hutchinson, N.J., Neary, B.P., and Dillon, P.J., 1991, Validation and use of Ontario's trophic status model for establishing lake development guidelines, *Lake and Reservoir Management*, 7: 13–23.

Jackson, D.A., Peres-Neto, P.R., and Olden, J.D., 2001, What controls who is where in freshwater fish communities — the roles of biotic, abiotic, and spatial factors, *Canadian Journal of Fisheries and Aquatic Sciences*, 58: 157–170.

Jackson, T.A., 1991, Biological and environmental control of mercury accumulation by fish in lakes and reservoirs of northern Manitoba, Canada, *Canadian Journal of Fisheries and Aquatic Sciences*, 48: 2449–2470.

Knapp, C.W, Graham, D.W., Steedman, R.J., and deNoyelles, F., Jr., 2003, Short-term impact of experimental deforestation on deep phytoplankton communities in four small Boreal forest lakes, *Boreal Environment Research*, 8:9–18.

Krueger, C. and Ebener, M., 2004, Rehabilitation of lake trout in the upper Great Lakes: past lessons and future challenges, in *Boreal Shield Watersheds: Lake Trout Ecosystems in a Changing Environment*, Gunn, J.M., Steedman, R.J., and Ryder, R.A., Eds., Lewis Publishers, Boca Raton, FL, Chap. 3.

Legault, M., Benoît, J., and Bérubé, R., 2004, Impact of new reservoirs, in *Boreal Shield Watersheds: Lake Trout Ecosystems in a Changing Environment*, Gunn, J.M., Steedman, R.J., and Ryder, R.A., Eds., Lewis Publishers, Boca Raton, FL, Chap. 5.

Lester, N.P. and Dunlop, W.I., 2004, Monitoring the state of the lake trout resource: a landscape approach, in *Boreal Shield Watersheds: Lake Trout Ecosystems in a Changing Environment*, Gunn, J.M., Steedman, R.J., and Ryder, R.A., Eds., Lewis Publishers, Boca Raton, FL, Cap. 16.

Loftus, K.H. and Regier, H.A., 1972, Introduction to the proceedings of the 1971 Symposium on Salmonid Communities in Oligotrophic Lakes, *Journal of the Fisheries Research Board of Canada*, 29: 613–616.

Magnuson, J.J., Meisner, J.D., and Hill, D.K., 1990, Potential changes in the thermal habitat of Great Lakes fish after global climate warming, *Transactions of the American Fisheries Society*, 119: 254–264.

Maitland, P.S., Regier, H.A., Power, G., and Nilsson, N.A., 1981, A wild salmon, trout, and char watch: an international strategy for salmonid conservation, *Canadian Journal of Fisheries and Aquatic Sciences*, 38: 1882–1888.

Marshall, T.R., Ryder, R.A., Edwards, C.J., and Spangler, G.R., 1987, Using the Lake Trout as an Indicator of Ecosystem Health: Application of the Dichotomous Key, Great Lakes Fishery Commission Technical Report 49, Great Lakes Fishery Commission, Ann Arbor, MI.

Matuszek, J.E. and Beggs, G.L., 1988, Fish species richness in relation to lake area, pH, and other biotic factors in Ontario lakes, *Canadian Journal of Fisheries and Aquatic Sciences*, 45: 1931–1941.

Matuszek, J.E., Goodier, J., and Wales, D.L., 1990, The occurrence of *Cyprinidae* and other small fish species in relation to pH in Ontario lakes, *Transactions of the American Fisheries Society*, 119: 850–861.

McMurtry, M.J., Wales, D.L., Scheider, W.A., Beggs, G.L., and Dimond, P.E., 1989, Relationship of mercury concentration in lake trout (*Salvelinus namaycush*) and smallmouth bass (*Micropterus dolomieui*) to the physical and chemical characteristics of Ontario lakes, *Canadian Journal of Fisheries and Aquatic Sciences*, 46: 426–434.

Mills, K.H., Chalanchuk, S.M., and Allan, D.J., 2000, Recovery of fish populations in Lake 233 from experimental acidification, *Canadian Journal of Fisheries and Aquatic Sciences*, 57: 192–204.

Minns, C.K., 1989, Factors affecting fish species richness in Ontario lakes, *Transactions of the American Fisheries Society*, 118: 533–545.

Molot, L.A. and Dillon P.J., 1997, Colour–mass balances and colour–dissolved organic carbon relationships in lakes and streams in central Ontario, *Canadian Journal of Fisheries and Aquatic Sciences*, 54: 2789–2795.

Molot, L.A., Dillon, P.J., Clark, B.J., and Neary, B.P., 1992, Predicting end-of-summer oxygen profiles in stratified lakes, *Canadian Journal of Fisheries and Aquatic Sciences*, 49: 2363–2372.

Payne, N.R., Korver, R.M., MacLennan, D.S., Nepszy, S.J., Shuter, B.J., Stewart, T.J., and Thomas, E.R., 1990, The Harvest Potential and Dynamics of Lake Trout Populations in Ontario, Lake Trout Synthesis, Population Dynamics Working Group, Ontario Ministry of Natural Resources, manuscript report, Toronto

Post, J.R., Sullivan, M., Cox, S., Lester, N.P., Walters, C.J., Parkinson, E.A., Jackson, L., and Shuter, B.J., 2002, Canada's recreational fisheries: the invisible collapse, *Fisheries*, 27(1): 6–17.

Powell, M.J. and Carl, L.M., 2004, Lake trout stocking in small lakes: factors affecting success, in *Boreal Shield Watersheds: Lake Trout Ecosystems in a Changing Environment*, Gunn, J.M., Steedman, R.J., and Ryder, R.A., Eds., Lewis Publishers, Boca Raton, FL, Chap. 12.

Rapport, D.J., Regier, H.A., and Hutchinson, T.C., 1985, Ecosystem behavior under stress, *American Naturalist*, 125: 617–640.

Rasmussen, J.B., Godbout, L., and Schallenberg, M., 1989, The humic content of lake water and its relationship to watershed and lake morphometry, *Limnology and Oceanography*, 34: 1336–1343.

Regier, H.A., 1979, Changes in species composition of Great Lakes fish communities caused by man, *Transactions of the North American Wildlife and Natural Resources Conference*, 44: 558–566.

Ryan, P.A. and Marshall, T.R., 1994, A niche definition for lake trout (*Salvelinus namaycush*) and its use to identify populations at risk, *Canadian Journal of Fisheries and Aquatic Sciences*, 51: 2513–2519.

Ryder, R.A., 1965, A method for estimating the potential fish production of north-temperate lakes, *Transactions of the American Fisheries Society*, 94: 214–218.

Ryder, R.A., 1982, The morphoedaphic index — use, abuse and fundamental concepts, *Transactions of the American Fisheries Society*, 111: 154–164.

Ryder, R.A. and Edwards, C.J., Eds., 1985, A Conceptual Approach for the Application of Biological Indicators of Ecosystem Quality in the Great Lakes Basin, Report to the Great Lakes Science Advisory Board of the International Joint Commission, Windsor, Ontario.

Ryder, R.A. and Orendorff, J.A., 1999, Embracing biodiversity in the Great Lakes ecosystem, in *Great Lakes Fisheries Policy and Management: A Binational Perspective*, Taylor, W.W. and Ferreri, C.P., Eds., Michigan State University Press, East Lansing, pp. 113–143.

Schindler, D.W., 1998, A dim future for boreal waters and landscapes, *BioScience*, 48(3): 157–164.

Schindler, D.W., Beaty, K.G., Fee, E.J., Cruikshank, D.R., DeBruyn, E.R., Findlay, D.L., Linsey, G.A., Shearer, J.A., Stainton, M.P., and Turner, M.A., 1990, Effects of climatic warming on lakes of the central boreal forest, *Science*, 250: 967–970.

Schindler, D.W., Curtis, P.J., Bayley, S.E., and Parker, B.R., 1997, DOC-mediated effects of climate change and acidification on boreal lakes, *Biogeochemistry*, 36: 9–28.

Schindler, D.W., Curtis, P.J., Parker, B.R., and Stainton, M.P., 1996, Consequences of climate warming and lake acidification for UV-B penetration in North American boreal lakes, *Nature*, 379: 705–708.

Schindler, D.W. and Gunn, J.M., 2004, Dissolved organic carbon as a controlling variable in Boreal Shield lakes, in *Boreal Shield Watersheds: Lake Trout Ecosystems in a Changing Environment*, Gunn, J.M., Steedman, R.J., and Ryder, R.A., Eds., Lewis Publishers, Boca Raton, FL.

Schindler, D.W., Newbury, R.W., Beaty, K.G., Prokopowich, J., Ruszczynski, T., and Dalton, J.A., 1980, Effects of a windstorm and forest fire on chemical losses from forested watersheds and on the quality of receiving streams, *Canadian Journal of Fisheries and Aquatic Sciences*, 37: 328–334.

Shuter, B.J., Jones, M.L., Korver, R.M., and Lester, N.P., 1998, A general life history based model for regional management of fish stocks: the inland lake trout (*Salvelinus namaycush*) fisheries in Ontario, *Canadian Journal of Fisheries and Aquatic Sciences*, 55: 2161–2177.

Shuter, B.J. and Meisner, J.D., 1992, Tools for assessing the impact of climate change on freshwater fish populations, *Geo-Journal*, 28(1): 7–20.

Snucins, E.J. and Gunn, J.M., 1995, Coping with a warm environment: behavioral thermoregulation by lake trout, *Transactions of the American Fisheries Society*, 124: 118–123.

Snucins, E. and Gunn, J., 2000, Interannual variations in the thermal structure of clear and colored lakes, *Limnology and Oceanography*, 45: 1639–1646.

Steedman, R.J., 2000, Effects of experimental clearcut logging on water quality in three small boreal forest lake trout (*Salvelinus namaycush*) lakes, *Canadian Journal of Fisheries and Aquatic Sciences*, 57(Suppl. 2): 92–96.

Steedman, R.J., 2003, Littoral fish response to experimental logging around small Boreal Shield lakes, *North American Journal of Fisheries Management*, 23:392–403.

Steedman, R.J. and France, R.L., 2000, Origin and transport of aeolian sediment from new clearcuts into boreal lakes, northwestern Ontario, Canada, *Water, Air, and Soil Pollution*, 122: 139–152.

Steedman, R.J., France, R.L., Kushneriuk, R.S., and Peters, R.H., 1998, Effects of riparian deforestation on littoral water temperatures in small boreal forest lakes, *Boreal Environment Research*, 3:161–169.

Steedman, R.J. and Kushneriuk, R.S., 2000, Effects of experimental clearcut logging on thermal stratification, dissolved oxygen, and lake trout (*Salvelinus namaycush*) habitat volume in three small boreal forest lakes, *Canadian Journal of Fisheries and Aquatic Sciences*, 57(Suppl. 2): 82–91.

Steedman, R.J., Kushneriuk, R.S., and France, R.L., 2001, Littoral water temperature response to experimental shoreline logging around small boreal forest lakes, *Canadian Journal of Fisheries and Aquatic Sciences*, 58: 1638–1647.

Vander Zanden, M.J., Wilson, K.A., Casselman, J.M., and Yan, N.D., 2003, Impacts of invasive species on food web dynamics, in *Boreal Shield Watersheds: Lake Trout Ecosystems in a Changing Environment*, Gunn, J.M., Steedman, R.J., and Ryder, R.A., Eds., Lewis/CRC, Boca Raton, FL.

Walseng, B., Yan, N.D., and Schartau, A.-K., 2003, Littoral microcrustaceans (Cladocera and Copepoda) indicators of acidification in Canadian Shield lakes, *Ambio*, 32(3):208–213.

Wania, F. and McLachlan, M.S., 2001, Estimating the influence of forests on the overall fate of semivolatile organic compounds using a multimedia fate model, *Environmental Science and Technology*, 36: 4860–4867.

Wright, R.F., 2001, Use of the dynamic model MAGIC to predict recovery following implementation of the Gothenburg Protocol, *Water, Air and Soil Pollution: Focus,* 00: 1–28.

Yan, N.D., Keller, W., Scully, N.M., Lean, D.R.S., and Dillon, P.J., 1996, Increased UV-B penetration in a lake owing to drought-induced acidification, *Nature,* 381: 141–143.

section VI

appendix one

Long-term monitoring sites on the Boreal Shield

Long-term monitoring programs are essential to identify the rate and direction of environmental change in the Boreal Shield ecozone. Canada's Ecological Monitoring and Assessment Network (EMAN; www.eman-rese.ca) currently provides a coordinating service to support data sharing, communication, and training in environmental and ecological monitoring and to assist in preparing state-of-the-resource assessment reports (e.g., Urquizo et al., 2000). Brief descriptions and maps of some of the key monitoring sites in the southern and southwestern Boreal Shield ecozone are given here. For more information on monitoring sites in the eastern part of the ecozone see, http://eqb-dqu.cciw.ca/eman/network/borshque.htm.

Experimental Lakes Area

The Experimental Lakes Area (ELA) (Figure A1.1) was established in 1968 and is located in northwestern Ontario approximately 250 km from Winnipeg and 50 km east-southeast of Kenora. The ELA includes 58 small lakes (with areas 1 to 84 ha) and their drainage basins, plus three additional stream segments set aside and managed through an agreement between the Canadian and Ontario governments. For an additional description of the site see the ELA special issue of the *Journal of the Fisheries Research Board of Canada* [28(2), 1971]. Descriptions of current projects and listings of research papers are available online at www.umanitoba.ca/institutes/fisheries/ELApubln.html.

Turkey Lakes Watershed

The Turkey Lakes Watershed (TLW) Study was initiated in 1980 to evaluate anthropogenic perturbation of Canadian Shield ecosystems (e.g., the effects of acidic deposition). The basin is located in the Algoma District of central Ontario about 50 km north of Sault Ste. Marie. It is an undeveloped and completely forested headwater basin with an area of 10.5 km^2; it contains a chain of four lakes (five distinct lake basins) that ultimately drain into Lake Superior via the Batchawana River (Figure A1.2). Comprehensive records of meteorological and surface water physical and chemical data have been maintained from the study's inception with biological data collected since the mid-1980s. For additional information on the physical, chemical, and biological characteristics of the TLW and research activities, participants, databases, and the like plus a searchable list of the more

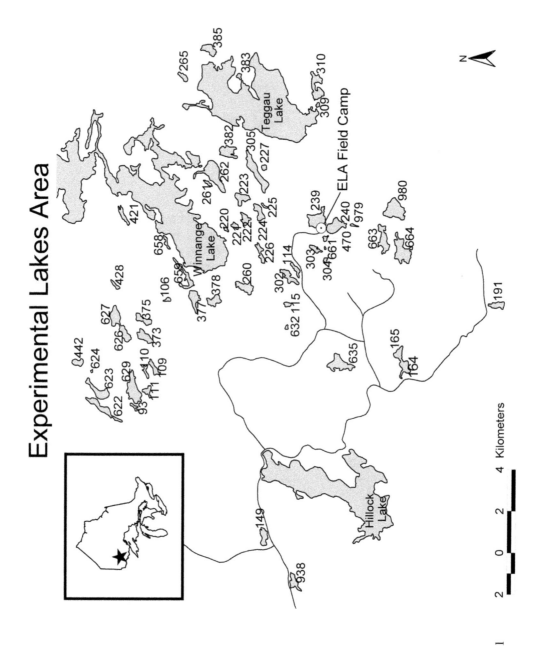

Figure A1.1 Experimental Lakes Area (ELA).

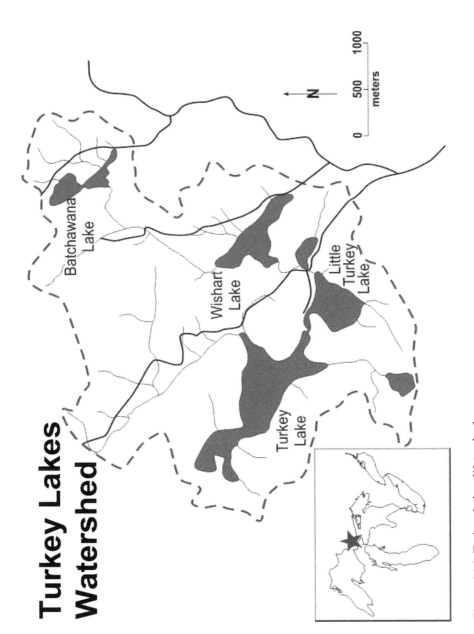

Figure A1.2 Turkey Lakes Watershed.

than 280 associated publications, see the Turkey Lakes Watershed study Web site at www.tlws.ca. There have been three special volumes of TLW publications: *Canadian Journal of Fisheries and Aquatic Sciences* [45(Suppl. 1), 1988]; *Ecosystems* [4(6), 2001]; and *Water, Air and Soil Pollution: Focus* [2(1), 2002].

Killarney Park

A chemical and biological monitoring program of 21 remote lakes in Killarney Park (Figure A1.3) was established to study the interaction of major stressors such as acidification, climate change, and metal deposition. Data records began in the 1970s, with some associated university research projects beginning in 1967. Killarney Park contains some of Canada's clearest waters, which are highly sensitive to climate-driven drought effects. Information on recent studies conducted under the Canada/Norway Northern Lakes Recovery Study at Killarney is provided in a special issue of *Ambio* [32(3), 2003].

North-Temperate Lakes Long Term Ecological Research

The North-Temperate Lakes Long Term Ecological Research (LTER) site, housed at the Center of Limnology — Trout Lake Station (University of Wisconsin–Madison) (Figure A1.4), is 1 of 24 LTER sites across the United States. Trout Lake Station was established in 1925 by Birge and Juday, two of North America's limnological pioneers. With long-term predictive regional ecology as the focus, intensive data collection began on seven LTER lakes in 1981 and continues. More information on the lakes, current research, and online data catalogue is available at http://limnosun.limnology.wisc.edu/.

Sudbury area lakes

Monitoring of lakes in the Sudbury, Ontario, area (Figure A1.5) began in the early 1970s, and additional study lakes were added in the 1980s. Monitoring includes regular sampling for chemistry, zooplankton, and phytoplankton on 15 lakes that vary in surface area from 5.8 to 315.8 ha. Studies of fish and benthic invertebrates are also completed periodically on some of these lakes. The lakes are within the (17,000 km²) area historically affected by the Sudbury smelter emissions, but vary greatly in their initial degree of damage, ranging from the highly acidic, metal-contaminated lakes close to Sudbury to acidified, undeveloped lakes in more remote areas (e.g., within Lady-Evelyn Smoothwater Park). Studies have primarily focused on recovery processes as the lakes respond to about 90% reduction in atmospheric deposition of sulfur and metals since 1960. More recently other stressors, including climate change, exotic species, and depletion of base cations, have been emphasized. The *Canadian Journal of Fisheries and Aquatic Sciences* [49(Suppl. 1), 1992] and http://laurentian.ca/biology/ecologyunit.html provide additional information.

Dorset Environmental Science Centre

Eight lakes with 20 inflowing tributaries, 8 outflows, and 20 small catchments (10 to 200 ha) in mixed deciduous-coniferous forests with extensive wetlands have been monitored by the Dorset Environmental Science Centre (Figure A1.6) since 1975 to assess the effects of environmental stresses (e.g., climate change, greenhouse gases, acid deposition, nutrient enrichment, mercury and other trace metals, ultraviolet radiation) on aquatic ecosystems. The lakes are located in south-central Ontario where the primary industry is tourism.

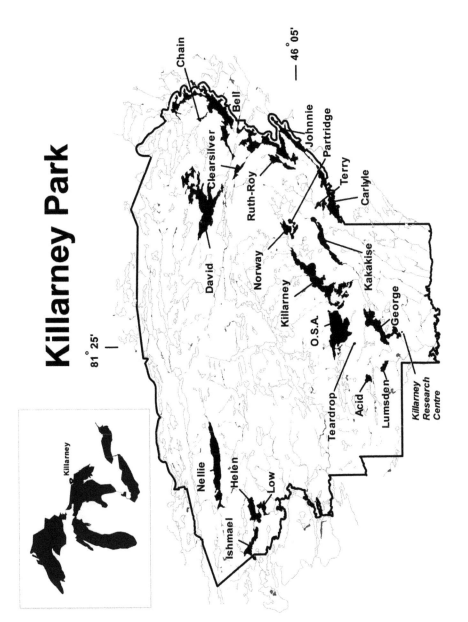

Figure A1.3 Study lakes used for the 1997–2002 Northern Lakes Recovery Study.

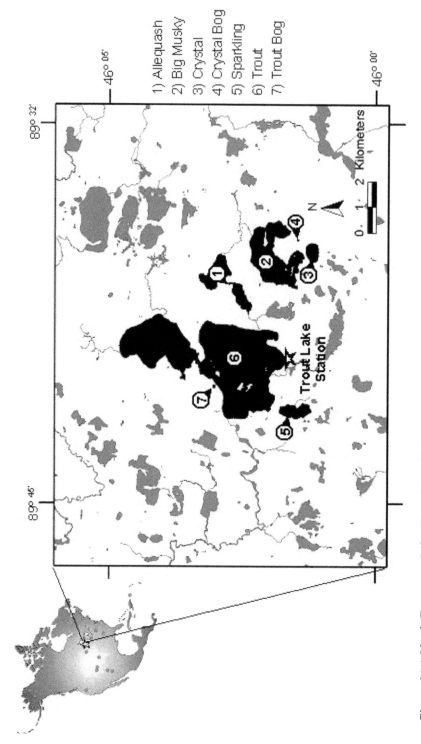

Figure A1.4 North-Temperate Lakes Long Term Ecological Research (LTER) Trout Lake Station.

Sudbury Area Lakes

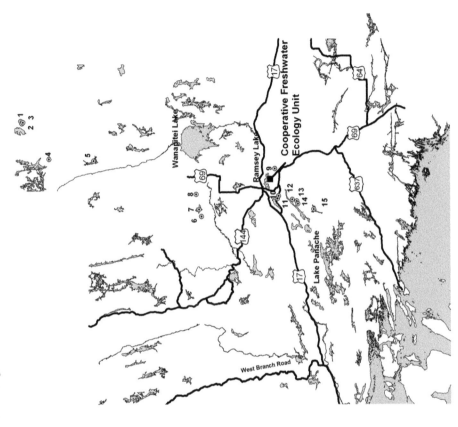

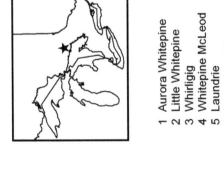

1 Aurora Whitepine
2 Little Whitepine
3 Whirligig
4 Whitepine McLeod
5 Laundrie
6 Sans Chambre
7 Nelson
8 Joe
9 Daisy
10 Middle
11 Hannah
12 Lohi
13 Clearwater
14 Swan
15 Wavy

Figure A1.5 Sudbury area lakes.

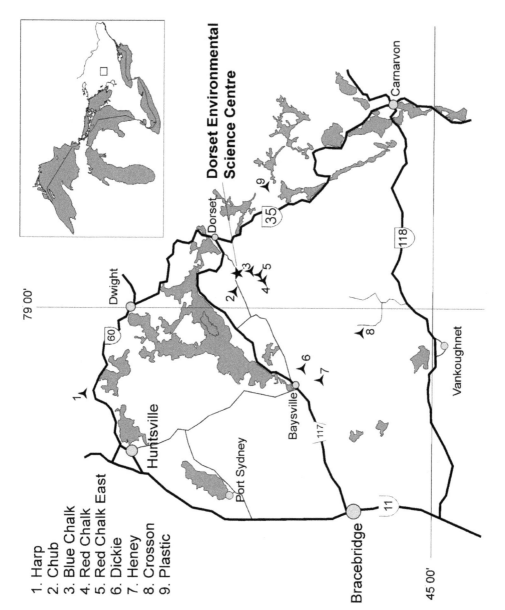

Figure A1.6 Dorset Environmental Science Centre.

Monitoring activities include meteorology, biogeochemical cycles, water chemistry, plankton, and benthos. For more information see www.trentu.ca.

Harkness Laboratory of Fisheries Research

Harness Laboratory of Fisheries Research is located on 5154.2-ha Lake Opeongo in Algonquin Park, Ontario (Figure A1.7). It is Canada's oldest freshwater research station with a prime focus on fish ecology. Since 1936, an ongoing census of angler's catch (lake trout and smallmouth bass) has been maintained continuously for Lake Opeongo. The data support analyses of long-term trends in growth, production, and population dynamics of two species (Shuter et al., 1987). Other long-term data sets focus on reproductive timing and production of young and age and growth patterns in other species for the last 10 to 20 years.

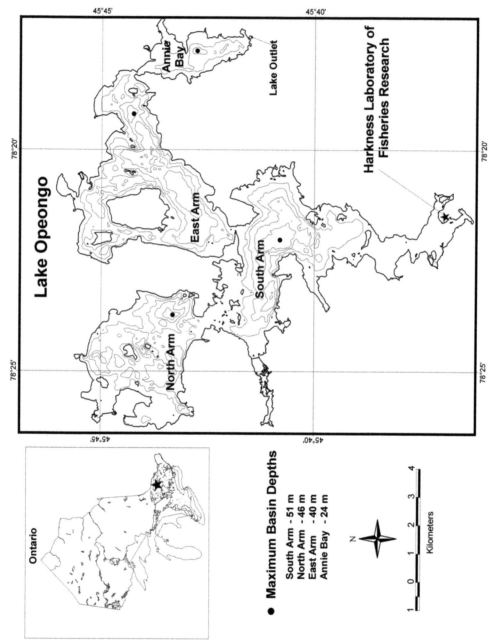

Figure A1.7 Harkness Laboratory of Fisheries Research.

References

Shuter, B.J., Matuszek, J.E., and Regier, H.A., 1987, Optimal use of creel survey data in assessing population of Lake Opeongo lake trout (*Salvelinus namaycush*) and smallmouth bass (*Micropterus dolomieu*), 1963–83, *Can. J. Fish. Aquat. Sci.*, **44**(Suppl. 2):229–238.

Urquizo, N., Bastedo, J., Brydges, T., and Shear, H., 2000, Ecological Assessment of the Boreal Shield Ecozone, Indicators and Assessment Office, Environment Canada, Ottawa. Available online at www.ec.gc.ca/soer-ree/english/default.cfm.

appendix two

Lake trout lakes of the Boreal Shield ecozone of North America

Lake trout lakes are listed alphabetically for the jurisdictions of Minnesota, New York (Adirondacks), Ontario (Northeastern, Northwestern, South Central) and Quebec. The Laurentian Great Lakes and 33 other large lakes (>10,000 ha), many of which cross jurisdictions, are not included. Alternate names are listed in parentheses, and a separate entry is given for each name. Lakes with the same name are ordered by latitude. This atlas is considered an accurate list of lake trout lakes in these jurisdictions at the time of publication; however many other lake trout lakes may be discovered in the future, particularly in parts of Quebec where less extensive surveys have been conducted. In addition, other lake trout lakes exist in the Shield bedrock regions of Alberta, Manitoba, Michigan, Saskatchewan, and Wisconsin. These areas are part of the Boreal Shield ecozone, but these lakes are not yet included in the atlas.

Lake names and descriptions were obtained from lake management agencies from Minnesota, New York, Ontario, and Quebec (see acknowledgment section for contact names) in 1998. The data was obtained from surveys conducted by government agencies and universities over a period of several decades. The most recent data was generally used but with the sampling period duration and the wide variety of methods and data sources, the descriptions for parameters such as Secchi depth or specific conductance should be considered as approximations. The current status (e.g. origin, reproduction, abundance) of the lake trout populations in many of these lakes is also uncertain.

With the large Ontario data set we attempted to update and verify the earlier data. Longitude and latitude of all Ontario lakes were checked by plotting each lake using 2000 Softmap: Ontario Top50 software (Softmap Technology, Inc.). Calculated centroids were used as the point of location of the lake. Location points were manually moved to the largest basin of the lake if the centroid missed the lake surface. Because of the large number of lakes with similar names, unique waterbody identification codes (OFIS codes) were added by adopting the codes assigned by Ontario Ministry of Natural Resources (OMNR) through their Ontario Fisheries Information System (OFIS). Blanks indicate that no OFIS code has yet been assigned. Arcview GIS 3.2 software was used to calculate surface areas for approximately 500 lakes with missing data from the OMNR Natural Resources Values Information System (NRVIS). At the time of publication (June 2003) there were still about 100 Ontario lakes from previous lake trout lists where the locational data could not be verified and the lakes were excluded from this atlas.

MINNESOTA

Lake name	Latitude	Longitude	Surface area (ha)	Maximum depth (m)	Mean depth (m)	Conductivity (µS/cm)	Secchi depth (m)
Ahmakose	48°01'34"	91°10'57"	15.2	22.9	8.3	29	5
Alder	48°02'43"	90°17'42"	215.3	21.9	9.1	38	5
Alpine	48°07'57"	90°59'25"	340.9	19.8	5.7	42	4
Alton	47°52'15"	90°54'33"	421.0	22.0	7.3	33	5
Amoeber	48°07'27"	91°08'50"	156.8	33.5	11.3	52	7
Basswood	48°03'29"	91°35'16"	9195.0	34.0		52	4
Bat	48°04'00"	90°54'52"	31.6	33.5	13.4	33	5
Bear	47°18'45"	91°17'29"	7.2	21.0		39	5
Big Trout	46°42'57"	94°09'28"	542.9	39.0	15.6	164	6
Birch	48°04'00"	90°32'36"	99.8	21.0	7.9	51	5
Blue Snow	48°03'32"	90°52'03"	19.9	15.2		22	4
Bluewater	47°25'17"	93°33'13"	147.3	36.6	15.1	225	5
Bone	47°45'25"	91°02'28"	18.8	15.5		78	3
Brule	47°56'17"	90°40'19"	1884.3	23.8	8.6	36	5
Burntside	47°56'42"	91°56'21"	2873.6	38.4	12.2	31	7
Canisteo Pit	47°19'34"	93°22'58"	392.2	88.7			3
Caribou	47°31'37"	93°38'24"	96.6	46.3	14.6	38	9
Cash	47°59'02"	90°44'55"	30.3	17.7	6.2		2
Cherokee	47°57'46"	90°47'55"	308.2	43.3	10.5	28	3
Cherry	48°08'11"	91°06'02"	58.8	30.2	12.6	53	7
Clearwater	48°04'52"	90°21'05"	522.8	39.9	15.1	42	7
Crooked	48°01'57"	90°55'08"	94.3	22.9	8.0	49	6
Cruiser	48°29'52"	92°48'19"	46.5	27.7	13.2	22	8
Crystal	48°02'56"	90°14'17"	84.9	27.7	12.1	33	5
Cypress	48°09'56"	91°06'59"	467.8	35.4	10.9	87	7
Daniels	48°04'50"	90°23'40"	185.9	27.4	10.8	43	8
Davis	47°59'13"	90°42'07"	147.4	19.5	8.6	21	4
Duncan	48°04'50"	90°27'59"	195.0	35.4	14.3	51	5
Dunn	48°04'57"	90°30'59"	36.1	20.1	7.5		4
East Bearskin	48°02'33"	90°21'49"	235.8	20.4	6.8	35	3

Lake	Latitude	Longitude					
Echo	47°35'14"	91°03'57"	18.6	18.6		60	7
Embarrass Pit	47°32'12"	92°16'55"	63.1	141.7		376	10
Ester	48°09'50"	91°04'14"	168.8	33.5	10.6	60	5
Explorer	48°02'58"	91°17'12"	23.8	22.9	10.2	43	7
Fat	48°17'29"	92°13'43"	41.3	15.2	6.6	21	5
Fay	48°04'45"	90°54'40"	26.4	19.8	9.9	40	4
Fern	48°03'28"	90°56'47"	22.8	21.3	10.0	45	5
Flour	48°03'07"	90°23'59"	134.0	24.4	8.6	51	6
Fraser	48°00'33"	91°12'09"	257.1	31.7	8.9	33	5
French	48°03'58"	90°56'47"	44.8	41.1	17.9	41	3
Frost	47°59'30"	90°49'03"	95.7	26.8	10.7	37	4
Gabimichigami	48°03'50"	91°01'34"	494.7	63.7	22.2	34	4
Gijikiki	48°09'08"	91°06'56"	45.5	25.0	8.8	51	6
Gillis	48°03'20"	90°55'48"	240.5	56.4	19.1	38	6
Gneiss	48°10'31"	90°48'28"	97.1	21.3	8.1	51	3
Gordon	47°59'20"	90°47'31"	49.1	29.0	7.5	35	4
Greenwood	47°59'58"	90°10'41"	817.1	34.1	9.9	29	5
Grindstone	46°07'20"	93°00'26"	213.2	51.8	22.7	86	5
Gun	48°19'20"	92°11'54"	71.8	41.8	10.9	23	8
Gunflint	48°05'53"	90°40'38"	1569.5	61.0	27.1	55	4
Hanson	48°08'23"	91°04'53"	113.6	30.5	12.7	60	4
Holt	48°07'08"	91°03'51"	44.3	22.3	8.1	40	5
Howard	48°04'38"	90°59'03"	61.5	38.1	14.7	44	2
Ima	48°00'56"	91°16'41"	308.1	35.4	11.9	35	4
Jap	48°05'57"	90°54'27"	47.9	21.3	9.3	40	7
Jasper	48°06'40"	91°01'20"	98.4	38.1	10.6	42	4
Jim	48°00'07"	90°17'27"	24.3	7.9	4.7		3
Karl	48°01'13"	90°46'13"	43.6	22.9	4.8	38	3
Kekekabic	48°03'58"	91°11'48"	675.3	59.4	22.8	43	5
Kemo	47°54'07"	90°25'55"	73.9	19.8	11.0		5
Kingfisher	48°06'59"	91°01'53"	14.3	12.8			4
Knife	48°05'38"	91°11'48"	2114.0	54.6		69	7
Little Kek	48°04'10"	91°09'54"	23.2	43.3	18.0	58	5
Little Knife	48°10'11"	91°06'06"	259.4	56.1	21.7	96	7
Little Saganaga	48°02'02"	90°59'35"	644.4	45.7	8.6	35	3

MINNESOTA (continued)

Lake name	Latitude	Longitude	Surface area (ha)	Maximum depth (m)	Mean depth (m)	Conductivity (µS/cm)	Secchi depth (m)
Little Trout	48°23'48"	92°31'18"	97.5	29.0	13.1	44	6
Little Trout	47°58'02"	90°28'53"	49.3	17.1	7.3	40	4
Little Trout	47°26'19"	93°32'42"	29.4	24.4	11.7	195	6
Long Island	48°00'37"	90°45'31"	338.6	25.9	8.1	32	2
Loon	48°04'40"	90°41'31"	435.7	65.5	21.6	45	5
Lunar (Moon)	48°08'30"	91°06'14"	23.2	18.9	6.7	46	4
Magnetic	48°06'20"	90°46'15"	175.9	27.4	12.2		5
Makwa	48°00'18"	91°03'35"	57.6	23.2	9.2	46	5
Maraboeuf	48°11'	90°49'	365.0	16.8		54	2
Mayhew	48°04'41"	90°35'31"	89.9	25.6	11.3	47	7
Mesaba	47°58'07"	90°56'38"	73.6	19.8	7.4		3
Misquah	47°58'54"	90°28'59"	23.3	18.6	9.5	55	3
Missionary	48°03'24"	91°16'35"	44.2	21.6	10.8	59	3
Moon (Lunar)	48°08'30"	91°06'14"	23.2	18.9	6.7	46	6
Moose	48°05'58"	90°05'13"	410.0	36.0	15.7	41	4
Moss	48°04'01"	90°28'48"	100.1	29.0	13.1	52	5
Mountain	48°06'15"	90°14'06"	835.7	64.0	17.6	45	5
Mukooda	48°20'09"	92°29'21"	308.5	23.8	12.2	58	7
North	48°06'01"	90°35'09"	1116.5	38.1	15.0	96	3
Ogishkemuncie	48°05'56"	91°03'18"	284.2	22.9	7.5	34	8
Ojibway	47°56'58"	91°32'34"	150.1	35.1	9.2	74	5
Owl	48°02'25"	90°54'33"	32.5	21.3	6.8	43	5
Oyster	48°13'33"	92°06'26"	291.9	39.6	13.3	30	5
Partridge	48°05'05"	90°29'19"	43.5	25.0	10.7	56	3
Peter	48°03'52"	90°59'24"	102.2	36.6	13.4	35	5
Pine	48°03'48"	90°13'01"	916.7	34.4	16.8	40	4
Pokegema	47°10'52"	93°34'37"	2665.2	34.1		203	7
Poplar	48°02'49"	90°30'32"	307.9	22.3		34	3

Lake	Latitude	Longitude					
Powell	48°03'42"	90°57'41"	20.8	22.9	8.0	40	4
Rabbit	48°09'39"	91°05'14"	41.9	32.0	12.9	51	5
Ram	47°57'30"	90°26'58"	26.9	12.2	6.0	41	7
Raven	48°02'13"	91°08'32"	82.6	17.1	7.4	31	6
Roosevelt	46°49'02"	93°56'59"	599.2	39.6			5
Rose	48°06'03"	90°26'48"	481.9	29.3	11.4	51	5
Saganaga	48°12'21"	90°55'31"	7119.9	85.3	11.1	40	5
Seagull	48°08'02"	90°55'26"	1608.8	44.2	10.1	65	4
Sema	48°05'11"	91°10'04"	33.0	21.9	4.3	44	3
Snipe	48°03'07"	90°49'30"	45.8	27.4	14.8	40	1
Snowbank	47°59'05"	91°26'16"	1334.9	45.7	19.2	65	6
South	48°05'43"	90°32'52"	485.0	43.0			9
State	47°58'38"	90°31'52"	21.4	15.2	5.9	29	
Strup	48°02'38"	91°11'12"	42.2	32.0	8.9	45	4
Swan	47°57'15"	90°31'35"	74.0	37.2	13.6	26	4
Takucmich	48°19'49"	92°09'41"	130.7	45.7	13.5	32	5
Thomas	47°59'21"	91°14'10"	602.0	33.5	8.4		3
Topaz	48°07'47"	91°07'26"	54.0	21.3	5.7	33	2
Town	47°58'31"	90°45'47"	31.9	22.0	9.3	30	4
Trout	47°58'23"	92°19'10"	3092.3	29.9	12.5	39	5
Trout	47°52'12"	90°10'21"	103.5	23.5	10.5	206	6
Trout	47°27'43"	93°33'17"	709.2	47.9	13.8	46	4
Tuscarora	48°02'28"	90°52'56"	333.5	39.6	10.9	35	4
Vernon	47°56'38"	90°34'36"	94.5	30.8	10.8	63	4
West Bearskin	48°03'59"	90°25'36"	200.5	23.8	10.2	43	5
West Fern	48°03'37"	90°58'33"	30.3	21.3	9.8	43	7
West Pike	48°05'11"	90°12'18"	285.2	36.9	10.0	37	6
Winchell	47°59'29"	90°35'50"	332.4	48.8	11.6	18	5
Wine	47°56'44"	90°56'06"	96.4	19.8			4
Wisini	48°02'23"	91°10'45"	38.0	41.8	13.1	27	4

NEW YORK (Adirondacks)

Lake Name	Latitude	Longitude	Surface Area (ha)	Maximum Depth (m)	Mean Depth (m)	Conductivity (uS/cm)	Secchi Depth (m)
Bear Pond	44°24'06"	74°17'11"	21.9	18.3	6.8	24	11
Bessie Pond	44°22'51"	74°23'14"	6.7	15.2	4.5	27	2
Big Hope Pond	44°30'43"	74°07'30"	8.9	11.5	5.8	23	3
Big Pine Pond	44°15'44"	74°08'52"	18.6	19.8	7.9	28	7
Bigsby Pond	43°50'11"	73°54'10"	474.0	22.9		23	6
Black Pond	44°26'12"	74°18'05"	29.0	13.7	6.2	42	3
Blue Mountain	43°51'07"	74°28'16"	535.1	31.1		40	8
Brandreth	43°25'32"	74°16'33"	355.1			25	
Brantingham	43°41'05"	75°16'33"	131.6	22.9	7.5	32	4
Bug	43°45'49"	74°43'47"	32.2	24.4	6.1	48	5
Canada	43°09'49"	74°32'27"	217.7	43.9		38	
Chapel Pond	44°08'28"	73°45'09"	7.4	23.8	9.3	28	7
Chazy	44°46'17"	73°48'29"	730.4	22.9	9.2	49	6
Church Pond	44°20'52"	74°20'12"	10.5	18.3	7.8	21	5
Clear	44°21'38"	74°16'16"	388.4	18.3	8.9	45	6
Clear Pond	44°28'50"	74°10'16"	39.1	16.8	7.3	35	4
Clear Pond	43°59'38"	73°49'40"	70.4	24.4	9.2	28	8
Connery Pond	44°18'23"	73°56'06"	32.8	15.2	5.3	39	3
Copperas Pond	44°19'45"	73°53'54"	8.9	23.8	9.9	33	6
Crane Pond	43°51'04"	73°39'47"	67.5	32.3	11.6	31	2
Dart	43°47'36"	74°52'16"	51.8	17.7	7.3	24	4
Deer Pond	44°16'34"	74°23'17"	47.0	19.5	10.4	28	7
Eagle	43°50'31"	74°29'17"	65.8	9.4	4.8	42	6
Eagles Nest	43°45'50"	74°43'42"	4.9	17.1	9.2	49	6
East Pond	43°55'47"	74°41'46"	22.3	17.1	6.8	25	6
Eighth Fulton Chain	43°46'12"	74°42'48"	122.6	24.7	11.9	54	5
Fawn	43°28'47"	74°27'56"	116.9	18.9	10.2	25	6
Fish Pond	44°23'42"	74°22'42"	47.0	15.3	7.0	27	6
Follensby Clear Pond	44°18'38"	74°20'45"	195.5	18.3	6.4	44	3
Follensby Pond	44°11'44"	74°22'41"	391.1			37	4
Gooseneck Pond	43°52'02"	73°35'59"	29.4	21.0		32	

Green	43°10'26"	74°30'39"	18.2	15.8	6.0	27	7
Green Pond	44°22'54"	74°18'03"	8.9	9.4	5.5	19	7
Green Pond	44°20'17"	74°20'37"	25.5	18.3	9.4	34	3
Gull Pond	44°12'59"	74°31'27"	117.3	23.2	10.2	25	7
Heart	44°10'47"	73°58'03"	10.7	16.8	5.1	20	5
Hoel Pond	44°21'29"	74°21'34"	181.7	24.4	8.1	23	7
Horseshoe Pond	44°19'02"	74°21'38"	34.4	7.9	4.7	20	3
Indian	43°45'20"	74°16'33"	1744.7	24.4		32	7
Jabe Pond	43°42'37"	73°31'52"	59.8	22.9	5.8	31	5
Jenkins Pond	44°07'55"	74°29'00"	122.2			29	9
Lake Clear	43°58'56"	74°55'24"	40.0	20.4	8.6	28	8
Lake Eaton	43°58'41"	74°27'00"	227.8	16.5		34	5
Lake Kushaqua	44°31'39"	74°06'13"	152.6	27.7	6.2	41	2
Lake Lila	43°59'35"	74°46'14"	572.2	18.3		24	3
Lake Marion	44°07'14"	74°43'04"	83.4	46.6	15.4	21	8
Lake Placid	44°18'17"	73°59'43"	1119.4	46.0		34	10
Ledge Pond	44°21'30"	74°25'23"	16.2	14.9	6.2	20	6
Limekiln	43°42'48"	74°48'47"	186.9	21.9	6.1	26	8
Little Green Pond	44°21'31"	74°17'52"	27.8	12.2	5.6	34	4
Little Simon Pond	44°09'42"	74°26'36"	57.5	32.0	11.0	26	9
Little Trout Pond	44°05'36"	74°39'15"	18.4	20.1	8.1	23	2
Long Pond	44°20'50"	74°24'03"	137.0	15.2	3.8	29	3
Long Pond	43°59'37"	75°11'33"	59.8	18.3	9.4	23	1
Loon	44°33'43"	74°04'36"	143.1	16.5	5.2	36	4
Lower Sargent Pond	43°51'34"	74°04'02"	52.2	9.1	6.4	29	4
Lydia Pond	44°23'45"	74°23'46"	7.7	11.6	6.5	38	4
Massawepie	44°15'00"	74°39'45"	176.7	22.9	6.9	43	4
Mc Kenzie Pond	44°19'27"	74°04'30"	97.0	16.2	6.9	33	6
Mirror	44°17'03"	74°58'54"	47.8	18.3	7.0	87	6
Mitchell Pond Lower	43°40'08"	74°45'17"	9.7	22.2	4.9	31	6
Mohegan	43°44'35"	74°39'02"	47.6	17.7	7.0	28	3
Moose Pond	44°21'57"	74°03'59"	56.7	21.3	8.7	35	5
Moose Pond	44°03'12"	74°11'55"	73.4	17.1		32	5
Moss	43°46'52"	74°51'11"	45.7	15.2	5.7	34	4

NEW YORK (Adirondacks) (continued)

Lake Name	Latitude	Longitude	Surface Area (ha)	Maximum Depth (m)	Mean Depth (m)	Conductivity (uS/cm)	Secchi Depth (m)
Negro	43°59'23"	74°53'13"	48.5	26.8	7.6	20	3
Newcomb	43°59'54"	74°06'01"	186.2	24.4		36	
Nine Corner	43°11'34"	74°32'35"	45.0	14.6	5.4	24	8
Oregon Pond	44°29'44"	74°07'16"	8.7	11.3	4.5	13	4
Owen Pond	44°19'23"	73°54'12"	7.6	9.4	3.7	35	4
Paradox	43°52'33"	73°43'46"	378.2	15.2	5.9	36	5
Piseco	43°59'27"	74°22'48"	1150.1	39.3		34	
Polliwog Pond	44°19'35"	74°21'22"	84.2	24.4	7.0	20	5
Queer	43°48'49"	74°48'25"	54.5	21.3	10.9	23	8
Raquette	43°53'12"	74°35'10"	2145.7	26.8		35	5
Rat Pond	44°21'10"	74°18'42"	11.7	8.8	3.7	22	2
Rich	43°58'31"	74°11'08"	167.3	19.8	9.1	33	4
Rock Pond	44°01'20"	74°25'28"	162.1	13.7	3.8	34	4
Rollins Pond	44°19'25"	74°24'36"	180.0	23.5	6.9	31	2
Sagamore	43°45'57"	74°37'43"	68.0	22.9	10.5	27	3
Saint Regis Pond	44°22'55"	74°17'35"	162.5	9.4	4.7	27	3
Salmon	43°57'51"	74°40'27"	131.7	27.1	8.3	22	7
Salmon	43°56'19"	74°56'36"	44.5	15.2	7.2	22	5
Salmon Pond	43°54'27"	74°22'42"	33.6	18.9	7.1	24	4
Sand Pond	43°56'53"	73°53'31"	25.8	12.2	5.0	43	3
Schroon	43°43'41"	73°48'40"	1682.0	44.2		69	4
Seventh	43°44'52"	74°45'54"	332.5	25.9	11.8	48	
Sixth Lake Fulton Chain	43°44'42"	74°46'56"	43.6	11.6	3.8	47	6
South	43°35'18"	74°37'17"	35.8	10.1	4.9	20	3
South Pond	43°55'43"	74°27'17"	172.6	16.5	4.8	28	4
Spitfire	44°25'23"	74°15'27"	105.2	9.4	4.8	36	4
Square Pond	44°29'28"	74°08'30"	17.2	33.5	10.0	39	6
Star	44°09'02"	75°02'03"	84.0	19.8	6.3	40	7
Stony Creek Ponds	44°12'41"	74°18'53"	74.9	12.5	3.8	40	4
Taylor Pond	44°29'36"	73°49'21"	346.5	24.4	13.4	24	6
Trout Pond	44°05'19"	74°39'29"	63.4	25.3	6.8	21	7

Lake name	OFIS code	Latitude	Longitude	Surface area (ha)	Maximum depth (m)	Mean depth (m)	Conductivity (µS/cm)	Secchi depth (m)
Tupper		44°12'13"	74°28'39"	1713.6	27.4	5.2	38	5
Twin Lake North		44°10'27"	75°02'35"	12.8	13.7	4.0	113	4
Twin Lake South		44°10'13"	75°02'43"	5.8	7.3		139	6
Upper Cascade		44°13'28"	73°52'28"	9.7	19.2	11.8	106	5
Upper Saranac		44°15'04"	74°17'50"	2019.8	31.3	4.3	39	
Upper Spectacle Pond		44°25'04"	74°18'01"	18.3	20.7	8.6	27	5
West		43°10'09"	74°32'03"	78.2	8.5	3.4	32	2
West Canada		43°35'41"	74°37'32"	96.4	21.3		23	
West Caroga		43°0'54"	74°29'28"	129.1	22.6	8.8	55	5
West Pine Pond		44°20'15"	74°25'25"	25.5	11.6	5.5	31	5
White		43°32'43"	75°08'41"	90.8	22.9	7.6	70	6

ONTARIO (Northeastern Ontario)

Lake name	OFIS code	Latitude	Longitude	Surface area (ha)	Maximum depth (m)	Mean depth (m)	Conductivity (µS/cm)	Secchi depth (m)
Aaron	17-6164-52142	47°04'24"	79°28'00"	21.3				
Abbott	17-4486-51793	46°46'06"	81°40'25"	28.5				
Acheson	17-4304-51587	46°34'48"	81°54'17"	159.2	33	11.1	44	6
Achigan (Bass)	16-7134-51997	46°55'15"	84°11'47"	279.6	44.2	12.4	27	5
Acid	17-4657-50979	46°02'09"	81°26'37"	19.6	29	10.9	24	2
Adelaide	16-7046-52151	47°03'38"	84°18'18"	137.3	47.2	12.2	37	7
Admiral (Duck)	17-3486-51441	46°26'13"	82°58'00"	84.3	16.2	6.6	32	3
Affleck (Upper Megisan)	17-3069-52350	47°14'30"	83°33'04"	131.6	9.1	3.8	45	2
Agawa Station	16-6849-52810	47°39'27"	84°32'14"	66.5	26	8.6	29	7
Aileen	17-5748-51919	46°52'43"	80°01'04"	157.6				
Airport	17-4143-51877	46°50'00"	82°07'11"	80.7	28.4	9.6	31	8
Alces	17-4129-51783	46°45'19"	82°08'23"	59.9	11	3.7	38	4
Alexander	17-3812-51598	46°35'13"	82°32'38"	36.7			23	
Alexander	17-4363-51070	46°06'56"	81°49'29"	37.7				
Alice	17-2851-52237	47°07'41"	83°49'37"	32.8				

ONTARIO (Northeastern Ontario) (continued)

Lake name	OFIS code	Latitude	Longitude	Surface area (ha)	Maximum depth (m)	Mean depth (m)	Conductivity (µS/cm)	Secchi depth (m)
Allan	17-5604-52073	47°01'12"	80°12'04"	67	25.3	8.8	25	5
Allen	17-3876-51569	46°33'30'	82°27'59"	23.9			29	
Allen	17-3542-51209	46°13'40"	82°53'27"	40.4				
Alma	17-5330-52032	46°59'01"	80°33'45"	98.0			43	
Almonte	16-6725-52872	47°43'47"	84°41'43"	73.5				
Aloft (Mountain)	16-6887-53553	48°19'29"	84°27'13"	106.1	28.1	7.9	49	6
Alphretta (Paradise)	17-5176-52025	46°58'39"	80°46'06"	487.4	35	8.2	37	11
Alto	17-3887-51504	46°29'42"	82°26'40"	13.1				
Anahareo	16-6707-53872	48°37'23"	84°41'03"	846.8			111	
Anelia	17-5279-52009	46°57'47"	80°37'57"	52.6				
Anima Nipissing	17-5827-52344	47°15'37"	79°54'14"	2049.7	76.2	13.7	45	9
Anjigami	16-6809-53004	47°51'04"	84°35'00"	1140.0			34	
Antrim	17-4523-51979	46°56'01"	81°37'33"	90.7	17.4	6.7	45	4
Anvil	17-5552-52523	47°25'23"	80°16'14"	226.2	22.3	7	35	4
Anvil (North Anvil)	17-3094-52159	47°04'17"	83°30'34"	92.8	18.3	6.8	36	4
Applesauce	17-3801-51606	46°35'26"	82°33'56"	53	39	8.3	24	5
Aquatuk	16-6588-60271	54°22'07"	84°33'17"	827.8				
Archambeau	17-3902-51615	46°36'39"	82°26'31"	49.0			29	
Archibald	17-3244-52261	47°09'54"	83°18'33"	47.1				
Armstrong	17-4560-51530	46°31'51"	81°34'05"	267.1	27.5	6.2		3
Army	17-3713-51911	46°51'49'	82°41'29"	64.8	17.1	7.6	36	5
Arnott	16-6730-54966	49°35'58"	84°36'21"	237.2				
Arrowhead (Lake 26)	17-3905-51513	46°30'30"	82°26'07"	64.1				
Augusta	16-6219-53229	48°02'59"	85°21'51"	69.5				
Avery	17-4823-52210	47°08'44"	81°13'37"	151.8	18	5.5	39	5
Axe	17-3146-51506	46°29'08"	83°24'56"	95.6				
Aylmer # 37 (Lawlor)	17-5216-51920	46°52'35"	80°43'56"	35.5			34	8
Bagpipe	17-4278-52035	46°59'21"	81°57'47"	39.8				
Ballard	16-7139-53404	48°10'57"	84°07'20"	90.6				
Banana	17-5029-52263	47°11'32"	80°57'45"	49.4	17	5.5	47	6

Lake	Map reference							
Banks	17-5452-52595	47°29'03"	80°24'01"	307.5	29.3	10	33	9
Bark	17-3874-51976	46°55'23"	82°28'21"	1255.1	29.3	7	41	6
Barmac	17-5684-52247	47°10'22"	80°05'51"	37.7	12.8	6.2	60	6
Barn (Lake 12)	17-4165-51920	46°52'43"	82°05'45"	20.6	15.3	4.5	33	4
Barnet	17-4865-52251	47°10'58"	81°10'34"	346.4	18.9	5.1	39	6
Barron (Lingo)	17-5152-52045	46°59'45"	80°47'59"	50.2	23		34	4
Barrow (Bouchard)	17-4157-51481	46°29'00"	82°05'55"	49.7	15.8	4.5	32	8
Barter	17-5674-52374	47°17'20"	80°06'25"	112.3	32.3	14.4	48	4
Bass (Cassidy)	17-5968-52457	47°21'33"	79°43'03"	60.9				
Bass (Achigan)	16-7134-51997	46°55'15"	84°11'47"	279.6	44.2	12.4	27	5
Bassoon	17-4704-51180	46°12'38"	81°24'00"	134.8			88	4
Basswood	17-3164-51328	46°19'47"	83°23'42"	2708.6	73.2	38.7	38	7
Batty (Whitefish)	17-3944-51408	46°24'48"	82°22'29"	244.3	37.2	13.6	28	7
Bauldry (Scott)	16-6745-53296	48°05'55"	84°39'24"	51.4	63.1	23.9	108	8
Bay (Coffee)	17-3525-51431	46°25'46"	82°55'15"	333.9	53.3	22	39	12
Bear	17-3921-51464	46°27'52"	82°24'18"	310.6	44.2	19.8		13
Bear	17-4652-51149	46°11'09"	81°27'00"	682.5	36.6	14.5	60	8
Bear (Kaotisinimigo)	17-6069-51648	46°37'49"	79°36'10"	549.9				
Beauty (Isabel)	17-5305-52643	47°32'00"	80°35'42"	199.9				
Beaver	17-3663-51518	46°30'30"	82°44'33"	23.1	25.6	7.7	71	4
Beaver (Mesomikenda)	17-4340-52773	47°38'24"	81°52'46"	1706.2	71.4	13.6	61	5
Becor	17-3085-52414	47°18'17"	83°32'11"	25.6	20.8			5.2
Beecher	17-3361-51423	46°25'02"	83°08'29"	154.9				
Beef (Pathfinder)	17-3570-51488	46°28'44"	82°51'40"	256.5	43.6	14.5	20	8
Beland	17-6155-52122	47°03'19"	79°28'34"	35.4	32	11.7	53	5
Bell (Gong)	17-3074-52158	47°04'42"	83°32'00"	384.5	21.4	6.9	37	4
Bell	17-3941-51575	46°33'53"	82°22'52"	112.8	38.1	11.3	32	8
Bell	17-4836-51079	46°07'42"	81°12'19"	347.4	26.8	8.1	29	4
Belle	17-3787-51299	46°18'49"	82°34'31"	33.8				
Bellows (Horseshoe)	17-3960-51305	46°19'20"	82°21'03"	274.3	28.1	8.9	30	6
Benner	17-5288-52236	47°10'02"	80°37'14"	58.7	26	8.8	30	8
Bergen	17-3373-51608	46°34'59"	83°07'23"	24.9				
Bergeron	17-5557-52556	47°27'11"	80°15'36"	32.4	35.1	10.9	39	6
Best	17-5881-52330	47°14'48"	79°50'10"	83.9	30.8	5.7	63	5
Bevans	17-3657-51687	46°39'37"	82°45'24"	31.8	23.5	9.8	38	11

ONTARIO (Northeastern Ontario) (continued)

Lake name	OFIS code	Latitude	Longitude	Surface area (ha)	Maximum depth (m)	Mean depth (m)	Conductivity (µS/cm)	Secchi depth (m)
Beyond	17-3675-51848	46°48'18"	82°44'10"	27.7	30.5	9.8	30	5
Bierce	17-3521-51843	46°47'51"	82°56'04"	71.2	16.5	5.4	33	6
Big	17-3480-51543	46°31'37"	82°58'53"	162.7				
Big Caribou (Caribou)	17-5718-50868	45°55'57"	80°04'06"	528.9	59.5	16.9	48	4
Big Chief	17-5681-52589	47°28'55"	80°05'45"	84.6	20.4	5.3		6
Big Horseshoe (Horseshoe)	17-3430-51737	46°42'01"	83°03'11"	208.9			33	
Big Missinaibi (Missinaibi)	17-3006-53592	48°21'32"	83°41'07"	7706.9	94	19.2	83	4
Big Skunk	16-6764-54957	49°35'25"	84°33'32"	321.5				
Big Squaw (Big Squirrel)	17-4281-52070	47°00'50"	81°56'49"	87.8	31	7.9	32	4
Big Squirrel (Big Squaw)	17-4281-52070	47°00'50"	81°56'49"	87.8	31	7.9	32	4
Big Trout	17-3962-51891	46°50'57"	82°21'40"	435.8	36.6	14.7	30	5
Big Trout	17-3459-51754	46°43'06"	83°01'02"	33.1			25	
Big Trout (Montreuil)	17-6607-51439	46°25'54"	78°54'28"	18.1				
Big Trout (Snapshot)	17-4025-51298	46°18'56"	82°15'53"	82.5	26.2	9.2	21	
Big Turkey (Turkey)	16-6958-52136	47°02'54"	84°25'10"	54	39.6	11.7	38	7
Bigwood	17-4936-51883	46°50'38"	81°05'34"	265.6			31	5
Bijou	17-3095-51641	46°36'18"	83°29'15"	62.3				
Bilton	17-3767-51706	46°40'55"	82°36'58"	52.5	21.7	5.2	44	4
Birch (Fetherston)	17-2736-51548	46°30'40"	83°57'03"	29.0				
Birch (Gough)	17-4246-51279	46°18'08"	81°58'46"	1080			37	
Black	17-3444-51662	46°38'12"	83°01'47"	57.2				
Black Beaver	16-6882-52534	47°24'32"	84°30'22"	165.9	40	12.1	23	11
Black Duck (Blackduck)	17-5914-52234	47°09'30"	79°47'25"	72.8	33.6	10.6	77	4
Black Trout	16-6609-53243	48°03'40"	84°50'16"	220.7	22.9	7	43	5
Blackduck (Black Duck)	17-5914-52234	47°09'30"	79°47'25"	72.8	33.6	10.6	77	4
Blackfish	16-7142-53440	48°12'56"	84°06'55"	216.4	33.9	12.3	52	3
Blackies	17-4669-52549	47°26'55"	81°26'22"	14.2				
Blackington	16-6616-53064	47°53'31"	84°50'23"	32.4	23.8	10	124	4
Blue (Paradise)	17-4802-54481	49°11'14"	81°16'17"	23.5				
Blue	17-6295-51774	46°44'24"	79°18'17"	189.4				

Blue (Kirkpatrick)	17-3400-51692	46°39'31"	83°05'20"	1097.5	70.2	17	29	7
Blue Sky	17-3648-51657	46°38'18"	82°45'15"	41.7	21.4	7.3	28	9
Bluesucker	17-5298-52239	47°10'10"	80°36'24"	144.5			43	
Bluewater	17-4415-51793	46°46'22"	81°46'25"	115			33	
Bobowash	17-3635-51568	46°33'10"	82°46'50"	64			30	
Bobwhite (Three Mile Lake)	17-4421-52015	46°58'00"	81°45'39"	66.5			26	
Bojack	16-7103-51885	46°49'07"	84°14'49"	28.7	8	4.9	47	3
Boland	17-3841-51846	46°48'23"	82°31'03"	56.2	34.7	9.2	24	8
Bone	16-7090-51872	46°48'23"	84°15'55"	121.2	32.6	7.7	53	6
Bonhomme	17-5216-51846	46°49'03"	80°42'44"	37.3				
Boomerang	17-3450-51921	46°52'19"	83°02'20"	34.7	43.9	7.6		4
Borden	17-3280-52990	47°49'39"	83°17'25"	1549.2	29	10.8		3
Borzoi	16-6922-53218	48°01'22"	84°25'14"	83.4				
Botha	17-4784-51994	46°57'17"	81°17'20"	38.2				
Bouchard (Barrow)	17-4157-51481	46°29'00"	82°05'55"	49.7	15.8	4.5		8
Bouck (Pine)	17-6046-52356	47°16'10"	79°37'06"	99.3	21.3	6.8		8
Boulder	16-6743-53628	48°23'47"	84°38'43"	88.0				
Boulton	17-5988-52076	47°00'29"	79°42'05"	46.1				
Boumage	17-3914-51928	46°52'58"	82°25'04"	113.8	16	6.4	140	5
Boundary (Lake 36)	17-3428-51647	46°37'10"	83°03'10"	18.8	24	7.7	18	5
Boundary (Commander)	17-3499-51427	46°25'24"	82°57'09"	87.7	26	13.9	39	8
Bowland	17-5120-52145	47°05'13"	80°50'31"	108	28	6.9	32	7
Brant	16-6835-52133	47°02'58"	84°35'00"	27.4			110	
Brewer (Lake 17)		46°51'09"	82°27'21"	0.9	15.5	6.3	25	4
Bridge	17-3344-51584	46°33'41"	83°09'42"	41.4			33	
Brigstocke (Lake 69)	17-5807-52370	47°17'05"	79°56'10"	34.2			36	
Broker	17-5002-51098	46°08'39"	81°59'49"	81	24.1	9.5	32	6
Brownbear (Hand)	17-5668-51731	46°42'36"	80°07'31"	71.6			93	
Bull	17-5317-52165	47°06'13"	80°34'57"	109.9	21	7.8	41	10
Bull (East Bull)	17-4085-51426	46°25'55"	82°11'12"	52.6	35	11.5	35	6
Burke	17-4630-50974	46°01'50"	81°28'32"	8.4	15.6	5.2	32	4
Burns	17-3382-51620	46°35'40"	83°06'51"	166.4	78	29.8	25	22
Burnt	17-3434-52003	46°56'24"	83°03'28"	14.9			23	
Burwash	17-4965-52193	47°07'52"	81°02'53"	1065.1	52.2	12.1	40	4
Bushcamp (Johnnie)	17-4826-51036	46°05'13"	81°13'30"	342.3	33.6	10	27	6

ONTARIO (Northeastern Ontario) (continued)

Boreal Shield Watersheds: Lake Trout Ecosystems in a Changing Environment

Lake name	OFIS code	Latitude	Longitude	Surface area (ha)	Maximum depth (m)	Mean depth (m)	Conductivity (µS/cm)	Secchi depth (m)
Byrnes	17-3092-51389	46°22′42″	83°28′42″	58.3	19.8	8.2	21	5
Cahill	17-6497-51358	46°21′41″	79°03′12″	93.2				
Camp 12 (Laurence)	17-3097-51982	46°54′44″	83°29′56″	10.9	13.4	7.6	54	4
Campover	17-3878-51671	46°39′30″	82°27′43″	59.1				
Canoe (Scarfe)	17-3429-51270	46°16′50″	83°02′20″	120.2				
Canyon	17-3792-51405	46°24′36″	82°34′14″	53.5	23.2	6.4	51	6
Capreol (Ella)	17-5101-51719	46°42′09″	80°52′03″	173.7	24	18.9	51	8
Carhess (Green)	17-4614-51736	46°43′14″	81°30′21″	44.4				
Caribou (Kakakiwibik)	16-6415-53841	48°35′46″	85°04′46″	423.6	31.7	6.9	117	3
Caribou (Rangers)	17-3909-51461	46°27′52″	82°25′51″	252.2	63.1	23.8	39	11
Caribou (Big Caribou)	17-5718-50868	45°55′57″	80°04′06″	528.9	59.5	16.9	48	4
Carpenter	16-7055-52294	47°11′15″	84°17′16″	40.9				
Cascaden	17-4617-51569	46°34′07″	81°30′42″	77.3	37	9.7	39	
Cassels (White Bear)	17-5970-52137	47°04′11″	79°43′24″	757.1	36.6	9.6	87	5
Cassidy (Bass)	17-5968-52457	47°21′33″	79°43′03″	60.9				
Castra	17-3323-51600	46°34′28″	83°11′21″	42.7	29	9.3	77	6
Caswell	17-5229-51902	46°51′51″	80°42′30″	39	24	7.9	38	9
Cedar (Kanichee)	17-5882-52189	47°07′05″	79°50′22″	252.5	33.6	7.3	68	4
Center	17-3544-51585	46°33′45″	82°53′34″	28.7			54	
Centre	17-5189-52102	47°02′51″	80°45′06″	121.6	25	7.1	43	8
Chaillon	16-6550-52845	47°41′57″	84°56′07″	63.3	38.4	10.9	48	3
Chamandy	17-4813-53013	47°52′11″	81°14′54″	70.0				
Chambers	17-5780-52177	47°06′30″	79°58′26″	209.8	29.9	4.7	71	5
Chambers 37	17-5805-52197	47°07′40″	79°56′18″	15.8	27	9	59	6
Chance	17-5340-52479	47°23′07″	80°32′56″	12.8	14.3	4.2	56	5
Charcoal	17-4332-51750	46°43′40″	81°52′27″	33.7	16.8	6.9	35	6
Charley	17-2728-51648	46°35′48″	83°57′33″	20.0				
Charon	16-5724-54965	49°37′10″	85°59′39″	304.9	30.5	10.8	164	3
Chiblow	17-3423-51340	46°20′50″	83°02′18″	2087.8	70.2	23.9	36	8
Chiniguchi	17-5240-51985	46°56′12″	80°41′48″	1295.7	44.2	13.6	33	18

Lake	ID	Latitude	Longitude					
Christman (Jimchrist)	17-3675-51601	46°35'01"	82°43'45"	58.6	18.3	7	69	6
Chrysler	17-4919-52573	47°28'16"	81°06'27"	86.1			42	
Chub	17-3172-51540	46°30'39"	83°23'02"	181.4		4		3
Chubb	17-2854-52191	47°05'41"	83°49'40"	55.5	27.5		43	6
Chuggin	17-5377-51779	46°45'20"	80°30'24"	29.5	16.5	5.8	57	4
Clarice	17-6090-53547	48°20'15"	79°31'47"	361	20			
Clayton	17-3924-51429	46°25'56"	82°24'19"	42.6		12.9	30	7
Clear	17-3909-51202	46°13'42"	82°24'46"	47.8	28			
Clear (Eva)	17-4106-54561	49°15'11"	82°13'41"	17.4				
Clear (Hess)	17-4589-51747	46°43'36"	81°32'14"	56.7				
Clear (Wakomata)	17-3191-51595	46°34'31"	83°21'38"	2468.7	73.2	28.3	40	8
Clear (Transparent)	17-6580-51459	46°27'00"	78°56'33"	69.3				
Clear (Lear)	17-3445-51313	46°19'04"	83°01'10"	171.4	39	13.8	31	8
Clear (Hampel)	17-5866-50871	45°56'02"	79°53'00"	85.5				
Clearwater	17-5002-53926	48°41'18"	80°59'50"	30.0				
Clearwater	17-5533-52097	47°02'26"	80°17'46"	118.6	47	13.8	66	11
Clearwater	16-7205-51738	46°41'02"	84°07'07"	12.4				
Cloudy	17-2746-51468	46°26'21"	83°56'03"	55.4	22.9	12.5	54	4
Clove (Dumbell)	17-3114-52360	47°15'07"	83°29'34"	31.9	15.2	6.8	39	3
Cobre	17-3627-51660	46°38'10"	82°47'40"	84.7	61.6	28	91	8
Coffee (Bay)	17-3525-51431	46°25'46"	82°55'15"	333.9	53.3	22	39	12
Colin Scott	17-5379-51863	46°49'49"	80°30'13"	43.9	43	16.4	38	15
Commander (Boundary)	17-3499-51427	46°25'24"	82°57'09"	87.7	26	13.9	36	8
Commando	17-4986-54342	49°03'44"	81°01'09"	13.4				
Como	17-3120-53099	47°54'45"	83°30'58"	1595.9	26	9.4	76	5
Conacher	17-3097-51710	46°40'04"	83°29'15"	197.4				
Constance	17-3290-51439	46°25'48"	83°13'28"	113.3	17.1	11.1	61	5
Cooper	17-3033-51548	46°31'15"	83°33'34"	85.6	23.8	10.2	33	5
Cooper	17-6138-52123	47°03'35"	79°30'22"	182.2	47.3	14.6	53	4
Corine	16-6739-54918	49°33'21"	84°35'43"	135.9				
Cork	17-2906-52364	47°15'25"	83°45'50"	22.4				
Cormier (Uranium)	17-3871-51523	46°31'03"	82°28'20"	137.8	33.5	7.9	36	8
Corner	17-3899-51448	46°26'33"	82°26'18"	83.0				
Crazy	17-3558-51622	46°34'45"	82°53'23"	22.2				
Cream	17-3540-51419	46°25'14"	82°54'06"	46.0			22	

ONTARIO (Northeastern Ontario) (continued)

Lake name	OFIS code	Latitude	Longitude	Surface area (ha)	Maximum depth (m)	Mean depth (m)	Conductivity (µS/cm)	Secchi depth (m)
Crooked	16-7208-51720	46°40'11"	84°06'35"	90.5	28	5.4	32	4
Crooked (Twist)	17-6578-51479	46°28'07"	78°56'42"	56.5				
Crooked (Swalwell)	17-5890-50995	46°02'35"	79°50'14"	75.3			35	5
Crosby	17-3399-51422	46°24'35"	83°05'33"	51.7	34.2	7.9		
Cross (Lake 12-1)	16-6398-54235	48°57'02"	85°05'25"	167.7				
Cross	17-5788-51912	46°52'13"	79°57'47"	1735.4	54.9	11.8	60	5
Crystal	17-5819-53339	48°09'18"	79°53'53"					
Crystal	17-5957-53022	47°52'05"	79°43'12"					
Crystal	17-3626-51758	46°43'26"	82°47'53"	25.5				
Crystal	16-7118-51658	46°36'52"	84°14'05"	24.9	13.5	4.7	36	7
Cucumber	17-5521-51877	46°50'40"	80°18'58"	80.5	22.9	10.3	74	5
Cummings	17-5605-51996	46°56'53"	80°12'22"	70.6	27.1	7.9	57	5
Cummings	17-3191-51490	46°28'46"	83°21'38"	501	37.8	17	62	4
Currie	17-5172-53366	48°11'02"	80°46'02"	196.6	26.8	9.5	72	2
Cut	17-6323-51802	46°45'50"	79°16'02"	70.4	39.7	13		10
Dana	17-4437-53579	48°22'25"	81°45'37"	300.0				
Dana (Pine)	17-5566-51724	46°42'16"	80°15'33"	110.5	25.9	7.4	45	3
Darragh	17-3039-51641	46°36'15"	83°33'31"	191.7	44.2	14.7	39	6
David	17-4776-51097	46°08'23"	81°17'33"	406.3	24.4	7	25	10
Davis	17-5241-52013	46°57'41"	80°40'35"	34.1	14	4.9	27	7
Dayohessarah	16-6447-54048	48°46'56"	85°01'56"	1113.6	26.6	10.1	106	7
Daystar	17-3465-51920	46°51'50"	83°00'54"	75.6	31.4	9.8	41	4
Dead Otter	16-5894-54131	46°51'58"	85°46'50"	133.6				7
Dean	17-5038-51901	46°52'23"	80°56'41"	25.6			39	
Deep	16-6702-53151	47°58'08"	84°43'11"	13.8	53.4	20.2	90	4
Dees	17-5371-52429	47°20'25"	80°30'31"	82	14	4.6	26	6
Deil (Devil's)	17-2752-51845	46°46'41"	83°56'40"	199.7	42.1	11.4	38	7
Denman (Little Chiblow)	17-3360-51357	46°21'13"	83°07'55"	644.4	38.1	17.3	81	6
Dennie	17-4313-51958	46°54'56"	81°54'08"	79.5	28	9.7	37	8
Depot	17-3786-51332	46°20'38"	82°34'38"	183.8	19.5	7.8	126	4

Name	Reference							
Desayeux (Yokum)	17-3037-51580	46°32'57"	83°33'48"	45.7	23	6.6	27	4
Deschamp	17-3903-51589	46°34'37"	82°25'31"	85.5	7.5	3	57	3
Deschamps	17-5603-51797	46°46'12"	80°12'35"	99.1	47.6	18	38	8
Devil's (Manitou)	17-5548-51889	46°51'13"	80°16'52"	343.5	42.1	11.4	38	7
Devil's (Deil)	17-2752-51845	46°46'41"	83°56'40"	199.7	34	9.6	68	8
Dewdney	17-5254-51902	46°52'43"	80°39'21"	176	22	5.5	84	3
Diabase	17-5747-52333	47°15'10"	80°00'23"	49.7	13.7	5.3	36	5
Diamond	17-2760-51448	46°25'18"	83°54'51"	154.9	38.1	9.6	35	4
Diamond	17-5580-52276	47°12'10"	80°14'32"	925.2	22.9	8.8	81	6
Diamond (Wigwas)	16-7240-51577	46°32'13"	84°04'38"	45	18.3	5.8	87	4
Dobie (Little Dobie)	17-3450-51565	46°32'45"	83°01'19"	43.2				
Dog	16-7139-53538	48°18'15"	84°06'28"	5184	74.7			
Dolly-B-Doo (Lake 15)	17-3633-51558	47°11'30"	83°38'29"	23.2	65.6	18.6	26	8
Dollyberry	17-3874-51449	46°32'30"	82°46'41"	143.7				
Dome (Lizotte)	17-5370-51830	46°27'02"	82°28'12"	21	60	15.4	39	9
Donald	17-3470-51698	46°48'01"	80°30'53"	498.2				
Dougall	17-5253-52060	46°40'30"	82°59'43"	40.1	53.4	13.3	37	9
Dougherty	17-4113-51624	47°00'41"	80°40'01"	412				
Dow	17-3746-51867	46°36'40"	82°09'29"	46.6	24	7.2	23	9
Dubbelewe	17-3523-51234	46°49'30"	82°38'44"	85.1				
Duborne (Lake of the Mountains)	17-3486-51441	46°15'01"	82°54'56"	933.6				
Duck (Admiral)	17-3114-52360	46°26'13"	82°58'00"	84.3	16.2	6.6	32	3
Dumbell (Clove)	17-3673-51503	47°15'07"	83°29'34"	31.9	15.2	6.8	39	3
Dunlop	17-3773-51525	46°29'49"	82°43'58"	1098.2	154.6	12.9	32	10
Duthorne	17-3432-51760	46°30'59"	82°35'47"	5.5				
Duval	17-5571-51960	46°43'16"	83°03'07"	150.1			37	
Eaglerock	16-6735-52992	46°55'35"	80°15'21"	167.6				7
Ear	17-4085-51426	47°49'39"	84°40'39"	60.5	27.5	7.7	96	5
East Bull (Bull)	17-3320-51621	46°25'55"	82°11'12"	52.6	35	11.5	32	6
East Caribou	17-3390-51571	46°35'37"	83°11'30"	161.2	30.5	11.4	38	6
East Twin	17-4828-52138	46°33'31"	83°06'05"	34.5				
Edna	17-5381-51884	47°04'47"	81°13'36"	233.1				
Edna	17-4518-52073	46°50'07"	80°29'41"	28.7			34	9
Elboga		47°01'13"	81°38'09"	27.9	16.2	6.1	122	4

ONTARIO (Northeastern Ontario) (continued)

Lake name	OFIS code	Latitude	Longitude	Surface area (ha)	Maximum depth (m)	Mean depth (m)	Conductivity (µS/cm)	Secchi depth (m)
Elbow (Onedee)	17-3357-51692	46°39′30″	83°08′48″	94.2			42	
Elbow		46°32′07″	83°01′00″	19.7				
Elinor	17-3946-51679	46°39′28″	82°22′35″	29.9	12.8	5.5	35	8
Elissa	17-5340-52501	47°24′21″	80°32′58″	24.1	9.8	2.4	57	3
Elizabeth	17-4512-51212	46°14′36″	81°38′05″	122.1	24.4		72	5
Ella (Capreol)	17-5101-51719	46°42′09″	80°52′03″	173.7	24	18.9	51	8
Elliot	17-3690-51389	46°23′33″	82°42′35″	615.1	38.1	16.3	105	5
Elmer	17-5125-53027	47°52′47″	80°49′57″	226.1	36	12.2	39	3.8
Emerald	17-3340-53046	47°52′30″	83°13′10″	50.3	30	13.2	250	10
Emerald	17-5515-51958	46°54′57″	80°19′24″	566.6	48.8	18.5	26	6
Emerald	17-3555-51294	46°18′14″	82°52′41″	435.9	91.5	39.7	29	14
Endikai (White)	17-3445-51613	46°35′25″	83°01′44″	591.7	48.2	29	92	6
Ericson (Lake 15)		48°55′03″	83°23′13″	11.2	18.4			
Ess (Lake 59)		46°48′15″	82°55′06″	33.7				
Esten	17-3726-51338	46°20′54″	82°39′16″	416.9	36	14.1	207	5
Eva (Clear)	17-4106-54561	49°15′11″	82°13′41″	17.4				
Evans	17-3717-51503	46°29′45″	82°40′16″	35.0				
Evelyn	17-5312-51934	46°53′40″	80°35′23″	110.7	24	6.5	88	5
Ezma	17-3580-51562	46°32′48″	82°51′05″	128.7			54	
Fairbank	17-4672-51457	46°27′57″	81°25′37″	703.1	44.5	18.3	66	8
Fearless	16-5940-53862	48°37′27″	85°43′28″	118.4				
Fern	17-2815-51794	46°44′05″	83°51′36″	86.6				
Ferrier	17-4769-53206	48°02′21″	81°18′39″	207.1	37.2	12.6	151	6
Fetherston (Birch)	17-2736-51548	46°30′40″	83°57′03″	29.0				
Fifty Dollar	17-3039-52174	47°05′29″	83°35′00″	27.4				
Finn	17-3358-51768	46°43′37″	83°08′55″	72.8			37	
First Justin (Justin)	17-5838-52437	47°20′46″	79°53′30″	38.1	32	6.4	28	3
Five Star (Lake 56)	17-3931-51855	46°49′08″	82°24′11″	92.1	42.7	14.2	54	11
Flack	17-3634-51607	46°35′16″	82°46′49″	951	70.1	20.8	38	11
Flagg	16-7024-52966	47°47′30″	84°17′52″	92.6	34.5	9.2	27	5

Name	Code	Latitude	Longitude					
Flamingo	17-3232-52322	47°13'16"	83°20'05"	42.9				16
Flatstone	17-5232-52811	47°41'05"	80°41'24"					6
Flipper	17-5532-54084	48°49'42"	80°16'29"	33.2	38.1	7.5	34	5
Florence	17-5335-52315	47°14'28"	80°33'58"	1006.5	23.2	11.5	38	
Florence	17-4381-51072	46°07'00"	81°48'01"	29.1	22.6	6.8	33	
Flying Bird (Pipe)	17-3088-51523	46°29'55"	83°29'25"	32.5			39	4
Flying Goose	17-3808-51397	46°24'07"	82°33'01"	36	28.4	7.4	30	7
Folson	17-4026-51425	46°25'52"	82°16'01"	198.9	23.1	7.8	105	
Forge	16-6888-53388	48°10'36"	84°27'31"	22.9				
Fortune	17-4339-53837	48°36'17"	81°53'49"	112.2				
Fourbass	17-6117-52162	47°05'25"	79°31'34"	387.5	40.9	12.8	78	3
Fox (Macauley)	17-4429-51586	46°35'21"	81°44'30"	471.4	42.4	10.7	69	4
Foxwell (Lake 9)		47°07'50"	83°43'08"	16.5	27			5.2
Foy	17-4809-51809	46°46'34"	81°15'04"	55.1	23.2	6.9	41	6
Fraleck	17-5089-51954	46°54'54"	80°52'57"	173.9			38	5
Franks	17-5267-51920	46°52'53"	80°38'40"	19.8				
Fraser	17-3876-52227	47°09'00"	82°28'57"	175	18	7.5		
Freddie's (Kindiogami)	17-3509-51882	46°49'57"	82°57'17"	465.2	21	7.5	34	8
Frederick	17-5228-52091	47°02'17"	80°41'55"	339.8			34	6
Friday	17-4734-51994	46°57'47"	81°20'34"	305.0				
Friday	17-3477-51958	46°53'45"	82°59'36"	37.6				
Friendly (Snyder)	17-3069-52112	47°01'41"	83°32'27"	33	12.2	4.3	29	3.6
Fullerton (Lanark)	17-3672-51583	46°34'18"	82°44'28"	56.1				
Gaff	17-3543-51567	46°33'12"	82°53'28"	31.2				
Galer	17-5371-53194	48°01'43"	80°30'06"					
Gamble	17-5339-52492	47°23'54"	80°33'01"	26.4	19	8.1	54	3
Gamitagama	16-6678-52804	47°39'27"	84°45'55"	195.2				
Garden	17-2933-51822	46°45'54"	83°42'24"	157.4	22.9	8.8	64	4
Gauvreau	17-6535-51376	46°22'34"	79°00'13"	39.3				
Gawasi	17-5446-51774	46°45'18"	80°25'24"	83.2				
Geiger (Lucas)	17-3860-51538	46°31'50"	82°29'12"	88.2	27	9.5	39	5
Geneva	17-4583-51789	46°45'52"	81°32'46"	356.4	25.3	6.3	41	5
George	17-4690-50971	46°01'50"	81°24'01"	188.5	36.6	16.4	29	9
Gibberry	17-3625-51552	46°32'14"	82°47'24"	58.7	54.9	18.6	26	11
Gibson	17-5140-53660	48°26'56"	80°48'39"	19.6				

ONTARIO (Northeastern Ontario) (continued)

Lake name	OFIS code	Latitude	Longitude	Surface area (ha)	Maximum depth (m)	Mean depth (m)	Conductivity (µS/cm)	Secchi depth (m)
Gibson	17-4083-51760	46°44'02"	82°12'00"	116.3			20	
Gilbert	17-4307-51870	46°50'05"	81°54'28"	73.1	28.4	7.4	27	4
Gimlet #1 (Upper Gimlet)	16-6786-52113	47°01'57"	84°38'50"	11				
Gimlet #2 (Gimlet)	16-6786-52100	47°01'17"	84°39'02"	21.9	19.8	7	35	5
Goetz	16-6751-53308	48°06'40"	84°38'59"	71.1	33.9	8.6	35	6
Goldie	17-2842-53251	48°02'50"	83°53'59"	1227	22	3.2	123	5
Gong (Bell)	17-3074-52158	47°04'42"	83°32'00"	384.5	21.4	6.9	37	4
Gord	17-3085-52432	47°18'54"	83°32'19"	99.5			71	
Gorrie	17-6104-52196	47°07'19"	79°32'42"	57.8	46	16.7	60	8
Gough (Birch)	17-4246-51279	46°18'08"	81°58'46"	1080			37	
Goulais	17-2975-52266	47°09'27"	83°40'01"	271.7	33.5	8.2	43	4
Gould	16-7033-52984	47°48'37"	84°17'07"	246.5	37.5	9.8	12	5
Grace	17-4535-51088	46°08'00"	81°36'04"	47.2	17.2	6.2	26	12
Granary (Magog)	17-3585-51264	46°16'32"	82°50'27"	284.9	30.5	16.3	34	5
Grandeur	17-3762-51303	46°19'04"	82°36'31"	106.8	23.2	8	107	4
Grant	16-6843-53260	48°03'40"	84°31'40"	89.6	33.9	8.8	110	5
Grassy (Wawiashkashi)	17-5514-51832	46°48'08"	80°19'09"	414.4	21.4	5.6	58	5
Grays	17-5467-52535	47°26'07"	80°22'48"	179.8	15.5	5.8	26	6
Great Mountain	17-4723-51114	46°09'26"	81°21'34"	198.3	37.5	9.9	26	7
Grebe	17-3813-51617	46°35'58"	82°33'06"	34.5			25	
Green (Carhess)	17-4614-51736	46°43'14"	81°30'21"	44.4				
Green	17-3631-51491	46°29'00"	82°46'58"	17				
Greenwater	17-5561-52598	47°29'28"	80°15'23"	55.1	22	8.4	32	7
Grey Owl	16-7118-52382	47°15'50"	84°12'10"	247.9	31.1		45	4
Grey Trout	17-3706-51682	46°39'23"	82°41'31"	88.5	17.5	7.8	15	9
Griffin	16-6971-52177	47°05'08"	84°24'15"	147.1	45.7	20.3	45	10
Grimard	17-3898-51411	46°25'16"	82°26'29"	20.0			35	
Guide (Penelope)	17-3091-51898	46°50'12"	83°30'12"	176	36.6	11.6		
Guilfoyle (Lake 43)	17-4054-55158	49°47'21"	82°18'50"	15.8			129	4
Guilmette (Trout)	17-6537-51099	46°07'35"	79°00'47"	65.4	32.3	10.2	32	2

Lake	UTM	Latitude	Longitude					
Gull (Lawer)	17-3006-52333	47°13'23"	83°38'00"	128.7	20.7	6	41	2
Gull	17-5618-51964	46°55'00"	80°11'26"	1312.8	54.9	20.3	47	6
Gull (Mewburn)	17-3504-51925	46°52'11"	82°57'32"	264.7	56.4	13.6	35	6
Gullbeak	17-3645-51403	46°24'13"	82°45'49"	232.5	35.1	12.3	51	5
Gullrock	17-5804-52398	47°18'33"	79°56'09"	229.4	12.7	4.1	27	7
Gusty	17-3645-51487	46°28'49"	82°45'55"	26.7	13.7	3.7	22	5
Haentschel	17-5057-52285	47°12'41"	80°55'18"	94.8	22	6.3	33	4
Halfway	17-4514-51934	46°53'40"	81°38'17"	247.2	21.4	8	54	4
Hammer	17-3828-51735	46°42'25"	82°31'59"	20.6			38	
Hammond (Twin)	17-5841-52630	47°30'57"	79°52'28"	190.4	24	6.2	94	3
Hampel (Clear)	17-5866-50871	45°56'02"	79°53'00"	85.5				
Hand (Sleith)	16-7178-53397	48°10'29"	84°04'17"	119	18.9	7.3	87	3
Hand (Brownbear)	17-5668-51731	46°42'36"	80°07'31"	71.6				
Hangstone	17-5845-51854	46°49'20"	79°53'27"	329.8	18	5.1	44	2
Hannah	17-4564-51144	46°11'02"	81°33'55"	388.4				
Harold	17-3902-51596	46°35'30"	82°25'59"	124.8	22	6.2	29	12
Harry	17-3892-53167	47°59'33"	82°29'15"	157			92	
Harry	17-3873-51582	46°34'10"	82°28'11"	41.7			29	
Hastie	17-3625-51213	46°14'04"	82°46'38"	119.3			23	
Hat	17-3509-51716	46°40'32"	82°56'35"	40.9			42	
Havilah (Ickta)	17-2927-51504	46°28'38"	83°42'05"	25	11.7	2.1	27	10
Hawk	16-7106-51960	46°53'11"	84°14'10"	38	21.3	5.7	22	6
Hawk	16-6813-53264	48°04'02"	84°33'59"	322.1	56.4	8.5	114	3
Hawley	16-6538-60420	54°30'14"	84°37'30"	1234.9	54	21.4	175	8
Hearst	17-5940-52379	47°17'23"	79°45'24"	53.9				
Helen	17-4563-51062	46°06'37"	81°33'43"	82.6	41.2	20.5	33	6
Helen	17-4925-52070	47°01'08"	81°05'52"	306.1			37	
Herridge	17-5893-52037	46°58'57"	79°49'30"	181.0				
Hess (Clear)	17-4589-51747	46°43'36"	81°32'14"	56.7				
Hess		46°22'26"	82°58'59"	85.4				
Hidden (Hider)	17-4152-52021	46°58'10"	82°06'52"	119.2	20.4	8.5	33	10
Hideaway	17-4820-51162	46°1'58"	81°12'40"	69.3			46	
Hider (Hidden)	17-4152-52021	46°58'10"	82°06'52"	119.2	20.4	8.5	33	10
High	17-4679-51161	46°12'21"	81°24'43"	22.6				
Highland	17-3298-51596	46°34'12"	83°13'16"	64			58	

ONTARIO (Northeastern Ontario) (continued)

Lake name	OFIS code	Latitude	Longitude	Surface area (ha)	Maximum depth (m)	Mean depth (m)	Conductivity (µS/cm)	Secchi depth (m)
Hilltop	16-6636-53071	47°53'55"	84°48'41"	24.9				6
Hobbs (Kennedy)	17-5680-51750	46°43'34"	80°06'38"	35.1	20	6.6	45	
Hogsback	17-3929-53528	48°19'16"	82°26'41"	26.5				
Horner (Stoney)	17-3062-51715	46°40'17"	83°31'59"	417.8			33	
Horseshoe (Big Horseshoe)	17-3430-51737	46°42'01"	83°03'11"	208.9			33	
Horseshoe (Bellows)	17-3960-51305	46°19'20"	82°21'03"	274.3	28.1	8.9	30	6
Horseshoe (Ringer)	17-4640-51180	46°12'58"	81°28'41"	61.2			61	
Horwood	17-4010-53144	47°58'05"	82°19'19"	5545.8	46	5.9	82	
Hough	17-3851-51403	46°24'32"	82°29'39"	163.2				2
Hound	17-3055-52250	47°09'05"	83°33'52"	34.5	18.9	6.3	53	3
Hubert	16-6933-52436	47°19'30"	84°26'30"	85.3	38	11.3	15	12
Hutton	17-5005-51840	46°48'42"	80°59'34"	77.4	15	4	44	4
Ian (Lake 7)		46°41'09"	82°19'26"	48.0			28	
Ickta (Havilah)	17-2927-51504	46°28'38"	83°42'05"	25	11.7	2.1	27	10
Irish	17-5318-51902	46°52'23"	80°34'46"	22.5			38	7
Iron	17-5491-52005	46°57'28"	80°21'17"	72.4	27	9.2	42	10
Iron	16-7109-51920	46°50'54"	84°14'17"	27.4				
Iron (South Iron)	17-5723-51858	46°49'31"	80°03'03"	31.6	17.7	7	33	4
Iron	17-2748-51543	46°30'26"	83°56'06"	216	45.7	10.9	50	5
Ironside	17-4953-51870	46°50'19"	81°03'42"	125.7	10.7	3.7	36	5
Iroquois Bay	17-4505-51058	46°06'27"	81°38'50"	495.8	97.3	34.4	101	6
Isabel (Beauty)	17-5305-52643	47°32'00"	80°35'42"	199.9				
Ishmael	17-4542-51063	46°06'38"	81°35'32"	72.8	19.8	11.3	34	5
Island (Michi)	16-5821-52883	47°44'44"	85°54'02"	126.3	24.5	9.9	54	5
Island	17-3084-52126	47°02'33"	83°31'11"	95.3	22.9	4.9	42	3
Island (Upper Island)	16-7101-51724	46°40'22"	84°15'00"	150.2	27.5	8.8	43	6
Island (Meniss)	16-7233-51592	46°33'03"	84°05'16"	44.1	23.8	9	36	4
Island (Mc Mahon)	17-2851-51570	46°32'03"	83°48'09"	224.8				
Jackpine	17-5808-52217	47°08'42"	79°56'04"	108	23.8	4.1	54	4
Jarvis	16-7175-51737	46°41'01"	84°09'19"	53.5	19.8	4.3	24	2

Jeanne	17-4099-51964	46°55'25"	82°11'16"	115.3			27	
Jerry	17-5489-54853	49°31'13"	80°19'27"	249.9	35	10.7	30	13
Jerry	17-4888-52702	47°35'13"	81°08'57"	212.4	22.6	8.7	27	9
Jerry	17-5263-52458	47°22'01"	80°39'11"	56.3	18.3	7	69	6
Jim Edwards	17-5431-52386	47°18'06"	80°25'51"	86.6			47	
Jimchrist (Christman)	17-3675-51601	46°35'01"	82°43'45"	58.6				
Jobammageeshig	17-3152-51548	46°31'23"	83°24'33"	357.1	33.6	10	27	6
Johnnie (Bushcamp)	17-4826-51036	46°05'13"	81°13'30"	342.3	33.6	8.9	56	5
Jumping Cariboo	17-5933-51926	46°52'41"	79°46'35"	408.7	32	6.4	28	3
Justin (First Justin)	17-5838-52437	47°20'46"	79°53'30"	38.1				
Kabiskagami	16-6792-53960	48°41'36"	84°33'53"	145.6	30.5	13.5	30	7
Kakakise	17-4750-51010	46°03'54"	81°19'11"	112.6	31.7	6.9	117	3
Kakakiwibik (Caribou)	16-6415-53841	48°35'46"	85°04'46"	423.6	33.6	7.3	68	4
Kanichee (Cedar)	17-5882-52189	47°07'05"	79°50'22"	252.5				
Kaotisinimigo (Bear)	17-6069-51648	46°37'49"	79°36'10"	549.9				
Kasasway (Upper Kasasway)	17-4304-53073	47°55'03"	81°55'54"	234.0				
Kashbogama	17-3861-52198	47°07'35"	82°30'08"	267	29	10.9	47	6
Kathleen	17-2820-53190	47°59'20"	83°55'18"	517	22	6.1	61	5
Katzenbach	16-6239-53245	48°03'49"	85°20'11"	243.8				
Kaufman	16-7159-51722	46°40'20"	84°10'32"	31.4	15.9	6.3	27	4
Kecil	17-4001-51234	46°15'48"	82°17'41"	425.9	26.2	12.1	36	5
Keelor	17-3500-51516	46°30'12"	82°57'26"	123.8	34.2	8.6	32	6
Kelly #27	17-4369-51883	46°46'46"	80°31'55"	17.1	17	7.6	42	6
Kennedy	17-5680-51750	46°50'51"	81°49'39"	254.5	52	15.8	45	7
Kennedy (Hobbs)	17-4325-51399	46°43'34"	80°06'38"	35.1	20	6.6	30	6
Kerr	17-4322-53002	46°24'40"	81°52'40"	126				
Ketchini	17-3897-52094	47°51'01"	81°54'23"	186.5	31	7.7	82	4
Kettle	17-5356-51847	47°01'32"	82°27'21"	299.6	33	7	37	4
Kettyle	17-4723-51015	46°48'43"	80°32'18"	53.3	23.5	8.9	36	13
Killarney	17-3509-51882	46°04'08"	81°21'20"	326.5	61	10.8	29	10
Kindiogami (Freddie's)	17-5787-52433	46°49'57"	82°57'17"	465.2				
King Dodds (Kittson)	17-3828-51348	47°20'23"	79°57'14"	78.5	15.3	4.8	66	4
Kings	17-3414-51701	46°21'32"	82°31'30"	87.5	22	9.1	43	4
Kirk		46°40'04"	83°04'25"	59	21.9	9.5	17	13

ONTARIO (Northeastern Ontario) (continued)

Lake name	OFIS code	Latitude	Longitude	Surface area (ha)	Maximum depth (m)	Mean depth (m)	Conductivity (μS/cm)	Secchi depth (m)
Kirkpatrick (Blue)	17-3400-51692	46°39′31″	83°05′20″	1097.5	70.2	17	29	7
Kitt	17-5812-52445	47°21′12″	79°55′22″	131.1	31.1	11.2	51	5
Kittson (King Dodds)	17-5787-52433	47°20′23″	79°57′14″	78.5	15.3	4.8	66	4
Klondyke	17-4157-51611	46°35′37″	82°05′57″	94.3	20.5	5.3	27	6
Kokoko	17-5733-52151	47°05′28″	80°02′04″	493.7	44.8	13.7	66	6
Kukagami	17-5344-51754	46°43′57″	80°33′03″	1864.8			44	7
Kumska	17-4975-51827	46°47′47″	81°02′52″	112.4	28.3	10.5	40	
Kwinkwaga	16-6225-54072	48°48′52″	85°19′43″	811.3	36.6	6.6	114	3
Kwitosse	17-4714-52923	47°47′06″	81°22′52″	19.2				
Labelle	17-3401-51997	46°55′27″	83°06′15″	32.8				
Lac aux Sables (Sable)	17-3988-51823	46°47′24″	82°19′06″	1162.3	32	9.2	32	6
Lac Cherie	17-2937-52086	47°00′02″	83°42′46″	226.7				
Laderoute	17-3767-51230	46°14′48″	82°35′57″	28.9				
Lady Dufferin	17-5221-52550	47°27′14″	80°42′44″	160.9	24.4	4.8	89	5
Lady Sydney	17-5599-52502	47°24′13″	80°12′20″	229.7	25.6	7.4	36	5
Lake 1		46°32′10″	82°18′51″	28.7			27	
Lake 1 (Survey)	17-3702-51629	46°36′33″	82°41′39″	18.2				
Lake 3		46°42′57″	82°10′45″	11.8				
Lake 5		46°57′25″	83°33′24″	13.5				
Lake 6		47°19′01″	83°33′09″	6.4	10.7			
Lake 7 (Ian)		46°41′09″	82°19′26″	48.0			28	2.2
Lake 8 (Lower)		47°08′01″	83°39′55″	68.1				
Lake 8 (Long)	17-3933-51911	46°52′27″	82°24′18″	78.5				
Lake 8		46°46′45″	83°00′59″	21.9			43	
Lake 9 (Foxwell)		47°07′50″	83°43′08″	16.5	27			5.2
Lake 9		46°42′14″	82°30′03″	9.6				
Lake 10		46°40′35″	84°07′39″	19.4				
Lake 11		46°30′28″	82°30′24″	29.2				
Lake 11 (Wright's,Rice)	17-4300-51062	46°06′26″	81°54′19″	27.1	26.5	12.9	35	8
Lake 12 (Barn)	17-4165-51920	46°52′43″	82°05′45″	20.6	15.3	4.5	33	4

Lake	Code	Latitude	Longitude					
Lake 12-I (Cross)	16-6398-54235	48°57'02"	85°05'25"	167.7				
Lake 13	17-4171-51922	46°52'48"	82°05'16"	4.6	14.9	6	26	7
Lake 15 (Dolly-B-Doo)		47°11'30"	83°38'29"	23.2				
Lake 15 (Ericson)	17-3722-51561	48°55'03"	83°23'13"	11.2	18.4			
Lake 15 (Wolf)	17-3735-51561	46°32'55"	82°39'59"	10.1				
Lake 16 (Polar Bear)		46°32'49"	82°38'56"	9.7				
Lake 17 (Brewer)		46°51'09"	82°27'21"	0.9				
Lake 20		46°32'46"	82°28'18"	14.5	5.9		38	2.5
Lake 21		47°10'39"	83°39'29"	24.7	33			7.7
Lake 21		47°04'57"	83°49'41"	14.8				
Lake 21		46°46'00"	83°02'27"	22.1				
Lake 21	17-4116-51795	46°45'54"	82°09'27"	35.3	16	5.8	36	5
Lake 23		46°32'17"	82°28'24"	13.9				
Lake 23 (Long)	17-4236-51138	46°10'30"	81°59'21"	64.8	35.1	15.3		7
Lake 24		47°09'36"	83°39'26"	16.6				
Lake 24	17-2803-52089	46°59'56"	83°53'22"	114	22	4.9		3
Lake 25		46°41'14"	82°38'24"	5.1				
Lake 26 (Arrowhead)	17-3905-51513	46°30'30"	82°26'07"	64.1	21.3			
Lake 27		47°15'40"	83°28'36"	34.0				
Lake 27		46°35'45"	82°33'29"	8.0				
Lake 30	17-4730-52369	47°17'13"	81°21'24"	6.5	25.6	11.2	143	5
Lake 30 (Moose)		47°11'04"	83°41'51"	31.8				
Lake 33		46°30'58"	82°27'10"	12				
Lake 34	17-3961-51866	47°11'00"	83°22'00"	16.3	7	3.6	26	4
Lake 36 (Little Trout)	17-3428-51647	46°49'36"	82°21'43"	64.3	24	7.7	33	5
Lake 36 (Boundary)	17-4476-52921	46°37'10"	83°03'10"	18.8				
Lake 36 (Secret)	17-3856-53100	47°46'57"	81°41'55"	26.2				
Lake 41 (Whigham)	17-4009-51893	47°56'05"	82°31'54"	10	15	5.1	35	4
Lake 42	17-4054-55158	46°51'06"	82°17'59"	72.8				
Lake 43 (Guilfoyle)	17-4929-52250	49°47'21"	82°18'50"	15.8				
Lake 44		47°10'49"	81°05'36"	25.3				
Lake 51		46°44'01"	82°41'24"	10.1				
Lake 54 (Sailor Trout)	17-3923-51852	46°43'20"	82°41'35"	18.5	15.5	6.6	60	8
Lake 55 (Little Touch)	17-3931-51855	46°48'49"	82°24'41"	20.8	42.7	14.2	54	11
Lake 56 (Five Star)		46°49'08"	82°24'11"	92.1				

ONTARIO (Northeastern Ontario) (continued)

Lake name	OFIS code	Latitude	Longitude	Surface area (ha)	Maximum depth (m)	Mean depth (m)	Conductivity (µS/cm)	Secchi depth (m)
Lake 59 (Ess)		46°48′15″	82°55′06″	33.7				
Lake 61 (Star)		46°49′17″	82°28′17″	30.2				
Lake 69 (Brigstocke)	17-5807-52370	47°17′05″	79°56′10″	34.2			35	
Lake 73		46°50′15″	82°37′11″	16.6				
Lake of the Mountains (Duborne)	17-3523-51234	46°15′01″	82°54′56″	933.6				
Lake One	16-7130-51713	46°39′49″	84°12′58″	38.4	16.5	6.6	35	4
Lake Panache (Panache)	17-4743-51217	46°15′22″	81°19′39″	8958.6			64	6
Lake Talon (Talon)	17-6497-51295	46°18′16″	79°03′22″	1404.9	60.1	12.9	56	4
Lanark (Fullerton)	17-3672-51583	46°34′18″	82°44′28″	56.1				
Lance	17-3069-52475	47°21′05″	83°33′23″	53.5	17.1	5.3	33	5
Landers	17-5390-52352	47°16′22″	80°29′01″	110.7	24	4.9	32	10
Larder	17-6010-53265	48°05′02″	79°38′05″	3703.7	33.5	12.3	191	4
Larry	17-3788-51673	46°39′00″	82°35′21″	9.0				
Laughing	17-2957-52027	46°56′49″	83°41′22″	78.5	19.2	3.8	29	2
Laundrie	17-5110-52189	47°07′32″	80°51′16″	375	20	4.9	29	
Laura	17-5301-51990	46°56′45″	80°36′13″	223.8	61	16.3	49	8
Laurence (Camp 12)	17-3097-51982	46°54′44″	83°29′56″	10.9	13.4	7.6	54	4
Lauzon	17-3592-51192	46°12′53″	82°49′13″	2242	82.3	23.8	54	7
Lawer (Gull)	17-3006-52333	47°13′23″	83°38′00″	128.7	20.7	6	41	2
Lawlor (Aylmer # 37)	17-5216-51920	46°52′35″	80°43′56″	35.5			34	8
Lear (Clear)	17-3445-51313	46°19′04″	83°01′10″	171.4	39	13.8	31	8
Leask	17-4933-52284	47°12′38″	81°05′10″	69.9	24.1	7.1	59	9
Leg Of Lamb	17-5100-53797	48°34′20″	80°51′51″	8.5				
Lepha	17-5714-52652	47°32′19″	80°03′06″	202.8	21	7.3	38	7
Lillybet	17-3511-51570	46°33′07″	82°56′30″	88.9	20.5	9.3	33	6
Limit	17-3886-52116	47°03′23″	82°27′58″	134.9	14	4.2		3
Lineus	17-2791-52545	47°24′31″	83°55′41″	25.9	37	10.5		3
Linger	17-5367-52152	47°05′31″	80°30′55″	69.8	18	3.1	30	7
Lingo (Barron)	17-5152-52045	46°59′45″	80°47′59″	50.2	23		34	4

Lake		Latitude	Longitude					
Little Agawa	16-6981-52469	47°20'51"	84°22'35"	131.2	41.5	10.9	81	7
Little Bear	17-4603-51135	46°10'34"	81°30'26"	140.5	22.6	12.2	39	6
Little Burwash	17-4930-52197	47°07'47"	81°05'33"	105.2	28.4	10.4	81	6
Little Chiblow (Denman)	17-3360-51357	46°21'13"	83°07'55"	644.4	38.1	17.3	68	6
Little Dayohessarah	16-6466-54051	48°47'01"	85°00'15"	55.5	30.1	10.4	41	5
Little Dennie	17-4303-51975	46°55'45"	81°54'55"	39.5	20	6.9	81	7
Little Dobie (Dobie)	17-3450-51565	46°32'45"	83°01'19"	43.2	18.3	5.8	32	6
Little Laundrie	17-5145-52182	47°07'17"	80°48'35"	58.2	24	6.2	52	6
Little Moon	17-3581-51407	46°24'26"	82°50'46"	103.6				
Little Panache	17-4721-51252	46°16'55"	81°21'43"	130.6	26.2	9.7	96	5
Little Pickerel	17-3224-51502	46°29'03"	83°18'36"	194.3	18.6	7.9	33	6
Little Pogamasing	17-4318-52011	46°57'05"	81°53'59"	360.8	24	8.4	48	6
Little Prune	17-4919-52250	47°10'49"	81°06'28"	31.9	31.4	7.8	56	5
Little Quinn	17-2855-52329	47°12'51"	83°50'25"	42.8	45.7	13.5	35	6
Little Quirke	17-3770-51544	46°32'20"	82°36'05"	396.5	31	12.6	32	10
Little Sister	17-3813-51665	46°38'35"	82°33'00"	61.6			60	
Little Squaw	17-4290-52054	47°00'22"	81°55'24"	31.7	15.5	6.6	35	8
Little Touch (Lake 55)	17-3923-51852	46°48'49"	82°24'41"	20.8	20	7.1	26	4
Little Trout	16-6169-53102	47°56'14"	85°26'10"	20.1	7	3.6	34	4
Little Trout (Lake 36)	17-3961-51866	46°49'36"	82°21'43"	64.3	12.8	5.4	38	5
Little Turkey	16-6969-52130	47°02'33"	84°24'31"	19.4				
Little Venetian		46°56'26"	81°12'33"	71.1			35	
Lizotte (Dome)	17-3874-51449	46°27'02"	82°28'12"	21			36	
Lloyd	17-4807-54476	49°10'58"	81°15'51"	11.4				
Lodestone	17-3078-52210	47°06'51"	83°32'06"	62.8	42.7	15.2	30	5
Lonely (Meredith)	17-2758-51519	46°29'23"	83°55'26"	302.7				
Long (Wiseman T)		46°58'51"	82°55'26"	22.2				
Long (Lake 8)	17-3933-51911	46°52'27"	82°24'18"	78.5				
Long	17-3486-51574	46°33'18"	82°58'31"	31.2				
Long	17-4936-51346	46°22'20"	81°04'54"	77.9				
Long (Lake 23)	17-4236-51138	46°10'30"	81°59'21"	64.8	35.1	15.3	137	7
Longhaul	17-3845-51805	46°46'12"	82°30'45"	16.7	15.1	4.3	29	3
Loon	17-4455-51177	46°12'54"	81°42'23"	151	20.4	9.8	60	5
Loon (Northland)	16-7200-51762	46°42'21"	84°07'19"	87.2	26	11.6	35	5
Loonskin	16-6761-53288	48°05'09"	84°38'18"	149.7	25.6	5.6	83	8

ONTARIO (Northeastern Ontario) (continued)

Lake name	OFIS code	Latitude	Longitude	Surface area (ha)	Maximum depth (m)	Mean depth (m)	Conductivity (µS/cm)	Secchi depth (m)
Lost	17-3242-52296	47°11′55″	83°19′16″	29.9				8
Low	17-4567-51055	46°06′12″	81°33′34″	33.8	28.4	14.4	74	8
Lowell	17-5912-52067	47°00′33″	79°47′59″	89.8	53.4	17.1	65	
Lower (Lake 8)		47°08′01″	83°39′55″	68.1				
Lower Bass	17-5601-52035	46°59′02″	80°12′31″	105.4	21.4	8	56	7
Lower Green	17-4455-52033	46°59′40″	81°43′10″	30.6			32	
Lower Griffin	16-6961-52174	47°04′52″	84°38′50″	27.7				
Lower Mace	17-3574-51509	46°29′50″	82°51′42″	196.2	25	9.7	33	5
Lower Matagamasi	17-5393-51873	46°50′10″	80°29′04″	131.8	18	6.9	34	7
Lower Shakwa	17-4261-51776	46°44′59″	81°58′00″	51	10.4	1.6	41	3
Lower Twin		47°02′49″	83°32′35″	23.0	18.3			5.6
Lucas (Geiger)	17-3860-51538	46°31′50″	82°29′12″	88.2	27	9.5	39	5
Lulu	17-5830-53532	48°19′43″	79°52′49″	51.0				
Lulu	17-5184-52500	47°24′22″	80°45′26″	80.7	23.2	7.4		
Lumsden	17-4665-50976	46°01′31″	81°25′59″	23.8	21.8	9	26	7
Mac Gregor	16-6850-52397	47°17′13″	84°33′12″	22.0			22	7
Macauley (Fox)	17-4429-51586	46°35′21″	81°44′30″	471.4	42.4	10.7	69	4
Macdonald (Mc Donald)	17-4573-51969	46°55′36″	81°33′36″	87.9			31	
Macfie	17-4702-52882	47°44′54″	81°23′48″	71.8				
Maconner	16-7182-53233	48°01′45″	84°04′26″	100.4	34.2	9.4	43	6
Madawanson	17-4096-51627	46°36′42″	82°10′47″	416.1	42.7	12.9	30	5
Maggie	17-5253-52254	47°11′31″	80°40′25″	66			34	
Magog (Granary)	17-3585-51264	46°16′32″	82°50′27″	284.9	30.5	16.3	34	5
Magrath	17-3088-51846	46°47′28″	83°30′22″	43.2	26	8.6	51	4
Makobe	17-5430-52549	47°26′44″	80°25′13″	2021.5	22.6	5.8	29	5
Mamainse	16-6811-52113	47°01′58″	84°36′57″	148.6				
Manitou (Devil's)	17-5548-51889	46°51′13″	80°16′52″	343.5	47.6	18	57	8
Manitowik	16-6953-53376	48°09′51″	84°22′30″	3130.3	119	37.6	88	4
Maquon	16-6766-52854	47°41′58″	84°38′45″	165.4				
Marcia	17-5380-52106	47°03′13″	80°30′14″	120.9			63	

Lake	Map ref.	Latitude	Longitude					
Marina	17-5258-52493	47°23'52"	80°39'31"	36.9	16.8	4.6	29	6
Marion	17-5330-51995	46°57'01"	80°34'27"	38.5	8.5	4.3	38	4
Marion	17-4319-51755	46°43'53"	81°53'28"	61.5	12.5	4	37	15
Marjorie	17-5292-51958	46°54'36"	80°37'14"	77.4	22	5.6	32	7
Marjory	16-6404-53215	48°02'02"	85°07'32"	61.0	47.2	11.4	45	5
Marsh	17-3689-51478	46°28'19"	82°42'34"	31.4	65.6	19.1	51	4
Marshland	17-3743-51307	46°19'04"	82°37'40"	107.8	27.4	9.6	39	6
Marten	17-5953-51723	46°41'58"	79°45'52"	1008.5	44.5	12.2	139	9
Mashagama	17-3213-51987	46°55'16"	83°20'41"	662.9	61	8.7	34	9
Maskinonge	17-5427-51794	46°46'25"	80°26'25"	1426.6	17.7	4.4	44	5
Matachewan	17-5243-53180	48°02'24"	80°42'01"	140.5	32	10.6	39	8
Matagamasi	17-5305-51809	46°45'36"	80°36'21'"	1392.7	85.3	22.5	78	7
Matchinameigus (Trout)	16-7135-53359	48°08'58"	84°07'27"	667.7	47	14.3	642	5
Mather's	17-3638-51369	46°22'30"	82°46'22"	106.8	25.2	9.3		5
Matinenda	17-3525-51387	46°23'11"	82°55'31"	3566.6				11
May	17-3852-51434	46°25'56"	82°29'02"	318				8
Mc Cabe	17-3797-51421	46°25'24"	82°33'57"	180.2				
Mc Caroll (Mc Carrel)	17-2753-51476	46°26'42"	83°55'26"	203.9	33.6	16.1	45	5
Mc Carrel (Mc Caroll)	17-2753-51476	46°26'42"	83°55'26"	203.9	33.6	16.1	45	5
Mc Carthy	17-3877-51310	46°19'31"	82°27'00"	657.9	28	8.8	148	12
Mc Clung	17-3272-51926	46°51'44"	83°15'24"	33.2				
Mc Connell	17-6268-51774	46°44'23"	79°20'23"	207.2				
Mc Cool	17-3918-51442	46°26'44"	82°24'28"	132.3				
Mc Cormick	16-6570-53423	48°12'58"	84°53'12"	153.2				
Mc Culloch	17-5226-52411	47°19'28"	80°42'02"	43.2	32.9	12.1	33	5
Mc Donald (Macdonald)	17-4573-51969	46°55'36"	81°33'36"	87.9	61.6	22	72	9
Mc Dougall	17-4544-52051	47°00'10"	81°35'35"	48.6	22	9.6	31	7
Mc Giffin	17-5401-52454	47°21'39"	80°28'04"	109.4	23.8	14.5	48	6
Mc Giverin	17-3678-51289	46°18'09"	82°42'59"	305.9	36	11.7	33	6
Mc Gown	17-5328-52011	46°57'51"	80°34'00"	102.2	32	10.2	53	4
Mc Grindle	17-4851-51943	46°54'25"	81°11'49"	132.9	9	1.7	34	2
Mc Guey	17-4331-51413	46°25'28"	81°52'12"	56.3	38.1	13.8	33	7
Mc Kensie (Mc Kenzie)	17-5586-52672	47°33'26"	80°13'04"	66.4	32.6	13.9	48	7
Mc Kenzie (Mc Kensie)	17-5586-52672	47°33'26"	80°13'04"	66.4	32.6	13.9	48	7
Mc Lander	17-4309-51420	46°25'45"	81°53'51"	58.9	19.2	7.1		4

ONTARIO (Northeastern Ontario) (continued)

Lake name	OFIS code	Latitude	Longitude	Surface area (ha)	Maximum depth (m)	Mean depth (m)	Conductivity (µS/cm)	Secchi depth (m)
Mc Laren	17-5786-52082	47°01'29"	79°57'56"	34.4				
Mc Lean	17-5778-52271	47°11'41"	79°58'22"	97.4				
Mc Mahon (Island)	17-2851-51570	46°32'03"	83°48'09"	224.8				
Mc Nab	17-6007-52240	47°09'48"	79°40'15"	44.4				
Megisan	17-3090-52347	47°14'56"	83°31'25"	620.3	18	6.4	49	4
Megisan (Torrance)	17-3096-52319	47°12'48"	83°30'53"	98.3	12.2	4.1		3
Memesagamesing (Saga)	17-5779-50969	46°01'18"	79°59'58"	1098.8	59.4	13.7	34	3
Memoir	17-3507-52629	47°30'13"	82°58'53"	131.6	39.3	14	92	6
Mendelssohn	17-5595-52642	47°31'41"	80°12'35"	460.5	33.5	10.7	40	6
Meniss (Island)	16-7233-51592	46°33'03"	84°05'16"	44.1	23.8	9	36	4
Meredith (Lonely)	17-2758-51519	46°29'23"	83°55'26"	302.7	42.7	15.2	35	5
Mesomikenda (Beaver)	17-4340-52773	47°38'24"	81°52'46"	1706.2	71.4	13.6	61	5
Meteor	17-4698-52386	47°18'09"	81°23'55"	249.6				
Mewburn (Gull)	17-3504-51925	46°52'11"	82°57'32"	264.7	56.4	13.6	35	6
Michaud	17-4821-51845	46°48'37"	81°14'03"	148.5	24	7	26	6
Michi (Island)	16-5821-52883	47°44'44"	85°54'02"	126.3	24.5	9.9	54	5
Mickey	17-5395-51925	46°53'06"	80°28'57"	59.4	19	6.8	36	4
Midlothian	17-5001-53061	47°54'31"	80°59'39"	367.3	32.3	8.2	60	5
Mijinemungshing	16-6718-52845	47°41'36"	84°42'37"	500.2				
Mike	17-4266-52165	47°06'17"	81°58'05"	40.0				
Millen	17-4113-51461	46°27'52"	82°09'20"	84.9	30.5	14.4	33	6
Miller	17-5208-52783	47°39'38"	80°43'25"	73.7	21.4	6.7		5
Millerd	17-5045-51248	46°16'41"	80°56'33"	176.6	18.9	4.6	45	6
Ministic	17-4564-51563	46°33'29"	81°34'23"	567.2	31.1	11.1	29	5
Mirror	17-3019-52319	47°12'45"	83°36'58"	44.5	21.3	8.5	50	6
Mishi	16-6222-53262	48°04'46"	85°21'32"	217.4				
Mishibishu	16-6184-53256	48°04'29"	85°24'35"	818.3				
Missinaibi (Big Missinaibi)	17-3006-53592	48°21'32"	83°41'07"	7706.9	94	19.2	83	4
Mississagi	17-3842-52259	47°10'41"	82°31'32"	238.3	19	4.7	56	4
Mistango	17-5396-54407	49°07'10"	80°27'26"	205.4				

Name	Reference	Latitude	Longitude					
Moccasin	17-3811-51776	46°44′35″	82°33′29″	163.9	31	10.6	30	9
Molybdenite	16-6520-53237	48°03′03	84°57′38″	93.6	49	8.1	18	4
Mongoose	16-7124-52271	47°09′43″	84°11′17″	117.6	30.5	9	37	3
Montreuil (Big Trout)	17-6607-51439	46°25′54″	78°54′28″	18.1				
Moon	17-3539-51400	46°24′45″	82°52′47″	506.4	41.8	16.2	38	6
Moon	17-4007-51778	46°44′54″	82°17′56″	48.4	10.1	4.8	30	4
Moonshine	17-3778-51798	46°45′46″	82°36′02″	14.8	20	5.7	54	7
Moose (Lake 30)		47°11′04″	83°41′51″	31.8				
Moose	17-4745-51661	46°39′10″	81°18′50″	204				
Morgan	17-4777-51687	46°40′23″	81°17′26″	42.8	32.5	12.1	26	5
Morrison	17-2857-52090	47°00′07″	83°49′07″	376.6				
Mosquito	17-4599-51542	46°32′36″	81°31′21″	97.1	13.1	5.9	30	4
Mount (Mountain)	17-3674-51703	46°40′30″	82°44′00″	209.3			45	
Mountain (Aloft)	16-6887-53553	48°19′29″	84°27′13″	106.1	28.1	7.9	49	6
Mountain	17-5885-52314	47°13′49″	79°49′38″	221.5	39.3	10.9	62	6
Mountain (Mount)	17-3674-51703	46°40′30″	82°44′00″	209.3			45	
Mountain	17-3207-51418	46°29′00″	82°45′28″	18.6				
Mountain (Woodrow)		46°24′35″	83°19′59″	49.1	32.7	11.2	42	5
Mousseau	17-5845-53285	48°06′23″	79°51′55″					
Mowat	17-5038-51902	46°51′51″	80°56′51″	93.9	44.8	12.1	32	7
Mozhabong	17-4175-51987	46°56′55″	82°05′18″	1520.4	21	10.1	37	2
Mud (Robertson)	16-7087-51842	46°46′52″	84°15′55″	147.6	25.6	4.4	53	5
Muldrew	17-4525-52091	47°02′07″	81°37′30″	242.3	19.8	11.3	50	4
Mumroe	17-5606-52675	47°33′36″	80°11′40″	75.9	42.7	8.1	119	3
Muskasenda	17-4773-53267	48°05′32″	81°18′16″	632	15.8	5.3	39	3
Muskwash	17-3412-51900	46°50′44″	83°04′57″	43.1	31.1	7.1	27	6
Mystery (South Anvil)	17-3079-52141	47°03′14″	83°31′46″	74.5				
Nansen	17-4108-54557	49°14′58″	82°13′31″	11.6				
Narvik	17-3353-51609	46°35′08″	83°09′24″	51.4			35	
Navy	17-3816-51903	46°51′27″	82°33′12″	97.2			40	
Nellie	17-4594-51088	46°08′00″	81°31′32″	260.5	54.9	19.2		30
Nellie	17-5152-54045	48°47′43″	80°47′34″	81.4				
Nelson	17-4928-51746	46°43′33″	81°05′38″	315.8	50.3	11.1	34	8
Nemegosenda	17-3426-53189	48°00′18″	83°06′38″	1803.2	93	26.1	128	5
Nepahwin	17-5026-51439	46°27′34″	80°59′05″	127.5	21		387	5

ONTARIO (Northeastern Ontario) (continued)

Lake name	OFIS code	Latitude	Longitude	Surface area (ha)	Maximum depth (m)	Mean depth (m)	Conductivity (µS/cm)	Secchi depth (m)
Net	17-5909-52183	47°06′47″	79°48′15″	771.3	47.5	7.8	77	6
Nettie	17-5751-53413	48°13′22″	79°59′18″	51.0				
Niccolite	17-5562-52573	47°28′05″	80°15′13″	33	12.2	4	45	5
Nina	17-3871-51590	46°34′36″	82°28′26″	22.3				
Ninegee	17-3097-52494	47°22′21″	83°31′16″	47.8	20.7	6.5	53	3
Noble	17-6110-51620	46°36′14″	79°33′53″	94.5	16	5	25	3
Nook	17-3887-51486	46°28′45″	82°26′32″	27	18			
Norris	17-5862-51895	46°51′19″	79°52′12″	85.8	18.3	7.3	56	3
Norse	17-3577-51344	46°21′18″	82°51′28″	33.6				
North Anvil (Anvil)	17-3094-52159	47°04′17″	83°30′34″	92.8	18.3	6.8	36	4
North Hubert	16-6930-52450	47°19′49″	84°26′41″	33.8	37.5	12.2	18	13
North Raft	16-6458-60417	54°30′14″	84°44′53″	661.9				
Northland (Loon)	16-7200-51762	46°42′21″	84°07′19″	87.2	26	11.6	35	5
Norway	17-4759-51035	46°05′14″	81°18′32″	63.3	33.6	15.1	29	11
Nushatogaini	17-3169-52406	47°17′19″	83°25′18″	186.2	37.5	10.3	50	4
O.S.A.	17-4691-51000	46°03′12″	81°23′53″	278.9	39.7	12	36	16
Obabika	17-5566-52101	47°02′19″	80°15′32″	3155.1	40.3	12.6	46	6
Ogas	16-6738-52944	47°46′59″	84°40′13″	197.8	42.1	13	46	5
Old Woman	16-6710-52766	47°37′21″	84°43′27″	267.6				
Olympus	17-3595-51584	46°34′29″	82°50′07″	7.4				
Onaping	17-4620-52125	47°08′19″	81°31′14″	4736.8			42	5
Onedee (Elbow)	17-3357-51692	46°39′30″	83°08′48″	94.2			42	
Opikinimika	17-4685-52458	47°21′30″	81°25′14″	658.1				
Orotona (Tee)	17-4054-51496	46°29′44″	82°13′57″	62.9	41.2	13.8	28	5
Osbourne	17-4881-51899	46°51′57″	81°09′49″	112.1	10	3.1	41	2
Oshawong	17-4728-52266	47°11′43″	81°21′45″	112.5	25.9	12.2	69	5
Ouellette	17-3851-51472	46°28′14″	82°29′29″	158.5	84	26.8	44	13
Palangio	17-4850-54325	49°02′50″	81°12′19″	24.5				
Panache (Lake Panache)	17-4743-51217	46°15′22″	81°19′39″	8958.6			64	6
Pancake	16-6807-52155	47°04′11″	84°37′36″	350.4	17.4	4.5	31	5

Paradise (Blue)	17-4802-54481	49°11'14"	81°16'17"	23.5	35	8.2	37	11
Paradise (Alphretta)	17-5176-52025	46°58'39"	80°46'06"	487.4	28	7.5	32	6
Parkin	17-5105-51914	46°52'43"	80°51'44"	107.4	10.4	2.8	36	2
Parsons	17-5308-52065	47°00'49"	80°35'38"	50.9	26.8	7.6	37	5
Path	17-4487-52064	47°00'30"	81°40'12"	185.9	43.6	14.5	20	8
Pathfinder (Beef)	17-3570-51488	46°28'44"	82°51'40"	256.5				
Pats	17-5735-51927	46°53'08"	80°02'06"	42.6	24.4	12.4	45	6
Patten	17-2880-51572	46°32'07"	83°45'49"	243.5	13.4	3.9	30	2
Patter (Patterson)	17-2936-52244	47°08'13"	83°43'55"	176.1	21.5		30	6.5
Patterson	16-7045-52307	47°12'30"	84°17'56"	46.8	13.4	3.9	44	2
Patterson (Patter)	17-2936-52244	47°08'13"	83°43'55"	176.1	28.4	10.7	26	5
Patterson (Stormy)	17-5947-51034	46°04'49"	79°46'26"	333.4				
Peach	17-4347-52545	47°26'35"	81°51'55"	41.1	18.3	8.1	38	5
Peak	17-3579-51304	46°18'51"	82°50'27"	93.1				
Pearl	17-3300-51609	46°34'54"	81°13'05"	72.8	11	6.4	129	7
Pedro	17-5352-51958	46°54'59"	80°32'15"	63.8	36.6	11.6	140	4
Penelope (Guide)	17-3091-51898	46°50'12"	83°30'12"	176				
Perron	17-6531-51368	46°22'10"	79°00'33"	49.2	29	8.1	43	4
Perry	17-5661-53749	48°31'41"	80°06'16"	107.9	50.6	18	41	6
Peshu	17-3371-52041	46°58'15"	83°08'25"	388.5	29	11.4	41	7
Petauguin	17-2961-51542	46°30'42"	83°39'26"	87.9	30.5	12.9	41	5
Peter	17-4836-51150	46°11'24"	81°12'50"	132.4	16.2	7.7	23	4
Phelbin	16-7191-51747	46°41'31"	84°08'03"	15.8	37.5	13.9	57	8
Philbrick	17-5585-52656	47°32'30"	80°13'18"	79.5			36	
Piano	17-4255-52114	47°03'15"	81°58'49"	51				
Picard (Pickard)	16-7082-51888	46°49'19"	84°16'11"	57.0				
Pickard (Picard)	16-7082-51888	46°49'19"	84°16'11"	57.0				
Pickle	16-7205-53464	48°14'05"	84°01'42"	222.3	44.5	14.3	68	4
Pilgrim	17-5259-52267	47°11'43"	80°39'30"	122.9	25	6.6	26	8
Pilon	17-4860-52213	47°08'50"	81°11'05"	76.5				
Pine (Bouck)	17-6046-52356	47°16'10"	79°37'06"	99.3	21.3	6.8	110	8
Pine (Prugh)	16-7133-51952	46°52'44"	84°12'30"	51.5	11.1	2.8	23	5
Pine (Dana)	17-5566-51724	46°42'16"	80°15'33"	110.5	25.9	7.4	45	3
Pipe (Flying Bird)	17-3088-51523	46°29'55"	83°29'25"	32.5	22.6	6.8	33	5
Pivot	16-6879-53337	48°07'28"	84°28'29"	122.9	16.8	5	39	6

ONTARIO (Northeastern Ontario) (continued)

Lake name	OFIS code	Latitude	Longitude	Surface area (ha)	Maximum depth (m)	Mean depth (m)	Conductivity (µS/cm)	Secchi depth (m)
Pocket		47°06′58″	84°15′02″	21.7				
Pogamasing	17-4364-52017	46°58′04	81°50′08″	1587	34	8.8	41	6
Point	17-2857-52221	47°07′00″	83°49′32″	75.4	28.7	5.7	45	4
Polar Bear (Lake 16)	17-3735-51561	46°32′49″	82°38′56″	9.7				
Pond	17-5525-51994	46°56′56″	80°18′33″	22.7	23.5	11.5	32	10
Portelance	17-3688-51840	46°47′56″	82°43′04″	76.5	13.7	5.4	26	5
Pothole		46°38′48″	82°31′35″	11.6			68	
Potvin	17-5396-51819	46°47′30″	80°28′49″	45.2	62	27.5	49	14
Poupore	17-3931-51815	46°46′55″	82°24′18″	188.2	21.9	6.9	36	5
Prairie Grass	17-3155-52376	47°16′02″	83°26′10″	238.8	14.6	4.2	60	3
Primeau	17-3033-51689	46°38′47″	83°34′11″	117.4			30	
Prugh (Pine)	16-7133-51952	46°52′44″	84°12′30″	51.5	11.1	2.8	23	5
Prune	17-4902-52259	47°11′11″	81°07′59″	179.5	21	7	54	6
Quimby	17-3659-51361	46°22′05″	82°44′33″	165.6	14.3	7.3	98	6
Quintet (1)	16-7098-52235	47°08′16″	84°12′42″	72.8	18.3	4.9	36	4
Quintet (2)		47°07′36″	84°13′40″	164.3	44.2	13.8	45	4
Quirke	17-3810-51468	46°29′23″	82°33′07″	2072.6				
Rabbit	17-6031-52053	46°59′47″	79°38′37″	2106	42.7	14.2	79	6
Rachel	17-5560-52001	46°57′06″	80°15′44″	83.4	16.8	5.5	53	5
Rackey	17-3445-51523	46°30′29″	83°01′35″	35.4	35.4	10.2	29	6
Radisson	17-5180-53389	48°12′58″	80°45′27″	543.5	54.3	14.4	68	5
Ramsey	17-5038-51474	46°28′57″	80°57′00″	875.1				
Rand	17-4885-51790	46°46′08″	81°08′58″	20.6				
Ranger	17-3053-51969	46°53′43″	83°33′17″	2254.1	73.2	21.7	65	9
Rangers (Caribou)	17-3909-51461	46°27′52″	82°25′51″	252.2	63.1	23.8	56	11
Raven	17-6079-53230	48°03′03″	79°33′03″	616.7	46.9	19	39	3
Rawhide	17-3764-51679	46°39′16″	82°36′44″	968.4	61	23	63	12
Rawson	17-5330-51961	46°55′03″	80°34′00″	157.1			39	5
Red Cedar	17-5812-51735	46°42′10″	79°56′18″	2421.7	34.7	5.8	68	3
Red Deer (Rottier)	17-3706-51687	46°39′54″	82°41′04″	104.8	16.8	5.9	29	8

Lake	Code	Latitude	Longitude					
Red Rock	16-6650-52869	47°43'41"	84°48'33"	29.4	33.6	11.3	57	6
Red Squirrel	17-5740-52232	47°09'30"	80°01'31"	387.3				
Redpine	17-3046-52007	46°55'49"	83°33'48"	20.5				12
Redrock	16-7092-51678	46°38'35"	84°16'54"	19.5				9
Regal	17-3402-51626	46°36'02"	83°05'00"	32.1	17	7.4	23	7
Regan	17-5170-52317	47°14'30"	80°46'03"	122.2	38	11	36	4
Ren	17-2779-53631	48°23'14"	83°59'52"	54.9	18.3	5.6	18	4
Reserve	16-7159-51704	46°39'19"	84°10'42"	111.6	27.5	9.5	40	5
Restoule	17-5950-51006	46°03'17"	79°46'28"	1237.2	29.3	7.7	53	8
Reuben	17-5994-52009	46°57'03"	79°41'52"	228.5	44.8	16.4	80	
Rib	17-5972-52300	47°13'08"	79°43'01"	682	26.5	12.9	35	
Rice (Wright's, Lake 11)	17-4300-51062	46°06'26"	81°54'19"	27.1			61	
Ringer (Horseshoe)	17-4640-51180	46°12'58"	81°28'41"	61.2	20	6.5	47	5
River	17-3527-51815	46°46'43"	82°55'55"	67.3			38	
Robb	17-3357-51721	46°41'04"	83°08'55"	183.8				
Robbie	17-4704-52886	47°45'09"	81°23'43"	19.0	23	9	35	3
Robert	17-3953-51641	46°37'30"	82°22'03"	47.3	21	10.1	37	2
Robertson (Mud)	16-7087-51842	46°46'52"	84°15'55"	147.6	22	8.3	30	5
Rochester	17-3843-51532	46°31'32"	82°30'49"	51.3	79.9	10.2	43	4
Rocky Island	17-3460-51979	46°55'45"	83°01'52"	5664.1	17	5.1	27	5
Rodd	17-5274-52242	47°10'22"	80°38'16"	33.6				
Rogers	17-3264-51648	46°36'57"	83°15'29"	20.8	22.5	9.2	36	9
Roland	17-4899-51755	46°44'05"	81°07'58"	11.3	53	14.5	102	4
Rollo	17-3767-53044	47°52'59"	82°38'29"	983.8			31	3
Rome	17-4759-52050	46°00'00"	81°18'55"	231.3				
Roosevelt	17-5975-52342	47°15'21"	79°42'42"	69.4				
Rooster	17-3838-51579	46°34'22"	82°30'40"	38.2	24.1	8.2	38	7
Rosemarie	17-3822-51676	46°39'20"	82°32'24"	87.6	16.8	5.9	29	8
Rottier (Red Deer)	17-3706-51687	46°39'54"	82°41'04"	104.8	30.5	11.6		6
Round	17-4232-51131	46°10'09"	81°59'43"	13.8	20.4	8.8	89	4
Round (Wawiagama)	17-5476-52058	47°00'01"	80°22'55"	617.2	19.8	7.3	42	5
Rushbrook	17-4302-51759	46°44'08"	81°54'44"	174	23.8	7.1	27	6
Russian	17-4067-51796	46°45'38"	82°13'16"	308.2				
Ruth-Roy	17-4806-51039	46°05'25"	81°15'02"	61.3				
Sable (Lac aux Sables)	17-3988-51823	46°47'24"	82°19'06"	1162.3	32	9.2	32	6

ONTARIO (Northeastern Ontario) (continued)

Lake name	OFIS code	Latitude	Longitude	Surface area (ha)	Maximum depth (m)	Mean depth (m)	Conductivity (µS/cm)	Secchi depth (m)
Saga (Memesagamesing)	17-5779-50969	46°01′18″	79°59′58″	1098.8	59.4	13.7	34	3
Sailor Trout (Lake 54)		46°43′20″	82°41′35″	18.5				
Sam	16-7008-51898	46°49′32″	84°21′50″	11.5				
Sam Martin	17-5156-51907	46°52′11″	80°47′42″	153.1	22.7	4.5	35	6
Samreid	17-3658-51570	46°33′17″	82°45′00″	81.8			37	
Sand	16-6855-52885	47°43′31″	84°31′33″	354.5				
Sandcherry	17-4771-51865	46°50′03″	81°17′52″	35.5	23.9	6.7	38	6
Sandy	16-7189-51682	46°38′00″	84°08′26″	14.6	7.8	3.3	32	3
Sasaginaga (Sass)	17-5981-52505	47°24′06″	79°41′54″	90.5	34.2	8.3	98	7
Sass (Sasaginaga)	17-5981-52505	47°24′06″	79°41′54″	90.5	34.2	8.3	98	7
Savage	17-4226-51406	46°24′45″	82°00′18″	322.1	36.6	7.8	28	4
Saw	17-3882-53338	48°08′58″	82°30′10″	55.7				
Saymo	17-3086-52062	46°59′00″	83°31′00″	848.5	51.9	17.7		7.7
Scarfe (Canoe)	17-3429-51270	46°16′50″	83°02′20″	120.2				
Scotia	17-4713-52126	47°04′41″	81°22′49″	827.2	63.1	17.8	51	7
Scott (Bauldry)	16-6745-53296	48°05′55″	84°39′24″	51.4	63.1	23.9	108	8
Seabrook	17-3256-52078	47°00′01″	83°17′47″	473.5	30.5	9	62	4
Seagram	17-5351-52163	47°06′03″	80°32′13″	102.2	19	7.2	30	7
Secret	17-5672-52107	47°02′53″	80°06′54″	7.8				
Secret (Lake 36)	17-4476-52921	47°46′57″	81°41′55″	26.2				
Selwyn	17-5102-51772	46°45′26″	80°52′36″	58.2	26.8	6	43	4
Semiwite	17-3712-51594	46°34′38″	82°40′55″	304.3	33.5	12.4	40	8
Sesabic	17-3197-52031	46°57′33″	83°22′11″	139.3				
Seymour	17-3298-51798	46°45′07″	83°13′49″	62	16	5.7	40	3
Shakwa	17-4248-51802	46°46′15″	81°59′12″	648.3	27.5	9.7	33	6
Shanguish	17-3867-52169	47°05′31″	82°29′38″	216.4	18	5		5
Shelden	17-3014-51622	46°35′21″	83°35′31″	129.8	22.6	9	35	6
Shillington	17-5265-52901	47°45′41″	80°38′34″	136	34.7	11.2	91	3
Shingwak	17-4746-51865	46°49′37″	81°20′00″	56.3				
Shoepack	17-3799-51829	46°47′21″	82°34′26″	40.2	22.3	7.6	40	6

Lake	Code	Latitude	Longitude					
Shoofly	17-4713-52302	47°13'45"	81°22'35"	185.7				
Sibbald	17-5312-54060	48°48'31"	80°34'27"	24.5				
Silvester	17-5267-51883	46°50'29"	80°38'43"	50.6			36	5
Sinaminda	17-4277-51933	46°53'09"	81°56'21"	1107.7	35.4	7.6	37	14
Sister	17-3799-51662	46°38'24"	82°34'09"	109.3	43	20.7	35	
Skintent	17-3117-51838	46°47'05"	83°27'56"	52.1				
Skirl	17-3283-51579	46°33'18"	83°14'21"	95.9				
Skookum	17-3008-51554	46°31'27"	83°35'39"	62.2	19.8	11.6	44	8
Skull	17-5550-52620	47°30'45"	80°16'11"	132.3	23.8	5.6		5
Skunk	17-3789-51715	46°41'14"	82°34'54"	53.4	10.7	3.9	45	5
Skunk	17-5640-51968	46°55'12"	80°09'54"	103.2	59	13.6	56	10
Sleith (Hand)	16-7178-53397	48°10'29"	84°04'17"	119	18.9	7.3	87	3
Slipper	17-6809-50176	45°17'18"	78°41'39"	51.7	34	14.2	26	5
Small Island	17-3046-52008	46°56'03"	83°33'49"	2.2				
Smith	17-4287-55286	49°54'28"	81°59'35"	29.3				6
Smith	17-5182-52469	47°22'41"	80°45'28"	255	25.9	6.9	26	11
Smoothwater	17-5243-52488	47°23'44"	80°40'48"	869.9	88.4	31.4	33	
Snapshot (Big Trout)	17-4025-51298	46°18'56"	82°15'53"	82.5	26.2	9.2	21	
Snowbird	17-3387-51651	46°37'19"	83°06'23"	35			29	
Snyder (Friendly)	17-3069-52112	47°01'41"	83°32'27"	33	12.2	4.3		3.6
Snyder (Strawberry)	17-3106-52001	46°55'46"	83°29'14"	14	12			5.5
Solace	17-5225-52251	47°10'33"	80°42'06"	316.8	57	11.8	30	13
Soul	17-4756-51648	46°38'18"	81°19'06"	22.8	34	13.6	32	8
South	17-3676-51456	46°27'10"	82°43'25"	28.3				
South Anvil (Mystery)	17-3079-52141	47°03'14"	83°31'46"	74.5	31.1	7.1	27	6
South Iron (Iron)	17-5723-51858	46°49'31"	80°03'03"	31.6	17.7	7	33	4
Spawning	17-5777-52085	47°01'38"	79°58'39"	197.5				
Spence	17-3479-51385	46°23'05"	82°58'41"	55.1				
Sportsman	17-3919-51837	46°48'10"	82°24'50"	44.7				
Spring	17-6316-51797	46°45'36"	79°16'37"	73.0				
Spring	17-3618-51491	46°29'01"	82°47'43"	18.8				
Spruce	16-7140-52235	47°07'54"	84°10'52"	71.4	25.6	1.4	102	5
Spud	17-3166-52189	47°06'21"	83°24'46"	45.7				
Squirrel	17-4285-52017	46°58'03"	81°56'33"	235.5	19	8.2	35	7
St. Anthony	17-5962-53128	47°57'55"	79°42'41"	488.9	30.5	10.1	66	3

ONTARIO (Northeastern Ontario) (continued)

Lake name	OFIS code	Latitude	Longitude	Surface area (ha)	Maximum depth (m)	Mean depth (m)	Conductivity (μS/cm)	Secchi depth (m)
St. Leonard	17-4507-51189	46°13'25"	81°38'19"	151.8	27.1	12.8	44	3
Star	17-3955-51626	46°36'40"	82°21'52"	56.8	15	4.8	36	3
Star (Lake 61)		46°49'17"	82°28'17"	30.2			35	
Stock	17-5358-53588	48°22'42"	80°31'13"	28.8	23.2		26	
Stone	17-3605-51473	46°27'59"	82°48'35"	30.4			33	
Stoney (Horner)	17-3062-51715	46°40'17"	83°31'59"	417.8				
Stormy (Patterson)	17-5947-51034	46°04'49"	79°46'26"	333.4	28.4	10.7	44	5
Stouffer	17-5228-52105	47°03'57"	80°40'58"	141	17		31	10
Stover	17-3349-51649	46°37'05"	83°09'20"	33.6	28.5	8.4	22	9
Strawberry (Snyder)	17-3106-52001	46°55'46"	83°29'14"	14	12			5.5
Strickland	16-6497-54012	48°44'55"	84°57'57"	226.2	32.1	5.2	76	5
Stringer	17-3364-51553	46°32'21"	83°08'18"	45.4				
Stull	17-5134-52341	47°15'43"	80°49'13"	256.9	34.3	7.5	34	5
Sugar	17-5670-52431	47°20'15"	80°06'34"	231.3	26.5	8	41	5
Sugarbush	17-4569-52032	46°59'18"	81°34'10"	54.4			27	
Summers	17-3662-51463	46°27'31"	82°44'50"	228.5	53.3	14.5	29	7
Sunny	17-5348-53340	48°09'38"	80°31'56"	75.7				
Sunrise	17-6014-52109	47°02'43"	79°39'58"	161.9	50.3	16.9	107	5
Survey (Lake 1)	17-3702-51629	46°36'33"	82°41'39"	18.2				
Surveyor	17-4158-51784	46°45'38"	82°06'23"	132.5	24.4	8.8	65	4
Susan	17-3866-51573	46°33'42"	82°28'47"	16.8				
Sutton	16-6500-60140	54°15'14"	84°41'51"	3763.9				
Swalwell (Crooked)	17-5890-50995	46°02'35"	79°50'14"	75.3	34.2	7.9	35	5
Swanson	17-4019-54633	49°18'59"	82°20'58"	36.2				
Talon (Lake Talon)	17-6497-51295	46°18'16"	79°03'22"	1404.9	60.1	12.9	56	4
Tea	17-3539-51453	46°26'50"	82°54'09"	59.5				
Teardrop	17-4678-50995	46°02'41"	81°24'46"	3.4	16.6	9.6	27	12
Teasdale	17-3849-51482	46°28'46"	82°29'56"	131.5	51.8	22.3	71	12
Tee (Ortona)	17-4054-51496	46°29'44"	82°13'57"	62.9	41.2	13.8	28	11
Telephone	17-4397-51960	46°55'00"	81°47'36"	13.6			27	5

Telfer	17-5165-51976	46°56′45″	80°47′18″	305	32.9	10.4	34	15
Ten Mile	17-3629-51531	46°31′21″	82°47′18″	932.4	91.5		36	11
Tenfish	17-3635-51676	46°38′52″	82°46′09″	96			29	9
Thieving Bear	17-5889-52256	47°10′46″	79°49′35″	114.8	29	7.4	60	5
Thor	17-4787-52200	47°08′05″	81°16′52″	266.6	53.3	14.9	110	4
Three	17-3398-51886	46°49′28″	83°06′03″	52.5	34.2	10.7	23	9
Three Mile Lake (Bobwhite)	17-4421-52015	46°58′00″	81°45′39″	66.5			30	
Threenarrows	17-4670-51065	46°06′47″	81°25′19″	810.1	51.9	14.5	29	7
Threetrails	17-6320-51770	46°44′07″	79°16′18″	34.1				
Tier	16-7009-51879	46°48′49″	84°22′04″	13.5				
Tikamaganda	16-7128-52654	47°30′29″	84°10′11″	438.8	33.6	13.1	26	6
Till	16-6559-52901	47°44′51″	84°55′06″	29.5	15.3	7.8	41	7
Timber	17-6506-51504	46°29′31″	79°02′14″	123.7				
Tip Top	17-3993-51687	46°40′22″	82°18′55″	88.2				
Toobee	17-3580-51785	46°44′46″	82°51′25″	88.2	17.7	6.9	72	5
Toocee (Toosee)	17-3484-51824	46°46′48″	82°59′31″	54.6	12.2	3.3	30	4
Toodee	17-3342-51796	46°45′05″	83°10′13″	138.7			35	
Toosee (Toocee)	17-3484-51824	46°46′48″	82°59′31″	54.6	12.2	3.3	30	4
Tooth	17-6130-52261	47°10′58″	79°30′31″	37.6	42.1	21.9	99	6
Torrance (Megisan)	17-3096-52319	47°12′48″	83°30′53″	98.3	12.2	4.1		3
Town Line	17-3338-51748	46°42′28″	83°10′27″	117.8				
Traill	17-5509-54115	48°51′23″	80°18′20″	75.7				
Transparent (Clear)	17-6580-51459	46°27′00″	78°56′33″	69.3				
Trapper	17-3062-52029	46°57′13″	83°32′50″	29.3	24.4	6.5	35	6
Treeby	16-6606-53045	47°52′36″	84°51′05″	129.5	27.8	9.3	70	4
Trethewey	17-5381-52549	47°26′52″	80°29′30″	565.3	27.4	7.5	35	6
Trethewey 27	17-5391-52563	47°27′39″	80°28′51″	22.1	11	4.7	47	4
Trollope	17-5969-53816	48°34′55″	79°41′27″	217.9	22.9	10.4		4
Trotter (Trout)	16-7291-51626	46°34′52″	84°00′39″	155.7	34.2	12.4	62	5
Troupe	16-6814-53519	48°17′56″	84°33′03″	103.2	25.9	6.9	82	6
Trout (Matchinameigus)	16-7135-53359	48°08′58″	84°07′27″	667.7	17.7	4.4	44	7
Trout	16-7228-52127	47°01′56″	84°04′03″	205.8				
Trout	16-7106-51670	46°37′34″	84°14′24″	252	51.8	16.3	29	7
Trout (Trotter)	16-7291-51626	46°34′52″	84°00′39″	155.7	34.2	12.4	62	5
Trout	17-3776-51312	46°19′31″	82°35′28″	40.6	12.8	5.6		5

ONTARIO (Northeastern Ontario) (continued)

Lake name	OFIS code	Latitude	Longitude	Surface area (ha)	Maximum depth (m)	Mean depth (m)	Conductivity (µS/cm)	Secchi depth (m)
Trout	17-6290-51301	46°18'56"	79°19'06"	1673.4	69.2	16.9	78	3
Trout	17-5317-51186	46°13'30"	80°35'55"	929.6	36.6	8.6	45	3
Trout (Guilmette)	17-6537-51099	46°07'35"	79°00'47"	65.4	32.3	10.2	32	2
Trump	17-3033-53528	48°18'00"	83°39'07"	190.9	28.7	8.1	36	7
Tujak	17-2951-52116	47°01'41"	83°41'47"	153.8	27.4	7.1		4.6
Tukanee	16-6315-53885	48°38'25"	85°12'40"	332.2	32	14.6	108	5
Tupper	16-7076-51899	46°49'57"	84°16'39"	55.8	12.2	3.9	21	5
Turkey (Big Turkey)	16-6958-52136	47°02'54"	84°25'10"	54	39.6	11.7	38	7
Turner	17-5695-52367	47°16'57"	80°04'52"	129.3	25.6	9.2	32	9
Turtle	17-6406-51290	46°18'01"	79°10'34"	253.3	56.4	9.9	59	4
Turtle	17-3716-51254	46°16'37"	82°40'22"	142.9	30.5	14	22	6
Turtleshell	17-5573-51927	46°53'21"	80°14'51"	161.1	42.7	12.5	54	7
Twab	16-7098-51955	46°52'53"	84°14'44"	29.9	14	4.3	65	6
Twenty Minute	17-6421-51475	46°28'04"	79°08'54"	32.6				
Twentythree Mile	17-3698-52223	47°08'37"	82°43'01"	42.5				
Twin (Hammond)	17-5841-52630	47°30'57"	79°52'28"	190.4	24	6.2	94	3
Twin	17-5304-52013	46°58'24"	80°36'29"	68.3			40	
Twin		46°44'43"	82°40'28"	15.6				
Twinhouse	15-6587-53594	48°22'09"	90°51'15"	105.8	15	5.5	23	3
Twist (Crooked)	17-6578-51479	46°28'07"	78°56'42"	56.5				
Tyson	17-4910-51070	46°07'01"	81°06'59"	1142.2	39.6	11.9	33	4
Upper Bark	17-3855-52054	46°59'01"	82°30'03"	732.3	38	8.6	42	4
Upper Bass	17-5579-52068	47°00'59"	80°14'14"	122.2	39	7.1	57	4
Upper Gimlet (Gimlet #1)	16-6786-52113	47°01'57"	84°38'50"	11				
Upper Green	17-3840-52215	47°08'12"	82°31'52"	504	19	9.2	47	6
Upper Green	17-4452-52066	47°00'44"	81°43'16"	71.3	27	9.2	28	6
Upper Island (Island)	16-7101-51724	46°40'22"	84°15'00"	150.2	27.5	8.8	43	6
Upper Kasaway (Kasasway)	17-4304-53073	47°55'03"	81°55'54"	234.0				
Upper Megisan (Affleck)	17-3069-52350	47°14'30"	83°33'04"	131.6	9.1	3.8	45	2

Lake	Code	Latitude	Longitude					
Upper Pancake	16-6806-52194	47°06'04"	84°37'10"	92.2	16.8	4	24	3
Upper Sheppard	16-7121-51939	46°51'38"	84°13'34"	35.5	11			4.25
Upper Twin	17-5947-52066	47°03'09"	83°32'51"	18.6	44.2	14.3	86	7
Upper Twin	17-3848-51810	47°00'27"	79°45'16"	185.4	12.5	3.7	26	3
Uppermost	17-3871-51523	46°46'28"	82°30'31"	12.9	33.5	7.9	36	8
Uranium (Cormier)	17-3644-51685	46°31'03"	82°28'20"	137.8	37.5	17.5	35	11
Vasseau	17-4810-51976	46°39'28"	82°46'32"	83.1			31	
Venetian	17-3223-52615	46°56'46"	81°14'52"	1019.7	28	7.6	95	3
Vezina	17-2803-51854	47°29'04"	83°21'30"	226.9	12.2	7	30	5
Vixen	17-3223-52536	46°47'14"	83°52'39"	75.3	32.6	7.8		5
Wabus	17-5097-51760	47°24'59"	83°21'36"	80.5	38.5	11	41	10
Waddell	17-3547-52257	46°44'20"	80°52'53"	67.3	35	11	96	5
Wagong	17-3191-51595	47°10'13"	82°54'59"	147.8	73.2	28.3	40	8
Wakomata (Clear)	17-4608-51161	46°34'31"	83°21'38"	2468.7	54.9	26.5	78	7
Walker	17-4657-52747	46°11'49"	81°30'38"	339.3				
Wanatangua	17-4663-52723	47°37'37"	81°27'22"	13.3				
Waonga	16-7174-52272	47°36'19"	81°26'53"	211.4	56.4	12.5	30	5
Wart	17-5842-51980	47°09'48"	84°07'40"	446.4	31.1	6.6	75	5
Wasaksina	17-5314-53425	46°55'54"	79°53'30"	601	36.6	8.9	114	6
Watabeag	17-3377-51571	48°14'13"	80°34'36"	2267.1				
Waterhole	17-4923-51272	46°32'51"	83°06'16"	34.4			33	
Wavy	16-6699-53200	46°18'09"	81°05'33"	306.7	32.9	15.6	92	7
Wawa	17-5476-52058	48°00'31"	84°43'46"	677.4	20.4	8.8	89	9
Wawiagama (Round)	17-5514-51832	47°00'01"	80°22'55"	617.2	21.4	5.6	58	4
Wawiashkashi (Grassy)	16-7183-51703	46°48'08"	80°19'09"	414.4				5
Weashkog	16-7186-51904	46°39'08"	84°08'49"	29.3				
Weckstrom	17-4523-51581	46°49'37"	84°08'59"	17.8	22	8.7	37	6
Weequed	17-4971-52299	46°35'35"	81°37'19"	217.9	35		54	
Welcome	17-4224-51851	47°13'09"	81°02'31"	681	29.6	9.5	32	6
Wenn	17-4246-51860	46°48'52"	82°00'51"	26.4	12.2	5	48	2
Wensley	17-4563-51199	46°49'33"	81°59'16"	45	25.6	9.9	54	5
West Bay	17-4771-51698	46°14'10"	81°33'29"	397.4			54	
West Morgan	17-3856-53100	46°40'28"	81°18'18"	87.2			25	
Whigham (Lake 41)	17-3974-51436	47°56'05"	82°31'54"	10				
Whiskey		46°26'23"	82°20'07"	992.8			293	

ONTARIO (Northeastern Ontario) (continued)

Lake name	OFIS code	Latitude	Longitude	Surface area (ha)	Maximum depth (m)	Mean depth (m)	Conductivity (µS/cm)	Secchi depth (m)
White	16-6003-54025	48°46'10"	85°38'07"	6090.3	48.2	29	92	6
White (Endikai)	17-3445-51613	46°35'25"	83°01'44"	591.7	36.6	9.6	87	5
White Bear (Cassels)	17-5970-52137	47°04'11"	79°43'24"	757.1	50.3	18.4	39	5
White Bear	17-3393-51732	46°41'57"	83°06'25"	284.9				
White Birch	16-7196-51746	46°41'27"	84°07'41"	8.5				
White Oak	17-5000-51272	46°17'55"	80°59'56"	273.6	43		33	10
White Owl	17-3798-52237	47°09'28"	82°35'11"	1081	19	4.7	45	4
White Pine	17-5128-52363	47°16'55"	80°49'50"	67	22	5.6	28	6
White Rock	17-3197-52134	47°03'05"	83°22'24"	80.2			37	
Whitefish	16-6898-53236	48°02'35"	84°27'00"	1821.1	54.9	15.2	31	3
Whitefish (Batty)	17-3944-51408	46°24'48"	82°22'29"	244.3	37.2	13.6	28	7
Whitewater	17-5756-52313	47°13'54"	80°00'05"	269.7	40.2	8.6	51	8
Whitman	17-2812-51867	46°47'44"	83°51'44"	35.1				
Whitney	17-5955-52284	47°12'15"	79°44'21"	57.2				
Wiggly	17-3936-51522	46°31'05"	82°23'08"	236.8	42	8.8	39	8
Wigwas (Diamond)	16-7240-51577	46°32'13"	84°04'38"	45	22.9	8.8	35	4
Wilkie	17-3598-51620	46°35'46"	82°49'25"	68.3	25	7.9	57	5
Williamson	17-3332-51424	46°25'11"	83°10'11"	225.4	33.5	12	33	4
Wilson	17-3405-51601	46°34'39"	83°04'53"	75.4	52	15.9	32	9
Windermere	17-2934-53143	47°57'07"	83°46'07"	3832	30	8	54	5
Windy	17-4664-51605	46°35'57"	81°26'19"	1112	65.6	10.7	55	
Winnie	17-4194-52131	47°04'15"	82°03'41"	144.4	19.9	7.8	41	7
Wiseman T (Long)	17-5279-51902	46°58'51"	82°55'26"	22.2			36	
Wolf	17-3722-51561	46°51'10"	80°37'55"	79.8			36	19
Wolf (Lake 15)	16-7105-51946	46°32'55"	82°39'59"	10.1				
Wolfe	17-3207-51418	46°52'25"	84°14'45"	139.3	21.3	4.6	27	7
Woodrow (Mountain)	17-5505-51928	46°24'35"	83°19'59"	49.1	32.7	11.2	42	5
Woods	17-4300-51062	46°53'17"	80°20'11"	23.1	17.7	8.2	77	8
Wright's (Rice, Lake 11)		46°06'26"	81°54'19"	27.1	26.5	12.9	35	8

Lake name	OFIS code	Latitude	Longitude	Surface area (ha)	Maximum depth (m)	Mean depth (m)	Conductivity (µS/cm)	Secchi depth (m)
Yokum (Desayeux)	17-3037-51580	46°32'57"	83°33'48"	45.7	23	6.6	27	4
Yorston	17-5349-52132	47°04'02"	80°32'08"	351.4	45	11	37	9
Yorston River Pool		47°06'41"	80°34'32"	35.1			43	
Zinger	17-5187-54742	49°25'20"	80°44'33"	200.0				

ONTARIO (Northwestern Ontario)

Lake name	OFIS code	Latitude	Longitude	Surface area (ha)	Maximum depth (m)	Mean depth (m)	Conductivity (µS/cm)	Secchi depth (m)
Abamategwia (Flatrock)	15-5795-55014	49°39'25"	91°53'43"	1357.7	33.2	9.4	40	3
Abbess	15-5970-53925	48°40'54"	91°40'47"	130.2	37.2	13	24	3
Abigogami	16-3670-54288	48°59'46"	88°49'13"	434.1	15.5	5.5	128	4
Abram	15-5765-55453	50°03'33"	91°55'23"	2395.7	36.6	12.1	66	2
Adele	15-6189-54492	49°11'02"	91°22'08"	333.8	30.5	9.9	24	6
Adventure	15-3695-56411	50°54'35"	94°51'11"	140.5	36	12.9	21	7
Aegean	15-3758-56302	50°48'44"	94°45'48"	482.1	59	10.3	25	5
Aerobus	15-4712-55760	50°20'16"	93°24'17"	1947.7	43.9	15.4	87	3
Agnes	15-6228-53450	48°14'32"	91°20'45"	2993.9	79.3	19.6	29	6
Aiabewatik	15-5385-54784	49°27'27"	92°28'14"	162.6	33	9	143	4
Ajax	15-4691-54639	49°19'43"	93°25'17"	510.6	29	10.2	50	4
Alexandra (Indian)	15-4623-55297	49°55'12"	93°31'34"	642.3				
Alonghill	15-4866-54374	49°05'49"	93°11'17"	532	64.1	10.7	105	7
Al's	15-4384-56032	50°34'46"	93°52'13"	218.4	18	11.3	42	5
Anders	16-3600-54270	48°58'54"	88°54'46"	161.6				
Angel (Gustauson)	15-5826-54970	49°37'23"	91°51'21"	144.6	20	7.6	33	8
Anishinabi	15-4674-55871	50°26'21"	93°27'41"	3399.5	88.4	31.4	108	8
Anstey (East)	15-4502-54405	49°07'00"	93°40'53"	76.1	50	13.3	72	7
Antoine	15-6095-53746	48°31'29"	91°31'38"	335.6				
Apps	15-4128-58107	52°26'34"	94°16'14"	972.4	40	10.9	30	4
Argo	15-5888-53453	48°15'44"	91°48'42"	956.7	55.5	20.2	26	7
Aronson	15-5023-54519	49°13'55"	92°58'01"	216.1	31	11	35	4
Arrow	15-7025-53379	48°09'53"	90°16'35"	3314.4	54.9	17.8	86	

ONTARIO (Northwestern Ontario) (continued)

Lake name	OFIS code	Latitude	Longitude	Surface area (ha)	Maximum depth (m)	Mean depth (m)	Conductivity (µS/cm)	Secchi depth (m)
Athelstane	15-7050-54058	48°46'31"	90°12'09"	1806.4	32.9	9.4	51	3
Atik	15-7097-53434	48°12'41"	90°10'37"	77.7	28.1	12.3	39	4
Atikwa (Deer)	15-4602-54790	49°27'48"	93°32'47"	5387.8	59.5	13.2	44	4
Atlantic	15-6100-55667	50°14'39"	91°27'24"	54.4	29.9	12.1	111	3
Audrey (ELA L. 596)	15-4384-55016	49°39'56"	93°51'14"	161.8	33	7.2	26	6
Augite	15-4487-55234	49°51'02"	93°42'57"	105.2	26	9.9	35	4
Azure	15-3869-58827	53°05'02"	94°41'09"	665.9	19.2	8.5	174	3
Back Lawrence	15-4763-54581	49°16'01"	93°19'49"	340.8	48.7	12.5	29	3
Bad (ELA L. 139)	15-4327-55016	49°39'54"	93°55'58"	167.5	29.9	8.2	32	5
Bad Vermilion	15-5236-53978	48°44'06"	92°40'58"	1813.1	49.4	17	74	5
Badwater	15-5772-53711	48°29'26"	91°57'18"	192.6	16.2	6.6	20	4
Balmain	15-4469-55244	49°52'10"	93°44'39"	119.7	33.5	12.7	33	4
Balmoral	15-5840-54616	49°18'19"	91°50'25"	294	22	6.1	21	3
Bamoos	16-5474-54071	48°49'02"	86°21'16"	182.5	69	26.9	35	5
Barbara	16-4420-54651	49°20'17"	87°47'54"	2360.8				
Baril	15-6650-54018	48°44'55"	90°45'57"	998.5	33.5	12.3	30	4
Bartley	15-4633-54470	49°10'49"	93°30'08"	124.2	36	9.9	65	5
Basket	15-5711-55030	49°40'06"	92°00'47"	4267.2	19.8	5.9	41	4
Bass (Toothpick)	15-4194-55517	50°06'55"	94°07'17"	296.9	21	7.9		4
Basswood	15-6055-53263	48°05'08"	91°34'01"	4840.2				
Bat	15-4855-54345	49°03'45"	93°11'23"	313.2	28.4	7.9	66	4
Batchewaung	15-6103-53911	48°39'52"	91°30'44"	1274.3	54.9	19.4	22	5
Beak	15-5542-54588	49°16'56"	92°15'15"	524.3	30	7.7	31	3
Beamish	15-3670-56298	50°48'32"	94°53'54"	560.9	19	5.2	32	4
Bear Trap	16-4621-54126	48°51'35"	87°31'06"	9.6				
Beaton (One Island)	15-4708-55595	50°11'32"	93°24'09"	160.6	27	12.9	81	6
Beatty	16-4566-55070	49°42'57"	87°36'07"	655.0				
Beauty	15-4119-55702	50°16'39"	94°14'01"	306	25.9	12.3	47	4
Beaverhouse	15-5662-53769	48°32'43"	92°06'03"	2000.8	63.4	23.6	28	3
Bee	15-4406-55214	49°50'38"	93°49'27"	49.6	22	8	92	4

Lake								
Bell	15-6500-55194	49°48'10"	90°54'42"	4169.6	34.4	9.5	44	6
Below Bow (Lake 155)	15-5986-54297	49°00'44"	91°39'02"	208.0	45.8	12.5	29	5
Bending	15-5627-54632	49°19'53"	92°08'10"	1135.7	19.8	7.5	36	4
Benstewart	15-5008-54467	49°10'27"	92°59'21"	55				
Berry	15-4302-54803	49°28'25"	93°57'50"	1234.6	29.3	11.7	32	6
Bert	15-4283-55451	50°03'21"	94°00'05"	313.6	28	12.1	69	6
Bethune	15-4528-54361	49°04'45"	93°38'51"	92.3	25	5.3	32	3
Bethune	15-6049-54231	48°57'14"	91°34'01"	165.9			32	
Big Bear (Big Bruin)	16-4846-54110	48°51'17"	87°12'34"	67.5			32	
Big Bruin (Big Bear)	16-4846-54110	48°51'17"	87°12'34"	67.5				
Big Canon (Big Canyon)	15-4455-55462	50°04'06"	93°45'45"	2284.1	56.4	20.2	50	5
Big Canyon (Big Canon)	15-4455-55462	50°04'06"	93°45'45"	2284.1	56.4	20.2	50	5
Big Island	15-4298-55335	49°57'18"	93°58'57"	134.1	31.5	12.3	26	6
Big Jet	15-4263-55335	49°57'02"	94°01'47"	96.3	18	7.9	20	6
Big Joe (Joe)	15-5685-54110	48°50'47"	92°03'35"	175.0			27	
Big Mc Caulay (Mc Caulay)	15-5752-53953	48°42'37"	91°58'08"	451.7	49.7	20.6	35	4
Big Sandy (Sandybeach)	15-5463-55189	49°49'11"	92°21'11"	3820.2	41	20.3	150	5
Big Vermilion (Vermilion)	15-5556-55428	50°02'11"	92°13'00"	8266.1	42.4	10.9	116	7
Bigshell	15-4010-56940	51°23'20"	94°25'05"	1024.9	28	8.4	41	3
Birch	15-6192-53229	48°03'26"	91°24'17"	174.4				
Black Sturgeon	16-3634-54687	49°21'29"	88°52'50"	4940.7			30	
Blackhole (Lower Moosehide, Moosehide)	15-5871-54555	49°14'54"	91°48'11"	178.8	23.8	11.8		3
Bluff	15-4056-55392	49°59'46"	94°19'37"	78.2	37.5	11	26	7
Bluffpoint	15-4725-54467	49°10'16"	93°22'53"	520.4	38.1	10.4	51	4
Boomerang	16-5454-54576	49°16'10"	86°22'31"	421.4	40	9	123	5
Bornite	15-4670-55751	50°19'44"	93°27'26"	497	22.5	10.7	80	5
Boulder (Cobble)	15-4552-55258	49°53'18"	93°37'07"	1146.9	32	10	49	3
Boulder (Moraine)	16-3644-54282	48°59'12"	88°51'05"	277.2	17.7	8.9	128	4
Box	15-4402-54257	48°59'04"	93°48'40"	23.0				
Boyer	15-5320-54804	49°28'41"	92°33'30"	1028.6	37.2	11.1	125	5
Bradley	15-3794-55972	50°30'50"	94°42'29"	158.5	60	11.3	24	4
Brent	15-5989-53503	48°18'11"	91°41'07"	945.0				
Bretz	15-4933-54429	49°08'34"	93°05'33"	95.5	26	10.5	54	6
Brightsand	15-6869-55111	49°43'31"	90°24'24"	1353.9				

ONTARIO (Northwestern Ontario) (continued)

Lake name	OFIS code	Latitude	Longitude	Surface area (ha)	Maximum depth (m)	Mean depth (m)	Conductivity (µS/cm)	Secchi depth (m)
Brown T (Margon)	16-4988-54124	48°52'26"	87°00'54"	14.6				
Brown Trout	15-4631-55092	49°44'32"	93°30'44"	258.2				6
Bruin	15-4349-55230	49°51'34"	93°54'24"	214.2	66.8	27.4	38	
Buckingham	15-6279-53787	48°32'48"	91°15'39"	299.3				3
Buddell	15-5840-56575	51°03'44"	91°48'01"	194.3	30	9.1	42	3
Buhl	16-5501-54593	49°17'09"	86°18'43"	150.1	34	13.1	47	3
Bukemiga	16-3411-55531	50°06'46"	89°13'33"	791.3	44.5	11.3	26	5
Bulging	15-3634-56451	50°56'39"	94°56'11"	1010.3	70	23.2	39	5
Bunny	15-3940-56284	50°47'58"	94°30'11"	132.9	33.5	11.4	23	4
Burchell	15-6747-53842	48°35'38"	90°37'58"	1026.6	74.7	24.3	80	5
Burke	15-6136-53287	48°06'16"	91°28'33"	263.1	39	16.5	54	5
Burnt Island	15-6584-54061	48°47'29"	90°50'25"	118.7	32	13.7	48	4
Burt	15-6075-53504	48°17'10"	91°32'56"	14.9				
Burton (ELA L. 259)	15-4436-55044	49°41'29"	93°46'56"	94.1	25.4	8.2	21	5
Buzzard (Winnange, ELA L.660)	15-4495-55118	49°45'08"	93°42'00"	2631.5	115.2	27.2	21	9
Cache	15-6439-53791	48°32'54"	91°02'51"	522.7				
Cairngorm	16-5042-54260	48°59'01"	86°56'28"	1032.1	61	14.7	27	4
Calder	15-4922-54538	49°14'19"	93°06'26"	131.1	43	17.1	83	5
Call	15-4115-56321	50°50'00"	94°15'00"	323.4	22.5	5.3	30	2
Cameron	15-4499-54575	49°16'11"	93°41'01"	1324.1	26.5	9.6	150	8
Cameron	15-7164-53390	48°10'11"	90°05'26"	30.8	17.7	9.2	43	5
Camp	15-6896-54026	48°44'57"	90°25'13"	95.7				
Campbell	15-4503-54318	49°02'24"	93°40'32"	76.1	38.5	13.5	98	6
Campus	15-5934-54510	49°12'50"	91°43'05"	425.7	22.1	8.2	27	3
Caribou	16-3495-55900	50°26'43"	89°07'04"	9287.5	55.5	7.2	36	5
Carl	15-4776-55506	50°06'33"	93°18'48"	79.6	25.9	10	30	4
Carleton	15-5117-54665	49°21'16"	92°50'26"	124.4	47	18.5	78	6
Carling	15-6217-56090	50°37'23"	91°16'30"	1488.6	40.9	8.8	62	4
Carp	15-6254-53267	48°05'47"	91°18'30"	466.0				

Lake	Code							
Castastrophe	15-3715-55289	49°54'01"	94°47'21"	279.8	36	9.7	71	3
Castle	16-3663-55543	50°07'55"	88°52'22"	626	18.3	9.5	60	6
Castle	15-7176-53418	48°11'41"	90°04'06"	33.2	35.1	11	67	2
Catlonite	16-4995-54471	49°10'52"	87°00'29"	326.2	16.3	6.1		6
Caution	15-4140-55352	49°57'56"	94°11'24"	149.4	23.8	6.7	131	4
Cavern	16-3778-54100	48°49'53"	88°39'59"	23.5	39.6	9.1	66	5
Caviar	15-4431-54685	49°22'11"	93°47'28"	3185.8	24.4	12.6	90	5
Cecil	15-6164-54885	49°32'26"	91°23'25"	1615.5	26.8	6.3	170	4
Cedarbough	15-5581-55367	49°58'46"	92°11'37"	188.4	18	7	135	5
Cedartree	15-4375-54620	49°18'35"	93°51'36"	548.4	27	8.1	35	3
Cherrington	15-3718-58269	52°34'00"	94°53'56"	2066.9	41.2	18	44	7.6
Church (Partridge)	15-6795-53776	48°31'39"	90°34'08"	59.73	16	4	29	4
Cirrus	16-5620-54179	48°54'55"	86°09'21"	318.3	85.4	22.4	18	4
Cirrus	15-5747-53824	48°35'35"	91°59'17"	2113.3	31	10.4	47	3
Clare	15-4315-55181	49°48'50"	93°57'11"	83	23.5	12.1	42	3
Clearwater (Delano)	15-4689-55111	49°45'14"	93°25'54"	821.4	73.2	27.1	22	9
Clearwater West	15-5755-54281	49°00'05"	91°58'09"	3609.8				
Cleveland	15-4364-57983	52°19'49"	93°56'23"	609.0	34.1	9.2	84	4
Cliff	15-4785-55574	50°10'09"	93°18'09"	2663.4				
Cliff	16-3700-54335	49°02'33"	88°46'43"	41.3	32	10	49	3
Cobble (Boulder)	15-4552-55258	49°53'18"	93°37'07"	1146.9	32.5	9.4	30	5
Cobourg	15-4596-54321	49°02'41"	93°33'05"	73.4	22.1	9	21	2
Cole	15-5900-53914	48°40'04"	91°46'12"	91.9	36	10.4	29	6
Collins (Trout)	16-3277-55707	50°15'52"	89°25'15"	886.3	24.3	7.4	36	3
Como	15-6228-53964	48°42'34"	91°19'52"	145.8				
Cone	15-5940-53465	48°16'20"	91°44'20"	66.6	32.3	8.3	44	4
Confederation	15-5210-56629	51°07'07"	92°41'29"	4174.3	42.7	11.6	116	3
Confusion	15-4187-56135	50°40'16"	94°09'19"	1426.2	65.8	25.3	45	6
Conifer	15-4270-56011	50°33'22"	94°01'30"	1125.1	44	15.9	183	5.5
Cooney	16-3260-55298	49°53'49"	89°25'24"	74				
Cosgrave	16-4324-54532	49°13'47"	87°55'42"	591.0	104	32.6	32	8
Coubran	16-5387-54105	48°50'38"	86°28'31"	145.6	42.7	8.5	63	5
Couture	15-6679-55535	50°06'53"	90°39'00"	508.3	43	18.7	29	4
Cox	15-4311-55431	50°02'21"	93°57'49"	92.9	37	11.6	42	4
Crabclaw (ELA L. 469)	15-4559-55166	49°48'04"	93°36'17"	577.6				

ONTARIO (Northwestern Ontario) (continued)

Lake name	OFIS code	Latitude	Longitude	Surface area (ha)	Maximum depth (m)	Mean depth (m)	Conductivity (µS/cm)	Secchi depth (m)
Crevasse	16-3297-55263	49°52'03"	89°22'09"	114.6	73	23.3	108	4
Crook	15-5634-54343	49°03'49"	92°07'59"	155.5	24	7	20	5
Crooked (Kaopskikamak)	15-5116-54405	49°07'07"	92°50'28"	1266.7	39	7.2	26	4
Crooked	15-5926-53397	48°12'46"	91°45'39"	3213.8	50.3	11.3	55	2
Cross	15-4115-55351	49°57'53"	94°14'01"	104.3	26.2	8.3	26	6
Crossroute	15-4674-54480	49°11'14"	93°26'44"	246.1	33.6	9.2	84	4
Crowrock	15-5858-54247	48°58'17"	91°49'50"	1502.2	50.3	7.4		
Cry	16-3557-55322	49°55'35"	89°00'40"	246.7	47.6	22.2	77	10
Crystal	15-6263-53964	48°42'49"	91°16'18"	612.7	47	15.4	31	8
Cypress (Ottertrack)	15-6419-53372	48°10'26"	91°05'30"	277.4				
Dad	15-4520-54331	49°03'03"	93°39'25"	75.3	41	13.6	96	7
Dakota	15-7039-53810	48°33'04"	90°14'11"	115.2				
Daniels	15-4399-55286	49°54'05"	93°50'07"	1599.4	78.6	23.1	24	6
Darky	15-5893-53500	48°17'59"	91°47'48"	497.8	54	15.9	18	4
Dash	15-4559-54377	49°05'38"	93°36'15"	275.8	62	13.2	54	5
Dashwa	15-5921-54203	48°55'47"	91°44'51"	1125.8	46.1	13.5	24	4
David	15-5450-53588	48°22'54"	92°23'33"	333.5	33	13.8	45	4
Davidson	16-4892-54151	48°53'19"	87°08'56"	44	19.2	4	33	3
Deception	16-3712-53869	48°37'27"	88°44'48"	23.8	19.2	7	165	5
Deer (Atikwa)	15-4602-54790	49°27'48"	93°32'47"	5387.8	59.5	13.2	44	4
Delaney	15-4250-55493	50°05'25"	94°02'53"	1281.8	81.4	22.8	40	6
Delano (Clearwater)	15-4689-55111	49°45'14"	93°25'54"	821.4	23.5	12.1	42	3
Derby	15-4508-54427	49°08'14"	93°40'27"	113.3	24	11.7	135	6
Dewan	15-6123-54618	49°18'01"	91°27'24"	93.8	23.5	4.3	59	6
Dibble	15-5704-54495	49°11'47"	92°02'07"	1093.3	31	5.7	28	3
Dicker	15-4332-55321	49°56'22"	93°55'51"	299.9	62.2	15.3	26	8
Dimple	15-5777-54539	49°14'02"	91°55'58"	373.5	25	12.3	27	5
Disraeli	16-3551-54440	49°08'12"	88°59'11"	444.8	32.9	10.6	168	4
Doan	15-6062-54453	49°09'11"	91°32'42"	441	29.6	7.2	26	4
Dogfly	15-4908-54386	49°06'06"	93°07'39"	738.2	51.2	16.1	56	4

Lake	Code	Latitude	Longitude					
Dogpaw	15-4342-54693	49°22'10"	93°54'03"	1994.4	24.4	10.3	80	4
Dogtooth	15-4148-55077	49°43'06"	94°10'48"	2727.7	42	10.8	41	5
Domain	15-3893-56463	50°57'41"	94°34'14"	549.5	36	7.4	29	3
Doman	15-5019-54252	48°58'53"	92°58'33"	48.6	22.6	10.4	35	6
Donald	15-3664-56554	51°02'27"	94°54'55"	1497	29	11.8	30	4
Doré	15-6143-53839	48°35'54"	91°26'57"	312.6	39.7	8	33	4
Douglas	15-4108-56510	51°00'19"	94°16'18"	976.5	23.5	7.5	44	2
Dow (Lake 4)		50°37'09"	94°43'07"	198.5				
Dowswell	15-3786-56088	50°37'12"	94°43'01"	197.1				
Dragon	15-3862-56199	50°43'18"	94°36'45"	217.6	19	8.8	26	4
Dragon	15-5828-54131	48°52'03"	91°52'22"	116.1				
Draper	15-6021-53763	48°32'15"	91°37'05"	180.1				
Dryberry	15-4395-54864	49°31'45"	93°50'13"	9545.9	21	5.6	29	3
Ducell	16-4895-54163	48°54'01"	87°08'13"	43.1	19	5.9	48	5
Dummy	15-3912-55526	50°07'06"	94°31'17"	99	33.6	14.1	24	4
Dumpy	15-4238-55746	50°19'09"	94°04'05"	350.1	33.2	15.6	93	4
Eagle Rock	15-5212-54524	49°13'28"	92°42'58"	294.1	17.4	5.1	72	9
Earngey	15-4363-55211	49°50'26"	93°53'14"	31.1	50	13.3	20	7
East (Anstey)	15-4502-54405	49°07'00"	93°40'53"	76.1	31	10.8	21	7
East Campus (Fish)	15-5968-54495	49°11'49"	91°40'08"	324.2	18.9	6.1	36	3
East Hardtack	15-6012-54215	48°56'31"	91°37'01"	137.3	102.7	27.5	35	6
East Hawk (Hawk)	15-4279-55138	49°46'27"	94°00'05"	892	20	6.5	24	2
East Plummes (Plummes)	15-6686-53595	48°22'05"	90°43'27"	208.9	25	7.9	233	2
Echo (Underbrush)	15-4114-56359	50°52'29"	94°15'05"	398.7	30.1	8.7	59	6
Echoing	15-5492-60415	54°31'20"	92°14'13"	5326.6	24.6	8.1		5
Eden	15-3599-56139	50°39'40"	94°58'44"	488.5	32			
ELA L. 111		49°44'33"	93°50'06"	9				
ELA L. 139 (Bad)	15-4327-55016	49°39'54"	93°55'58"	167.5	29.9	8.2	32	5
ELA L. 161 (Hillock)	15-4350-55001	49°39'10"	93°54'00"	981.1	80.2	27.3	32	8
ELA L. 188		49°36'03"	93°47'42"	37	>25			
ELA L. 223		49°41'53"	93°42'29"	27	14			
ELA L. 224		49°41'25"	93°42'58"	26	27			
ELA L. 228 (Teggau)	15-4531-55044	49°41'27"	93°39'01"	1352.1	162.8	47.5	25	6
ELA L. 239		49°39'45"	93°43'22"	54	30			
ELA L. 254 (Ethelma)	15-4306-55046	49°41'43"	93°57'48"	387.2	33.6	14.4	33	6

ONTARIO (Northwestern Ontario) (continued)

Lake name	OFIS code	Latitude	Longitude	Surface area (ha)	Maximum depth (m)	Mean depth (m)	Conductivity (μS/cm)	Secchi depth (m)
ELA L. 256 (Veronica)	15-4385-55043	49°41'28"	93°51'10"	120.4	27.5	12.8	26	7
ELA L. 258 (Ella)	15-4420-55046	49°41'35"	93°48'16"	69.9	36.5	13.4	23	4
ELA L. 259		49°41'48"	93°46'54"	97	20			
ELA L. 259 (Burton)	15-4436-55044	49°41'29"	93°46'56"	94.1	25.4	8.2	21	5
ELA L. 260		49°41'46"	93°46'00"	34	14			
ELA L. 262		49°42'38"	93°41'33"	84	>30			
ELA L. 305		49°41'35"	93°41'13"	52	33			
ELA L. 310		49°39'45"	93°38'21"	50	>20			
ELA L. 322		49°37'19"	93°40'18"	74	19			
ELA L. 373	15-4424-55102	49°44'39"	93°47'56"	27	21			
ELA L. 374 (Porcus)	15-4390-55083	49°43'30"	93°50'43"	533.1	53.3	21.1	27	6
ELA L. 375	15-4432-55103	49°44'40"	93°47'15"	19	27			
ELA L. 376 (Manomin)	15-4466-55121	49°45'33"	93°44'21"	768.1	80.1	25.2	30	3
ELA L. 377	15-4443-55076	49°43'15"	93°46'25"	27	18			
ELA L. 378	15-4440-55065	49°42'38"	93°46'35"	26	12			
ELA L. 379 (Sheila)	15-4424-55064	49°42'33"	93°47'57"	159.5	47	18.1	27	7
ELA L. 382		49°42'17"	93°40'40"	37	13			
ELA L. 385		49°42'47"	93°36'39"	25	14			
ELA L. 399 (Mirror)	15-4301-55137	49°46'33"	93°58'20"	47.1	35	8.7	24	7
ELA L. 442	15-4411-55137	49°46'30"	93°49'05"	16	18			
ELA L. 464 (Highwind)	15-4345-55065	49°42'21"	93°54'59"	717.1	62.2	20.4	32	5
ELA L. 467		49°39'33"	93°45'45"	210	21			
ELA L. 468 (Roddy)	15-4474-55027	49°40'35"	93°43'45"	292.9	29	9.7	30	6
ELA L. 469 (Crabclaw)	15-4559-55166	49°48'04"	93°36'17"	577.6	37	11.6	42	4
ELA L. 563 (Kushog)	15-4309-55096	49°44'59"	93°57'34"	221.3	26.3	8.1	30	4
ELA L. 592 (Horseshoe)	15-4380-55114	49°45'14"	93°51'39"	35.1	18.2	6.6	23	5
ELA L. 596 (Audrey)	15-4384-55016	49°39'56"	93°51'14"	161.8	33	7.2	26	6
ELA L. 606 (Feist)	15-4398-55176	49°48'34"	93°50'17"	486.9	91.5	42.8	30	11
ELA L. 615 (Windermere)	15-4417-55173	49°48'31"	93°48'30"	95.8	49.1	11.6	30	6
ELA L. 622		49°46'03"	93°50'35"	37	31			6

ELA L. 623	15-4419-55113	49°45'49"	93°50'19"	36	21	14.5	18	6
ELA L. 625 (Fish)		49°45'12"	93°48'24"	80	40.1			
ELA L. 629	15-4414-55011	49°44'55"	93°50'07"	63	18	4.9	20	3
ELA L. 634 (Pidlubney)	15-4477-55182	49°39'41"	93°48'44"	55.2	20.5	6.7	30	4
ELA L. 647 (Upper Stewart, Stewart)		49°48'58"	93°43'38"	265.5	19.2			
ELA L. 649 (Lower Stewart, Stewart)	15-4454-55169	49°48'16"	93°45'32"	189.7	29.6	8	36	3
ELA L. 653 (Geejay)	15-4437-55142	49°46'43"	93°46'47"	186.5	46.3	18	30	1
ELA L. 664		49°37'23"	93°44'37"	67	>30			
ELA L.1002 (T-Bone)	15-4357-55095	49°44'11"	93°53'33"	51.2	24	9.9	30	4
ELA L.262 (Rogers)	15-4505-55066	49°42'38"	93°41'33"	84	>30		20	
ELA L.626 (Gibson)	15-4426-55113	49°45'12"	93°47'49"	25.7	13	7	22	9
ELA L.627 (Jorgenson)	15-4433-55118	49°45'28"	93°47'15"	37.4	7.5	4.7	21	5
ELA L.660 (Winnange, Buzzard)	15-4495-55118	49°45'08"	93°42'00"	2631.5	115.2	27.2	21	9
Elbow	15-6460-54008	48°44'35"	91°00'40"	583.6	58.6	22.5	62	6
Elevation	15-6577-53565	48°20'44"	90°52'20"	105.7	23	8.6	23	5
Ella (ELA L. 258)	15-4420-55046	49°41'35"	93°48'16"	69.9	36.5	13.4	23	4
Ellis	16-4879-54129	48°52'09"	87°09'58"	147.6	36.9	9.2	36	5
Elm	16-3710-54376	49°04'50"	88°46'14"	63.7	71.7	11.5	233	6
Elsie	15-5909-54484	49°11'09"	91°45'08"	1456.3	39	17.7	24	7
Emarton	15-4907-56260	50°47'12"	93°07'58"	125.8	24.4	13.8	93	2
Embryo	15-4044-56455	50°57'22"	94°21'40"	624.8	37.5	10.3	38	3
Emerald	15-6315-53307	48°07'26"	91°14'21"	250.8				
Emery	15-5824-54442	49°08'56"	91°52'15"	457.2	61.9	23.9	29	5
Entwine	15-5231-54423	49°08'33"	92°41'12"	1610.6	70.1	21.9	20	4
Esker (Three Moose)	16-5078-54343	49°03'50"	86°53'35"	7229				
Esnagami	16-5113-55744	50°19'46"	86°50'09"	1165.5	30.2	6.9	114	4
Esox	15-4810-54373	49°05'25"	93°15'09"	387.2	56.1	20.5	68	5
Ethelma (ELA L. 254)	15-4306-55046	49°41'43"	93°57'48"	1707.8	33.6	14.4	33	6
Eva	15-6340-53970	48°42'56"	91°10'24"	1203.5	54.9	13.5	33	6
Eye	15-5915-54149	48°52'57"	91°45'05"	629.7	18.3	5.2	25	7
Factor	15-5688-53944	48°42'16"	92°03'35"	56	49.4			2
Fahey	15-4964-54421	49°07'40"	93°02'15"			11.9	53	6

ONTARIO (Northwestern Ontario) (continued)

Lake name	OFIS code	Latitude	Longitude	Surface area (ha)	Maximum depth (m)	Mean depth (m)	Conductivity (µS/cm)	Secchi depth (m)
Faircloth	16-4168-56676	51°09'21"	88°11'00"	168.4	28.4	10.2	153	3
Fallingsnow	16-2930-53355	48°08'27"	89°46'58"	143.3	32.6	13.9	42	7
Favel	15-4299-55381	49°59'36"	93°58'40"	500.2	39	16.1	41	5
Fawcett	15-5090-55503	50°06'24"	92°52'25"	64.6	19.2	9.2		5
Feather (Quill)	15-4483-54282	49°00'23"	93°42'25"	161.1	24.4	7.2		5
Feist (ELA L. 606)	15-4398-55176	49°48'34"	93°50'17"	486.9	91.5	42.8	30	11
Ferguson	15-6367-53715	48°29'01"	91°09'19"	483.9				
Fernow	16-5680-55298	49°55'06"	86°03'11"	687.3				
Fish (East Campus)	15-5968-54495	49°11'49"	91°40'08"	324.2	31	10.8	20	7
Fish (ELA L. 625)	15-4419-55113	49°45'12"	93°48'24"	80	40.1	14.5	18	6
Fisher	15-4599-54900	49°33'44"	93°33'21"	299.7	25.1	8.4	60	5
Fishhook (Washeibemaga)	15-5346-54737	49°25'03"	92°31'34"	509	25.9	8.5	60	4
Flatrock (Abamategwia)	15-5795-55014	49°39'25"	91°53'43"	1357.7	33.2	9.4	40	3
Flegg	15-4343-55259	49°53'00"	93°54'45"	299.8	40	15.7	24	9
Flora	15-5762-54014	48°45'46"	91°57'53"	100.5	32.2	15.1	93	4
Fly	15-5228-56544	51°02'42"	92°40'45"	757.6	23.5	6.5	30	2
Fly	15-4341-55407	50°01'01"	93°55'12"	162.1	26	6.1	39	4
Fog	15-4645-54542	49°14'01"	93°29'28"	273.8	22.5	11.7	101	4
Forgotten	15-4028-55463	50°03'53"	94°21'21"	219.1	35	16.5	36	5
Fork	15-6696-53978	48°42'43"	90°41'38"	155.3				
Forsberg	15-5949-54175	48°54'20"	91°42'13"	208	30.2	9.7	22	5
Forty Mile	15-5993-55460	50°03'38"	91°36'48"	237	21	8.6	90	3
Fourbay	15-6523-55501	50°05'19"	90°52'27"	324.5	46.6	11	47	3
Fourstar	15-4866-54636	49°19'39"	93°11'07"	106.9	23.4	9.4	26	3
Fox	15-3721-55362	49°57'43"	94°47'05"	358.9				
Foxtail	15-4923-54691	49°22'38"	93°06'59"	51.6	25	7.7	20	5
French	15-6363-53918	48°39'56"	91°08'22"	301.9	25.9	12.5	37	3
Furlonge	15-4723-54386	49°06'07"	93°22'22"	172.6	31.7	9.5	182	4
Game	15-4546-55219	49°50'58"	93°37'58"	183	38.2	15	41	3
Gardnar	15-5375-55206	49°50'30"	92°28'45"	151.8	21.7	7.9	66	4

Name	Code	Latitude	Longitude					
Gargoyle	15-6384-54394	49°05'44"	91°06'09"	495.2	22.9	4.3	38	3
Garter (Mc Leans)	16-4671-54133	48°52'29"	87°26'41"	87.1	52.2	18.7	39	4
Geejay (ELA L. 653)	15-4437-55142	49°46'43"	93°46'47"	186.5	46.3	18	30	1
Geordie	16-5379-54088	48°49'57"	86°29'28"	20.0				
George	15-4196-55443	50°02'51"	94°07'23"	62.8	24.7	10.2	32	5
Ghost	15-5240-55204	49°50'14"	92°40'00"	461	29	16	81	6
Gibi	15-4182-54953	49°36'46"	94°07'51"	575.1	54.3	13.6	86	5
Gibson (ELA L.626)	15-4426-55113	49°45'12"	93°47'49"	25.7	13	7	22	9
Glacier	15-6325-53417	48°12'56"	91°13'40"	114.5				
Glenn	15-3832-56403	50°54'22"	94°39'45"	672.4	22	6.9	30	3
Gooseneck	15-3704-55440	50°02'13"	94°48'37"	172	29.9	8.2	50	4
Gordon	15-4490-55266	49°53'27"	93°42'38"	904.7	89.9	36.7	35	13
Graham	15-4464-54294	49°00'39"	93°44'04"	31.5				
Granite	15-3654-55075	49°42'34"	94°52'03"	141.7	19.8	11.2	56	4
Grant	15-4920-54470	49°10'41"	93°06'45"	365.8	35.5	7.3	75	6
Grave	15-4981-54403	49°07'01"	93°01'30"	129.5	42	12.9	65	6
Green	15-5355-54750	49°25'43"	92°30'35"	35.6	25.3	13.8	116	6
Green (Little Green)	15-4684-55017	49°40'09"	93°26'14"	132.8	32	19.2	35	7
Greenhedge	16-4701-54424	49°08'07"	87°24'35"	338.7				
Greenwater	15-6904-53837	48°34'43"	90°25'52"	3060.3	54.9	18.1	65	5
Greenwich	16-3638-54070	48°48'36"	88°51'33"	484.3	61	19.5	25	4
Grehan	16-5271-54815	49°29'25"	86°37'28"	232.4	62	22.7	202	4
Grey Trout	15-5670-54304	49°01'26"	92°05'02"	1297.8	48.8	17	23	7
Grimshaw	15-4951-54235	48°58'11"	93°04'30"	72.5				
Grouse	15-6812-53794	48°32'37"	90°32'35"	87.3	38.4	15.7	50	5
Gulliver	15-6218-54485	49°10'53"	91°19'20"	1478	50.6	10.1	23	3
Gullwing	15-5293-55287	49°54'15"	92°35'15"	2426.8	27.5	16.9	132	2
Gunflint	15-6725-53298	48°06'05"	90°41'42"	726.4				
Gunter	16-2948-55306	49°53'36"	89°51'37"	212.4	31	7.3	24	4
Gustauson (Angel)	15-5826-54970	49°37'23"	91°51'21"	144.6	20	7.6	33	8
Haggart	15-3615-56383	50°52'01"	94°58'51"	1930.7	59	15.8	32	3
Hammell	15-4269-56605	51°05'25"	94°02'13"	876.7	30.5	10	36	5
Hammerhead	15-3706-56518	51°00'35"	94°50'58"	505.6	29.5	9	33	2
Hansen	15-3805-56428	50°55'20"	94°42'02"	561.3	25.5	8.8	27	3
Hardtack	15-5996-54211	48°56'08"	91°38'22"	180.7	36	10.4	23	5

ONTARIO (Northwestern Ontario) (continued)

Lake name	OFIS code	Latitude	Longitude	Surface area (ha)	Maximum depth (m)	Mean depth (m)	Conductivity (µS/cm)	Secchi depth (m)
Hare	16-5447-54037	48°47'11"	86°23'31"	50.8	29.3	14.4	39	4
Harris	15-4845-54485	49°11'07"	93°12'03"	1804.9	61.1	13.7	54	4
Hatchet	15-4098-56485	50°58'59"	94°17'06"	333.8	36.3	8	39	3
Hawk (East Hawk)	15-4279-55138	49°46'27"	94°00'05"	892	102.7	27.5	36	6
Hawkcliff	15-4624-54987	49°38'35"	93°31'03"	535.8	58.8	18.2	23	2
Hawkeye	16-3199-53954	48°41'21"	89°26'29"	435.8	33.6	11.7	48	4
Hector	15-4746-54627	49°19'07"	93°21'01"	468.4	27	10.4	29	2
Helder	15-4153-55778	50°20'52"	94°11'25"	474.8	44.5	18.4	34	5
Helena	15-4597-54397	49°06'41"	93°33'08"	105.2	57	26.8	77	9
Herb	15-5913-54428	49°07'58"	91°44'55"	75.3	24.9	11.4	23	9
Herb	15-4222-55250	49°52'25"	94°04'54"	38.1	20	8.1		4
Heuston	15-5760-54440	49°08'46"	91°57'28"	349.4	56.4	18.1	17	6
Hickerson	15-4866-54214	48°56'52"	93°11'01"	60.2	27.5	10	42	3
High	15-3457-55073	49°42'02"	95°08'00"	778.5	20.8	11.9	72	5
Highwind (ELA L. 464)	15-4345-55065	49°42'21"	93°54'59"	717.1	62.2	20.4	32	5
Hill	15-4666-54568	49°15'51"	93°27'35"	655.3	18.9	7.5	35	4
Hillock (ELA L. 161)	15-4350-55001	49°39'10"	93°54'00"	981.1	80.2	27.3	32	8
Hilma	16-3752-53979	48°43'25"	88°41'50"	36.1	11.9	4.8	111	8
Hilton	15-4503-59412	53°37'10"	93°45'06"	551.5				6
Holger	15-5995-55749	50°19'11"	91°36'11"	306.2				
Hollingsworth	16-3616-56066	50°35'20"	88°57'16"	602.4	26	6.4	28	2
Holmes	15-5587-54132	48°52'08"	92°11'43"	147.4	34.2	15.4	23	5
Holstein	15-4650-54480	49°11'18"	93°28'39"	126.4	34	9.4	63	5
Hood	15-6735-53727	48°29'08"	90°39'14"	120.4	17.1	7.7	30	4
Hope	15-4445-54729	49°24'24"	93°45'59"	246.7	52.8	15	99	7
Horn	15-4287-55480	50°04'55"	93°59'48"	103.2	24.3	10.7	41	6
Hornberg	15-4514-54404	49°07'14"	93°39'34"	56.7				
Hornblende	16-4719-54180	48°54'52"	87°22'47"	20.6				
Hornick	16-3326-55334	49°55'56"	89°19'55"	305.1	61	24.7	38	6
Horseshoe	15-4476-55297	49°55'09"	93°43'48"	85	29	11	29	5

Lake	Code							
Horseshoe	15-4323-55298	49°55'08"	93°56'44"	157.5	36	9.1	23	4
Horseshoe (ELA L. 592)	15-4380-55114	49°45'14"	93°51'39"	35.1	18.2	6.6	23	5
Horseshoe	15-5833-54665	49°20'41"	91°51'15"	104.2	16	4.2	21	3
Huronian	15-6641-53950	48°41'21"	90°46'39"	359.8	39.6	11.1	28	6
Huston	15-3498-55846	50°23'43"	95°06'52"	190.6	30.5	11.8	26	6
Icarus	15-6814-53436	48°13'06"	90°33'25"	725.2	42.7	11.6	40	3
India	15-3909-55505	50°05'55"	94°31'31"	96.4	24.3	9.5	23	4
Indian (Alexandra)	15-4623-55297	49°55'12"	93°31'34"	642.3				
Indian	15-5965-54891	49°32'21"	91°39'57"	3962.8	36	9.4	55	4
Innes	16-3750-54094	48°49'36"	88°42'04"	231.2	42.4	8.8	93	2
Inspiration	16-3567-55633	50°12'19"	89°00'20"	556.1	88.5	11.4	29	6
Irene	15-6015-54420	49°07'42"	91°36'34"	1429	81.1	25.3	23	6
Isabelle	15-4698-55819	50°23'25"	93°25'30"	154.7	22	10.7	75	7
Isinglass	15-4500-54658	49°20'39"	93°41'09"	696.6	48.2	18.7	155	10
Islets	15-5533-54518	49°13'11"	92°16'19"	189.3	38	11.4	24	5
James	15-4953-54441	49°09'22"	93°03'12"	100.4	24	7	48	3
Jason	15-4247-55403	50°00'44"	94°03'04"	33.4	17.7	8.5	30	4
Jason	16-4741-54101	48°50'47"	87°21'13"	54.9	18.3	5.1		3
Jean (Little Kashabowie)	15-5958-53772	48°32'34"	91°42'10"	1468.2	74.1	16.1		5
Jeff	15-6177-53358	48°09'49"	91°24'59"	118.5				
Jessie	16-4028-54500	49°11'49"	88°20'04"	1001.9				
Jim	15-6008-53925	48°40'46"	91°37'44"	195	21	8.2	20	4
Jim	15-3982-55510	50°06'29"	94°25'04"	267.1	25.5	8	42	6
Joe (Big Joe)	15-5685-54110	48°50'47"	92°03'35"	175.0			27	
Jorgenson (ELA L.627)	15-4433-55118	49°45'28"	93°47'15"	37.4	7.5	4.7	21	5
Joyce	15-6113-53468	48°15'50"	91°30'53"	433.0				
Kag (Kagianagami)	16-4393-56423	50°55'12"	87°51'38"	7588	45	8.9	126	6
Kagianagami (Kag)	16-4393-56423	50°55'12"	87°51'38"	7588	45	8.9	126	6
Kahabeness	15-4987-54454	49°09'46"	93°01'11"	194.7	32	7.7	33	4
Kahshahpiwi	15-6126-53450	48°14'40"	91°28'52"	549.1				
Kaiarskons (Kaiashkons)	15-4710-54335	49°03'34"	93°23'16"	2147.1	58	24.5	84	4
Kaiashkons (Kaiarskons)	15-4710-54335	49°03'34"	93°23'16"	2147.1	58	24.5	84	4
Kamikau	15-6986-53540	48°18'35"	90°19'16"	119.1	19.2	7.1	28	5
Kaminni	15-5013-54633	49°19'21"	92°58'46"	329.4	32	7.3	23	2
Kaopskikamak (Kay)	15-6111-54547	49°14'08"	91°28'23"	362.2	29.9	6.9	24	4

ONTARIO (Northwestern Ontario) (continued)

Lake name	OFIS code	Latitude	Longitude	Surface area (ha)	Maximum depth (m)	Mean depth (m)	Conductivity (µS/cm)	Secchi depth (m)
Kaopskikamak (Crooked)	15-5116-54405	49°07'07"	92°50'28"	1266.7	39	7.2	26	4
Kaoskauta	15-4895-54697	49°22'52"	93°08'02"	256.8	41.7	8.6	20	5
Kapesakosi	15-4760-54731	49°24'56"	93°19'48"	261.9	21.2	7.6	28	5
Kasakokwog	15-5896-53865	48°37'29"	91°47'34"	768.9	59.5	21.4	26	5
Kashabowie	15-6913-53983	48°42'22"	90°23'57"	2289.6	35	7.4		3
Katimiagamak (Kay)	15-4479-54411	49°07'28"	93°42'46"	1032	76.3	23.1	98	8
Kawashegamuk (Long)	15-5477-54774	49°26'59"	92°20'32"	1834.7	32.9	14	141	8
Kawawiag	15-5104-54512	49°12'49"	92°51'30"	468.2	35.1	8.4	32	6
Kay (Kaopskikamak)	15-6111-54547	49°14'08"	91°28'23"	362.2	29.9	6.9	24	4
Kay (Katimiagamak)	15-4479-54411	49°07'28"	93°42'46"	1032	76.3	23.1	98	8
Keefer	15-6162-53525	48°18'14"	91°26'46"	230.0				
Kennewapekko	15-5320-54689	49°22'17"	92°33'48"	216.4	37.2	13.1	39	6
Kenny	15-5334-54745	49°25'26"	92°32'10"	85.8	34.5	16	113	10
Kenny	16-4479-54007	48°45'32"	87°42'32"	230.0				
Kershaw	16-2939-55252	49°50'43"	89°52'00"	214.5				
Kett	15-6017-53337	48°08'57"	91°38'05"	146.2				
Keys	15-4268-55431	50°02'08"	94°01'11"	574.7	90	37	32	7
Kilburn	15-3954-56165	50°41'09"	94°28'51"	1359.8	61.9	8.6	33	2
Killala (North Killala)	16-5341-54366	49°04'48"	86°31'36"	976.1	90	36.7	105	5
Kilvert	15-4257-55053	49°41'10"	94°01'33"	974	45.8	12	35	6
King	16-4358-54036	48°47'01"	87°52'24"	126.6	36	11.7	42	4
Kingfish	15-5758-54387	49°05'52"	91°57'41"	426.9	37.2	9.3	20	5
Kinmoapiku	15-5723-54610	49°17'42"	92°00'24"	212.6	38	7.5	27	5
Kinnyu	15-5281-54545	49°14'42"	92°36'30"	231.5	24.4	6.6	20	2
Kishkutena	15-4440-54320	49°02'08"	93°46'47"	1066.8	24.4	8.6	87	5
Klinestiver	16-5388-54811	49°28'47"	86°27'59"	157.6	40	10.9		3
Knife	15-6340-53288	48°06'31"	91°12'39"	608.1				
Kowastigiman	15-5158-58183	52°30'50"	92°45'37"	110.7				
Kushog (ELA L. 563)	15-4309-55096	49°44'59"	93°57'34"	221.3	26.3	8.1	30	4
Lac des Iles	16-3103-54530	49°12'35"	89°36'59"	1498.7	27.1	6.9	52	2

Lac la Croix	15-5630-53553	48°21'49"	92°08'25"	6461.2	23.5	7.5	44	2
Lake 4 (Dow)	15-5729-54366	50°37'09"	94°43'07"	198.5	51.9	14.7	23	6.3
Lake 5PB10-125 (Peterson, Lake 125)		49°04'47"	92°00'07"	282				
Lake 5PB10-93 (Monte, Lake 93)	15-5727-54414	49°07'23"	92°00'13"	136.9				
Lake 5PB17-82	15-6070-54327	49°02'23"	91°32'10"	91	20.7	7	23	4
Lake 8 Block 5	15-5916-53473	48°16'27"	91°45'56"	20.2	18.9	9.2	23	5
Lake 20		49°07'44"	92°08'14"	57	32	10.9	18	5
Lake 26		49°07'21"	92°08'42"	30	37	11.8	22	8
Lake 39		49°05'42"	92°09'54"	37	23	8.8	19	7.5
Lake 42		49°05'04"	92°09'35"	26	17	6.4	15.6	7
Lake 42 (Secret)	15-5627-54364	49°04'46"	92°08'19"	149.4	28.8	14.1	20	6
Lake 93 (Lake 5PB10-93, Monte)	15-5727-54414	49°07'23"	92°00'13"	136.9				
Lake 103		49°02'11"	92°02'41"	81	18		19	3.7
Lake 125 (Lake 5PB10-125, Peterson)	15-5729-54366	49°04'47"	92°00'07"	282	51.9	14.7	23	6.3
Lake 155 (Below Bow)	15-5986-54297	49°00'44"	91°39'02"	208.0	25.9	11.1	63	2
Lake of Bays	15-6207-55385	49°59'20"	91°18'59"	4415.4	76.2	22.2	44	4
Lawrence	15-4729-54570	49°16'17"	93°22'31"	1870.7	22	5.7	23	3
Leano	15-3985-56261	50°46'53"	94°26'21"	310.6	33.6	14.4	39	5
Lee	15-4176-56429	50°56'10"	94°10'02"	282.5	36.6	15.3	27	6
Leitch	15-4170-56459	50°57'39"	94°10'51"	170.8	55.8	21	40	7
Lemman	15-4142-55736	50°18'22"	94°12'24"	1360.9	28.9	10	32	4
Lift	15-4460-55234	49°51'42"	93°45'08"	96.1	23	10.6	35	4
Lilac	15-5464-53494	48°17'44"	92°22'54"	298.4	22	8.5	32	3
Linge	15-4101-56795	51°15'14"	94°17'23"	843.2	78.6	41.2	87	6
Linklater	15-4385-55243	49°52'21"	93°51'27"	628.9	17.4	5.5		3
Little Bow	15-4452-55268	49°02'10"	91°36'39"	91.7	46.7	17.3	35	
Little Gordon	15-4684-55017	49°53'33"	93°45'24"	176	32	19.2	21	8
Little Green (Green)	15-5687-54337	49°40'09"	93°26'14"	132.8	28.4	23.8	24	7
Little Greytrout	15-6024-54365	49°03'15"	92°03'39"	150.1	36.3	11.3	24	5
Little Gull	15-6005-54371	49°04'27"	91°35'52"	325.4	27.1	13.1		6
Little Irene		49°04'51"	91°37'27"	54.7				6

ONTARIO (Northwestern Ontario) (continued)

Lake name	OFIS code	Latitude	Longitude	Surface area (ha)	Maximum depth (m)	Mean depth (m)	Conductivity (µS/cm)	Secchi depth (m)
Little Jet	15-4264-55321	49°56'13"	94°01'31"	37.1	15	6.1	20	5
Little Joe	15-4682-55795	50°22'07"	93°26'50"	221.3	24	11.2	72	4
Little Kashabowie (Jean)	15-5958-53772	48°32'34"	91°42'10"	1468.2	74.1	16.1		5
Little Mennin (Spruce)	15-5612-54761	49°26'35"	92°09'21"	278.6	47	10.3	30	5
Little Moosehide (Upper Moosehide)	15-5886-54545	49°14'21"	91°47'00"	218.2	22.3	10.7	38	4
Little Moraine (Trout)	16-3657-54310	49°01'14"	88°50'16"	234.7	32	13.2	106	10
Little North	15-6802-53327	48°07'28"	90°34'38"	39	7	3	108	4
Little Pic	16-5264-54678	49°21'52"	86°38'13"	120.3	24.4	6.5	128	4
Little Raleigh	15-5800-54783	49°27'13"	91°53'46"	74.1	19.2	6.6	29	5
Little Sandford	15-6016-54393	49°06'06"	91°36'21"	98.4	29.1	11	29	8
Little Santoy	16-5063-54057	48°48'24"	86°54'53"	122	56.7	15.5	123	4
Little Scattergood	15-5192-54632	49°19'23"	92°44'09"	61.2	24.4	9.5	38	4
Little Sparkling	15-6993-55238	49°50'02"	90°13'35"	317.3	35	9.8	33	4
Little Trout	15-4833-56555	51°03'10"	93°14'21"	3521.7				
Little Trout	15-4222-55387	49°59'51"	94°05'08"	69.2	27	10.5	33	5
Little Vermilion	15-5581-55389	50°00'13"	92°11'23"	2244.8	24.3	10.1	167	4
Little Woman	15-5163-56621	51°06'43"	92°46'02"	565.1	37	9	93	6
Loch Erne	15-6946-53876	48°36'49"	90°21'26"	165	32	13.8	104	7
Loch Lomond	16-3272-53473	48°15'29"	89°19'44"	1696.2	70.5	21.3	58	4
Long (Kawashegamuk)	15-5477-54774	49°26'59"	92°20'32"	1834.7	32.9	14	141	6
Long-Legged (Stork)	15-4172-56262	50°47'20"	94°10'22"	6915.5	35.4	8.9	41	3
Loonhaunt	15-4632-54287	49°00'59"	93°30'00"	2070.4	47.3	11.3	53	4
Louisa	15-6250-53348	48°09'25"	91°19'20"	717.9	76.3	23.2	34	5
Lower Bow	15-6071-54377	49°03'36"	91°34'09"	382.8	34.1	7.6	23	2.7
Lower Manitou	15-5023-54567	49°15'49"	92°58'11"	8284.8	81	19.9	66	4
Lower Moosehide (Moosehide, Blackhole)	15-5871-54555	49°14'54"	91°48'11"	178.8	23.8	11.8	30	3
Lower Stewart (Stewart, ELA L. 649)	15-4454-55169	49°48'16"	93°45'32"	189.7	29.6	8	36	3

Name	Code	Latitude	Longitude					
Lucy	16-4876-55740	50°19'13"	87°10'28"	730.4	35.1	6.5	167	3
Luella	16-3792-56718	51°11'22"	88°43'19"	768.5	55.8	16.3	32	3
Lyne (Lynx)	16-4734-54128	48°52'01"	87°21'49"	164.3	55.8	16.3	32	3
Lynx (Lyne)	16-4734-54128	48°52'01"	87°21'49"	164.3	25.3	10.6	24	5
Mabel	15-5942-54463	49°09'29"	91°42'54"	472.8	29	10.9	100	6
Mac Intosh	16-3752-54002	48°44'42"	88°41'48"	42.4				
Mackay	15-3590-58314	52°37'45"	95°04'36"	790.7	37.2	8.2	62	5
Mackenzie (Mc Kenzie)	16-3530-55684	50°15'20"	89°03'20"	509.5	18		37	3
Mackenzie	16-3699-53994	48°44'06"	88°46'12"	53.4	22.5	3.7	51	3
Madoson	16-5611-54166	48°54'00"	86°09'57"	63.7				
Magnetic	15-6663-53296	48°06'19"	90°45'33"	99.0	30.1	10.3	42	4
Malaher	15-3580-57547	51°55'27"	95°03'14"	557.9	50	16.1	56	8
Mameigwess	15-5853-54894	49°33'42"	91°49'25"	5309.7	39	11.4	30	5
Mang	15-5081-54510	49°12'31"	92°53'43"	165.9	80.1	25.2	30	3
Manomin (ELA L. 376)	15-4466-55121	49°45'33"	93°44'21"	768.1				
Margon (Brown T)	16-4988-54124	48°52'26"	87°00'54"	14.6	28.1	7	29	6
Marion	15-6298-53936	48°41'02"	91°14'12"	436.6	41.5	11.9	20	4
Mark	15-4239-55321	49°56'35"	94°03'55"	265.5	67	7.1	25	4
Mather	15-3590-56330	50°50'02"	95°00'09"	533.1	30.5	9.9	26	4
May	15-4240-55414	50°01'19"	94°03'40"	51.8	36	16.5	26	5
Mc Alpine	16-6008-53882	48°38'04"	91°37'22"	301.1	37.2	12.7	42	2
Mc Aree	15-5787-53500	48°18'02"	91°56'21"	845	49.7	20.6	35	4
Mc Caulay (Big Mc Caulay)	15-5752-53953	48°42'37"	91°58'08"	451.7	49.7	8.5		5
Mc Crea	15-6908-56374	50°51'35"	90°17'19"	3741.2	67.1	16.4	38	6
Mc Cusker	15-3845-57222	51°38'43"	94°40'08"	2777.7				
Mc Dougall	15-6158-53674	48°26'50"	91°26'22"	371.6				
Mc Ewen	16-6348-53492	48°17'23"	91°10'55"	661.6				
Mc Gonigel (Pluto)	15-5972-55017	49°39'48"	91°39'14"	159.5	24.4	13.5	30	4
Mc Gruer	16-6472-58717	52°58'35"	90°48'28"	77.9	38	10.2	66	8
Mc Innes	15-4501-57850	52°12'27"	93°43'30"	6533.9	53.4	15.3	44	3
Mc Intyre	15-6027-53429	48°14'11"	91°36'57"	553.3				
Mc Kay	16-5406-54956	49°36'48"	86°26'17"	3131.6	48.8	9.4	159	4
Mc Kenzie (Mackenzie)	16-3530-55684	50°15'20"	89°03'20"	509.5	37.2	8.2	62	5
Mc Laurin	16-3511-55649	50°13'12"	89°05'09"	304.2	81	21.7	32	7
Mc Leans (Garter)	16-4671-54133	48°52'29"	87°26'41"	87.1	52.2	18.7	39	4

ONTARIO (Northwestern Ontario) (continued)

Lake name	OFIS code	Latitude	Longitude	Surface area (ha)	Maximum depth (m)	Mean depth (m)	Conductivity (µS/cm)	Secchi depth (m)
Mc Niece	15-6151-53432	48°13'53"	91°27'45"	56.5	37	8.8	24	3
Meddick	15-4291-58135	52°28'09"	94°02'38"	1133.3	61	16.7	35	9
Medicine	15-4432-55232	49°51'39"	93°47'23"	211.5	50	19.7	39	3
Medicine Stone	15-4227-56424	50°55'41"	94°06'11"	1199.9	32.3	8.4	27	3
Meggisi	15-5287-54593	49°17'20"	92°36'18"	1180.9	18.4	5.2	30	3
Mermaid	15-4220-55436	50°02'29"	94°05'24"	58.2	48	23.1	198	5
Merpaw	16-3637-57235	51°38'57"	88°58'09"	108.2	27	6.5	39	15
Metionga	15-6824-55113	49°43'02"	90°28'03"	2029.5				3
Middle Bow		49°03'38"	91°31'27"					
Middle Kilburn	15-3917-56201	50°43'28"	94°32'04"					
Miner	16-3803-54067	48°48'13"	88°37'43"	4.9	22.9	8.1	87	6
Mirror (ELA L. 399)	15-4301-55137	49°46'33"	93°58'20"	47.1	35	8.7	24	7
Missus	15-4815-54449	49°09'41"	93°15'05"	235.5	32	11.6	54	6
Mister	15-4811-54430	49°08'39"	93°15'12"	211	30	10.4	62	6
Mix	15-5274-57646	52°02'15"	92°36'21"	231.5				
Mold	15-5282-55495	50°05'01"	92°36'28"	49.4				
Monte (Lake 5PB10-93, Lake 93)	15-5727-54414	49°07'23"	92°00'13"	136.9	17	8.1	23	4
Moose	15-5181-54639	49°19'47"	92°45'01	117.2				
Moose	15-7171-53313	48°06'16"	90°04'38"	222.8	12.2	4.3	39	4
Moosehide (Lower Moosehide, Blackhole)	15-5871-54555	49°14'54"	91°48'11"	178.8	23.8	11.8	30	3
Mooseland	15-7155-54743	49°23'17"	90°01'15"	1091.4	159.8	22.1	63	2
Moraine (Boulder)	16-3644-54282	48°59'12"	88°51'05"	277.2	17.7	8.9	128	4
Moss	16-4132-53876	48°38'25"	88°10'28"	229.7	49	21.9	56	4
Mount	15-5600-54290	49°00'56"	92°10'25"	1081.3	37.2	12.3		3
Mountain	16-3388-55667	50°13'51"	89°15'20"	94	24.5	8.4	23	6
Mountain (Northwest Mountain)	15-4112-55373	49°58'59"	94°14'31"	168.4	34.2	17.1	39	4
Mountain	15-7072-53309	48°06'38"	90°13'43"	459.8				

Lake								
Mowe	15-6672-53552	48°19'25"	90°44'16"	612.8	33	12.3	42	3
Musclow	15-3646-56968	51°24'13"	94°56'40"	2434	45.1	18.1	47	3
Myrt	15-6681-53680	48°26'59"	90°43'31"	272.6	15.1	5	28	6
N.L	15-3910-56213	50°44'07"	94°32'41"	272.1				
N.L	15-3540-56417	50°54'37"	95°04'37"	304.7				
N.L	15-3695-56428	50°55'26"	94°51'25"	73.4				
Nalla	16-3680-53921	48°40'10"	88°47'34"	48.9	21	8.3	41	2
Namaygoos	15-5718-54592	49°17'04"	92°00'38"	126.3	24	8.8	23	7
Namego	15-3934-55525	50°07'13"	94°29'34"	206.6	41	17.2	29	8
Nameiben	16-3458-55570	50°08'51"	89°09'30"	126.4	65	5.9	63	5
Narrow	15-5913-53921	48°40'40"	91°45'32"	51.8	15	8	21	2
Nemakwis	15-5136-58183	52°30'18"	92°48'05"	306.2				
Nest	15-6079-53338	48°08'54"	91°32'39"	79.4				
Nevison	15-5828-54113	48°51'07"	91°52'13"	151.8	19	7.9	21	4
Niobe	15-6228-53982	48°43'25"	91°19'57"	317.3	19.5	6.4	46	4
Niven	15-5734-54018	48°46'28"	92°00'01"	290.7	33	8.7	65	6
No Man	15-6302-53333	48°08'31"	91°15'01"	25.1	8.2	3.8	66	8
Nolan	16-3742-54014	48°45'16"	88°42'44"	22.9	29.3	10	96	8
Nora	15-5842-54507	49°12'18"	91°50'35"	1592.6	55.2	16.2	27	5
North	16-6844-53328	48°07'17"	90°31'45"	1091.2	36	15.3	93	6
North Arabi	15-6566-54931	49°34'34"	90°50'04"	68.6	29	10.6	44	3
North Killala (Killala)	16-5341-54366	49°04'48"	86°31'36"	976.1	90	36.7	105	5
North Mawn	15-6866-54968	49°35'41"	90°25'12"	192.7	27.8	9	24	5
North Trout	15-4518-58612	52°54'31"	93°43'14"	742.7				
North Whalen	16-3133-55504	50°04'26"	89°36'02"	445.3	14	3.7	27	3
Northern Light	15-6723-53466	48°15'06"	90°40'45"	6869.8	39.7	8.4	38	3
Northwest Mountain (Mountain)	15-4112-55373	49°58'59"	94°14'31"	168.4	34.2	17.1	39	4
Norway	15-6261-54379	49°04'38"	91°16'16"	505.4	62.5	21.9	28	10
Notellum	15-4264-55879	50°26'25"	94°02'12"	261	42	12.5	36	5
Noxheiatik	15-5416-54762	49°26'39"	92°25'56"	90	30.5	9.3	65	6
Nydia	15-6418-53991	48°43'42"	91°04'33"	549.6	48.5	9.9	33	7
Nym	15-6136-53934	48°41'56"	91°27'28"	1838.1	37.2	8.7	27	5
Obonga	16-3346-55390	49°58'54"	89°18'32"	3578.4	71.7	17.4	50	3
Olifaunt	15-6144-53765	48°32'05"	91°27'02"	561.7				

ONTARIO (Northwestern Ontario) (continued)

Lake name	OFIS code	Latitude	Longitude	Surface area (ha)	Maximum depth (m)	Mean depth (m)	Conductivity (µS/cm)	Secchi depth (m)
Olive	15-4096-56873	51°19'57"	94°17'35"	783	17	4.8	21	3
Oliver	16-3081-53489	48°15'52"	89°35'13"	199.1	38.1	22.7	64	6
Onamakawash	16-3145-55756	50°18'20"	89°36'10"	1385.6	47.3	8.6	53	3
One Island (Beaton)	15-4708-55595	50°11'32"	93°24'09"	160.6	27	12.9	81	6
Onepine	15-3875-57398	51°48'00"	94°37'54"	1210.7	15	6.7	29	2
Onnie	15-4109-56434	50°56'01"	94°16'01"	161.9	26	8.4	36	3
Opichuan	16-4452-56767	51°14'09"	87°47'09"	310.2	27	7.6	114	3
Optic	15-3932-56417	50°55'48"	94°31'11"	343.4	24	9.5	39	4
Osaquan	15-5906-54745	49°25'05"	91°45'02"	371.4	22	4.8	33	3
O'Sullivan	16-4960-55853	50°25'43"	87°03'13"	4292.5	45.7	8.4	174	4
Other Man	15-6375-53394	48°11'35"	91°09'03"	178.4	32.6	10.9	104	8
Otisse	16-4890-54142	48°52'58"	87°08'56"	13	21	7.1	41	3
Otterskin	15-4531-54543	49°14'12"	93°38'26"	476.1	67	20.2	143	11
Ottertrack (Cypress)	15-6419-53372	48°10'26"	91°05'30"	277.4				
Otukamamoan (Trout)	15-5100-54230	48°57'35"	92°51'31"	5165.8	53.7	16.9	39	4
Owl	16-5000-54272	48°59'42"	87°00'49"	996.0				
Owl	15-5940-53910	48°40'00"	91°43'19"	29.4	10.2	6.2	21	2
Paddy	15-5701-54572	49°15'58"	92°02'03"	458.8	26	8.1	23	6
Paguchi (Trout)	15-6056-54920	49°34'05"	91°32'58"	2447.6	30.2	9.1	65	2
Pagwachuan	16-5657-55074	49°43'05"	86°05'10"	2714.8	54.9	18.8	169	6
Paint	16-4496-55086	49°43'51"	87°41'39"	336.7	43	5.1	213	5
Pangloss	16-3358-55142	49°45'48"	89°16'40"	337.4	40	7.3	57	5
Panorama	15-4398-54235	48°57'40"	93°49'39"	493.9	38	10.3	80	6
Partridge (Church)	15-6795-53776	48°31'39"	90°34'08"	59.73	41.2	18	44	7.6
Passover	15-4826-54857	49°31'03"	93°14'22"	378.7	48.8	10.1	35	3
Patty	15-4313-55449	50°03'16"	93°57'35"	45.3	22.6	8.1	27	4
Paull	15-3834-56252	50°46'09"	94°39'13"	448.7	30	8	27	4
Peak	15-5264-54828	49°29'30"	92°38'08"	331.2	25.3	11.8	111	4
Pekagoning	15-5590-54442	49°08'41"	92°11'31"	1319.6	38.4	10.6	26	5
Pelican	15-5737-55520	50°07'15"	91°58'10"	2341.5	35.1	10.4	66	2

Name	Code							
Penassi	15-6300-55346	49°57'01"	91°11'00"	1441.3	27.1	6.6		2
Penassi	15-4887-54610	49°18'05"	93°09'11"	134.1	23	7.6	23	2
Pennock Lake West	16-2832-53599	48°21'25"	89°55'36"	19.1	15.3	4		4.7
Perch	15-3903-55478	50°04'26"	94°31'56"	260.2	29.9	8.4	48	4
Pete	16-2870-53613	48°22'15"	89°52'37"	17	10.7	5.8	23	4
Peterson (Lake 5PB10-125, Lake 125)	15-5729-54366	49°04'47"	92°00'07"	282	51.9	14.7		6.3
Pettit	15-5537-54222	48°57'15"	92°16'16"	1197.1	25	10.6	31	3
Pic	15-4141-55473	50°04'36"	94°12'01"	114.4	22.9	9.1	41	5
Pickerel	15-6146-53867	48°37'42"	91°26'04"	6058.8	74.7	17.7	31	4
Pidlubney (ELA L. 634)	15-4414-55011	49°39'41"	93°48'44"	55.2	20.5	4.9	20	3
Pineneedle	15-4085-56132	50°39'55"	94°17'15"	1037.7	67.1	19.3	45	6
Pipestone	15-4595-54375	49°05'47"	93°33'08"	3890.7	53.4	18.6	87	4
Piskegomang Brook	15-4699-55014	49°40'48"	93°23'50"	127.7	21	7.1	26	6
Plough	15-6351-53343	48°08'46"	91°10'56"	48.6				
Plummes (East Plummes)	15-6686-53595	48°22'05"	90°43'27"	208.9	20	6.5	35	2
Pluto (Mc Gonigel)	15-5972-55017	49°39'48"	91°39'14"	159.5	24.4	13.5	30	4
Poacher	15-6191-53284	48°05'11	91°24'14"	123.0				
Pond (Trapper)	15-5816-53498	48°17'53"	91°53'58"	76.9	25.3	11.5	26	4
Poohbah	15-5975-53595	48°22'44"	91°40'58"	1529.7		20.8	35	
Popeye	15-5821-54650	49°20'03"	91°52'14"	103.8	24	9.1	20	3
Populus	15-4557-54838	49°30'22"	93°36'35"	665.6	39	8.2	86	4
Porcus (ELA L. 374)	15-4390-55083	49°43'30"	93°50'43"	533.1	53.3	21.1	27	6
Prairie	16-5211-54299	49°01'31"	86°42'32"	169.1	36	14.6	179	7
Priam	15-4745-54662	49°21'08"	93°21'19"	289.1	25	7	48	3
Pringle	16-3708-54345	49°03'10"	88°46'00"	182.1	29	10.2	49	7
Prospect	16-5105-54115	48°51'30"	86°51'24"	35.4				
Quetico	15-5780-53794	48°33'51"	91°56'51"	4261.3	61	13.1	23	3
Quill (Feather)	15-4483-54282	49°00'23"	93°42'25"	161.1	24.4	7.2		5
Rail	15-4365-55203	49°49'41"	93°52'56"	71.5				
Raleigh	15-5786-54755	49°25'27"	91°54'34"	1698.9	29.3	10.4	26	6
Ram	15-6095-53783	48°32'44"	91°30'51"	83.1				
Rawlinson	15-5380-54589	49°17'01"	92°28'39"	234.1	32.1	11.1	25	2
Rawn	15-6293-53806	48°34'08"	91°14'57"	335.9	30.5	10	27	2
Rayner	16-5363-55355	49°58'23"	86°29'38"	435.3				

ONTARIO (Northwestern Ontario) (continued)

Lake name	OFIS code	Latitude	Longitude	Surface area (ha)	Maximum depth (m)	Mean depth (m)	Conductivity (µS/cm)	Secchi depth (m)
Red Deer	15-4168-55418	50°01'31"	94°09'22"	779.9	67.7	26	32	6
Red Paint	15-6171-54337	49°02'35"	91°23'49"	515.9	48.5	12.2	47	5
Redfox	15-6569-53648	48°25'04"	90°52'56"	62.4	24	13.3	62	6
Richard	15-4345-55213	49°50'31"	93°54'28"	64.1	23.8	7.1	45	6
Richardson	15-5670-55575	50°10'08"	92°03'21"	190.4	22.9	12.1	71	6
Rieder	15-5748-60832	54°53'37"	91°49'46"	488	32	9.1	144	3
Riverview	15-6208-54398	49°06'01"	91°20'49"	237.3	67.1	17.6	18	6
Robbie Burns	15-4455-55326	49°56'32"	93°45'40"	200.3	29	9.4	26	4
Robinson	15-5995-53387	48°11'35"	91°39'48"	420.9	35.1	12.7	23	7
Roddy (ELA L. 468)	15-4474-55027	49°40'35"	93°43'45"	292.9	29	9.7	30	6
Roderick	15-4041-57141	51°34'19"	94°23'05"	2271.4	31.1	9.4	29	3
Rogers (ELA L.262)	15-4505-55066	49°42'38"	93°41'33"	84	>30		20	
Roland	15-5855-53461	48°15'39"	91°50'56"	230.4	32	12	17	4
Ronny	15-4086-55352	49°57'56"	94°16'16"	73.2	20.5	7.6	26	4
Ross	15-6528-53582	48°21'45"	90°56'07"	312	26	10.1	33	3
Ross	15-4870-55428	50°02'18"	93°10'40"	324.1	20.1	7.5	30	6
Rostoul	15-3759-56455	50°57'44"	94°45'57"	545.7				
Roughstone	15-5539-54298	49°01'13"	92°15'10"	296.5	24.2	9.9	24	6
Route	15-5270-55450	50°03'25"	92°37'03"	1135.4	43.3	13.7	72	2
Rowan	15-4593-54629	49°19'43"	93°33'02"	5471	39.7	10.3	58	5
Rowdy	15-3949-56006	50°33'30"	94°28'36"	1128.3	55.9	9.2	35	3.8
Royd	15-3759-56575	51°03'16"	94°46'01"	801	60.1	14.3	29	3
Rudge	15-6659-53963	48°41'42"	90°44'04"	492.1	35.7	8.9	23	7
Ruffle	16-5208-54346	49°03'50"	86°42'38"	481.9	36	15.9	105	13
Rutter	15-5581-54358	49°04'20"	92°12'23"	337.5	21.4	5.5	28	3
Saganaga	15-6547-53441	48°14'41"	90°55'19"	3524.9				
Saganagons	15-6508-53496	48°18'17"	90°55'22"	2508.0				
Sakwite	15-5073-54306	49°01'18"	92°54'25"	379.3	24.5	8.8		3
Sanctuary	15-7008-54035	48°45'14"	90°16'05"	66.8	26.4	11	24	5.2
Sandbeach	15-5586-54227	48°57'08"	92°11'53"	681.5	54.9	20.7	47	4

Lake	Map reference	Latitude	Longitude					
Sandford	15-5955-54404	49°06'47"	91°41'16"	2921.7	114.7	36.8	27	7
Sandhill	15-4545-54478	49°10'59"	93°37'23"	80.2	30.5	12.5	150	7
Sandstone	15-7053-53459	48°14'03"	90°14'13"	935.2	23.8	7.3	75	5
Sandybeach (Big Sandy)	15-5463-55189	49°49'11"	92°21'11"	3820.2	41	20.3	150	5
Santoy	16-5087-54120	48°51'46"	86°52'55"	1002				
Sark	15-6173-53543	48°20'29"	91°25'27"	283.4	19.3	11.3	119	4
Sasakwei (Summit)	15-5250-54810	49°29'00"	92°39'16"	126.3	38.1	6.1	30	4
Scattergood	15-5206-54600	49°17'19"	92°42'33"	281.7	53	16.6	132	5
Schistose	15-4562-54445	49°09'36"	93°36'57"	344.7	36.2	15.3	48	7
Schultz	15-4694-55607	50°11'53"	93°25'46"	312.2	27.8	9.2	26	3
Scotch (Upper Scotch)	15-6345-54482	49°10'26"	91°09'16"	1291.6	22	11.1	43	4
Scotty	15-4244-55773	50°20'43"	94°03'20"	771.7	41.2	6.4	53	5
Seagrave	15-5576-56820	51°17'05"	92°10'13"	1390.8	28.8	14.1	20	6
Secret (Lake 42)	15-5627-54364	49°04'46"	92°08'19"	149.4	68	22.1	19	5
Sedgwick	15-5706-54405	49°06'39"	92°01'56"	654.2				
Seeley	16-5429-54066	48°48'46"	86°24'56"	87.2	28.6	9.4	27	4
Seggemak	15-5370-54703	49°23'07"	92°29'07"	169.5	42.7	10.9	62	4
Selim (Whitesand)	16-4720-54109	48°51'09"	87°22'54"	269.4	47.5	9	38	6
Serpent	15-6089-54300	49°00'52"	91°30'42"	415.8				
Sesikinaga	15-5595-56666	51°08'06"	92°07'00"	1354.1				
Shade	15-6165-53376	48°10'57"	91°26'50"	213.1	30	7.3	15	4
Shaw	15-6035-54589	49°16'29"	91°34'52"	169.9	41.1	9.8		5
Shawanabis	16-3240-55701	50°15'30"	89°28'10"	849.9	37.8	7.6	23	4
Shebandowan	15-6988-53913	48°38'40"	90°18'29"	5905.2	47	18.1	50	7
Sheila (ELA L. 379)	15-4424-55064	49°42'33"	93°47'57"	159.5				
Sheridan	15-6253-53286	48°06'14"	91°19'20"	83.1				
Shingwak	15-4491-54608	49°18'28"	93°41'55"	830	22.9	4.7	27	3
Shiny	15-6279-55307	49°55'00"	91°13'12"	295.4	26	9	21	4
Shrub	15-4351-55357	49°58'24"	93°54'22"	558.7				
Side	15-6090-53375	48°11'09"	91°32'05"	25.0				
Sidious	15-4096-56216	50°44'28"	94°16'53"	213.4	16.8	6.4	33	3
Silence	15-6207-53435	48°14'33"	91°22'08"	230.7	91.5	39	36	7
Silver	15-4153-55253	49°52'28"	94°10'42"	2649.1				
Silver	15-4564-54361	49°04'42"	93°35'49"	36.5				
Silver	16-3737-53877	48°37'50"	88°42'50"	78.4	27.5	8.3	138	5

ONTARIO (Northwestern Ontario) (continued)

Lake name	OFIS code	Latitude	Longitude	Surface area (ha)	Maximum depth (m)	Mean depth (m)	Conductivity (µS/cm)	Secchi depth (m)
Silvertip	15-5536-54324	49°02'34"	92°16'38"	721.5	54.9	16.8	30	6
Silvery	15-4354-55159	49°47'33"	93°53'53"	447.2	39.1	13.2	23	7
Six Mile	15-6522-55386	49°58'56"	90°52'37"	1074.2	29.6	8.4	62	
Sixth	15-3791-55635	50°12'54"	94°41'42"	106.1	25.4	8.8	54	4
Slender	15-4489-54311	49°01'59"	93°41'56"	240.4	45.8	16.2	54	6
Smoothrock	16-3266-55997	50°31'28"	89°26'49"	9824.5	47.2	5.4	39	3
Snook	15-3795-55615	50°11'49"	94°41'26"	263.1	37.2	14.2	33	2
Soho	15-5921-53889	48°38'56"	91°44'49"	270.7	57.3	16.7	20	4
South	16-4271-56531	51°01'36"	88°02'22"	542	33.6	8.6	110	5
South	15-6163-54669	49°20'45"	91°23'52"	52.3	15.9	6.4	33	8
South	15-6824-53301	48°06'02"	90°31'46"	183.8				
South		48°05'41"	90°40'23"	450.5	42.9	19.2		29.7
South Otterskin	15-4544-54517	49°13'10"	93°37'45"	213.7	36	11.8	179	7
South Trout	15-4541-58591	52°52'58"	93°40'17"	657.5	29.8	12.6	114	4
South Wapageisi	15-5449-54565	49°15'42"	92°23'00"	611.9	18.3	6	38	3
Sox	16-4689-54139	48°52'32"	87°25'46"	84.2	47.6	11.7	28	4
Sparkle	15-6161-54458	49°09'28"	91°24'08"	216.5	22	8	23	5
Sparkling	15-7031-55216	49°48'51"	90°10'37"	1266	30.5	9.7	21	2
Sphene	15-4753-54261	48°59'19"	93°20'14"	505.8	38.5	8.4	89	4
Spoonbill	15-3697-57331	51°44'05"	94°53'08"	3229.1	45.8	7.9	33	5
Spring	15-5469-53527	48°19'37"	92°22'02"	74	30.2	17.2	36	7
Springpole	15-5600-56874	51°20'04"	92°08'56"	2861.3	35.1	6.3	72	6
Spruce (Little Mennin)	15-5612-54761	49°26'35"	92°09'21"	278.6	47	10.3	30	5
Squeers	15-6804-53764	48°31'07"	90°33'15"	384.1	33.6	11.5	42	7
Star	15-5322-54462	49°10'11"	92°33'30"	35.4	26	12.2	17	3
Stetham	15-6708-53937	48°40'32"	90°40'53"	222.8	27	7.7	33	4
Stewart (Lower Stewart, ELA L. 649)	15-4454-55169	49°48'16"	93°45'32"	189.7	29.6	8	36	3
Stewart (Upper Stewart, ELA L. 647)	15-4477-55182	49°48'58"	93°43'38"	265.5	19.2	6.7	30	4

Stoat	15-4614-54940	49°35'36"	93°32'23"	118.8	35	11.9	36	5
Store	15-5944-55905	50°27'41"	91°40'35"	469.5	16.5	4.9	74	4
Stork (Long-Legged)	15-4172-56262	50°47'20"	94°10'22"	6915.5	35.4	8.9	41	3
Stormy	15-5503-54708	49°23'03"	92°18'06"	2441.2	36.6		102	6
Strange	15-5108-54527	49°13'44"	92°51'05"	45.3	32	10.7	51	6
Strong	15-4851-54232	48°57'54"	93°12'23"	350.5	25.3	8.6	40	3
Sullivan	15-4696-54465	49°10'21"	93°25'00"	337.5	30.5	9.5	50	5
Sumach	15-4321-56018	50°33'59"	93°57'07"	743.3	27	12.8	68	4
Summit (Sasakwei)	15-5250-54810	49°29'00"	92°39'16"	126.3	19.3	11.3	119	4
Sun	16-5303-54739	49°25'07"	86°34'58"	171.2	36.9	12.3	111	4
Sunbeam	15-6885-53406	48°11'31"	90°27'48"	88	12	6.3	41	5
Sunbow	15-6843-53443	48°13'38"	90°31'04"	552.3	48.8	13	54	3
Sunday	15-6170-53296	48°06'37"	91°25'38"	419.6	14.9	7.7	60	5
Sunset	16-2818-53430	48°12'20"	89°56'11"	65.6	59.5	20.6	29	5
Sunshine	15-5219-54678	49°21'50"	92°41'22"	574.4	36.9	7.4	186	3
Superb	16-5004-55935	50°29'38"	86°59'57"	625.4	45	10.5	18	5
Surprise	16-3120-55738	50°17'21"	89°38'14"	366.4	29	8.6	41	4
Sword	15-3724-55393	49°59'30"	94°46'35"	364.6	70.1	19.7	33	8
Sydney	15-3980-56120	50°39'18"	94°26'46"	5868.2				
Syenite	16-4866-54161	48°53'44"	87°11'00"	30.0				
T	15-4278-55293	49°54'57"	94°00'45"	148.4	50	12.4	21	6
Tadpole	15-4785-54790	49°27'59"	93°17'59"	531	36	9.6	24	4
Talon River	15-3705-56160	50°40'45"	94°49'12"	1634.2	18.3	5.8	32	4
Taylor	15-5285-54705	49°23'17"	92°36'27"	291.6	24	9.9	30	4
T-Bone (ELA L.1002)	15-4357-55095	49°44'11"	93°53'33"	51.2	162.8	47.5	25	6
Teggau (ELA L. 228)	15-4531-55044	49°41'27"	93°39'01"	1352.1	33	8.4	32	4
Telescope	15-4007-56429	50°55'50"	94°24'33"	829	38	10.5	24	6
Tent	15-4255-55296	49°54'49"	94°02'22"	102.3	36.6	12.2	69	7
That Man	15-6276-53314	48°07'24"	91°17'20"	155.9	39.3	11.7	91	6
This Man	15-6332-53357	48°09'48"	91°12'26"	314.1	37	10.4	34	5
Thompson	15-5527-53591	48°23'04"	92°17'13"	954.8				
Three Moose (Esker)	16-5078-54343	49°03'50"	86°53'35"					
Thunder	15-5241-55137	49°46'40"	92°39'55"	1123	23.5	11.1	84	4
Thunder (Thundercloud)	15-5352-54684	49°22'05"	92°30'31"	176.2	38.4	17.9	44	5
Thundercloud (Thunder)	15-5352-54684	49°22'05"	92°30'31"	176.2	38.4	17.9	44	5

ONTARIO (Northwestern Ontario) (continued)

Lake name	OFIS code	Latitude	Longitude	Surface area (ha)	Maximum depth (m)	Mean depth (m)	Conductivity (µS/cm)	Secchi depth (m)
Tilly	15-6508-53878	48°37'36"	90°57'25"	409	31	9.5	26	5
Tinto	15-7029-53799	48°32'28"	90°15'17"	154.8	30	9.1	74	5
Titmarsh	15-6832-53585	48°21'13"	90°31'41"	959.5	49.4	13.6	30	4
Tompkins	15-4464-54275	49°00'26"	93°44'31"	23.7				
Toothpick (Bass)	15-4194-55517	50°06'55"	94°07'17"	296.9	21	7.9		4
Trapper (Pond)	15-5816-53498	48°17'53"	91°53'58"	76.9	25.3	11.5	26	4
Treelined	15-3993-55754	50°19'26"	94°24'54"	636.7				
Trout (Collins)	16-3277-55707	50°15'52"	89°25'15"	886.3	36	10.4	29	6
Trout	15-3634-55652	50°13'29"	94°54'56"	136.7				
Trout	15-3706-55191	49°48'43"	94°47'55"	242.9				
Trout	15-4185-55144	49°46'44"	94°07'53"	206.7	45.8	11	29	7
Trout	15-4651-55115	49°45'07"	93°29'08"	357	40.5	17.1	29	6
Trout (Paguchi)	15-6056-54920	49°34'05"	91°32'58"	2447.6	30.2	9.1	65	2
Trout (Little Moraine)	16-3657-54310	49°01'14"	88°50'16"	234.7	32	13.2	106	10
Trout (Otukamamoan)	15-5100-54230	48°57'35"	92°51'31"	5165.8	53.7	16.9	39	4
Trout	15-5485-53484	48°17'23"	92°20'32"	305.7	33	14.3	44	3
Trout Fly (Troutfly)	16-3692-57289	51°41'59"	88°53'29"	1306.5	24.4	8.5	159	7
Troutfly (Trout Fly)	16-3692-57289	51°41'59"	88°53'29"	1306.5	24.4	8.5	159	7
Trout Fly (Troutfly)	15-5396-55220	49°51'08"	92°26'56"	200.8	32	12	86	2
Troutfly (Trout Fly)	15-5396-55220	49°51'08"	92°26'56"	200.8	32	12	86	2
Tuck	15-6028-53374	48°10'47"	91°37'31"	287.2				
Tully	15-5974-55920	50°28'13"	91°37'29"	1088.4	29.9	9.5	68	5
Tunnel	16-3378-55707	50°16'19"	89°16'31"	618.9	21.5	6.6	29	3
Turtle	15-5770-54220	48°56'59"	91°56'57"	1153.4	20.1	7.8	23	3
Twinhouse	15-6587-53594	48°22'09"	90°51'15"	105.8	15	5.5	23	3
Underbrush (Echo)	15-4114-56359	50°52'29"	94°15'05"	398.7	25	7.9	24	2
Uneven	16-2947-55385	49°57'43"	89°51'43"	1156.9	48	10.7	29	3
Uphill	15-5175-54670	49°21'23"	92°45'32"	350.2	50	20.1	54	5
Upper Bow		49°04'07"	91°29'03"	345.5	16.8	5.3	23	2.7
Upper Hatchet	15-4065-56473	50°58'18"	94°19'55"	260.4	48.5	18	30	6

Lake	Code							
Upper Lawrence	15-4724-54519	49°13'16"	93°22'37"	613.8	41.1	12.3	42	6
Upper Manitou	15-5149-54725	49°24'05"	92°47'00"	5191.1	56	17.2	75	6
Upper Medicine Stone	15-4255-56385	50°53'40"	94°03'31"	1072.8	41.2	15.7	30	3
Upper Moosehide (Little Moosehide)	15-5886-54545	49°14'21"	91°47'00"	218.2	22.3	10.7	38	4
Upper Scotch (Scotch)	15-6345-54482	49°10'26"	91°09'16"	1291.6	27.8	9.2	26	3
Upper Stewart (Stewart, ELA L. 647)	15-4477-55182	49°48'58"	93°43'38"	265.5	19.2	6.7	30	4
Vale	16-3421-55611	50°11'00"	89°12'42"	215.8	27	9.5	60	3
Valhalla	15-4223-56844	51°18'36"	94°06'25"	675	22	7	30	4
Valjean	15-5835-54705	49°22'35"	91°50'39"	240.2	19.2	7	42	4
Van Nostrand	15-6419-54342	49°02'50"	91°03'12"	234.1	23.8	8.1	51	4
Vane	15-4817-54223	48°57'27"	93°15'02"	346.4	29.6	12.8	33	3
Vermilion	15-3884-55442	50°02'43"	94°33'29"	783	54	18.7	45	5
Vermilion (Big Vermilion)	15-5556-55428	50°02'11"	92°13'00"	8266.1	42.4	10.9	116	7
Veronica (ELA L. 256)	15-4385-55043	49°41'28"	93°51'10"	120.4	27.5	12.8	26	7
Vickers	15-5017-54457	49°09'18"	92°58'48"	1387.7	42.1	10.9	33	3
Victoria	15-6048-54975	49°37'28"	91°32'36"	936.9	28.1	9.5	42	3
Vista	15-4943-54343	49°03'48"	93°04'36"	562.5	68.9	13.1	62	5
Voltaire	16-3401-55086	49°42'51"	89°13'22"	274.5	67.1	17.9	78	4
Wabaskang	15-4880-55820	50°23'31"	93°10'08"	6192.7	28.7	7.6	153	5
Wabindon	15-7076-53374	48°09'29"	90°12'21"	183.9	22	7.4	60	5
Walker (Walkers)	16-4779-54086	48°49'55"	87°18'02"	150.7	32	6.8	45	4
Walkers (Walker)	16-4779-54086	48°49'55"	87°18'02"	150.7	32	6.8	45	4
Walkover	16-3657-55587	50°10'05"	88°52'51"	153.2	45	8.2	61	4
Walleye	15-4820-54820	49°29'43"	93°14'35"	304.3	25.9	9.6	27	4
Walmsley	15-5300-54789	49°27'35"	92°35'17"	201.3	20.7	11.2	113	7
Walotka	16-3483-54344	49°02'44"	89°04'31"	97	23	9.8	45	7
Walt	15-5904-54434	49°08'19"	91°45'26"	102.1	28	10.2	20	7
Walter	15-6058-53782	48°32'21"	91°34'29"	307.8				7
Wanipigow River	15-3491-56500	50°59'48"	95°09'12"					
Wapageisi	15-5472-54633	49°19'20"	92°21'02"	1651.8	48.8	11.1	45	3
Wapisipi	15-5136-58165	52°29'26"	92°48'28"	154.5				
Warclub	15-4475-54833	49°30'00"	93°43'26"	301	44.8	13.1	44	4
Washagomis	15-5225-56766	51°14'56"	92°40'29"	395.3	28.7	9.7	59	5

ONTARIO (Northwestern Ontario) (continued)

Lake name	OFIS code	Latitude	Longitude	Surface area (ha)	Maximum depth (m)	Mean depth (m)	Conductivity (µS/cm)	Secchi depth (m)
Washeibemaga (Fishhook)	15-5346-54737	49°25'03"	92°31'34"	509	25.9	8.5	60	4
Wasp	15-5956-54324	49°02'43"	91°41'43"	200.7	24.1	6.5	21	5
Watershed	15-6815-53784	48°32'09"	90°32'28"	172.4	12.2	4.2	44	4
Waweig	16-3500-55543	50°07'16"	89°05'35"	1210.8	42.7	12.4	58	4
Weikwabinonaw	15-6958-53562	48°19'50"	90°21'32"	1255.9				
Weiseieno	15-4793-54737	49°25'03"	93°17'11"	226.4	36.3	10.5	27	4
White Otter	15-5816-54402	49°06'16"	91°52'12"	8249	56.4	22.1	29	5
Whitesand (Selim)	16-4720-54109	48°51'09"	87°02'54"	269.4	42.7	10.9	38	4
Whitewater	16-3455-56307	50°48'35"	89°11'35"	8948.0				
Whitewater	15-5275-54864	49°31'57"	92°37'19"	110.2	17	10	102	7
Wiggins	16-3738-54079	48°48'44"	88°43'05"	21	18.9	10.7	53	
Wigwasan	16-3383-55532	50°06'14"	89°15'40"	764.5	32	11.2	26	4
Willard	15-4299-55210	49°50'35"	93°58'03"	575.7	42.7	14.7	36	4
Wilson	15-3558-55954	50°29'36"	95°01'54"	311.6	42.7	8.8	21	4
Windermere (ELA L. 615)	15-4417-55173	49°48'31"	93°48'30"	95.8	49.1	11.6	30	6
Windigoostigwan	15-6508-53978	48°42'12"	90°58'28"	865.6	59	16.7	34	3
Wine	15-4760-55892	50°27'25"	93°20'25"	1565.1	29.9	10.7	91	
Winkle	15-5010-54289	49°00'53"	92°59'11"	382	27.5	7.6	56	7
Winnange (Buzzard, ELA L.660)	15-4495-55118	49°45'08"	93°42'00"	2631.5	115.2	27.2	21	9
Woman	15-5179-56721	51°12'17"	92°44'38"	2035.2	33.2	10.3	65	4
Wonderland	15-4198-55480	50°04'48"	94°07'09"	893.6	95.8	45.6	37	9
Worth	15-3736-55400	50°00'10"	94°45'53"	114.6	29	9.9	44	5
Wreck	15-4185-55443	50°02'51"	94°08'14"	26.1	19	12		6
Wrist	15-3772-56364	50°52'12"	94°44'39"	326	63	18.3	23	5
Wyder	15-3895-55970	50°30'54"	94°33'16"	257.3	31	8.1	41	4
Yellowhammer	15-6896-53457	48°14'15"	90°26'42"	184.8	18.3	4.9	39	2

Lake name	OFIS code	Latitude	Longitude	Surface area (ha)	Maximum depth (m)	Mean depth (m)	Conductivity (µS/cm)	Secchi depth (m)
Yoke	15-4664-54431	49°08'43"	93°27'38"	633.8	29	9.3	63	5
Yucca	16-5268-54713	49°23'12"	86°37'45"	177.4	34.8	13.9	133	4
Yum Yum	15-6139-53394	48°12'21"	91°28'35"	82.1				
Zarn	15-6023-55439	50°02'44"	91°34'17"	1255.7	34.2	10.3	73	5
Zeemel	15-6768-58283	52°34'40"	90°23'09"	354.6	30.2	6.4	56	3

ONTARIO (South Central Ontario)

Lake name	OFIS code	Latitude	Longitude	Surface area (ha)	Maximum depth (m)	Mean depth (m)	Conductivity (µS/cm)	Secchi depth (m)
Allen	17-7159-49996	45°07'13"	78°15'14"	97.8	29.8	8.5	84	5
Alluring	17-7226-50740	45°47'10"	78°08'15"	30.6	36	13.8	40	4
Alsever	18-2667-50631	45°41'00"	77°59'33"	195	31.1	9.4	43	3
Angle (Fifteen Mile)	17-6596-50236	45°20'58"	78°57'56"	86.2	33.5	13.2	34	6
Animoosh	17-7225-50724	45°46'18"	78°08'18"	62.8				
Anson	17-6664-49909	45°03'13"	78°53'13"	74.5				
Anstruther (Eagle)	17-7210-49580	44°44'36"	78°12'26"	633.8	39	12.6	39	5
Arabis (Murphys)	18-2810-50600	45°39'41"	77°48'48"	68.8	20.7	7.8	45	8
Art (Spruce)	17-7007-49926	45°03'40"	78°27'12"	126.3	24.7	7.3	34	16
Ashby	18-3150-49957	45°05'40"	77°21'30"	259.5	36.6	4.3	53	6
Ashby White (Ashden)	18-3141-50021	45°08'54"	77°21'46"	137.6	23.8	9.9	190	6
Ashden (Ashby White)	18-3141-50021	45°08'54"	77°21'46"	137.6	23.8	9.9	190	6
Astrolabe	18-3569-50514	45°36'11"	76°50'02"	22.1				
Aylen	18-2779-50548	45°36'08"	77°50'10"	2014.6	72.8	27.2	51	7
Bacon	17-6281-50820	45°52'53"	79°20'58"	47.8				
Bailey	18-3147-49993	45°07'29"	77°21'24"	7.4	24	9.8	56	6
Balfour	18-2721-50571	45°37'52"	77°55'27"	81.4	25.3	8.7	50	4
Baptiste	18-2642-49981	45°05'55"	77°59'25"	2226.3	32	4.7		4
Bark	18-2782-50364	45°26'26"	77°50'24"	3798.5	87.5	24.3	46	5
Bark (Barker)	18-3129-49983	45°06'58"	77°22'59"	140.8	38.1	4.1	53	6
Barker (Bark)	18-3129-49983	45°06'58"	77°22'59"	140.8	38.1	4.1	53	6
Barnard	18-3058-49997	45°07'34"	77°28'12"	39.1	48.2	18.9	49	8

ONTARIO (South Central Ontario) (continued)

Lake name	OFIS code	Latitude	Longitude	Surface area (ha)	Maximum depth (m)	Mean depth (m)	Conductivity (µS/cm)	Secchi depth (m)
Bass (Dyson)	17-6059-50076	45°12'59"	79°39'06"	90.7	17.1	10		6
Bass (Limerick)	18-2934-49742	44°53'46"	77°36'13"	828.8	30	9.7	212	5
Bass (Steenburg)	18-2880-49669	44°49'41"	77°40'35"	280.7	20.1	5.3		4
Bass (Cherry)	17-7181-49488	44°39'41"	78°14'56"	22.4	18	5.5	60	4
Bass Haunt (Basshaunt)	17-6994-49995	45°07'23"	78°27'50"	48.8				
Basshaunt (Bass Haunt)	17-6994-49995	45°07'23"	78°27'50"	48.8				
Bat (Bear)	17-6861-50020	45°08'55"	78°37'57"	46.8				
Bay	17-6400-50405	45°30'20"	79°12'28"	136.0				
Bay	18-2749-49893	45°01'26"	77°51'29"	84.9	21.9	10.8	227	10
Bay (Big Mink)	17-7292-50149	45°15'25"	78°04'52"	196.2	23.8	6	48	4
Bear	17-6796-50231	45°20'29"	78°42'30"	94.7	36.6	9.4	34	6
Bear (Bat)	17-6861-50020	45°08'55"	78°37'57"	46.8				
Beaver (Mc Innis)	17-7149-49577	44°44'33"	78°17'09"	154.9				
Beech	17-6815-49938	45°04'37"	78°41'40"	136	26.8	7.1	59	4
Bella	17-6539-50341	45°26'45"	79°01'50"	327.9	40	16.2	37	5
Bernard	17-6257-50660	45°44'10"	79°23'00"	2057.7	47.9	15.1	55	6
Big Bear (Glamor)	17-7075-49817	44°57'39"	78°22'10"	194.7	22	10	102	7
Big Clear (Picard)	17-7069-49611	44°46'37"	78°23'23"	75.8	35.1	10.1	185	5
Big Clear (Clear)	18-3843-49346	44°33'28"	78°27'21"	169	61	20.4	76	8
Big Crow	17-6991-50783	45°49'45"	78°26'03"	440	27.1	8.2	40	4
Big Deer (Wahwashkesh)	17-5747-50629	45°43'22"	80°02'07"	1719.9	45.4	12.1	46	4
Big Gibson	17-7182-51266	46°15'36"	78°10'05"	78.3				
Big Hawk	17-6779-50032	45°09'31"	78°44'15"	365.2	59	18	32	7
Big Mair	18-3497-50033	45°10'09"	76°54'44"	20.1	22	8.8	167	5
Big Mink (Tyne)	17-6410-50941	45°59'15"	79°10'43"	49.4				
Big Mink (Bay)	17-7292-50149	45°15'25"	78°04'52"	196.2	23.8	6	48	4
Big Ohlmann (Rock)	18-3420-49908	45°03'17"	77°00'23"	32	42.1	20.5	71	4
Big Otter (Otter)	17-5810-50147	45°16'31"	79°58'15"	506.1	44.8	10.8	32	6
Big Porcupine	17-6867-50357	45°27'03"	78°36'24"	235.3	31.7	7.5	27	8
Big Rideau (Rideau)	18-4040-49580	44°46'33"	76°12'59"	6479	100	12.3	208	5

Lake								
Big Salmon (Salmon)	18-3807-49325	44°32'15"	76°30'06"	148.2	42.4	15.3	87	8
Big Trout	17-6849-50701	45°45'34"	78°37'27"	1518.7	31.1	8.3	32	5
Big Trout	17-6635-49768	44°55'47"	78°55'37"	204	32	9.2	29	3
Big Twin (Eighteen Mile)	17-6595-50273	45°22'58"	78°57'45"	135.7				
Biggar	17-6610-50894	45°56'29"	78°55'06"	381.5	32	9.7	38	3
Bigwind	17-6532-49909	45°03'21"	79°03'28"	106.5	36.6	7.8		
Birch	18-3782-49339	44°32'47"	76°32'29"	196.1				
Bitter (Hardwood)	17-6904-50047	45°10'17"	78°34'35"	42.9	13.1	6.1	32	5
Black	17-6720-49649	44°49'09"	78°49'28"	67.7	17.4	7.9	39	5
Black (Eyre)	17-6962-50147	45°15'37"	78°29'55"	65.6	28.7	11.8	150	4
Black (Fortescue)	17-7023-49679	44°50'18"	78°26'33"	78.7	17.4	9.7	47	4
Black (Gilbank)	17-5985-50163	45°17'43"	79°44'38"	18.8				
Blackstone	17-5878-50090	45°13'51"	79°52'52"	531.7	24.3	9.4	33	4
Blue	17-6804-50712	45°46'22"	78°40'43"	44.3				
Blue	17-5788-50141	45°16'42"	79°59'43"	4.0	10.6	3	37	7
Blue	17-6966-50033	45°09'28"	78°29'54"	22.8	31	16	212	13
Blue	18-3390-49726	44°53'27"	77°02'20"	28.6	26.2	7.5	191	4
Blue	18-3753-49258	44°28'41"	76°34'00"	10.4	22	9.4	29	8
Blue Chalk (Clear)	17-6620-50068	45°11'54"	78°56'19"	50.2	64.1	18.1	46	7
Bob	17-6745-49756	44°54'59"	78°47'02"	219.8	25.6	7.5		7
Bobs	18-3741-49489	44°41'02"	76°35'12"	2449.2	22	6.4		5
Bonnechere	17-6889-50367	45°27'53"	78°35'03"	104.5			27	
Bonnie	17-6367-49998	45°08'25"	79°15'40"	42.0	17.1	5.5	34	2
Boot	17-7193-50542	45°36'30"	78°11'27"	94.3	29.3	7.8	40	5
Booth	17-7191-50596	45°39'52"	78°11'53"	493.7	71.1	23.1	50	6
Boshkung	17-6789-49924	45°03'51"	78°43'49"	715.8				
Bottle	17-7154-49597	44°45'36"	78°16'41"	162.2				
Bow	17-6926-49688	44°50'54"	78°33'47"	50.3				
Brewer	17-7100-50518	45°35'25"	78°18'25"	35.4				
Brophy	17-5726-50407	45°31'05"	80°04'12"	18.5				
Bruce	17-6836-50496	45°34'41"	78°38'42"	25.9	12.8	6.3	32	4
Brule	17-6705-50558	45°38'15"	78°48'44"	84	27	11.5	29	4
Brule (Wensley)	18-3388-49903	45°02'59"	77°02'52"	570.6	56.4	22.2	125	7
Buck (Mc Cann)	17-6427-50610	45°41'24"	79°10'01"	96.0				
Buck	17-6254-50307	45°25'04"	79°23'40"	265.5	23.5	9.9	32	2

ONTARIO (South Central Ontario) (continued)

Lake name	OFIS code	Latitude	Longitude	Surface area (ha)	Maximum depth (m)	Mean depth (m)	Conductivity (µS/cm)	Secchi depth (m)
Buck	17-6573-50283	45°23'30"	78°59'28"	39.5	26.8	15.1	35	8
Buck	18-2851-50189	45°17'35"	77°44'24"	42.7				
Buck (Shabomeka)	18-3313-49730	44°53'27"	77°08'19"	267.5	32	12.4		5
Buck (South)	18-3856-49326	44°32'19"	76°26'19"	755	40.9	11.9	194	5
Buckshot (Indian)	18-3369-49845	44°59'31"	77°04'18"	439.3	29	9.7	98	4
Burns	18-3360-50201	45°19'03"	77°05'27"	143.2	36.6	11.3	146	7
Burnt Island	17-6840-50576	45°38'56"	78°38'47"	854.3	32.9	10.8	32	8
Burntroot	17-6801-50811	45°51'25"	78°40'16"	1085.7	25	6.5	32	4
Butt	17-6600-50619	45°41'32"	78°56'54"	456.2	49.1	16.6	27	7
Buzzard (Trout)	17-7216-49499	44°40'14"	78°12'16"	89.7				
Cache	17-6886-50457	45°32'26"	78°35'06"	287.2	32.6	6.7	37	6
Calumet	17-6895-50843	45°53'20"	78°33'30"	54.8	37.2	13.7	48	5
Camp	17-6637-50337	45°26'19"	78°54'31"	189.3	42.7	15	32	6
Camp (Little Mackie)	18-3457-49937	45°04'54"	76°57'32"	235.5				
Canisbay	17-6881-50495	45°34'30"	78°35'20"	152.2	25.9	8	34	6
Canning	17-6856-49789	44°56'41"	78°38'25"	244	22.3	6.7	80	5
Canoe	17-6782-50464	45°33'14"	78°43'03"	344.5	36.6	12.8	30	7
Canoe	18-3773-49386	44°35'10"	76°32'58"	291.4	47.3	22.9	202	9
Canonto	18-3587-49905	45°03'23"	76°47'45"	224				
Cardwell (Long)	17-6182-50207	45°19'53"	79°29'30"	203.6				
Carl Wilson	17-6852-50986	46°01'21"	78°36'29"	421.5	23.2	7.9	45	4
Carson	18-2845-50435	45°30'53"	77°45'39"	272.7	41.5	12.5	53	4
Cashel (Little Salmon)	18-2987-49765	44°54'54"	77°33'01"	170.1	28.8	7.1	251	3
Catchacoma	17-7116-49585	44°45'13"	78°19'19"	687.6	43.9	20.3	56	4
Catfish	17-6896-50895	45°56'20"	78°33'48"	640.7	22	5.2	33	4
Cauchon	17-6765-51031	46°03'34"	78°43'21"	235.9	41.8	12.4	38	4
Cauliflower (Clydawadka)	17-7147-50305	45°23'48"	78°15'24"	224.8				
Cavendish (Mc Ginnis)	17-7152-49568	44°44'04"	78°16'49"	23.5	25.9	10.6	155	4
Cedar	17-6953-50993	46°01'22"	78°28'11"	2543.1	58.5	14.1	39	4
Centre	18-3065-50915	45°57'05"	77°29'44"	71.8				

Name	ID	Lat	Long					
Centre	17-7326-49882	45°00'56"	78°02'59"	144	17.7	4.1	47	4
Chandos	18-2645-49676	44°49'08"	77°58'17"	1387.4	47.9	13.3	200	5
Charleston	18-4196-49317	44°32'24"	76°00'06"	2517.1	91.5	17.4	206	5
Charlotte	18-3098-50298	45°23'45"	77°25'52"	112.1	24	7.4	192	6
Cherry (Bass)	17-7181-49488	44°39'41"	78°14'56"	22.4	18	5.5	60	4
Chickaree	17-6931-50703	45°45'41"	78°30'58"	25.9	16.5	7	37	6
Clean (Clear)	17-6940-50133	45°14'45"	78°31'54"	160.4	43.3	14.8	35	12
Clear (Watt)	17-5942-51055	46°05'54"	79°46'52"	265.5				
Clear (Schamerhorn)	17-6358-50403	45°30'19"	79°15'39"	109.6				
Clear (Lake Clear)	18-3287-50341	45°26'28"	77°11'24"	1727.1	42.7	11.2	214	6
Clear (Solitaire)	17-6558-50282	45°23'31"	79°00'37"	122.4	29	10.9	33	7
Clear	17-5943-50124	45°15'49"	79°47'54"	220	44.5	16.3	29	10
Clear (Clean)	17-6940-50133	45°14'45"	78°31'54"	160.4	43.3	14.8	35	12
Clear (Blue Chalk)	17-6620-50068	45°11'54"	78°56'19"	50.2	22	9.4	29	8
Clear (Redstone)	17-6797-50055	45°10'53"	78°42'46"	96.3	29.6	11.8	29	9
Clear (Margaret)	17-6666-50011	45°08'43"	78°52'58"	60.5	32.3	11.2	30	7
Clear (Hudson)	17-7255-49943	45°04'00"	78°08'05"	73.7	23	11.1	33	8
Clear	17-6561-49891	45°02'25"	79°01'06"	100.3	61	20.4	76	8
Clear (Big Clear)	18-3843-49346	44°33'28"	76°27'21"	169	29.9	10.5	25	4
Clearwater	17-6389-49628	44°48'25"	79°14'37"	72.1	21.3	8	26	8
Clinto (Hardwood)	17-6671-50198	45°18'49"	78°52'00"	137.6				
Clydawadka (Cauliflower)	17-7147-50305	45°23'48"	78°15'24"	224.8				
Cobden (Muskrat)	18-3511-50603	45°40'57"	76°54'42"	1219.3				
Coghlan	18-2691-50329	45°24'44"	77°57'12"	88.9	16.2	7.3	39	5
Coon	17-7024-50461	45°32'30"	78°24'24"	24.9	15.3	7.3	30	3
Costello	17-7089-50526	45°35'50"	78°19'13"	33.5	18.9	8.7	54	3
Cotter (Otter)	18-3117-49924	45°03'45"	77°23'29"	305.6	29.3	8.6	52	4
Couchain	18-2871-50627	45°41'13"	77°44'02"	54.4				
Cowper (Spider)	17-5733-50114	45°15'30"	80°03'56"	480.5	34.5	8.1	32	6
Cox	17-7176-49514	44°41'02"	78°15'09"	90.2	15.6	4.4	44	6
Crab (Nunikani)	17-6779-50074	45°12'02"	78°44'00"	115.6	24.1	7.9	29	7
Cradle	17-6890-50379	45°28'18"	78°34'56"	16.8	33.6	11.2	24	5
Crane	17-5830-50066	45°12'36"	79°56'33"	560.9				
Crooked (Little Proudfoot)	17-6386-50621	45°42'02"	79°13'11"	20.4				
Crooked (Mayo)	18-2962-49902	45°02'13"	77°35'15"	196.5	32.9	8.2	206	10

ONTARIO (South Central Ontario) (continued)

Lake name	OFIS code	Latitude	Longitude	Surface area (ha)	Maximum depth (m)	Mean depth (m)	Conductivity (µS/cm)	Secchi depth (m)
Cross (Lyell)	18-2697-50311	45°23'51"	77°56'41"	200.5	36.6	11.1	40	4
Crotch	17-7265-50594	45°39'33"	78°05'40"	113.1	20.4	7.1	39	4
Crotchet	17-6623-49804	44°57'37"	78°56'32"	111.5	36	11.5	81	7
Crow	18-3719-49514	44°42'16"	76°37'04"	436.4	38.1	14.5	187	6
Crozier (Mc Fadden)	17-6685-50221	45°20'03"	78°50'53"	54.3	30.5	11.6	37	7
Crystal	18-3065-49990	45°07'25"	77°27'36"	55.3	35	9.8	32	4
Crystal	17-6995-49588	44°45'30"	78°28'47"	483.2	32.9	11.2	175	5
Daisy	17-6602-50581	45°39'40"	78°56'34"	128.4	24.4	6	28	6
Dark (Pusey)	17-7193-49920	45°02'54"	78°12'53"	56.7				
Darlingtons	17-5757-50240	45°22'01"	80°01'58"	21.2				
Davis	17-6812-49620	44°47'27"	78°42'37"	93.8	29.9	10.3	89	3
Deer (Deete)	17-6108-50740	45°48'46"	79°34'43"	368.3	15.6	4.6	30	2
Deer	17-7288-49909	45°02'14"	78°05'47"	189.7	20.1	8.3	45	6
Deete (Deer)	17-6108-50740	45°48'46"	79°34'43"	368.3	15.6	4.6	30	2
Delano	17-6877-50428	45°30'55"	78°35'52"	26.1	18.6	6.6	33	3
Delphis	17-7066-49983	45°06'35"	78°22'24"	26.4	28	10.1	29	7
Delta (Lower Beverley)	18-4097-49387	44°35'52"	76°08'12"	766.5	25.9	9.2		3
Desert	18-3738-49330	44°32'33"	76°35'11"	382	68.3	22.4	180	6
Devil	18-3847-49367	44°34'50"	76°27'24"	1061.5	44.5	14.4	164	6
Devil (Mephisto)	18-2966-49782	44°55'58"	77°34'40"	166.6	42	14.7	225	7
Devils (Lutterworth)	17-6707-49705	44°52'10"	78°50'22"	61.0				
Devil's (Salerno)	17-6985-49705	44°51'21"	78°29'01"	144.6	13.4	6.3	150	5
Diamond	18-3013-50268	45°22'06"	77°32'14"	80.8	39.6	13.8	98	4
Diamond	17-7334-49952	45°04'30"	78°02'05"	165.9	25	8.3	90	7
Dickey	18-2831-49628	44°47'14"	77°44'34"	209	54	18.4	140	5
Dickson	17-7171-50733	45°46'47"	78°12'19"	974.7	16.8	5.5	39	4
Dixon	18-2939-49797	44°56'35"	77°36'42"	30.3				
Doe	17-6239-50430	45°31'52"	79°24'49"	1221.6				
Dog	18-3938-49203	44°25'52"	76°20'11"	964	49.7	5.8	117	3
Dotty (Long)	17-6571-50360	45°27'42"	78°59'26"	152.8	29			

Drag	17-7042-49937	45°04'36"	78°24'26"	1018	55.5	20.1	66	5
Draper	18-3795-49263	44°28'56"	76°30'51"	96.0	16.4	9.9	41	5
Duck (Little Seguin)	17-5937-50257	45°22'45"	79°48'02"	86.9	17.1	10		
Dudman (Loon)	17-7067-49878	45°00'57"	78°22'35"	227.2	24.4	7	54	6
Dyson (Bass)	17-6059-50076	45°12'59"	79°39'06"	90.7	31.1	9.4	243	6
Eagle	17-6973-50011	45°08'16"	78°29'24"	234.6	39	12.6	39	5
Eagle (Wollaston)	18-2756-49689	44°50'28"	77°50'09"	363	35.4	12.1	115	5
Eagle (Anstruther)	17-7210-49580	44°44'36"	78°12'26"	633.8	23.2	6.6	84	6
Eagle	18-3651-49486	44°40'51"	76°42'00"	661.6				5
East (Two Islands)	17-7067-49934	45°03'54"	78°22'44"	57.2				
East Tower	17-6380-50862	45°55'03"	79°13'12"	8.8	29.9	6.7		6
Eels	17-7269-49741	44°53'10"	78°07'33"	945.8				
Eeyore	18-2714-51136	46°08'21"	77°57'35"	31.4	22	9.3	41	3
Effingham (Little Wes)	18-3122-49855	44°59'06"	77°21'36"	236.9				
Eighteen Mile (Big Twin)	17-6595-50273	45°22'58"	78°57'45"	135.7				
Elzevir	18-3143-49522	44°42'05"	77°20'40"	14				
Emsdale	17-6408-50417	45°30'58"	79°11'48"	65.9	18.3	11.2	37	2
Erables	17-6715-50963	46°00'03"	78°47'15"	383	32	10.8	114	6
Esson (Otter)	17-7149-49887	45°01'17"	78°16'15"	244.8			49	
Eustache	17-7235-50905	45°55'46"	78°06'44"	28.6	17.4	7.9	39	5
Eyre (Black)	17-6962-50147	45°15'37"	78°29'55"	65.6	23.8	5.2	41	4
Fairholme	17-5856-50474	45°34'35"	79°54'11"	66.3	69.5	22.1	44	3
Fairy	17-6425-50209	45°19'52"	79°10'39"	711.5	24.4	9.1	87	6
Faraday (Trout)	18-2699-49935	45°03'31"	77°55'17"	114	49.1	17.7	86	9
Farquhar	17-7196-49958	45°05'13"	78°12'00"	336.1				
Fassett	17-6471-50976	46°01'04"	79°05'58"	156.7				
Fatty's (Oxbow)	17-6587-50339	45°26'19"	78°58'00"	169.2	35.1	1.6	34	6
Fifteen Mile (Angle)	17-6596-50236	45°20'58"	78°57'56"	86.2	33.5	13.2	55	5
Fishtail	17-7203-50025	45°08'38"	78°11'58"	256	38.1	17.2		
Flat Iron	18-3139-50699	45°45'35"	77°23'34"	30.0				
Flaxman (Long)	17-5920-50206	45°20'03"	79°49'36"	62.7	27.5	11.4	22	5
Fletcher	17-6738-50244	45°21'11"	78°46'35"	255.8	23.2	7.9		
Fools	17-6920-50600	45°40'11"	78°32'10"	35	12.2	7.5	33	7
Fortescue (Black)	17-7023-49679	44°50'18"	78°26'33"	78.7	28.7	11.8	150	4
Found	17-6845-50465	45°32'59"	78°38'08"	11.3				

ONTARIO (South Central Ontario) (continued)

Lake name	OFIS code	Latitude	Longitude	Surface area (ha)	Maximum depth (m)	Mean depth (m)	Conductivity (µS/cm)	Secchi depth (m)
Fowke (Spring)	17-6023-50734	45°48′31″	79°40′46″	243.6	53.4	14.5	53	4
Fox	18-3164-49965	45°05′20″	77°19′47″	26.1				
Foys	18-2760-50735	45°46′49″	77°52′41″	73	18.3	7.1	43	5
Fraser	17-7088-50445	45°31′28″	78°19′33″	48.2	28	7.8	30	5
Galeairy	17-7110-50392	45°28′34″	78°17′59″	887.9	22.9	6.1	32	4
Garter	18-3771-49403	44°36′27″	76°32′52″	62.0				
Garvin	18-3363-50253	45°21′50″	77°05′22″	14.6				
Gilbank (Black)	17-5985-50163	45°17′43″	79°44′38″	18.8	17.4	9.7	47	4
Gillies	17-4736-50056	45°12′20″	81°20′07″					
Gin	18-2988-49924	45°03′31″	77°33′27″	101.2	13.7	4.6		5
Glamor (Big Bear)	17-7075-49817	44°57′39″	78°22′10″	194.7	22	10	102	7
Gliskning (Joe)	17-7180-50452	45°31′40″	78°12′30″	42.5				
Go Home	17-5910-49842	45°00′26″	79°50′43″	666	32.6	8.6	44	5
Godda	17-7155-50569	45°38′07″	78°14′05″	39.5	13.7	7	45	5
Gold	17-7162-49553	44°43′12″	78°16′18″	317.6	47	15.6	45	5
Gooderham (Pine)	17-7066-49762	44°54′39″	78°22′53″	87.4	18.9	6.8	34	5
Gouinlock	17-6815-51013	46°02′39″	78°39′10″	66.3	25.9	10.9	60	7
Gould	18-3747-49260	44°28′47″	76°34′21″	221.7	61.6	18	188	7
Grace	17-7178-49945	45°04′15″	78°13′59″	226.2	38.7	15.3	62	6
Grand	18-2820-50842	45°52′36″	77°48′06″	778.9	39.6	7.4	52	5
Grandview	17-6530-50069	45°12′03″	79°03′04″	74.4				
Granite	18-3533-49916	45°03′55″	76°51′44″	26.7	11.3	6.7	164	7
Graphite (Long)	17-6485-50660	45°44′05″	79°05′33″	68.4	14.3	4.8	27	6
Grass (Sweny)	17-6400-50597	45°40′51″	79°12′07″	134.7	36.6	11.1	32	5
Green	18-3492-50132	45°15′30″	76°55′15″	82.0				
Green (Mair)	18-3559-49965	45°06′30″	76°49′54″	49				
Green (Little Green)	18-3513-49801	44°57′35″	76°53′03″	29.1	21	12.5	300	8
Greenleaf	18-2715-50845	45°52′41″	77°56′55″	61.6	67.4	16.7	54	4
Green's	17-6434-50338	45°26′42″	79°09′57″	19.9				
Grimsthorpe	18-3108-49715	44°52′26″	77°23′35″	93.4	23	7.6	44	3

Name	Code	Latitude	Longitude					
Grindstone	18-3461-49861	45°00'59"	76°57'16"	164.2	20	7.5		5
Gull	17-6755-49685	44°51'14"	78°46'36"	995	49.1	16.5	140	5
Halfway (Long)	18-2967-50318	45°24'43"	77°35'50"	81.6	26.5			
Haliburton	17-7045-50074	45°11'56"	78°23'44"	1012.9	54.9	17.4	53	3
Halls	17-6774-49973	45°06'38"	78°44'47"	543.2	80.5	28.1	38	8
Hambone	17-6576-50594	45°40'21"	78°58'38"	40.9	28.7	9.6	25	10
Happy Isle	17-6941-50691	45°44'53"	78°30'42"	535.8	41.1	13	35	10
Harburn (Island)	17-7007-50095	45°12'45"	78°26'43"	24.7	17.7	5.4	31	4
Hardup (Poverty)	17-6582-50283	45°23'32"	78°58'43"	32.0				
Hardwood (Clinto)	17-6671-50198	45°18'49"	78°52'00"	137.6	21.3	8	26	8
Hardwood (Bitter)	17-6904-50047	45°10'17"	78°34'35"	42.9				
Harness	17-6928-50423	45°30'32"	78°31'52"	68.4	24.7	8.2	31	6
Harp	17-6460-50265	45°22'45"	79°08'05"	70.0				
Haskins	18-2820-50495	45°34'09"	77°47'40"	28.9	29	11.6	41	15
Havelock	17-6857-50178	45°17'34"	78°37'54"	196.3				
Hay	17-7203-50289	45°22'46"	78°11'31"	681.7	4.6	2.2	42	3
Head	17-6903-50431	45°31'08"	78°33'50"	87.4	15.5	4.8	33	5
High	17-6179-50100	45°14'09"	79°29'53"	64.2				
Hilly (Monck)	17-7281-49872	45°00'21"	78°06'26"	143.3	12.2	4.5	33	4
Hiram	17-6980-50572	45°38'28"	78°27'37"	40.7	13.4	5.5	37	3
Hogan	17-6942-50832	45°52'00"	78°29'59"	1303.2	31	6.7	40	4
Holland	18-2806-49968	45°05'36"	78°47'24"	29.8	26.8	11.8	61	9
Hollow (Kawagama)	17-6767-50182	45°17'39"	78°44'43"	2818.8	73.2	21.8	33	7
Horn	17-6179-50580	45°40'12"	79°29'18"	471.8	34.7	11.3	29	5
Horseshoe	17-5903-50174	45°18'27"	79°50'40"	370	20.4	6.5	53	6
Horseshoe	17-6831-49843	44°59'24"	78°40'37"	289.8	22.9	7		8
Hudson (Clear)	17-7255-49943	45°04'00"	78°08'05"	73.7	23	11.1	33	8
Hungry	17-6538-50445	45°32'22"	79°01'49"	15.1				
Hungry	18-3503-49630	44°48'28"	76°53'13"	255	32.3	8.1	83	2
Indian	18-3947-49383	44°35'34"	76°19'38"	269.2	25.9	10	159	4
Indian (Buckshot)	18-3369-49845	44°59'31"	77°04'18"	439.3	29	9.7	98	4
Indian (Proudfoot)	17-6372-50605	45°41'17"	79°14'21"	126.8	29	10.3	36	5
Island (Harburn)	17-7007-50095	45°12'45"	78°26'43"	24.7	17.7	5.4	31	4
Jack	17-7358-49530	44°41'35"	78°01'26"	1221.3	42.7	8.4		5
Jamieson	18-2885-49933	45°03'50"	77°41'11"	46.1	23.8			

ONTARIO (South Central Ontario) (continued)

Lake name	OFIS code	Latitude	Longitude	Surface area (ha)	Maximum depth (m)	Mean depth (m)	Conductivity (µS/cm)	Secchi depth (m)
Jeffrey	18-2733-49889	45°01'16"	77°52'26"	42.1	38.7	12.5	231	8
Joe	17-6781-50506	45°35'06"	78°42'47"	137.6	24.4	6.1	31	3
Joe (Gliskning)	17-7180-50452	45°31'40"	78°12'30"	42.5				
Joeperry (Wolf)	18-3188-49757	44°54'52"	77°17'53"	168.8	22.9	7.3	40	4
John	18-2816-49795	44°56'15"	77°46'04"	20.0				
Johnson	17-6865-50150	45°15'55"	78°37'20"	158	2.7			
Joseph (Lake Joseph)	17-5998-50031	45°10'35"	79°43'48"	5155.6	93	24.7	68	6
Kabakwa	17-6740-49978	45°06'51"	78°47'14"	113.6	17			
Kamaniskeg	18-2891-50334	45°25'50"	77°41'49"	2908.8	40	9.3	50	6
Kashagawigamog	17-6900-49852	44°59'41"	78°35'18"	817.3	39.7	13	79	6
Kashwakamak	18-3389-49691	44°51'32"	77°02'20"	1172	21.9			
Kawagama (Hollow)	17-6767-50182	45°17'39"	78°44'43"	2818.8	73.2	21.8	33	
Kearney	17-6629-51334	46°20'12"	78°52'58"	90.8				
Kearney	17-7001-50500	45°34'35"	78°26'09"	32	18.3	9.4	43	4
Kelly (North)	17-6868-50133	45°15'02"	78°37'11"	105.2				
Kenneth	17-6904-50423	45°30'36"	78°33'42"	39.7	17.4	5	36	3
Kennisis	17-6858-50095	45°12'57"	78°38'29"	1360		23.5		
Kilbourne	18-3232-49846	46°59'40"	77°14'33"	15.2				
Kimball	17-6818-50232	45°20'30"	78°40'39"	212.5	67.1	22.8	30	8
King	18-3135-49969	45°06'10"	77°22'10"	35.9				
Kingscote	17-7185-50089	45°12'51"	78°13'59"	213.7	26.2	7.5	38	6
Kioshkokwi	17-6636-51050	46°04'14"	78°53'26"	1126.8	45.8	12.5	36	4
Kirkwood	17-6914-50393	45°28'57"	78°33'03"	33.9	17.4	7.1	28	6
Kishkebus	18-3289-49744	44°54'16"	77°10'02"	83.2	32.9	12.3		3
Klaxon	17-6862-50035	45°09'43"	78°37'50"	41.0				
Knowlton	18-3717-49235	44°27'11"	76°36'48"	182	33.9	9.8	234	5
Koshlong	17-6983-49826	44°58'10"	78°29'00"	409	42	10.2		5
Kulas	18-2863-50387	45°28'18"	77°44'06"	1374.6	73.2			
Kushog	17-6741-49945	45°05'40"	78°47'34"	639.5	38.1	9.9	47	6
L. Bear (Little Glamor)	17-7076-49835	44°58'40"	78°22'01"	63.5	9.2	4.4	79	4

Lake								
L. Rideau (Upper Rideau)	18-3941-49484	44°40'54"	76°20'10"	1362.8	22	8.1	225	3
La Muir	17-6869-50771	45°49'27"	78°35'39"	756.8	39.3	10.4	38	7
Lake Clear (Clear)	18-3287-50341	45°26'28"	77°11'24"	1727.1	42.7	11.2	214	6
Lake Joseph (Joseph)	17-5998-50031	45°10'35"	79°43'48"	5155.6	93	24.7	68	6
Lake of Bays	17-6567-50131	45°15'14"	79°00'21"	6904.1	70.1	22.2	41	8
Lake of Many Islands	17-6022-50683	45°45'44"	79°41'46"	219.8	29	7	56	2
Lake of Two Rivers	17-6966-50501	45°34'45"	78°28'49"	292.2	37.5	15.4	38	3
Lake Rosseau (Rosseau)	17-6112-50030	45°10'26"	79°35'05"	6374.4	90	23.2	48	6
Lake St. Peter (St. Peter)	17-7333-50217	45°18'43"	78°01'23"	231.3	28	7.9	51	3
Lake Vernon (Vernon)	17-6337-50206	45°19'42"	79°17'38"	1505.1	37.2	12.7	41	3
L'Amable	18-2783-49885	45°01'05"	77°48'45"	178.4	35	23	125	8
Laurel (Laurie)	17-6847-51035	46°03'46"	78°36'36"	65.2	29	9.4	43	6
Laurie (Laurel)	17-6847-51035	46°03'46"	78°36'36"	65.2	29	9.4	43	6
Lavallee	18-2687-49805	45°56'34"	77°55'51"	81	32	14.5	236	5
Lavieille	17-7137-50829	45°52'05"	78°14'45"	2425.7	48.8	14.4	47	5
Lawrence	17-6935-50408	45°29'47"	78°31'18"	78.9	14.6	4.5	30	6
Len (Trout)	18-3086-50023	45°08'54"	77°26'06"	157.6	33.6	14.5	120	5
Leo	18-3986-49222	44°26'52"	76°16'26"	38.5				
Limerick (Bass)	18-2934-49742	44°53'46"	77°36'13"	828.8	30	9.7	212	5
Limestone	18-3556-50120	45°14'55"	76°50'21"	25.0				
Limestone	18-2974-49942	45°04'26"	77°34'25"	52.9	36.6	9.1	255	7
Lipsy	17-6857-50049	45°10'33"	78°38'36"	69	52			
Little	17-7233-51100	46°06'31"	78°06'42"	35.5	12	5.3	30	5
Little Anstruther (Serpentine)	17-7247-49674	44°49'35"	78°09'26"	18.2				
Little Black	18-3808-49320	44°31'34"	76°30'11"	21	22.6	8.6	64	4
Little Bob	17-6749-49712	44°52'21"	78°47'00"	73.8	14.3	6.7	49	6
Little Boshkung	17-6795-49895	45°02'17"	78°43'10"	126.8	48.5	14.3	43	6
Little Cauchon	17-6812-51026	46°03'15"	78°39'48"	269.3				
Little Clear	18-3804-49340	44°33'04"	76°30'14"	21.9				
Little Coon	17-6883-50339	45°26'07"	78°35'37"	38.6	16.5	7.2	29	5
Little Dickson	17-7196-50760	45°48'21"	78°10'11"	118.1	25	7.3	41	7
Little Flower (Mayflower)	17-6396-50276	45°23'24"	79°12'57"	8.1				
Little Glamor (L. Bear)	17-7076-49835	44°58'40"	78°22'01"	63.5	9.2	4.4	79	4
Little Green (Green)	18-3513-49801	44°57'35"	76°53'03"	29.1	21	12.5	300	8

ONTARIO (South Central Ontario) (continued)

Lake name	OFIS code	Latitude	Longitude	Surface area (ha)	Maximum depth (m)	Mean depth (m)	Conductivity (µS/cm)	Secchi depth (m)
Little Hawk (Pipikwabi)	17-6802-50021	45°09'17"	78°42'18"	344.3	93	31.5	32	9
Little Island	17-6847-50441	45°31'35"	78°38'12"	81.1	22.9	6.9	30	6
Little Joseph (Little Lake Joseph)	17-6034-50057	45°12'15"	79°41'09"	288.3	38	15.4	38	5
Little Kennisis	17-6890-50140	45°15'13"	78°35'39"	229.6				
Little Lake Joseph (Little Joseph)	17-6034-50057	45°12'15"	79°41'09"	288.3	38	15.4	38	5
Little Mackie (Camp)	18-3457-49937	45°04'54"	76°57'32"	235.5				
Little Manitouwaba	17-5996-50289	45°24'30"	79°43'39"	79.3	18.3	6.4	22	5
Little Mayo	18-2954-49923	45°03'21"	77°35'53"	25.3	17.4	6.6	330	6
Little Merrill	18-3109-49756	44°54'39"	77°23'42"	57				
Little Proudfoot (Crooked)	17-6386-50621	45°42'02"	79°13'11"	20.4				
Little Redstone	17-6910-50096	45°12'30"	78°34'16"	226.2	62.2	12.8	33	5
Little Salmon (Cashel)	18-2987-49765	44°54'54"	77°33'01"	170.1	28.8	7.1	251	3
Little Salmon	18-3793-49329	44°32'30"	76°31'05"	38				
Little Seguin (Duck)	17-5937-50257	45°22'45"	79°48'02"	86.9	16.4	9.9	41	5
Little Weagan	17-6362-50640	45°43'05"	79°14'58"	17.5				
Little Wes (Effingham)	18-3122-49855	44°59'06"	77°21'36"	236.9	22	9.3	41	3
Little Whitefish	17-5947-50141	45°16'32"	79°47'31"	376.2				
Little Wren	17-6686-50055	45°11'05"	78°51'14"	12.1				
Littledoe	17-6787-50546	45°37'24"	78°42'09"	120.1	20.7	5.6	28	5
Livingstone	17-6786-50252	45°21'38"	78°43'04"	189.1	36.6	12.7		
Lobster	17-7190-50465	45°32'21"	78°11'40"	132.9	25	5.6	36	4
Long (Graphite)	17-6485-50660	45°44'05"	79°05'33"	68.4	14.3	4.8	27	6
Long (Oliphant)	17-6412-50632	45°42'34"	79°11'15"	175.6	33.5	13.9	27	7
Long (Dotty)	17-6571-50360	45°27'42"	78°59'26"	152.8	29			
Long (Halfway)	18-2967-50318	45°24'43"	77°35'50"	81.6	26.5			
Long (Sixteen Mile)	17-6595-50248	45°21'38"	78°55'48"	135.7				
Long (Flaxman)	17-5920-50206	45°20'03"	79°49'36"	62.7	27.5	11.4	22	5
Long (Cardwell)	17-6182-50207	45°19'53"	79°29'30"	203.6				

Long (Tedious)	17-6906-50039	45°09'54"	78°34'27"	33.3	24.5	9.4	39	5
Long	17-7076-49910	45°02'39"	78°21'49"	84.5	17.4	8.1	35	3
Long	17-7242-49521	44°41'14"	78°10'41"	95.5	23	7.9	77	6
Long Mallory	18-3294-49850	44°59'43"	77°09'21"	63.3	22.9	5	35	3
Long Schooner	18-3443-49962	45°06'06"	76°58'40"	217	16.5	4.5	34	4
Longbow	17-6650-50692	45°45'31"	78°52'17"	100	26.2	9	31	7
Longer	17-6808-50747	45°48'00"	78°40'07"	170.3	14	4.9	35	7
Loon (Pevensey)	17-6387-50586	45°40'10"	79°13'12"	155.2	23			
Loon	17-6576-50317	45°25'22"	78°59'07"	31.5				
Loon (Purdy)	18-2858-50244	45°20'34"	77°43'59"	132.8				
Longline	17-6592-50126	45°15'03"	78°58'17"	27.2				
Loon (Dudman)	17-7067-49878	45°00'57"	78°22'35"	227.2				
Loon Call	17-7256-49580	44°44'31"	78°09'01"	90.2				
Lorimer	17-5806-50435	45°32'16"	79°58'30"	458.7	24.4	8	58	4
Lorne	17-6505-50931	45°58'47"	79°03'32"	99.3	25	6.4	31	6
Lost Dog (Namegos)	17-6478-50906	45°57'16"	79°05'28"	49.5	31.1	11.1	32	7
Loucks	17-7202-49511	44°40'53"	78°13'19"	34.5	18.6	7.5	36	6
Loughborough	18-3873-49226	44°26'48"	76°24'26"	738.1	38.4	14.5	271	3
Louie	17-6785-50286	45°23'23"	78°43'12"	33.2				
Louisa	17-6968-50383	45°28'20"	78°28'57"	567	61	17	34	6
Lower Beverley (Delta)	18-4097-49387	44°35'52"	76°08'12"	766.5	25.9	9.2		3
Lower Fletcher (Skin)	17-6699-50230	45°20'25"	78°49'37"	61	30.5	13.1	34	5
Lower Hay	17-7188-50317	45°24'35"	78°12'19"	419.9	32.3	9.3	42	3
Lower Paudash	17-7365-49835	44°58'16"	78°00'13"	461.3	21.3	6.6	109	6
Lower Raven	17-6440-50613	45°41'31"	79°09'03"	38.9	15.6	6.1	27	10
Loxley	17-6827-51039	46°03'57"	78°38'11"	51.6	29	10.6	36	6
Loxton	17-6384-50899	45°57'03"	79°12'51"	48.7				
Loyst	18-3441-49393	44°35'32"	76°57'50"	16.2	31.1	10.3		7
Lucky	18-3412-49900	45°02'49"	77°00'56"	99.6	27.8	11.9	110	9
Lutterworth (Devils)	17-6707-49705	44°52'10"	78°50'22"	61.0				
Lyell (Cross)	18-2697-50311	45°23'51"	77°56'41"	200.5	36.6	11.1	40	4
Lynx	17-6873-50890	45°55'55"	78°35'01"	94.4	25.3	8.4	37	5
Macdonald	17-6916-50119	45°14'11"	78°33'33"	137.8	39.6	10.8	32	10
Mackie	18-3435-49932	45°04'28"	76°59'05"	156.8	22.9	8.6		6
Madawaska	17-7050-50228	45°19'51"	78°23'03"	123.4				

ONTARIO (South Central Ontario) (continued)

444   *Boreal Shield Watersheds: Lake Trout Ecosystems in a Changing Environment*

Lake name	OFIS code	Latitude	Longitude	Surface area (ha)	Maximum depth (m)	Mean depth (m)	Conductivity (µS/cm)	Secchi depth (m)
Mair (Green)	18-3559-49965	45°06'30"	76°49'54"	49				
Manitou (Wilkes)	17-6548-50975	46°00'56"	79°00'02"	1423.2	33.6	12.2	39	3
Maple	17-6685-50984	46°01'10"	78°49'28"	324.8	18.3	6.7	34	3
Maple	17-5941-50243	45°22'04"	79°47'52"	201.9				
Maple (Ninatigo)	17-6833-49961	45°05'41"	78°40'26"	335.3	36.6	11.8	60	5
Marble	18-3309-49674	44°50'25"	77°08'17"	175.6	23.5			
Margaret (Clear)	17-6666-50011	45°08'43"	78°52'58"	60.5	32.3	11.2	30	7
Marsden (Marsh)	17-6961-50115	45°13'43"	78°30'13"	230.2	24	3.8	36	3
Marsh (Marsden)	17-6961-50115	45°13'43"	78°30'13"	230.2	24	3.8	36	3
Martineau	18-2838-50593	45°39'21"	77°46'28"	18.3				
Mary	17-6369-50115	45°14'23"	79°15'44"	1065.4	56.4	24.7	43	3
Mayflower (Little Flower)	17-6396-50276	45°23'24"	79°12'57"	8.1				
Mayo (Crooked)	18-2962-49902	45°02'13"	77°35'15"	196.5	32.9	8.2	206	10
Mazinaw (Upper Mazinaw)	18-3257-49760	44°56'14"	77°12'45"	1590.4	144.9	41.2	71	6
Mc Cann (Buck)	17-6427-50610	45°41'24"	79°10'01"	96.0				
Mc Cauley	17-7254-50470	45°32'32"	78°06'42"	132.9				
Mc Causland	18-3297-49700	44°51'56"	77°09'15"	36				
Mc Coy	17-5765-50143	45°16'48"	80°01'26"	65.7				
Mc Crae	17-5947-49760	44°55'34"	79°48'09"	208.6				
Mc Craney	17-6637-50481	45°34'08"	78°54'06"	361	61.3	14.2	28	5
Mc Fadden (Crozier)	17-6685-50221	45°20'03"	78°50'53"	54.3	30.5	11.6	37	7
Mc Garvey	17-6909-50339	45°26'05"	78°33'35"	70.4	20	4.8	23	4
Mc Gee	17-7244-49474	44°38'49"	78°10'12"	30.0				
Mc Ginnis (Cavendish)	17-7152-49568	44°44'04"	78°16'49"	23.5	25.9	10.6	155	4
Mc Innis (Beaver)	17-7149-49577	44°44'33"	78°17'09"	154.9				
Mc Intosh	17-6735-50597	45°40'17"	78°46'11"	316.8	29.6	7.4	24	5
Mc Kaskill	17-7300-50677	45°43'12"	78°02'20"	278	19.2	6.2	45	6
Mc Kenzie	17-7331-50275	45°21'48"	78°01'28"	311.6	27.8	9	46	4
Mc Master	18-2843-50369	45°27'14"	77°45'27"	113.1				
Mc Sourley (Tee)	18-2708-51225	46°13'09"	77°58'15"	33.0				

Lake	ID	Latitude	Longitude					
Mephisto (Devil)	18-2966-49782	44°55'58"	77°34'40"	166.6	42	14.7	225	7
Merchant	17-6920-50712	45°46'17"	78°31'46"	411.6	32	8.9	34	7
Mill	17-5784-50242	45°22'07"	79°59'54"	685.5				
Minden (Mountain)	17-6807-49834	44°58'28"	78°42'46"	319.4	31.4	13.5	54	9
Mink	17-6710-51031	46°03'45"	78°47'36"	234.5	45.8	15	38	3
Miskokway (Simikoka)	17-5602-50551	45°38'57"	80°13'40"	237.5	41.2	14.5	21	4
Miskwabi	17-7109-49919	45°03'01"	78°19'25"	263.8				
Mississagagon	18-3358-49706	44°52'31"	77°04'20"	507				
Mississagua	17-7123-49536	44°42'33"	78°19'04"	587.6	39.7	17.8	44	4
Monck (Hilly)	17-7281-49872	45°00'21"	78°06'26"	143.3	12.2	4.5	33	4
Monmouth	17-7216-49764	44°54'30"	78°11'32"	76.6	16			
Moore	17-7340-50356	45°26'15"	78°00'34"	308.1	12.8	3.9	46	4
Moore	17-6734-49625	44°47'56"	78°48'32"	182.5	24.4	7.7	58	2
Moose	17-6995-50027	45°09'02"	78°27'42"	289.4	43.9	16.6	57	4
Mosque (Mosquito)	18-3494-49857	45°00'38"	76°54'22"	138.2	34.4	7	89	6
Mosquito (Mosque)	18-3494-49857	45°00'38"	76°54'22"	138.2	34.4	7	89	6
Mountain (Minden)	17-6807-49834	44°58'28"	78°42'46"	319.4	31.4	13.5	54	9
Mountain	18-2646-49772	44°54'42"	77°58'55"	25.6				
Murphys (Arabis)	18-2810-50600	45°39'41"	77°48'48"	68.8	20.7	7.8	45	8
Murray	18-3811-50023	45°09'59"	76°30'46"	22.4				
Muskrat (Cobden)	18-3511-50603	45°40'57"	76°54'42"	1219.3				
Namakootchie	17-6744-50426	45°31'14"	78°46'18"	18.2				
Namegos (Lost Dog)	17-6478-50906	45°57'16"	79°05'28"	49.5	31.1	11.1	32	7
Napier	18-3739-50059	45°11'51"	76°36'18"	14.6				
Nepawin	17-6970-50733	45°47'16"	78°27'52"	34.9	17	5.8	38	6
Ninatigo (Maple)	17-6833-49961	45°05'41"	78°40'26"	335.3	36.6	11.8	60	5
North	17-6403-50645	45°43'19"	79°11'44"	52.6	41.1	14.9	28	9
North (Kelly)	17-6868-50133	45°15'02"	78°37'11"	105.2				
North Depot	17-7101-51050	46°04'20"	78°16'53"	116.1	19.8	7.9	44	4
North Grace	17-6940-50350	45°26'41"	78°31'07"	102.7	25.3	6.5	29	5
North Pigeon	17-6756-49720	44°52'54"	78°46'37"	50.1	39			
North Tea (Waskigomog)	17-6524-50894	45°56'25"	79°02'48"	1364	31.1	10	37	3
Nunikani (Crab)	17-6779-50074	45°12'02"	78°44'00"	115.6	24.1	7.9	29	7
Oblong	17-7017-50059	45°10'46"	78°25'53"	91.1	27.5	10	51	8
Oliphant (Long)	17-6412-50632	45°42'34"	79°11'15"	175.6	33.5	13.9	27	7

ONTARIO (South Central Ontario) (continued)

Lake name	OFIS code	Latitude	Longitude	Surface area (ha)	Maximum depth (m)	Mean depth (m)	Conductivity (µS/cm)	Secchi depth (m)
Opeongo	17-7048-50649	45°42'55"	78°22'06"	5921.9	49.4	14.6	43	6
Oriole	18-3587-50119	45°14'54"	76°48'02"	14.8				
Otter (Big Otter)	17-5810-50147	45°16'31"	79°58'15"	506.1	44.8	10.8	32	6
Otter (Cotter)	18-3117-49924	45°03'45"	77°23'29"	305.6	29.3	8.6	52	4
Otter (Esson)	17-7149-49887	45°01'17"	78°16'15"	244.8	32	10.8	114	6
Otter	18-4111-49593	44°47'01"	76°07'26"	602.2	36.6	10	270	3
Otter (Shark)	17-7217-49461	44°38'10"	78°12'16"	31.2	17.4	7		6
Otterslide	17-6876-50640	45°42'25"	78°35'36"	304.1	20.1	6.2	35	3
Owl	17-6846-50524	45°36'09"	78°37'57"	47.7	18.9	6.3	34	6
Oxbow (Fatty's)	17-6587-50339	45°26'19"	78°58'00"	169.2	35.1	1.6		
Oxtongue	17-6628-50250	45°21'10"	78°55'00"	249.3	26.8	8.9	35	4
Palmerston (Trout)	18-3549-49862	45°00'41"	76°50'59"	539.1	56.4	20.4	147	6
Papineau	18-2795-50249	45°20'28"	77°48'54"	830.8	61	19.4	94	6
Paudash	17-7323-49829	44°57'51"	78°03'25"	754.7	46.4	12	110	5
Paugh	18-2894-50519	45°35'25"	77°41'59"	710.5	51.8	14.1	52	4
Pen	17-7049-50366	45°27'10"	78°22'25"	378.6	34.8	9.2	33	4
Pen (Peninsula)	17-6488-50221	45°20'36"	79°06'20"	864.8	34.1	9.7	56	4
Peninsula (Pen)	17-6488-50221	45°20'36"	79°06'20"	864.8	34.1	9.7	56	4
Percy (Pine)	17-7080-50094	45°12'33"	78°21'05"	341				
Pevensey (Loon)	17-6387-50586	45°40'10"	79°13'12"	155.2	26.2	9	31	7
Phipps	17-6902-50388	45°28'43"	78°33'58"	22.9	18.6	6	27	5
Picard (Big Clear)	17-7069-49611	44°46'37"	78°23'23"	75.8	35.1	10.1	185	5
Piglet	17-7281-51157	46°09'31"	78°02'42"	17.7				
Pine (Percy)	17-7080-50094	45°12'33"	78°21'05"	341				
Pine	17-6521-49919	45°03'54"	79°04'17"	77.1	22.6	9.3		
Pine (Silent)	17-7321-49771	44°54'44"	78°03'41"	118.3	23.1	6.5	33	5
Pine (Gooderham)	17-7066-49762	44°54'39"	78°22'53"	87.4	18.9	6.8	34	5
Pinetree	17-7094-50474	45°33'02"	78°19'02"	106.6	26.5	6.9	27	6
Pipikwabi (Little Hawk)	17-6802-50021	45°09'17"	78°42'18"	344.3	93	31.5	32	9
Pog	17-6987-50494	45°34'26"	78°27'11"	27.4	16.5	5.2	39	4

Name	Code	Latitude	Longitude					
Portage	17-5942-50071	45°12'52"	79°48'01"	97.5	25.9	12.1	56	8
Potspoon	18-3743-49398	44°36'09"	76°35'00"	88.2	12.2	2.9	28	4
Potter	17-6733-50536	45°36'56"	78°46'29"	83				
Poverty (Hardup)	17-6582-50283	45°23'32"	78°58'43"	32.0				
Prospect	17-6470-49834	44°59'27"	79°08'06"	69.4				
Prottler	17-7086-50365	45°27'13"	78°19'54"	48	23.5	8.1	32	5
Proudfoot (Island)	17-6372-50605	45°41'17"	79°14'21"	126.8	29	10.3	36	5
Proulx	17-7024-50723	45°46'37"	78°23'46"	383.5				
Purdy (Loon)	18-2858-50244	45°20'34"	77°43'59"	132.8	23			
Pusey (Dark)	17-7193-49920	45°02'54"	78°12'53"	56.7				
Radiant	17-7100-50965	45°59'39"	78°17'19"	643.4	36.9	7.7	40	6
Ragged	17-6839-50376	45°28'41"	78°38'12"	629.4	37.8	5.8	31	6
Raglan (Raglan White)	18-3030-50161	45°16'21"	77°30'40"	136.8	20.1			
Raglan White (Raglan)	18-3030-50161	45°16'21"	77°30'40"	136.8	20.1			
Rain	17-6619-50545	45°37'52"	78°55'07"	168.4	23.5	5.5	27	5
Rainy	18-3160-49741	44°53'56"	77°19'47"	560.7	44			
Rankin	17-5855-50156	45°17'27"	79°54'32"	137.8				
Rathbun	17-7213-49618	44°46'38"	78°12'11"	115.2				
Raven	17-6686-50085	45°12'22"	78°51'06"	560.2	41.8	8.1	28	3
Rebecca	17-6537-50324	45°25'49"	79°02'04"	210.5	29	7.9	40	5
Red Chalk	17-6612-50058	45°11'20"	78°56'56"	58.3	32	10.9	30	4
Red Horse	18-4143-49324	44°32'29"	76°04'57"	301.7	37.2	10.2	263	3
Red Pine	17-6806-50080	45°12'23"	78°42'00"	383.9	40	10.6	32	8
Redstone	17-6934-50060	45°10'25"	78°32'09"	1130.3	82.4	21.9	38	8
Redstone (Clear)	17-6797-50055	45°10'53"	78°42'46"	96.3	29.6	11.8	29	9
Reid	18-3476-49924	45°04'09"	76°56'04"	102.8	19.8	7.5	208	5
Rideau (Big Rideau)	18-4040-49580	44°46'33"	76°12'59"	6479	100	12.3	144	5
Robinson	18-2854-49764	44°54'40"	77°43'05"	29.1	34.8	6.6	47	5
Robitaille	18-2764-50628	45°41'00"	77°52'12"	147.6	18.9	6.5	36	5
Rock	17-7039-50418	45°30'30"	78°23'50"	509.1	30.5	7.9	71	4
Rock (Big Ohlmann)	18-3420-49908	45°03'17"	77°00'23"	32	42.1	20.5	27	4
Rosebary	17-6616-50692	45°45'36"	78°55'18"	189.4	17.4	6		4
Ross	17-7014-50111	45°13'35"	78°26'03"	24.6				
Rosseau (Lake Rosseau)	17-6112-50030	45°10'26"	79°35'05"	6374.4	90	23.2	48	6
Round	18-3030-50572	45°38'21"	77°31'23"	3068.8	54.9	13.2	66	4

ONTARIO (South Central Ontario) (continued)

Lake name	OFIS code	Latitude	Longitude	Surface area (ha)	Maximum depth (m)	Mean depth (m)	Conductivity (µS/cm)	Secchi depth (m)
Round Island	17-7183-50679	45°43'57"	78°11'43"	156.3	17	5.4	35	5
Round Schooner	18-3434-49981	45°07'05"	76°59'20"	192.2	32	15.2	119	7
Salerno (Devil's)	17-6985-49705	44°51'21"	78°29'01"	144.6	13.4	6.3	150	5
Salmon	17-7019-49656	44°49'14"	78°26'44"	171.8	30.5	11.3	180	8
Salmon (Big Salmon)	18-3807-49325	44°32'15"	76°30'06"	148.2	42.4	15.3	87	8
Sand	17-6425-50540	45°37'35"	79°10'18"	568.2	59.4	22.8	32	3
Sawyer	17-6636-50569	45°38'54"	78°54'02"	47.5	19	7.8	27	8
Schamerhorn (Clear)	17-6358-50403	45°30'19"	79°15'39"	109.6				
Sec	18-3016-50772	45°49'12"	77°33'26"	165.3	13	5.1	44	4
Serpentine (Little Anstruther)	17-7247-49674	44°49'35"	78°09'26"	18.2				
Shabomeka (Buck)	18-3313-49730	44°53'27"	77°08'19"	267.5	32	12.4		5
Sharbot (West Sharbot)	18-3668-49582	44°46'06"	76°41'14"	705.6	32	8.1	188	6
Shark (Otter)	17-7217-49461	44°38'10"	78°12'16"	31.2	17.4	7		6
Sheldon	17-6703-49682	44°50'46"	78°50'29"	52.9				
Sherborne (Trout)	17-6737-50047	45°10'32"	78°47'23"	251.8	35.1	9.5	31	7
Shiner	18-3525-50118	45°14'47"	76°52'44"	4.5				
Shirley	17-7239-50633	45°41'05"	78°07'02"	480.9	26.8	7.4	37	4
Silent (Pine)	17-7321-49771	44°54'44"	78°03'41"	118.3	23.1	6.5	33	5
Silver (Tiffin)	17-5943-50084	45°13'28"	79°47'56"	74.7				
Silver	18-3735-49648	44°49'39"	76°35'57"	249.0				
Silver (Triangle)	17-7178-49498	44°40'06"	78°15'06"	27.2	14	4.7	77	5
Simikoka (Miskokway)	17-5602-50551	45°38'57"	80°13'40"	237.5	41.2	14.5	21	4
Simpson	18-3110-50029	45°09'24"	77°24'12"	25.1	18.6	10.4	200	6
Sisco	17-6492-50918	45°57'53"	79°04'30"	35.1	31.1	13.3	30	8
Six Mile	17-6000-49730	44°54'14"	79°43'52"	1418.5				
Sixteen Mile (Long)	17-6595-50248	45°21'38"	78°57'48"	135.7				
Skeleton	17-6216-50117	45°15'08"	79°27'59"	2155.5	64.7	28.9	40	9
Skin (Lower Fletcher)	17-6699-50230	45°20'25"	78°49'37"	61	30.5	13.1	34	5
Skootamatta	18-3220-49677	44°50'11"	77°15'04"	742				

Lake	UTM code	Latitude	Longitude					
Slipper	17-6809-50176	45°17'18"	78°41'39"	51.7	34	14.2	26	5
Smoke	17-6811-50426	45°30'45"	78°40'50"	607.1	54.9	16.2	36	6
Smyth (Surprise)	17-6357-50902	45°57'15"	79°14'57"	157.5				
Solitaire (Clear)	17-6558-50282	45°23'31"	79°00'37"	122.4	29	10.9	33	7
Source	17-6834-50477	45°33'31"	78°39'13"	271.1	44.2	8	33	8
South (Buck)	18-3856-49326	44°32'19"	76°26'19"	755	40.9	11.9	194	5
South Wildcat	17-6894-50214	45°19'20"	78°35'02"	96.7	58	14.3	68	5
Soyers	17-6883-49881	45°01'18"	78°36'22"	330.6	48.8	8.1	32	6
Spider (Cowper)	17-5733-50114	45°15'30"	80°03'56"	480.5	34.5	14.5	53	4
Spring (Fowke)	17-6023-50734	45°48'31"	79°40'46"	243.6	53.4	4.3	300	4
Spring	18-3060-50034	45°09'34"	77°28'08"	46.9	23.2	7.3	34	16
Spruce (Art)	17-7007-49926	45°03'40"	78°27'12"	126.3	24.7	11	48	4
St. Andrews	18-2913-50800	45°50'26"	77°41'12"	89.8	24.4	15.8	34	8
St. Nora	17-6706-50023	45°09'15"	78°49'46"	263.9	39	7.9	51	3
St. Peter (Lake St. Peter)	17-7333-50217	45°18'43"	78°01'23"	231.3	28	9.3	32	3
Star	17-5975-50201	45°19'47"	79°45'22"	158	22.6	5.3		4
Steenburg (Bass)	18-2880-49669	44°49'41"	77°40'35"	280.7	20.1			
Stewart	17-5971-49992	45°08'31"	79°45'51"	152.3	21.3	5.3	44	5
Stocking	17-6818-50162	45°16'39"	78°40'56"	63.7	15.3	8.6	50	5
Stoplog	17-7195-49479	44°39'11"	78°13'50"	48.3	24.4	15.4	63	4
Stormy	17-7043-49833	44°58'35"	78°24'48"	78.1	42.7	11.2	32	7
Stubbs (Trout)	18-2856-50410	45°29'20"	77°44'23"	137	19.5	8	50	
Sucker	17-6035-50114	45°15'04"	79°40'50"	103.6	18.3	5.8	26	
Sucker	17-7171-49602	44°45'52"	78°15'18"	150	17.1		29	
Sugar	17-5971-50251	45°22'35"	79°45'38"	151.1			35	
Sunbeam	17-6792-50595	45°39'36"	78°41'38"	81.3				
Sunday	17-7025-50524	45°35'48"	78°24'20"	45.9	13.1	4		5
Surprose (Smyth)	17-6357-50902	45°57'15"	79°14'57"	157.5				
Swan	17-6789-50403	45°29'42"	78°42'38"	81.4				3
Sweny (Grass)	17-6400-50597	45°40'51"	79°12'07"	134.7	19.2	6.3	29	6
Sylvia	17-7074-50446	45°31'37"	78°20'38"	37.3	36.6	11.1	32	5
Tallan	17-7329-49698	44°50'45"	78°03'09"	45.1	12.2	3.7	30	5
Tanamakoon	17-6855-50457	45°32'31"	78°37'42"	106.4	26	8.6	195	4
Tasso	17-6615-50354	45°27'32"	78°56'08"	170	25	8.1	36	5
Tea	17-6359-50589	45°40'20"	79°15'20"	34.7	22.3	6.2	25	3

ONTARIO (South Central Ontario) (continued)

Lake name	OFIS code	Latitude	Longitude	Surface area (ha)	Maximum depth (m)	Mean depth (m)	Conductivity (µS/cm)	Secchi depth (m)
Tea	17-6768-50415	45°30'33"	78°44'02"	156.1	14.8	5.4	31	3
Tedious (Long)	17-6906-50039	45°09'54"	78°34'27"	33.3				
Tee (Mc Sourley)	18-2708-51225	46°13'09"	77°58'15"	33.0				
Tepee	17-6774-50515	45°35'44"	78°43'30"	85	25	6.4	30	5
Thanet	18-2809-49622	44°46'53"	77°46'04"	109.2	26.8	8.7	110	6
Thirty Island	18-3730-49363	44°34'16"	76°35'56"	196	32			
Three Legged	17-5769-50127	45°15'58"	80°01'16"	177.7	40.6	14.3	29	6
Three Mile	17-6625-50948	45°59'20"	78°54'07"	408.4	36.6	9.8	34	5
Tiffin (Silver)	17-5943-50084	45°13'28"	79°47'56"	74.7				
Tim	17-6534-50681	45°45'17"	79°01'33"	182.6	19.8	6.8	22	5
Timberwolf	17-6711-50607	45°40'53"	78°48'16"	167.8	20.4	7	32	3
Tom Thompson	17-6768-50553	45°37'48"	78°43'45"	144	27.8	8.6	24	4
Triangle (Silver)	17-7178-49498	44°40'06"	78°15'06"	27.2	14	4.7	77	5
Trout	18-2809-51095	46°06'20"	77°50'04"	30.0				
Trout	17-6812-50651	45°43'22"	78°40'07"	559.6	13.1	4	29	4
Trout	17-5651-50482	45°35'09"	80°09'55"	293.3				
Trout (Stubbs)	18-2856-50410	45°29'20"	77°44'23"	137	42.7	15.4	63	4
Trout	17-5801-50304	45°25'25"	79°58'39"	225.8	27.1	6.8	24	6
Trout (Sherborne)	17-6737-50047	45°10'32"	78°47'23"	251.8	35.1	9.5	31	7
Trout (Len)	18-3086-50023	45°08'54"	77°26'06"	157.6	33.6	14.5	120	5
Trout (Faraday)	18-2699-49935	45°03'31"	77°55'17"	114	24.4	9.1	87	6
Trout (Palmerston)	18-3549-49862	45°00'41"	76°50'59"	539.1	56.4	20.4	147	6
Trout (Buzzard)	17-7216-49499	44°40'14"	78°12'16"	89.7				
Turtle	17-5994-50178	45°18'34"	79°43'34"	121	14.6	5.6	40	5
Twelve Mile	17-6807-49881	45°01'33"	78°42'10"	336.5	27.5	12	52	5
Two Islands (East)	17-7067-49934	45°03'54"	78°22'44"	57.2	23.2	6.6	84	5
Tyne (Big Mink)	17-6410-50941	45°59'15"	79°10'43"	49.4				
Upper Mazinaw (Mazinaw)	18-3257-49760	44°56'14"	77°12'45"	1590.4	144.9	41.2	71	6
Upper Raven	17-6432-50623	45°42'06"	79°09'39"	24.0				

Lake	Map ref	Latitude	Longitude					
Upper Rideau (L. Rideau)	18-3941-49484	44°40'54"	76°20'10"	1362.8	22	8.1	225	3
Valiant	17-7236-51207	46°12'19"	78°06'09"	38.6	19.8	6	33	4
Vernon (Lake Vernon)	17-6337-50206	45°19'42"	79°17'38"	1505.1	37.2	12.7	41	3
Victoria	17-7326-50560	45°37'15"	78°00'59"	891.6	47.6	14.8	41	7
Vireo	18-2681-50678	45°43'46"	77°58'58"	84	20.4	6.5	38	7
Wabun	18-3562-50096	45°13'36"	76°49'57"	44.5	26.8	13.3	170	7
Wadsworth	18-2982-50348	45°26'04"	77°34'33"	198.8	26.2	7.2	99	4
Wahwashkesh (Big Deer)	17-5747-50629	45°43'22"	80°02'07"	1719.9	45.4	12.1	46	4
Wanamaker	18-2982-49952	45°04'56"	77°33'50"	74.5	13.1	3.9		6
Waseosa	17-6350-50291	45°24'16"	79°16'28"	155.7			37	
Waskigomog (North Tea)	17-6524-50894	45°56'25"	79°02'48"	1364	31.1	10		3
Waterloo	17-7166-51172	46°10'32"	78°11'37"	181.7	22.3	7.6		4
Watt (Clear)	17-5942-51055	46°05'54"	79°46'52"	265.5				
Wensley (Brule)	18-3388-49903	45°02'59"	77°02'52"	570.6	56.4	22.2	125	7
Weslemkoon	18-3085-49890	45°01'38"	77°25'27"	1955.4	54.9	9.3	46	5
West Sharbot (Sharbot)	18-3668-49582	44°46'06"	76°41'14"	705.6	32	8.1	188	6
Whiskyjack	17-6785-50835	45°53'04"	78°42'25"	34.9			43	
White Partridge	17-7244-50795	45°50'03"	78°06'11"	574.4	47.3	15.3	50	6
White Trout (White Trout Bay)	17-6713-50125	45°14'51"	78°49'02"	2967.9				
White Trout Bay (White Trout Bay)	17-6713-50125	45°14'51"	78°49'02"	2967.9				
Whitebirch	17-6692-51029	46°03'34"	78°48'41"	80.8	27.5	8.1	37	3
Whitefish	17-7010-50468	45°32'37"	78°25'02"	205.6	24.4	6.2	39	4
Whitefish	17-5960-50159	45°17'28"	79°46'44"	294.6	33.6	11.2	42	7
Whitefish	18-3101-49752	44°54'23"	77°24'18"	56.1				
Whitegull	17-6985-50590	45°39'29"	78°27'05"	72.1				
Whyte	18-2959-49931	45°03'47"	77°35'26"	37.9	38.1	13.5	410	6
Wilbermere	17-7201-49873	45°00'25"	78°12'23"	40.0				
Wilkes (Manitou)	17-6548-50975	46°00'56"	79°00'02"	1423.2	33.6	12.2	39	3
Wilkins	18-2727-50635	45°41'15"	77°55'01"	237.8	35.1	11.9	39	5
Windigo	17-7092-51120	46°07'35"	78°17'23"	160.7	22	10	38	4
Wolf	17-5717-50405	45°30'58"	80°04'54"	62.6				
Wolf	17-6911-50149	45°16'45"	78°34'50"	53.4				

ONTARIO (South Central Ontario) (continued)

Lake name	OFIS code	Latitude	Longitude	Surface area (ha)	Maximum depth (m)	Mean depth (m)	Conductivity (µS/cm)	Secchi depth (m)
Wolf (Joeperry)	18-3188-49757	44°54'52"	77°17'53"	168.8	22.9	7.3	40	4
Wollaston (Eagle)	18-2756-49689	44°50'28"	77°50'09"	363	31.1	9.4	243	5
Woodland	17-6146-49730	44°54'10"	79°32'54"	81.7				
Wright	17-7108-50705	45°45'29"	78°17'18"	77.7				
Young	17-6139-50064	45°12'15"	79°32'56"	109.4	24			

QUEBEC

Lake name	Latitude	Longitude	Surface area (ha)	Maximum depth (m)	Mean depth (m)	Conductivity (µS/cm)	Secchi depth (m)
Abattis, Des	46°21'33"	76°36'00"	1114.0	40.0		40	5
Achigan	46°14'53"	77°06'15"	673.0	30.0		32	5
Achigan	45°56'26"	73°58'28"	526.0	27.0		23	5
Achigan, De l'	46°23'	75°47'	153.0				
Adams	47°45'29"	73°00'30"	184.0			22	
Aéroplane, De l'	46°17'53"	75°06'27"	16.0			14	5
Aigle, À l'	46°46'27"	74°43'48"	88.0				
Aigle, De l'	46°06'36"	75°27'31"	98.0			91	7
Aigles, Des	49°06'08"	71°43'28"	185.0				
Aigles, Des	47°01'18"	73°26'29"	381.1	49.6	13.6	13	8
Ailleboust, D'	49°34'35"	71°27'55"	986.0				
Albert (Bob Grant)	47°44'55"	73°31'22"	490.0				
Aldor	47°12'15"	78°08'01"	85.0				
Alex	49°15'25"	71°27'26"	613.0				
Alex, Petit	49°22'28"	71°25'50"	56.0				
Alfred	46°08'53"	74°58'53"	16.0	24.4	9.8	20	

Alfred, Petit	48°03'58"	75°46'44"	18.0			25	
Alma	46°35'16"	77°51'03"	93.0				
Amyot	47°57'58"	76°51'30"	129.0				
Andrew	47°44'47"	73°25'25"	62.0				
Angus	46°38'41"	76°22'16"	67.0	17.0		33	5
Anne	45°49'28"	74°18'31"	117.0	23.0		23	5
Anne, Lac	49°33'00"	68°59'22"	75.0				6
Anonyme	51°38'14"	74°44'01"	428.0	31.8			
Anonyme	51°11'10"	75°41'39"	691.0	21.0			
Antoine	46°22'12"	76°59'08"	435.0				
Antoinette	49°55'59"	74°25'29"	137.0	39.6		32	4
Antostogan	46°57'13"	76°38'01"	1315.0				
Anvers (Grandclair)	47°47'56"	73°37'06"	111.0			29	4
Archambault	46°18'28"	74°14'33"	1380.0	39.0	14.4		
Argent, D'	47°17'01"	79°08'06"	205.0	24.0			
Argent, D' (Polonais)	48°44'27"	71°40'33"	1119.0	56.4	16.2	140	5
Argent, D' (Silver)	46°38'30"	72°35'49"	60.0	40.0		187	4
Argente	45°51'44"	74°27'53"	52.0	43.0		90	4
Argile	45°51'37"	75°33'41"	451.0	27.0	13.3		2
Assomption, L'	46°27'41"	74°03'13"	91.0				
Atocas	47°02'57"	75°15'45"	181.0				
Aubin	46°58'25"	73°00'19"	134.1	22.9	6.8	18	
Aubry	47°03'59"	78°59'32"	199.0			15	6
Augier	48°39'03"	76°45'49"	264.0			30	4
Aumond	46°34'33"	77°32'26"	965.0	22.0			
Aux Dorés	49°51'14"	74°20'40"	4040.0	40.5			
Aux Dorés	47°58'57"	72°16'06"	60.0				
B.-L.	47°13'40"	78°37'12"	12.0				
Badeaux	46°59'47"	73°29'06"	80.0				
Bailloquet	48°12'59"	70°34'18"	80.0			45	
Banc De Sable, Du	48°54'55"	71°27'34"	175.0				
Banc De Sable, Grand	49°35'48"	71°53'41"	173.0				
Banc De Sable, Petit	49°42'	71°41'					
Bangall	46°00'	75°49'	161.0				
Baptiste	46°23'09"	77°08'10"	189.0				

QUEBEC (continued)

Lake name	Latitude	Longitude	Surface area (ha)	Maximum depth (m)	Mean depth (m)	Conductivity (µS/cm)	Secchi depth (m)
Barbé	48°15'	70°44'	21.0				
Bardy	47°36'16"	73°25'20"	70.0	23.0		21	3
Baribeau	46°20'56"	74°09'38"	111.0	21.0			
Bark, Petit	46°55'12"	76°20'47"	223.0				
Barrage, Du	45°44'	75°50'	16.0				
Barrete	46°58'19"	74°38'33"	78.0				
Barry (Burton)	46°16'23"	78°03'43"	34.0			20	11
Barthou	47°52'30"	76°56'40"	158.0				
Bat	47°02'44"	79°00'52"	256.0				
Baude	47°04'36"	73°17'58"	325.0	36.5	15.7	18	8
Bay	47°24'57"	78°17'18"	1311.0	37.0	7.6	21	6
Beauchene	46°39'26"	78°57'30"	3807.0	70.0	16.0	40	8
Beauchene, Petit	46°42'12"	78°52'40"	243.0		13.0	53	8
Beaupre	47°27'20"	73°38'11"	109.0	49.0		18	
Beautiful	46°58'31"	74°44'44"	57.0				
Becquerel	48°45'02"	73°13'34"	26.0				
Beland	48°50'51"	73°19'35"	210.0				
Belisle	47°04'31"	74°35'18"	155.0				
Bell	46°03'34"	76°37'58"	300.0	40.0	12.0	90	6
Bellevue	47°11'09"	72°13'51"	91.0				
Bernabe	48°55'04"	71°36'57"	273.0				
Betchie	49°59'53"	69°59'41"	1847.0				
Bevin	45°56'36"	74°34'50"	282.0	33.0		73	4
Bibitte	46°13'47"	74°39'19"	47.0	37.0			
Bidiere (Nicolas)	47°25'51"	75°04'41"	495.0				
Binet	46°11'49"	75°19'25"	36.0	15.8		22	3
Biscornet	48°44'33"	73°11'33"	12.0			24	2
Bison, Du	48°31'	73°03'	40.0				
Bitobi	46°06'	75°57'	199.0				
Blair	47°01'37"	74°37'54"	34.0			29	4

Lake	Latitude	Longitude					
Blais	45°56'	75°10'	47.0				4
Blanc	46°49'15"	72°16'20"	212.0	21.0		19	8
Blanc	46°24'05"	77°14'05"	143.0	26.0	11.9	35	
Blanche, De la	46°05'12"	74°27'16"	41.0			27	
Blanchin	48°05'59"	77°03'21"	264.0				
Blazer	47°30'27"	73°06'40"	75.0				
Bleu	46°36'58"	78°23'51"	1689.0			26	
Blue Sea	46°13'14"	76°03'17"	1437.0	60.0	19.0	165	7
Bob Grant (Albert)	47°44'55"	73°31'22"	490.0				
Bois Franc	46°45'11"	78°36'50"	1373.0				
Bois Franc, Du	46°55'09"	76°23'42"	215.0	25.0			3
Bois, Des	45°55'26"	75°24'41"	54.0				
Bois, Lac des	50°54'00"	69°06'26"	107.0				
Bois-Long	50°55'00"	68°57'00"	722.0				
Boissadel	47°00'51"	77°19'37"	142.0	59.0		45	5
Boisseau	46°10'54"	74°49'05"	65.0				
Boisvert	47°24'05"	75°13'21"	153.0				
Boivin	47°56'56"	72°17'42"	54.0	105.0	35.0		7
Bondy	47°05'02"	75°51'08"	531.0	25.0	9.0	32	3
Bonin	49°16'28"	74°39'09"	453.0				
Bonlac, Le	47°07'22"	72°08'29"	67.0				
Borry (Burton)	46°17'15"	78°05'00"	13.0				
Bouchard	46°06'12"	76°52'56"	54.0			21	11
Boucher	49°28'45"	68°55'55"	347.0	30.0	9.1		2
Boucher	47°03'07"	72°47'33"	365.0				
Boucher, Rivière	49°10'00"	69°06'00"					
Boue, De la	46°00'	76°36'	83.0				
Boulanger	47°21'24"	73°00'50"	23.0	32.1	11.5		4
Bouleau Blanc	46°54'22"	74°47'51"	99.0			44	
Bouleau, Du	47°01'43"	78°44'11"	83.0			31	
Bouleaux, Des	46°42'34"	78°55'58"	158.0		10.1	48	4
Bouleaux, Petit, Des	46°42'02"	78°53'42"	31.0		6.0	60	6
Bourbeau	49°57'36"	74°19'02"	606.0	36.6	18.0	135	7
Bouteille	52°55'00"	67°35'00"	1320.0				
Bowker	46°25'14"	73°12'53"	231.0	59.1	23.9	54	10

QUEBEC (continued)

Lake name	Latitude	Longitude	Surface area (ha)	Maximum depth (m)	Mean depth (m)	Conductivity (µS/cm)	Secchi depth (m)
Branssat	46°25'16"	77°02'23"	306.0				
Brascoupe	46°34'02"	76°11'15"	679.0	62.0		60	6
Brehaut	47°07'41"	73°35'48"	469.0	38.0		23	
Bright Sand	51°58'26"	65°44'35"	593.0				
Briquet	46°48'	76°45'	316.0				
Britannique	45°45'	75°16'	184.0				
Brochet, Au	49°40'00"	69°36'00"	4481.0				
Bruce	46°43'50"	77°20'09"	749.0				
Brule	52°18'01"	63°50'23"	8884.0				
Brulé	45°50'	75°37'	21.0				
Brule (Pigeon)	46°11'41"	76°52'30"	114.0				
Brunelle	48°14'21"	73°21'22"	163.0			16	
Burden	47°10'41"	79°07'45"	41.0				
Burton (Barry)	46°16'23"	78°03'43"	34.0			20	11
Burton (Borry)	46°17'15"	78°05'00"	13.0			21	11
Busted	47°00'53"	77°34'55"	368.0				
Byrd	47°02'52"	76°53'03"	2719.0				
Cabanac	47°05'52"	76°51'43"	36.0			27	4
Cabot	46°24'36"	73°57'35"	36.0	46.0		37	
Cache	46°21'12"	74°39'17"	301.0	24.0	10.8	36	10
Caché	49°50'29"	74°24'12"	339.0	22.5		28	4
Cache, De la	50°41'58"	68°37'43"	596.0				4
Café	46°41'	76°50'	27.0				
Cain	46°54'16"	75°23'43"	142.0	30.0	18.4	45	4
Cairine	46°21'53"	76°14'01"	28.0				
Caisse	46°27'39"	74°00'41"	52.0	15.0			3
Callas, Des	48°50'03"	73°38'03"	40.0				
Cameron	46°06'31"	74°39'31"	347.0	44.0		61	4
Canal, Du	47°00'33"	78°36'50"	52.0				
Canot, Du	47°46'01"	75°07'01"	96.0				

Lake	Latitude	Longitude					
Carabine	47°05'59"	74°30'32"	111.0				
Cardinal, Du	45°51'	75°48'	54.0				
Caribou	49°08'30"	69°54'20"	185.0	38.0		35	4
Caribou	46°17'09"	74°42'11"	117.0	13.0		60	3
Caribou (Mousseux)	46°02'00"	74°27'08"	49.0	66.0		19	8
Carignan	47°15'41"	72°46'39"	520.0		16.0	19	5
Carignan, Petit	47°10'45"	72°50'40"	228.0				
Carmel	47°19'07"	73°24'42"	32.0				
Carmin (Rouge)	46°04'	74°59'	104.0				
Carufel	49°48'35"	72°06'49"	17.0				
Castelveyre	47°09'40"	74°07'30"	479.0	24.0		22	5
Caugnawana	46°32'23"	78°18'27"	746.0			30	
Caugnawana, Petit	46°37'08"	78°26'28"	150.0	16.0	5.4	26	4
Cavendish	47°59'37"	76°17'48"	44.0				
Cayamant	46°06'56"	76°16'15"	725.0	59.0	12.3	65	5
Cayamant, Petit	45°59'	76°20'	383.0				
Cazes, De	48°51'54"	73°41'23"	189.0				
Cecile	48°30'45"	73°49'14"	163.0				
Cedres, Des	48°14'07"	70°44'15"	270.0				
Cedres, Des	46°18'14"	76°06'47"	793.0	38.7	14.1	85	5
Cèdres, Petit, Des	46°18'	76°04'	282.0				
Cerf, Du	46°16'35"	75°29'49"	1267.0	120.0	33.3	69	5
Cerf, Petit, Du	46°20'	75°30'	339.0	30.0	11.0	70	5
César	47°48'21"	75°15'01"	145.0	28.0		27	
Chailly	46°26'28"	78°37'02"	70.0				
Chaine De Lac	46°24'34"	73°11'34"	85.0	24.1	10.0	61	5
Chantier	47°01'08"	74°19'48"	91.0	24.0			
Chapleau	46°13'44"	74°56'34"	497.0	30.0		47	6
Charette	48°38'13"	76°22'01"	220.0			16	6
Charlebois	46°04'59"	74°03'06"	52.0	15.0			
Charley	47°00'38"	74°55'37"	31.0				
Charlie	46°46'19"	77°03'20"	189.0				
Chartier	47°25'09"	77°30'25"	984.0				
Chasseurs, Des	47°02'54"	76°17'20"	233.0	39.0	13.7	30	
Chaud	46°27'13"	74°46'07"	621.0	34.0		20	4

QUEBEC (continued)

Lake name	Latitude	Longitude	Surface area (ha)	Maximum depth (m)	Mean depth (m)	Conductivity (µS/cm)	Secchi depth (m)
Chavannes	47°09'29"	74°06'16"	199.0	18.0		23	5
Chavannes	46°51'16"	77°09'09"	774.0				
Chemin, Du	45°56'43"	75°41'37"	4.0				
Chénier	46°12'14"	75°05'40"	26.0	9.7			4
Chenon	47°19'39"	78°28'05"	725.0			15	6
Cheroy	46°29'51"	78°18'36"	142.0			30	6
Chien, Du	48°24'	73°27'	136.0				
Chienne, À la	47°02'01"	73°31'05"	551.6	48.0	15.8	17	3
Chigoubiche	49°04'55"	73°30'57"	3268.0				
Chopin	46°54'09"	74°49'05"	52.0			44	
Choquette	47°05'23"	78°58'27"	62.0				
Chubb	47°09'12"	78°48'35"	114.0			26	
Cinq Milles	47°13'21"	78°42'23"	176.0			24	5
Clair	46°32'08"	78°45'50"	150.0			39	
Clair	46°21'20"	78°25'31"	16.0				
Clair	46°20'35"	74°22'44"	34.0				
Clair	45°44'14"	74°19'33"	47.0	22.0			5
Clair	45°44'06"	75°48'02"	88.0			49	7
Clair (Clairvaux)	46°05'32"	74°03'41"	34.0				9
Clair (Devenyns)	47°05'25"	73°50'14"	2163.0	60.0	16.1	25	
Clair, Petit	47°33'00"	72°59'03"	29.0				
Clair, Petit	47°09'43"	73°34'23"	65.0	34.0	12.7	18	5
Claire	47°21'20"	73°26'33"	40.0	17.0			
Clairvaux (Clair)	46°05'32"	74°03'41"	34.0				7
Clais	48°06'13"	76°42'24"	36.0				
Clarice	48°20'17"	79°31'45"	50.0	20.0		40	
Claw	46°42'37"	76°38'33"	93.0	7.0		30	6
Clement	47°01'06"	78°58'25"	448.0				
Clément	46°54'	77°41'	29.0				
Clignancourt	49°07'58"	73°28'31"	25.0				

Name							
Cloutier	47°40'11"	73°12'20"	101.0	24.0		67	4
Cloutier	46°10'48"	73°38'53"	197.0				
Coffin	47°46'14"	76°02'51"	57.0			35	
Collins (Cullin)	46°21'06"	77°50'38"	117.0				
Connely	49°19'	71°57'	295.0				
Connely	45°53'54"	73°57'51"	124.0	20.0		75	3
Cooks	47°07'47"	78°48'04"	176.0			34	2
Corbeau, Du	46°44'00"	76°28'05"	153.0	50.0		40	4
Corbeau, Du	46°12'02"	75°28'37"	207.0	32.0		59	6
Corbeau, Du	46°07'03"	75°12'14"	36.0			39	
Corbeau, Petit	46°44'28"	76°26'43"	65.0	18.0		31	5
Cormon	47°27'	75°37'	456.0				
Cornes, Des	46°43'30'	75°09'19"	409.0	47.0	10.0	48	7
Corrigan	46°25'36"	77°25'13"	260.0	30.0		35	3
Cottentre	46°41'56"	78°42'58"	186.0			34	
Coude, En	48°02'11"	73°36'45"	92.0				
Couveuse, À la	46°35'35"	73°03'31"	73.0	28.0	11.0	26	6
Cox	46°55'45"	76°54'05"	106.0				
Cran, Du	48°51'32"	72°52'20"	53.0				
Crevier	47°06'13"	75°41'44"	419.0			59	4
Cris, Des	47°29'	75°26'	49.0				
Cristal	47°21'25"	73°48'08"	220.6	36.0	13.0	14	3
Croche	46°37'39"	76°28'23"	513.0	29.0		35	
Croche	45°55'	75°41'	117.0				
Croche (Forster)	46°22'50"	76°34'22"	119.0	20.0		40	5
Croche-Sud	46°59'43"	75°26'29"	124.0				
Croissant	51°23'43"	65°45'31"	526.0				
Croucher	46°57'07"	77°30'54"	57.0				
Cruiser	48°59'03"	71°27'39"	155.0				
Crystal	46°38'39"	73°38'19"	75.0	40.0			
Cuillere	48°40'19"	76°42'54"	262.0			15	
Cullin (Collins)	46°21'06"	77°50'38"	117.0			35	
Cypres, Des	49°11'17"	71°50'43"	113.0				
Cypres, Des	47°47'59"	74°14'18"	47.0				
Daigle	52°49'00"	67°12'00"	235.0				

QUEBEC (continued)

Lake name	Latitude	Longitude	Surface area (ha)	Maximum depth (m)	Mean depth (m)	Conductivity (µS/cm)	Secchi depth (m)
Dalou	48°52'54"	73°33'10"	9.0				
Dame, De la	45°50'	75°20'	117.0				
Danford	45°57'	76°08'	135.0				
Daviault	52°48'04"	67°04'30"	536.0				
David	46°35'	75°14'	220.0	25.0		51	4
David	46°29'08"	76°26'51"	754.0	55.0		50	8
Deces, Du	49°38'12"	69°18'43"	176.0				
Decharge	46°07'04"	74°48'07"	114.0	45.0		37	5
Decoste	47°13'47"	73°26'31"	116.0	45.1	16.1	19	6
Demers	48°52'23"	73°21'49"	67.0				
Denain	47°53'32"	76°59'36"	596.0			19	6
Desautels	49°24'00"	73°19'49"	365.0				
Desert	46°35'28"	76°18'24"	2979.0	52.0		36	6
Desmolier	46°10'01"	74°59'54"	18.0	15.2		112	
Deux Decharges, Des	51°58'00"	70°40'00"	1578.0				
Deux Iles, Des	46°59'20"	76°47'33"	104.0	27.4			6
Devenyns (Clair)	47°05'25"	73°50'14"	2163.0	60.0	16.1	25	9
Diable, Du	46°43'09"	78°56'09"	78.0	18.0	6.2	51	5
Diable, Du	46°06'26"	75°10'59"	85.0	64.0	25.1	425	8
Diamant	47°16'37"	73°58'13"	192.0	23.2		28	
Dinant	47°45'53"	73°34'07"	75.0				
Dix Milles, Des	47°53'44"	74°48'31"	1717.0				
Dix-Milles	46°47'27"	77°44'50"	1611.0				
Dodds	45°42'48"	75°36'34"	53.0				
Doolittle	46°41'	76°48'	805.0				
Doré	46°01'	75°02'	109.0				
Doré, À	48°15'06"	70°48'12"	30.0				
Doreil	46°48'32"	70°08'55"	88.0				
Double	50°48'00"	70°24'00"	1830.0				
Doucet	49°18'35"	71°55'19"	256.0				

Lake							
Douglas	46°44'01"	78°53'56"	36.0			64	5
Doyon	46°10'47"	77°16'51"	55.0	30.0	7.3	35	5
Draper	47°59'26"	76°52'29"	39.0				
Du Loon (Huard)	46°33'26"	78°27'39"	70.0				
Dubuc	49°19'00"	69°52'00"	946.0				
Ducarny	47°24'	75°27'	47.0				
Dufaux	47°32'05"	73°36'55"	78.0	16.8	4.7	20	5
Dufferin	48°49'27"	72°43'27"	64.0	26.0	8.2		
Dufresne	51°24'00"	65°45'00"	526.0				
Dugal	47°05'59"	78°39'33"	85.0	48.0		31	6
Duhamel	46°08'30"	74°38'15"	52.0	29.0		245	4
Dulain	49°45'	71°35'	370.0				
Dumas	46°03'31"	75°14'10"	39.0	13.4			
Dumont	46°03'41"	76°27'20"	1772.0	41.2	10.4	42	8
Dumoulin	47°32'05"	73°00'10"	275.0				
Duplessis	47°20'19"	75°14'30"	279.0				
Dupont	46°39'45"	76°34'08"	21.0	46.6	9.7	18	3
Dupuis	47°22'56"	73°46'03"	560.1				
Dutau	48°42'	73°11'	28.0	41.0		30	
Eau Claire, À l'	46°40'26"	79°03'13"	54.0				
Eau Claire, À l'	46°32'34"	73°03'25"	749.0				
Ecarte	47°11'53"	78°35'53"	321.0				
Echelle, L'	46°57'29"	78°38'02"	109.0	100.0	33.3	46	7
Echo	45°54'43"	75°27'06"	508.0	45.7		41	3
Écluse, De l'	45°50'38"	75°24'13"	375.0	56.0			
Écorces, Des	45°59'39"	74°32'25"	648.0		14.5	50	4
Écorces, Des	45°53'46"	75°19'08"	124.0			26	
Eddie	46°34'37"	78°25'45"	88.0				
Edja	46°11'	76°00'	199.0				
Elgin	46°57'54"	74°38'17"	52.0				
Embarras	46°53'44"	76°30'19"	1241.0	48.8		28	4
Emeraude	48°55'39"	71°39'30"	37.0				
En Coeur	47°10'43"	79°02'29"	161.0				
Épervanche	51°50'00"	70°03'00"	177.0				
Epinettes Noires	49°05'54"	71°40'11"	53.0				

QUEBEC (continued)

Lake name	Latitude	Longitude	Surface area (ha)	Maximum depth (m)	Mean depth (m)	Conductivity (µS/cm)	Secchi depth (m)
Epingle	52°18'00"	69°06'00"	366.0	35.0	8.7	9	4
Equerre	46°16'30"	74°58'54"	70.0	20.0		33	3
Eric	51°54'52"	65°34'20"	3212.0				
Ernest	46°10'41"	75°12'18"	117.0	24.4		43	4
Eternite	48°13'29"	70°33'18"	438.0	38.0	12.2	25	4
Eternite, Petit	48°13'59"	70°28'36"	40.0				
Ethel	51°50'48"	65°43'43"	326.0				
Etienniche	49°31'12"	71°21'32"	993.0	35.0	6.9		
Etire	51°02'00"	70°51'40"	167.0				
Fabre	46°22'27"	75°08'43"	47.0	24.4	10.2	38	
Falle	47°30'49"	73°38'42"	60.0	24.7	5.1	20	
Faucon	45°48'	75°20'	186.0				
Fer-À-Cheval	45°57'	75°43'	7.0				
Fils, Du	46°37'02"	78°07'28"	1746.0	33.0		24	
Fleche	50°02'05"	68°11'21"	770.0	36.0	8.9		
Foie	48°19'00"	74°17'06"	168.0	37.0	11.5	14	7
Forant	46°24'33"	77°10'15"	746.0				
Forges, Des	48°44'23"	76°44'09"	572.0			20	5
Forster (Croche)	46°22'50"	76°34'22"	119.0	20.0		40	5
Fou	47°11'34"	72°20'42"	132.0				
Fourche, De la	45°51'38"	75°20'15"	44.0	29.0	42.6	40	4
Fournier	51°32'12"	65°25'00"	4869.0				
Francais, Des	46°08'18"	73°37'57"	109.0	23.0		80	5
Frances	45°52'56"	76°31'26"	23.0				
Franquet	47°26'15"	73°01'33"	62.0				
Fremont	47°39'05"	73°39'38"	176.0	22.9	5.8	20	
Froid	46°40'21"	74°31'29"	219.0				
Froid, Petit	46°41'07"	74°32'12"	54.0	21.0		10	3
Fulham	47°45'50"	75°57'29"	93.0				
Gagamo	46°40'06"	76°32'24"	342.0	25.0		35	4

Lake							
Gagnon	46°06'52"	75°07'33"	1857.0	72.5	24.6	34	6
Gaillarbois, Lac	51°58'34"	67°25'12"	4761.0	8.1		8	9
Galarneau	46°08'18"	76°48'04"	637.0	40.0		32	9
Galloway	50°01'14"	74°13'20"	54.0	45.0	15.0	135	
Galt	47°09'10"	79°07'43"	80.0				
Gaston	48°21'14"	73°34'57"	39.0				
Gaston	48°00'58"	74°38'41"	186.0	32.0		92	9
Gatineau	46°33'04"	75°39'20"	83.0				
Gaucher	47°24'40"	73°11'03"	166.0	42.4	11.9		
Gauthier	46°59'45"	77°53'48"	163.0				
Gauthier	46°10'27"	74°31'18"	44.0	14.0		26	3
Gauvreau	45°39'	75°59'	83.0				
Gazeau	50°55'44"	71°37'48"	369.0				
Genest	47°07'03"	71°41'23"	13.0				
George, Grand	46°46'35"	78°43'58"	238.0			35	5
George, Petit	46°45'53"	78°41'56"	65.0			40	3
Georges	46°17'47"	73°59'48"	101.0	12.0			
Gerland	46°27'49"	77°04'13"	119.0				
Gervais	46°16'17"	74°40'50"	103.0	53.0		46	6
Gilbert	48°28'	73°25'	22.0				
Gilman	49°54'40"	74 20'21"	163.0	24.0	6.9	105	3
Gilmore	46°43'26"	76°26'56"	83.0	25.0		70	4
Go	47°29'14"	73°59'19"	88.0	58.0	26.6	13	6
Goeland	49°47'19"	71°41'06"	1181.0			20	
Goeland, Du	46°46'17"	78°20'22"	699.0	24.0	6.0	33	4
Goelands, Aux	49°18'49"	74°36'18"	212.0	21.0	7.0	75	5
Goglu, Du	45°54'43"	75°20'42"	13.0	35.0	6.1		
Gordon	46°36'	78°24'	190.0				
Gorman	47°01'19"	75°18'01"	207.0	99.0	30.3		
Gouin	49°32'30"	70°14'22"	1450.0	29.0			
Gour	46°09'47"	73°47'48"	104.0			23	5
Gralas	47°15'48"	77°03'09"	39.0			25	5
Grand	45°40'33"	75°39'23"	373.0	43.9	17.9	89	
Grand (Macousine)	47°46'17"	72°07'05"	171.0			19	
Grand Portage, Du	49°07'37"	69°50'20"	52.0				

QUEBEC (continued)

Lake name	Latitude	Longitude	Surface area (ha)	Maximum depth (m)	Mean depth (m)	Conductivity (µS/cm)	Secchi depth (m)
Grandclair (Anvers)	47°47'56"	73°37'06"	111.0				
Grande Loutre	51°07'30"	71°33'00"	343.0				
Grandes Baies, Des	46°22'11"	75°06'55"	368.0	40.0		25	3
Grandes Pointes, Aux	49°07'34"	71°33'25"	587.0			23	4
Grant	47°15'11"	79°00'51"	194.0				
Gravel	46°48'04"	75°23'01"	166.0	110.0		38	5
Green (Vert)	46°29'35"	76°17'27"	122.0	52.0		40	6
Grenouille	48°14'	70°42'	21.0				
Gris	46°35'17"	78°20'23"	28.0				
Grosbois	47°06'38"	72°49'14"	129.0	23.0	8.0	27	6
Grosse Ile	48°44'11"	73°31'45"	40.0				
Grundy	47°41'52"	73°24'37"	32.0				
Guay	47°12'45"	79°01'34"	508.0				
Guenette	47°54'21"	74°20'11"	142.0				
Guinecourt	50°55'00"	69°16'00"	1070.0				
Guynne	47°30'19"	73°03'40"	96.0	21.0		25	
Guynne, Petit	47°30'26"	73°01'15"	36.0				
Hachette	47°01'50"	74°33'46"	18.0			26	
Hachette	46°59'06"	74°35'41"	34.0			24	
Haine (Heron)	47°22'16"	74°52'09"	168.0			19	
Hamel	48°13'47"	70°35'08"	6.0				
Hamilton	46°35'32"	78°13'59"	438.0			29	
Harbison	47°32'06"	73°39'10"	36.0	21.3	5.6	20	
Harkin	46°43'	76°45'	73.0				
Harrington	45°51'29"	74°33'20"	106.0	27.0			
Hawkesbury	47°19'39"	77°44'31"	352.0				
Hazeur	46°40'25"	73°50'22"	117.0	30.0		28	6
Head	47°46'23"	74°36'30"	52.0				
Hecaté	48°48'55"	73°37'42"	33.0	20.0	7.2		4
Heney	46°01'47"	75°55'25"	1233.0	32.5	14.7	131	3

Lake	Latitude	Longitude					
Henri	49°59'45"	71°50'53"	318.0			19	
Heron (Haine)	47°22'16"	74°52'09"	168.0				
Hickey	45°56'39"	76°37'10"	231.0				
Hilaire	46°03'	76°23'	78.0				
Hoswart	48°38'57"	76°42'11"	107.0				
Houle	47°33'13"	73°30'51"	24.0				
Howard	46°35'51"	75°39'49"	98.0	23.0	9.0	50	4
Huard (Du Loon)	46°33'26"	78°27'39"	70.0				
Huddersfield	45°55'25	76°32'21"	225.0				
Ignace	46°37'26"	76°24'53"	104.0	21.0		29	
Ile Blanche, À l'	49°15'53"	71°44'17"	41.0			30	4
Ile, De l'	49°09'49"	69°54'09"	75.0				
Iles, Des	49°15'22"	71°19'44"	34.0				
Iles, Des	48°12'14"	73°38'02"	30.0				
Iles, Des	47°17'15"	78°38'10"	80.0				
Iles, Des	46°30'	75°30'	1620.0	37.0	9.0	28	5
Iles, Des	46°05'59"	74°01'45"	484.0	23.0	12.0	71	7
Iles, Des	45°58'	74°55'	212.0			27	
Indienne, De l'	46°01'34"	76°36'09	355.0				
Iroquois	46°03'	75°10'	142.0			39	
Irvine	46°50'54"	74°44'30"	41.0				
Isabel	45°47'	76°06'	153.0			30	
Jacques	46°23'41"	78°11'17"	31.0				
Jalobert	48°05'44"	75°49'16"	606.0				
Jean-Péré	47°03'49"	76°37'37"	2719.0	34.0		37	3
Jenssen	48°32'55"	73°07'25"	153.0	28.0		31	
Jobin	46°42'52"	76°20'30"	23.0				
John Bull	46°30'36"	76°53'30"	163.0				
Johnson	45°46'35"	76°23'22"	106.0				
Johnston	45°43'	75°59'	80.0				
Joinville	46°17'56"	75°12'10"	259.0	50.3	15.4	34	6
Joli	47°51'00"	73°13'44"	104.0	23.0		18	
Joli (Mc Grey)	46°25'51"	73°17'59"	65.0				
Jolivet	48°46'04"	73°22'39"	31.0				
Joncas	47°08'39"	77°39'48"	433.0			31	

QUEBEC (continued)

Lake name	Latitude	Longitude	Surface area (ha)	Maximum depth (m)	Mean depth (m)	Conductivity (µS/cm)	Secchi depth (m)
Jules, À	49°44'29"	69°08'46"	140.0				
Kallio	51°05'35"	72°55'33"	290.0	22.0	6.7	50	5
Kamiltaukas, Lac	49°38'00"	68°56'00"	167.0				
Kapahkueshikanapishkatsh	49°33'04"	72°44'20"	237.0				
Kar-Ha-Kon	46°20'37"	75°21'19"	91.0	15.0			5
Kensington	46°24'	75°43'	445.0				
Kettle	47°06'53"	74°45'36"	70.0				
Kikwissi	46°57'49"	78°33'12"	2225.0	40.0	15.8	26	
Kondiaronk	46°55'52"	76°45'18"	1590.0				
La Haie	48°28'23"	78°42'03"	39.0	39.0		35	6
La Loutre	45°59'42"	74°39'18"	65.0	32.0		35	
La Rue	52°40'00"	67°12'00"	680.0				
Labelle	46°13'00"	74°51'18"	738.0	61.0		63	6
Lac 7223	46°46'51"	78°41'23"	39.0	14.0		34	5
Lac 7232	46°49'42"	78°53'13"	62.0	17.0		44	
Lac 10539	47°05'25"	78°58'02"	36.0				
Lac 17893	48°38'00"	75°58'33"	161.0				
Lac 18812	48°32'48"	76°36'26"	117.0				
Lac 39378	46°22'58"	78°23'41"	10.0				
Lac 39755	46°35'48"	78°17'02"	13.0				
Lac 46543	48°00'10"	74°38'18"	19.0	19.4	6.4		
Lac A8388	48°00'36"	74°37'24"	35.0				
Lac C0678	48°53'21"	76°26'24"	23.0				
Lac C1677	48°40'45"	76°02'26"	14.0				5
Lac C1987	48°39'06"	76°47'15"	63.0				
Lac C2027	48°33'31"	76°39'22"	46.0				
Lac H	50°43'00"	70°04'30"	150.0				
Lac I	50°36'15"	70°07'00"	165.0				
Lachance	47°27'27"	73°05'16"	39.0				
Lafleur	46°24'18"	76°32'57"	39.0	34.0		29	

Lafontaine	47°29'30"	71°15'52"	44.0				
Lafontaine	46°04'	75°10'	109.0				
Laforest	46°06'	76°31'	101.0				3
Lajoie	46°25'19	74°16'56"	57.0				4
Laliberté	49°34'34"	71°31'23"	475.0	40.0		37	
Lamb	46°28'44"	77°14'26"	629.0				
Lambert	46°33'38"	73°12'23"	44.0			23	
Lanterne	47°02'36"	74°56'39"	13.0				
Laperrière	47°17'54"	79°26'24"	49.0				
Larin	46°04'24"	74°27'23"	24.0	24.0			
Larue	46°26'	76°57'	124.0	50.0		92	7
Laurel	45°52'18"	74°28'38"	60.0				
Lavoie	52°07'17"	63°42'21"	712.0	65.0			
Ledge	47°37'40"	74°11'39"	192.0			20	2
Legare	46°57'35"	73°57'50"	1134.0			22	6
Leluau	47°28'52"	74°52'19"	316.0	36.0	14.0	23	
Lemere	47°06'15"	72°47'24"	75.0				
Lemoine	49°22'11"	71°19'28"	240.0			25	
Lenoir	46°51'25"	74°31'57"	117.0				5
Lesage	46°18'39"	75°02'32"	471.0	40.0	12.0	47	5
Lescarbot	47°48'30"	72°04'15"	472.8	22.1	6.7	22	4
Lescot	47°13'53"	78°26'03"	559.0			27	5
Lesueur	47°26'42"	75°18'47"	671.0	51.5	18.3		
Letemplier	49°27'14"	68°46'53"	391.0				
Letemplier, Petit	49°25'00"	68°45'00"	115.0				
Lievre	46°53'21"	77°03'22"	357.0	17.0		30	2
Lievre	46°32'30"	77°41'24"	166.0			31	
Lisleroy	46°27'15"	78°36'00"	44.0			28	
Little Grant (Sterne)	46°35'31"	77°56'11"	13.0				
Livernois	47°08'34"	73°20'01"	205.8	43.0			4
Long	50°14'30"	72°57'10"	369.0				
Long	48°42'	73°30'	23.0				
Long	47°13'21"	72°18'33"	91.0				
Long	46°50'16"	72°08'18"	308.0				
Long	46°32'13"	72°56'52"	83.0				
Long							

QUEBEC (continued)

Lake name	Latitude	Longitude	Surface area (ha)	Maximum depth (m)	Mean depth (m)	Conductivity (µS/cm)	Secchi depth (m)
Long	46°43′29″	78°27′29″	47.0			28	
Long (Nord)	47°07′48″	73°29′43″	41.0	21.3	7.1	19	
Longworth	47°43′52″	74°21′36″	67.0				
Lost (Perdu)	46°25′12″	77°47′34″	70.0	43.0		29	6
Louis	47°25′37″	78°07′14″	54.0			32	
Louise	45°46′09″	74°25′08″	440.0	44.0		48	6
Loutre, De la	45°58′	75°40′	39.0				
Louvain	47°48′30″	73°38′19″	104.0				
Low Ball	51°41′00″	67°08′00″	330.0	24.0		16	5
Lozeau	52°06′00″	63°49′44″					
Lucernede, De l'Ecluse	45°43′55″	75°46′50″	122.0	43.0		46	5
Lymburner	49°57′23″	74°04′01″	104.0	37.0		40	6
Lynch	46°31′12″	74°44′26″	171.0	18.0		25	5
Lynch	46°24′41″	77°05′43″	1642.0	66.0		25	8
Lytton	46°38′43″	76°06′43″	246.0	35.0		50	5
Mabille	51°59′56″	62°56′16″	1699.0				
Machisque	50°51′44″	71°48′49″	2130.0			10	3
Macouaga	46°48′25″	74°41′48″	78.0			26	
Macousine (Grand)	47°46′17″	72°07′05″	171.0			19	
Madeleine	47°01′56″	75°50′19″	319.0				
Maganasipi	46°32′03″	78°23′23″	919.0			29	
Main, De la	49°04′07″	71°28′39″	20.0				
Maison De Pierre	46°53′20″	74°41′58″	1200.0				
Major	46°50′14″	75°30′30″	363.0	45.0	15.0	33	3
Malakisis	46°42′02″	78°39′24″	117.0			30	4
Malfait	49°19′03″	72°02′46″	197.0				
Malouin	46°32′32″	77°58′01″	546.0	45.0	10.9	26	
Maltais	46°59′18″	77°55′51″	251.0				
Managowich	49°08′54″	71°47′49″	25.0				
Maniluve	46°49′19″	77°08′09″	171.0				

Name							
Manitou	46°03'33"	74°22'12"	404.0	21.0	7.8	27	4
Manouanis	50°50'00"	70°19'00"	3420.0				7
Marguerite	47°01'43"	75°48'12"	622.0			26	7
Marie Lefranc	46°08'10"	74°59'45"	668.0	79.6	32.9	88	7
Marie-Louise	46°14'01"	74°58'18"	103.0	30.0		80	
Marin	46°32'11"	78°49'18"	401.0			40	
Marion	46°52'34"	77°46'39"	593.0				
Mars	47°26'24"	78°06'56"	443.0	25.0	5.9	23	5
Marsac	46°52'34"	79°05'25"	427.0				
Marteau, Du	48°14'20"	73°57'53"	174.0				
Maryse	47°14'07"	78°35'53"	81.0				
Maskinonge	46°18'59"	73°23'03"	1018.0	28.0		44	2
Masson	47°30'42"	73°08'23"	37.0				
Masson	46°34'41"	77°39'05"	52.0	21.0		32	4
Masson	46°02'41"	74°02'03"	355.0	46.0		83	4
Matane	48°42'03"	68°58'22"	142.0	52.0	26.6	50	4
Matchi-Manitou	48°00'52"	77°02'57"	3781.0	70.0	18.3	18	3
Matonipi	51°51'12"	69°48'15"	3289.0	48.0	12.0	14	5
Matonipis	51°49'53"	69°40'01"	3004.0				
Mauran	47°09'57"	75°51'52"	62.0				
Mauves, Des	46°11'22"	74°56'37"	451.0	30.4	15.2	36	
Mayer	46°34'09"	76°45'23"	41.0	18.0		19	
Mazana	47°07'33"	74°31'14"	782.0	20.0			2
Mc Arthur	45°42'	75°40'	127.0				
Mc Cann	46°25'14"	77°18'08"	285.0	27.0		36	6
Mc Connel	46°58'46"	77°39'02"	176.0				
Mc Cuaig	45°50'04"	76°26'59"	65.0				
Mc Donald	45°51'41"	74°34'49"	383.0	18.0		65	4
Mc Fee	45°42'49"	75°37'29"	93.0				
Mc Gillivray	46°04'28"	77°06'33"	899.0				
Mc Gregor	45°38'55"	75 38'53"	492.0	40.0		120	4
Mc Grey (Joli)	46°25'51"	73°17'59"	65.0	23.0		18	
Mc Kay	45°59'46"	76°38'11"	83.0				
Mc Kinley	49°13'27"	69°52'35"	24.0				
Mc Latchie	46°51'43"	77°02'53"	202.0				

QUEBEC (continued)

Lake name	Latitude	Longitude	Surface area (ha)	Maximum depth (m)	Mean depth (m)	Conductivity (µS/cm)	Secchi depth (m)
Mc Lennan	47°25'53"	75°47'08"	989.0				
Mc Tavish	47°41'50"	73°23'54"	42.0				
Meilleur	48°43'55"	73°35'24"	352.0				
Meloche	47°25'52"	75°03'18"	24.0				
Meloneze	49°32'20"	70°57'36"	225.0				
Melvin	47°26'21"	78°04'19"	60.0			27	
Membre	46°57'22"	77°27'59"	18.0				
Memewin	46°28'07"	78°41'45"	899.0			3	
Mer Bleue	46°14'	76°14'	544.0				
Mercier	46°11'54"	74°38'28"	117.0	34.0		104	7
Merlin	48°44'44"	73°14'25"	36.0			19	
Mesplet	48°46'22"	75°46'38"	2849.0	45.0	12.4	26	2
Mewab	46°45'	76°44'	36.0				
Midway, Lac	52°28'40"	67°01'40"	813.0				
Milieu, Du	47°10'12"	73°22'47"	41.0				
Mimi	48°44'30"	73°16'44"	23.0				
Minerve	46°13'28"	75°01'42"	337.0	46.0	4.4	63	5
Minet	47°31'12"	73 40'23"	57.0	13.7			
Misère, Lac de la Petite	49°40'00"	69°19'00"	30.0				
Miss Randolf	47°12'29"	78°40'40"	78.0	38.0		24	
Missionnaire Nord, Du	46°55'01"	72°33'33"	426.0	61.0	25.0	24	10
Missionnaire Sud	46°53'42"	72°32'24"	167.0				
Mitchel, Petit	46°18'44"	74°43'40"	18.0				
Mitchell	46°19'03"	74°42'57"	41.0	17.0			
Moise	48°47'51"	73°17'37"	67.0				
Moiseau	46°05'14"	76°52'59"	202.0				
Molard	48°35'	73°25'	241.0				
Monroe	46°20'23"	74°30'36	155.0	25.0			
Monseigneur-Bourdages	49°15'00"	70°06'00"	100.0				
Montagnais, Des	49°40'05"	69°05'58"	52.0				

Montagne Noire	46°12'	74°18'	280.0	34.0	13.1	35	4
Montagne Noire, De la	46°11'49"	74°16'18"	280.0	34.0	13.1		4
Montagne, De la	49°08'00"	69°49'11"	21.0				
Montagne, De la	48°34'58"	73°05'04"	102.0			32	
Montauban	46°53'05"	72°10'09"	456.0		50.8	340	7
Montjoie	46°17'	75°08'	1210.0	100.0		40	8
Moore	46°19'28"	77°44'08"	65.0	33.0			
Morand	48°22'	73°26'	62.0				
Morency	45°55'44"	74°02'02"	26.0	20.0			4
Morhiban, De	51°50'00"	62°54'00"					
Mousseux (Caribou)	46°02'00"	74°27'08"	49.0	13.0		60	3
Murray	46°27'	75°49'	311.0				
Nagard	46°21'34"	77°08'30"	70.0			29	8
Namego	47°56'49"	76°58'48"	122.0	50.0			
Nargile	47°28'25"	75°14'49"	36.0				
Nash	47°44'55"	74°26'15"	34.0			14	
Neault	48°19'25"	73°35'27"	35.0				
Necessite	47°04'45"	73°57'39"	124.0			12	3
Neiges, Des	47°28'17"	71°02'08"	733.0			17	5
Nemiscachingue	47°23'21"	74°31'51"	1660.0			26	4
Neuf Milles, Des	47°05'04"	77°09'54"	800.0	32.0		22	3
Nicolas (Bidiere)	47°25'51"	75°04'41"	495.0				
Nicole, Lac	51°32'48"	69°34'22"	255.0				
Nilgault	46°36'06"	77°15'08"	1629.0	30.8	20.4		
Noel	46°32'05"	73°28'46"	26.0				
Noir	47°30'17"	71°14'49"	10.0				
Nominingue	46°25'57"	74°58'55"	2198.0	36.0	12.0	35	5
Nominingue, Petit	46°21'58"	75°01'23"	655.0	33.0		19	3
Nord (Long)	47°07'48"	73°29'43"	41.0	21.3	7.1		
Norman	51°59'50"	63°41'44"	1577.0				
North	47°12'55"	78°36'46"	10.0			27	
Nougon	46°41'50"	77°30'00"	54.0			27	
Numero 8	51°51'08"	71°09'00"	230.0				
Numero 17	50°16'37"	70°36'35"	108.0				
Numero 18	50°13'00"	71°31'15"	130.0				

QUEBEC (continued)

Lake name	Latitude	Longitude	Surface area (ha)	Maximum depth (m)	Mean depth (m)	Conductivity (µS/cm)	Secchi depth (m)
Nylon, Du	48°41'04"	73°30'09"	104.0				
Ogascanane	47°05'30"	78°25'00"	3082.0	40.0	10.6	28	5
Onistagane	50°42'50"	71°19'41"	4584.0				
Oriole	46°32'35"	77°50'53"	41.0				
Oriskani	47°31'46"	73°41'59"	103.0	27.4	6.1	21	
Oscar	47°16'42"	75°12'16"	113.0				
Ostaboningue	47°08'07"	78°52'16"	3315.0	80.0		28	
O'Sullivan	47°34'	76°00'	1090.0				
Otjick	46°26'18"	76°45'16"	52.0	28.0		38	6
Otter (Petits) 1	46°43'06"	78°49'16"	36.0				
Ouaouati	47°09'34"	75°50'36"	176.0				
Ouareau	46°16'54"	74°08'26"	1492.0	54.0		20	5
Ouimet	45°50'27"	74°09'32"	54.0	22.0		36	3
Ouimet	46°10'07"	74°35'13"	158.0	23.0		61	3
Ouinegomic	47°56'44"	73°14'07"	70.0			20	
Pambrun	51°42'00"	70°44'00"	918.0				
Parker	47°26'48"	72°50'28"	135.0	23.0	9.0	22	5
Patelin	46°18'23"	74°59'07"	89.0	18.0			
Patrick	46°06'19"	73°58'44"	153.0	30.0		35	3
Patterson	46°08'	76°12'	109.0				
Paul	46°38'04"	76°23'58"	57.0	21.0		33	5
Paul	46°12'	76°45'	57.0				
Paul	46°10'26"	74°58'47"	285.0				
Pauli	50°22'03"	72°34'37"	326.0	45.7	16.5	80	7
Paul-Joncas	46°32'30"	78°00'12"	238.0				
Peier (Peter)	46°10'21"	76°42'36"	70.0			27	
Pemichangan	46°03'38"	75°51'01"	1544.0			125	7
Perchaude	51°32'30"	70°29'30"	354.0				
Perdrix Blanche	46°37'03"	76°33'35"	127.0	16.0		25	4
Perdrix, De la	46°45'03"	77°21'01"	57.0				

Lake	Latitude	Longitude					
Perdu	50°43'23"	70°12'15"	5594.0	43.0		29	6
Perdu (Lost)	46°25'12"	77°47'34"	70.0				
Perles, Aux	49°11'21"	69°49'56"	104.0				6
Perley	45°53'58"	75°25'38"	32.0	26.5	12.3	20	9
Perodeau	46°45'42"	75°09'14"	238.0	75.0	20.1		6
Persée	48°45'01"	73°18'33"	39.0			28	
Petawaga	47°03'09"	75°53'46"	2290.0				
Peter (Peier)	46°10'21"	76°42'36"	70.0				
Petit Caribou	46°17'58"	74°43'06"	11.0	30.0		23	4
Petit Collin	46°42'23"	74°04'14"	41.0	9.0			
Petits, (Otter) 1	46°43'06"	78°49'16"	36.0				
Pierre	51°30'30"	65°09'15"	818.0	33.0	6.4	28	
Pierre	47°24'01"	78°21'33"	321.0	22.0		110	5
Pierre	46°11'27"	73°41'04"	57.0				
Pierre, À	47°47'16"	73°50'44"	65.0	28.0			
Pierre-Antoine	47°28'51"	73°12'05"	70.0				
Pigeon (Brule)	46°11'41"	76°52'30"	114.0			32	
Piles, Des	46°38'48"	72°47'39"	401.0				
Pilet, Du	48°47'10"	72°20'51"	78.0	37.0	12.0	29	4
Pimodan	46°23'26"	75°17'34"	357.0			25	
Pin Rouge	46°23'48"	73°57'29"	41.0				
Pins Rouges, Des	47°23'48"	74°54'45"	324.0	32.4	8.2	23	4
Pins Rouges, Des	47°07'43"	73°54'45"	298.8	25.0	9.0	22	8
Pins Rouges, Des	46°36'18"	73°06'53"	104.0	50.0	18.0	85	6
Plages, Des	45°59'18"	74°53'45"	471.0	19.0	10.7	65	5
Plat	45°55'19"	75°23'41"	52.0			27	4
Poigan	47°12'19"	76°19'27"	1577.0				
Poisson Blanc	49°04'56"	71°04'56"	98.0				
Poisson Blanc	45°58'19"	75°44'24"	8521.0	124.0	34.0	50	6
Polonais (Argent, D')	48°44'27"	71°40'33"	1119.0				
Pommeroy	47°04'27"	78°39'36"	1191.0				
Poole	46°34'09"	78°27'38"	10.0				
Pope	46°36'	75°42'	267.0	24.0	7.4	59	4
Porc, Du	49°24'00"	68°58'00"	67.0				
Potherie Superieure	47°07'00"	73°47'00"	462.5	35.5	10.7	18	8

QUEBEC (continued)

Lake name	Latitude	Longitude	Surface area (ha)	Maximum depth (m)	Mean depth (m)	Conductivity (µS/cm)	Secchi depth (m)
Poulter	47°05'31"	76°44'15"	2176.0			25	5
Prairies, Des	50°34'53"	70°17'13"	1334.0				
Preston	46°06'29"	75°03'08"	702.0	125.0	53.7	510	10
Preston, Petit	46°02'	75°03'	127.0				
Primont	50°21'56"	72°22'08"	313.0				
Providence	48°10'39"	73°22'14"	26.0				
Provisions, Aux	45°57'05"	75°41'49"	49.0				
Provost	46°23'59"	74°15'50"	117.0				3
Pythonga	46°22'37"	76°25'44"	1865.0	68.2	13.0	45	6
Quemandeur, Du	48°20'46"	73°39'07"	41.0				
Quenouilles, Aux	46°09'44"	74°22'31"	248.0	19.0		32	3
Quiblier, Petit	47°24'07"	77°52'01"	62.0			26	
Quinn	46°29'08"	75°44'17"	373.0	43.0		47	6
Raccourcio, Lac du	49°20'00"	69°07'00"	205.0				
Raimbault	46°46'19"	74°28'24"	74.0				
Rainboth	46°51'48"	75°24'45"	80.0	28.0		21	2
Rame	46°43'18"	78°02'37"	461.0			26	3
Raquette, À la	46°23'39"	73°38'56"	41.0	20.0		23	4
Rat-Musqué, Petit du	45°53'22"	75°15'28"	23.0	10.3		25	
Rats, Aux	47°29'15"	73°09'31"	410.0	34.0		40	
Raven	47°28'14"	78°33'40"	18.0				
Raymond	46°30'	75°06'	20.0	60.0	26.0	38	5
Raymond	46°16'19"	77°03'57"	151.0	24.0		30	5
Recifs, Aux	51°33'30"	70°57'30"	293.0				
Redan	46°50'	76°45'	169.0				
Resolin	46°27'38"	77°08'13"	394.0				
Retard, Du	49°41'26"	69°07'04"	127.0				
Richard	48°36'01"	73°00'58"	10.0				
Richard	46°34'16"	74°04'45"	47.0			46	8
Richardson, À la Truite	46°11'25"	74°54'16"	80.0	24.0			4

Name	Latitude	Longitude					
Richaume	50°57'08"	72°04'24"	316.0				
Rioux	48°00'04"	76°19'14"	49.0				
Rita	49°12'57"	74°23'47"	179.0	31.0	11.0	10	5
Rivieres, Des	46°37'23"	76°08'55"	236.0	16.0		60	5
Robinson	47°10'02"	72°20'57"	104.0				
Roc-Causacouta	47°08'57"	73°34'44"	335.2	28.5	6.2	42	4
Roche Blanche, De la	47°05'40"	73°37'58"	78.0	19.8	6.7	31	4
Rocher, Du	46°49'19"	76°26'35"	280.0				
Rocher, Du	45°53'00"	75°20'23"	31.0				
Rochester	46°48'14"	77°12'30"	119.0	54.8			7
Rochon	46°43'16"	75°13'31"	326.0	40.0		105	6
Roddick	46°15'	75°53'	578.0				6
Roger	48°45'53"	76°13'59"	202.0	45.0	20.1	44	
Rognon	46°16'12"	75°03'50"	83.0				
Rohault	49°22'15"	74°20'01"	5802.0				5
Roland	46°55'53"	76°25'19"	153.0	21.3			
Romain	47°25'23"	77°42'32"	197.0				
Romain	46°35'11"	76°50'38"	117.0	25.0		19	
Rond	49°25'09"	75°58'36"	695.0				
Rond	47°11'50"	73°39'52"	109.0				
Rond	46°39'	76°15'	553.0	23.0		40	4
Rose	47°32'47"	73°25'29"	32.0				
Ross	47°15'20"	78°26'03"	350.0			22	
Rouge (Carmin)	46°04'	74°59'	104.0				
Rouge, Grand	46°54'34"	74°38'02"	202.0			26	
Rouge, Petit	46°56'55"	74°38'27"	62.0			25	
Roy	49°45'23"	69°53'54"	1526.0	19.0			
Royal	46°34'29"	76°22'05"	339.0	11.0		34	5
Royal, Petit	46°35'05"	76°21'22"	26.0			26	3
Ruisseaux, Des	46°35'20"	76°51'39"	80.0				
Rupert	46°59'49"	74°45'25"	457.0				
Sables, Aux	46°53'01"	72°21'48"	531.0	41.0	22.0	17	7
Sables, Aux	46°29'35"	73°23'19"	47.0	18.3	6.7	28	
Sables, Des	47°16'08"	74°00'17"	250.5		10.6	19	
Sables, Des	46°38'37"	78°27'19"	417.0	41.0		29	3

QUEBEC (continued)

Lake name	Latitude	Longitude	Surface area (ha)	Maximum depth (m)	Mean depth (m)	Conductivity (µS/cm)	Secchi depth (m)
Sables, Des	46°02'28"	74°18'07"	293.0	23.0		89	4
Sablier, Du	46°48'	76°43'	42.0				
Sacacomie	46°30'59"	73°13'31"	974.0	79.0	30.0	16	16
Saint-Amand	47°14'00"	79°08'30"	93.0			30	5
Saint-Amand	46°52'	77°12'	101.0				
Saint-Amour	47°18'39"	76°22'09"	834.0				
Saint-Charles	45°46'	75°52'	155.0				
Saint-Denis	46°18'	75°06'	330.0	66.5	24.3	40	5
Sainte-Marie	46°24'24"	75°02'20"	62.0	7.0	3.0	80	5
Sainte-Marie	45°57'59"	74°17'46"	140.0	15.0		45	6
Saint-Francois	45°52'53"	74°21'31"	70.0	12.0		30	4
Saint-Joseph	46°54'32"	71°38'37"	1111.0			21	
Saint-Joseph	46°24'44"	75°01'49"	18.0	7.0	3.0		
Saint-Joseph	45°58'28"	74°19'52"	145.0	30.0			
Saint-Michel	47°02'15"	72°46'30"	85.0			40	4
Saint-Patrice	46°18'	77°18'	2860.0	40.0	12.8	38	4
Saint-Pierre	45°43'09"	75°42'39"	356.0			105	4
Saint-Sixte	45°49'	75°15'	158.0				
Sairas	46°50'37"	77°08'28"	47.0				
Salone	47°21'20"	73°55'54"	919.0	25.0	5.1		
Samlock	45°58'	75°42'	60.0				6
Sangsues	46°28'05"	77°55'54"	1119.0			20	
Sapin Croche	49°52'00"	71°45'00"	381.0				
Sarasin	46°07'12"	74°12'08"	44.0	19.0			
Saseginaga	47°06'14"	78°34'04"	3755.0	40.0		28	
Saul	46°54'56"	74°44'07"	52.0			26	
Sault	48°30'17"	78°41'11"	34.0	38.0		54	
Sauvage	46°03'01"	74°31'21"	52.0	12.0			
Sauvageau	48°18'	73°19'	420.0				
Savary	46°44'38"	76°24'51"	451.0			38	

Lake							
Savary, Petit	46°43'11"	76°24'21"	93.0				4
Schierbeck	47°24'	75°45'	137.0				4
Scudder Pond	46°32'14"	78°17'57"	10.0			20	
Seagull	46°28'14"	77°05'22"	62.0	37.0	9.8	9	6
Sechelles	52°20'00"	69°12'00"	2717.0	30.0	12.4		6
Sedillot	49°31'31"	69°06'32"	530.0	59.0		69	
Seize Iles, Des	45°53'44"	74°27'51"	365.0				
Seneca	46°33'09"	78°21'29"	44.0	40.5	15.3	270	4
Sept-Freres, Des	46°20'17"	75°09'55"	311.0				
Sept-Iles	47°30'38"	71°14'23"	80.0				
Sept-Milles	46°42'00"	77°40'31"	635.0				
Serpent	49°49'00"	71°37'00"	855.0				6
Serpent	46°09'05"	75°28'51"	236.0	67.0	19.9	72	5
Shaughnessey	46°16'09"	74°56'14"	18.0	14.0		140	5
Silver (Argent, D')	46°38'30"	72°35'49"	60.0	56.4	16.2	45	8
Simon	45°57'56"	75°04'47"	2849.0	117.4	48.2	41	6
Simon (Trois-Montagnes)	46°09'39"	74°46'25"	334.0	68.0			
Six Iles	47°09'57"	78°40'39"	562.0				
Sleigh	46°45'35"	72°45'43"	101.0				
Smith	46°12'20"	78°49'43"	484.0	18.0	5.6	30	6
Smith	47°33'32"	77°08'28"	164.0				
Snooks	46°35'33"	73°37'09"	57.0	22.9		28	
Soliere	45°56'17"	77°31'15"	127.0	12.0			
Sopwith	46°09'01"	76°31'00"	161.0				
Sourd, Du	46°34'48"	75°16'27"	834.0	53.9	13.6	37	5
Souris, Des	45°47'	72°59'39"	202.0	24.0	8.0	29	8
Sparling	46°35'14"	76°26'	36.0				
Spearman	47°41'09"	78°35'08"	386.0			27	
Spellman	46°35'31"	73°22'59"	28.0			28	
Sterne (Little Grant)	46°20'23"	77°56'11"	13.0				4
Stevens	47°38'	77°43'51"	83.0	35.0		40	
Stramond	46°05'55"	76°01'	2365.0		15.0		5
Stubbs	46°32'23"	76°38'55"	285.0			45	
Sudrie	46°12'43"	77°17'08"	73.0				
Superieur	46°12'43"	74°28'27"	166.0	20.0		46	4

QUEBEC (continued)

Lake name	Latitude	Longitude	Surface area (ha)	Maximum depth (m)	Mean depth (m)	Conductivity (µS/cm)	Secchi depth (m)
Swastika	46°39'29"	74°34'01"	60.0	21.0		25	
Sylvere	46°20'58"	74°04'13"	179.0	31.0		34	8
Tabac#1	51°09'00"	67°01'00"	95.0				
Tabac, Du	47°40'56"	74°25'24"	83.0				
Tagwagan	47°01'45"	78°36'59"	10.0			29	
Tap	46°19'53"	77°46'41"	83.0	22.0		26	6
Taylor	46°29'30"	76°44'28"	171.0	40.0		32	8
Tee	46°47'03"	79°02'18"	448.0				
Terrasses, Des	48°38'21"	75°55'11"	272.0				
Tete D'Orignal, De la	46°36'34"	78°38'02"	334.0			30	
Tetepisca	51°05'20"	69°21'51"	2953.0				
Thompson	47°10'23"	75°27'29"	75.0				
Thompson	46°35'46"	78°32'34"	65.0			33	
Thorne	45°41'07"	76°21'39"	65.0	31.4	13.6		5
Tilley	46°39'00"	76°35'56"	60.0				
Tom	48°45'53"	73°35'00"	12.0				
Tomasine	46°43'30"	76°18'57"	461.0	35.1	9.9	30	4
Tortue, À la	46°25'56"	76°15'20"	433.0	25.0	7.1	42	
Touladi	49°48'22"	72°08'04"	91.0			21	
Touladi	49°35'08"	72°06'24"	42.0				
Touladi	47°45'08"	73°02'04"	36.0				
Touridi	47°40'52"	73°59'40"	148.0	30.0	12.0	21	3
Trader	46°41'	76°45'	145.0				
Transparent	47°29'42"	77°23'49"	189.0	37.0		30	
Travail	46°55'	77°41'	59.0				
Traversy	47°16'22"	72°40'58"	23.0	52.0	21.2		5
Trefle	47°01'45"	74°55'54"	10.0			26	7
Tremblant	46°14'53"	74°38'12"	945.0	91.0	23.0	28	7
Trente Et Un Milles, Des	46°11'15"	75°48'28"	4973.0	87.8	26.1	120	7
Trois-Montagnes (Simon)	46°09'39"	74°46'25"	334.0	68.0		41	6

Lake							
Trout	46°29'30"	77°38'19"	57.0	21.0		26	3
Troyes	47°10'43"	74°12'16"	1251.0	38.0			
Truite Grise	49°19'04"	73°18'42"	36.0				
Truite, À la	49°29'52"	72°48'10"	495.0				
Truite, À la	47°59'07"	73°38'10"	62.0				
Truite, À la	47°01'43"	78°37'33"	311.0				
Truite, À la	46°46'49"	79°06'38"	174.0			27	
Truite, À la	46°40'24"	78°44'40"	135.0	47.0		33	5
Truite, À la	46°16'38"	78°16'35"	3937.0	65.0	11.0	140	7
Truite, À la	46°05'15"	75°45'39"	70.0				
Turnbull	47°26'20"	74°50'48"	315.0				
Vaillant	48°09'35"	74°13'37"	70.0				
Valtrie	46°42'15"	74°33'28"	18.0	6.0		19	4
Vases, Des	49°06'06"	71°57'35"	77.0				
Vau, De	50°56'00"	72°14'19"	591.0				
Ventadour	47°45'51"	72°02'17"	386.0	42.5	10.8	20	5
Vermont	48°56'49"	71°09'10"	849.0				
Vers, Aux	46°09'04"	77°08'38"	179.0				
Vers, Petit	46°08'39"	77°07'11"	30.0				
Vert	46°03'19"	76°52'53"	176.0	52.0		77	6
Vert (Green)	46°29'35"	76°17'27"	122.0				
Viau	48°47'28"	73°18'31"	57.0				
Viceroi	45°51'	75°06'	445.0				
Vieille, De la	46°45'38"	76°13'41"	417.0	36.6		40	4
Vigie, De la	48°21'50"	73°34'27"	73.0				
Vignerod	47°19'11"	73°24'04"	161.0	35.0		31	6
Villedonne	46°51'34"	78°43'44"	290.0				
Villiers	47°07'09"	74°00'59"	1382.0	37.0		20	5
Vital	51°28'05"	65°17'50"	1134.0	106.2	14.1	60	4
Waconichi	50°09'04"	73°59'59"	8184.0				
Wacouno	51°34'57"	65°42'42"	1557.0				
Wallace	47°34'13"	73°18'05"	45.0				
Wapus	47°25'	75°43'	181.0				
Weldie	46°43'31"	76°39'02"	137.0	28.0		25	4
Wenworth	45°49'44"	74°26'57"	73.0	24.0		45	4

QUEBEC (continued)

Lake name	Latitude	Longitude	Surface area (ha)	Maximum depth (m)	Mean depth (m)	Conductivity (µS/cm)	Secchi depth (m)
Wetetnagami	48°55'35"	76°1'47"	1981.0			16	4
Willard	46°53'50"	74°35'35"	52.0			24	
Winawiash	47°23'16"	78°08'24"	1461.0	46.0	6.9	26	4
Windy	46°46'54"	78°48'30"	295.0			39	
Winsch	50°02'03"	74°13'01"	122.0	35.0	11.7	130	12
Wright	46°20'19"	76°47'56"	321.0	40.0		25	7
Xavier	46°08'51"	74°44'16"	104.0	48.0		20	7
Yellow	46°39'42"	76°36'15"	54.0				
Young	46°48'59"	77°11'04"	93.0				
Ypres	47°57'36"	76°54'10"	640.0				
Yser	47°54'36"	76°51'54"	1147.0	40.0	10.8		6
Yza	49°12'59"	71°23'27"	65.0				
Zavitz	46°34'51"	78°27'19"	34.0				

appendix three

Common and scientific names for fish species in selected Boreal Shield lake trout lakes from Ontario*

Blacknose dace – *Rhinichthys atratulus*
Bluegill – *Lepomis macrochirus*
Bluntnose minnow – *Pimephales notatus*
Brook stickleback – *Culaea inconstans*
Brook trout – *Salvelinus fontinalis*
Brown bullhead – *Ameiurus nebulosus*
Burbot – *Lota lota*
Central mudminnow – *Umbra limi*
Cisco (lake herring) – *Coregonus artedi*
Common shiner – *Luxilus cornutus*
Creek chub – *Semotilus atromaculatus*
Fathead minnow – *Pimephales promelas*
Finescale dace – *Phoxinus neogaeus*
Golden shiner – *Notemigonus crysoleucas*
Iowa darter – *Etheostoma exile*
Johnny darter – *Etheostoma nigrum*
Lake chub – *Couesius plumbeus*
Lake trout – *Salvelinus namaycush*
Lake whitefish – *Coregonus clupeaformis*
Largemouth bass – *Micropterus salmoides*
Logperch – *Percina caprodes*
Longnose sucker – *Catostomus catostomus*
Mottled sculpin – *Cottus bairdi*
Ninespine stickleback – *Pungitius pungitius*
Northern pike – *Esox lucius*
Northern redbelly dace – *Phoxinus eos*

* This is a list of species captured using the Nordic netting multimesh gillnets in 40 small (17-5863 ha) lake trout lakes from across the Boreal Shield ecozone of Ontario, 2000–2002. Data from Ed Snucins, Cooperative Freshwater Ecology Unit, Laurentian University, Sudbury, Ontario. Names follow Robin, C.R., Bailey, R.M. Bond, C.E., Brooker, J.R., Lachner, E.A., Lea, R.N., and Scott, W.B. 1991. Common and scientific names of fishes from the United States and Canada. American Fisheries Society, Special Publication #20, Bethesda, Maryland.

Pearl dace – *Margariscus margarita*
Pumpkinseed – *Lepomis gibbosus*
Rainbow smelt – *Osmerus mordax*
Rock bass – *Ambloplites rupestris*
Round whitefish – *Prosopium cylindraceum*
Slimy sculpin – *Cottus cognatus*
Smallmouth bass – *Micropterus dolomieu*
Spoonhead sculpin – *Cottus ricei*
Spottail shiner – *Notropis hudsonius*
Trout-perch – *Percopsis omiscomaycus*
Walleye (yellow pickerel) – *Stizostedion vitreum*
White sucker – *Catostomus commersoni*
Yellow perch – *Perca flavescens*

appendix four

Conversion factors

Convert column 1 into column 2 by multiplying by	Column 1 SI unit	Column 2 non-SI unit	Convert column 2 into column 1, by multiplying by
		Mass	
2.20×10^{-3}	gram, g (10^{-3} kg)	pound, lb	453.59
3.52×10^{-2}	gram, g (10^{-3} kg)	ounce (avoirdupois), oz	28.35
2.205	kilogram, kg	pound, lb	0.454
0.01	kilogram, kg	quintal (metric), q	100
1.10×10^{-3}	kilogram, kg	ton (2000 lb), ton	907.2
1.102	megagram, Mg (tonne)	ton (United States), ton	0.907
1.102	tonne, t	ton (United States), ton	0.907
		Length	
0.621	kilometer, km (10^3 m)	mile, mi	1.609
1.094	meter, m	yard, yd	0.914
3.28	meter, m	foot, ft	0.305
1.0	micrometer, μm (10^{-6} m)	micron, μ	1.0
3.94×10^{-2}	millimeter, mm (10^{-3} m)	inch, in.	25.4
10	nanometer, nm (10^{-9} m)	angstrom, Å	0.1
		Area	
2.471	hectare, ha	acre	0.405
2471	square kilometer, km² [(10^3 m)²]	acre	4.04×10^{-3}
0.386	square kilometer, km² [(10^3 m)²]	square mile, mi²	2.590
2.47×10^{-4}	square meter, m²	acre	4.05×10^3
10.76	square meter, m²	square foot, ft²	9.29×10^{-2}
1.55×10^{-3}	square millimeter, mm² [(10^{-3}m)²]	square inch, in.²	645
		Volume	
9.73×10^{-3}	cubic meter, m³	acre-inch	102.8
35.31	cubic meter, m³	cubic foot, ft³	2.83×10^{-2}
6.10×10^4	cubic meter, m³	cubic inch, in.³	1.64×10^{-5}
2.84×10^{-2}	liter, l (10^{-3} m³)	bushel, bu	35.24
1.057	liter, l (10^{-3} m³)	quart (liquid), qt	0.946
3.53×10^{-2}	liter, l (10^{-3} m³)	cubic foot, ft³	28.3
0.264	liter, l (10^{-3} m³)	gallon (United States)	3.785
33.78	liter, l (10^{-3} m³)	ounce (fluid), oz	2.96×10^{-2}
2.11	liter, l (10^{-3} m³)	pint (fluid), pt	0.474
		Yield and Rate	
0.893	kilogram per hectare, kg ha⁻¹	pound per acre, lb acre⁻¹	1.12
7.77×10^{-2}	kilogram per cubic meter, kg m⁻³	pound per bushel, bu⁻¹	12.87
1.49×10^{-2}	kilogram per hectare, kg ha⁻¹	bushel per acre, 60 lb	67.11
1.59×10^{-2}	kilogram per hectare, kg ha⁻¹	bushel per acre, 56 lb	62.89
1.86×10^{-2}	kilogram per hectare, kg ha⁻¹	bushel per acre, 48 lb	53.75
0.107	liter per hectare, l ha⁻¹	gallon (United States) per acre	9.35

Convert column 1 into column 2 by multiplying by	Column 1 SI unit	Column 2 non-SI unit	Convert column 2 into column 1, by multiplying by
893	tonnes per hectare, t ha^{-1}	pound per acre, lb acre^{-1}	1.12×10^{-3}
893	megagram per hectare, Mg ha^{-1}	pound per acre, lb acre^{-1}	1.12×10^{-3}
0.466	megagram per hectare, Mg ha^{-1}	ton (2000 lb) per acre, ton acre^{-1}	2.24
2.24	meter per second, m sec^{-1}	mile per hour	0.447

Pressure

9.90	megapascal, MPa (106 Pa)	atmosphere	0.101
10	megapascal, MPa (106 Pa)	bar	0.1
1.00	megagram per cubic meter, Mg m^{-3}	gram per cubic centimeter, g cm^{-3}	1.00
2.09×10^{-2}	pascal, Pa	pound per square foot, lb ft^{-2}	47.9
1.45×10^{-4}	pascal, Pa	pound per square inch, lb in.$^{-2}$	6.90×10^{3}

Temperature

1.00 (K − 273)	Kelvin, K	Celsius, °C	1.00 (°C + 273)
(9/5 °C) + 32	Celsius, °C	Fahrenheit, °F	5/9 (°F − 32)

Energy, Work, Quantity of Heat

9.48×10^{-4}	joule, J	British thermal unit, Btu	1.05×10^{3}
0.239	joule, J	calorie, cal	4.19
10^{7}	joule, J	erg	10^{-7}
0.735	joule, J	foot-pound	1.36
2.387×10^{-5}	joule per square meter, J m^{-2}	calorie per square centimeter (Langley)	4.19×10^{4}
10^{5}	newton, N	dyne	10^{-5}
1.43×10^{-3}	watt per square meter, W m^{-2}	calorie per square centimeter minute (irradiance), cal cm^{-2} min^{-1}	698

Electrical Conductivity

10	siemen per meter, S m^{-1}	millimho per centimeter, mmho cm^{-1}	0.1

Water Measurement

9.73×10^{-3}	cubic meter, m^{3}	acre-inches, acre-in.	102.8
9.81×10^{-3}	cubic meter per hour, m^{3} h^{-1}	cubic feet per second, ft^{3} sec^{-1}	101.9
4.40	cubic meter per hour, m^{3} h^{-1}	US gallons per minute, gal min^{-1}	0.227
8.11	hectare-meters, ha-m	acre-feet, acre-ft	0.123
97.28	hectare-meters, ha-m	acre-inches, acre-in.	1.03×10^{-2}
8.1×10^{-2}	hectare-centimeters, ha-cm	acre-feet, acre-ft	12.33

| | Concentrations[a] | |
	F1	F2
Aluminum (Al^{+3})	0.11119	0.03715
Ammonium (NH_4^+)	0.05544	0.05544
Barium (Ba^{+2})	0.01456	0.00728
Beryllium (Be^{+3})	0.33288	0.11096
Bicarbonate (HCO_3^-)	0.01639	0.01639
Boron (B)	—	0.09250
Bromide (Br^-)	0.01251	0.01251
Cadmium (Cd^{+2})	0.01779	0.00890
Calcium (Ca^{+2})	0.04990	0.02495
Carbonate (CO_3^{-2})	0.03333	0.01666
Chloride (Cl^-)	0.02821	0.02821
Chromium (Cr)	—	0.01923
Cobalt (Co^{+2})	0.03394	0.01697
Copper (Cu^{+2})	0.03148	0.01574
Fluoride (F^-)	0.05264	0.05264
Germanium (Ge)	—	0.01378
Gallium (Ga)	—	0.01434
Gold (Au)	—	0.00511
Hydrogen (H^+)	0.99209	0.99209
Hydroxide (OH^-)	0.05880	0.05880
Iodide (I^-)	0.00788	0.00788
Iron (Fe^{+2})	0.03581	0.01791
Iron (Fe^{+3})	0.05372	0.01791
Lead (Pb)	—	0.00483
Lithium (Li^+)	0.14411	0.14411
Magnesium (Mg^{+2})	0.08226	0.04113
Manganese (Mn^{+2})	0.03640	0.01820
Molybdenum (Mo)	—	0.01042
Nickel (Ni)	—	0.01703
Nitrate (NO_3^-)	0.01613	0.01613
Nitrite (NO_2^-)	0.02174	0.02174
Phosphate (PO_4^{-3})	0.03159	0.01053
Phosphate (HPO_4^{-2})	0.02084	0.01042
Phosphate ($H_2PO_4^-$)	0.01031	0.01031
Potassium (K^+)	0.02557	0.02557
Rubidium (Rb^+)	0.01170	0.01170
Silica (SiO_2)	—	0.01664
Silver (Ag)	—	0.00927
Sodium (Na^+)	0.04350	0.04350
Strontium (Sr^{+2})	0.02283	0.01141
Sulfate (SO_4^{-2})	0.02082	0.01041
Sulfide (S^{-2})	0.06238	0.03119
Titanium (Ti)	—	0.02088
Uranium (U)	—	0.00420
Zinc (Zn^{+2})	0.03060	0.01530

[a] Milligrams per liter $\times$ F1 = milliequivalents per liter (mEq); milligrams per liter $\times$ F2 = millimoles per liter.

Index

A

H

I

T